ADVANCES IN THE THEORY OF ATOMIC AND MOLECULAR SYSTEMS

Progress in Theoretical Chemistry and Physics

VOLUME 19

Honorary Editors:

W.N. Lipscomb *(Harvard University, Cambridge, MA, U.S.A.)*
Yves Chauvin *(Institut Français du Pétrole, Tours, France)*

Editors-in-Chief:

J. Maruani *(formerly Laboratoire de Chimie Physique, Paris, France)*
S. Wilson *(formerly Rutherford Appleton Laboratory, Oxfordshire, U.K.)*

Editorial Board:

V. Aquilanti *(Università di Perugia, Italy)*
E. Brändas *(University of Uppsala, Sweden)*
L. Cederbaum *(Physikalisch-Chemisches Institut, Heidelberg, Germany)*
G. Delgado-Barrio *(Instituto de Matemáticas y Física Fundamental, Madrid, Spain)*
E.K.U. Gross *(Freie Universität, Berlin, Germany)*
K. Hirao *(University of Tokyo, Japan)*
R. Lefebvre *(Université Pierre-et-Marie-Curie, Paris, France)*
R. Levine *(Hebrew University of Jerusalem, Israel)*
K. Lindenberg *(University of California at San Diego, CA, U.S.A.)*
M. Mateev *(Bulgarian Academy of Sciences and University of Sofia, Bulgaria)*
R. McWeeny *(Università di Pisa, Italy)*
M.A.C. Nascimento *(Instituto de Química, Rio de Janeiro, Brazil)*
P. Piecuch *(Michigan State University, East Lansing, MI, U.S.A.)*
S.D. Schwartz *(Yeshiva University, Bronx, NY, U.S.A.)*
A. Wang *(University of British Columbia, Vancouver, BC, Canada)*
R.G. Woolley *(Nottingham Trent University, U.K.)*

Former Editors and Editorial Board Members:

I. Prigogine (†)
J. Rychlewski (†)
Y.G. Smeyers (†)
R. Daudel (†)
H. Ågren (*)
D. Avnir (*)
J. Cioslowski (*)
W.F. van Gunsteren (*)

I. Hubač (*)
M.P. Levy (*)
G.L. Malli (*)
P.G. Mezey (*)
N. Rahman (*)
S. Suhai (*)
O. Tapia (*)
P.R. Taylor (*)

† : deceased; * : end of term

For other titles published in this series, go to
www.springer.com/series/6464

Advances in the Theory of Atomic and Molecular Systems

Conceptual and Computational Advances in Quantum Chemistry

Edited by

PIOTR PIECUCH
Michigan State University, East Lansing, MI, USA

JEAN MARUANI
CNRS, Paris, France

GERARDO DELGADO-BARRIO
CSIC, Madrid, Spain

and

STEPHEN WILSON
University of Oxford, UK

 Springer

Editors

Piotr Piecuch
Department of Chemistry
Michigan State University
East Lansing, MI 48824
USA
piecuch@chemistry.msu.edu

Jean Maruani
Laboratoire de Chimie Physique
CNRS and UPMC
11 Rue Pierre et Marie Curie
F-75005 Paris
France
jean.maruani@upmc.fr
jemmaran@gmail.com

Gerardo Delgado-Barrio
Instituto de Física Fundamental
CSIC
Serrano 123
E-28006 Madrid
Spain
gerardo@imaff.cfmac.csic.es

Stephen Wilson
Physical & Theoretical Chemistry Laboratory
University of Oxford
South Parks Road
Oxford OX1 3QZ
United Kingdom
quantumsystems@gmail.com

ISSN 1567-7354
ISBN 978-90-481-2595-1   e-ISBN 978-90-481-2596-8
DOI 10.1007/978-90-481-2596-8
Springer Dordrecht Heidelberg London New York

Library of Congress Control Number: 2009929297

Printed on acid-free paper

Springer is part of Springer Science+Business Media (www.springer.com)

Progress in Theoretical Chemistry and Physics

A series reporting advances in theoretical molecular and materials sciences, including theoretical, mathematical and computational chemistry, physical chemistry, and chemical physics

Aim and Scope

Science progresses by a symbiotic interaction between theory and experiment: theory is used to interpret experimental results and may suggest new experiments; experiment helps to test theoretical predictions and may lead to improved theories. Theoretical Chemistry (including Physical Chemistry and Chemical Physics) provides the conceptual and technical background and apparatus for the rationalization of phenomena in the chemical sciences. It is, therefore, a wide ranging subject, reflecting the diversity of molecular and related species and processes arising in chemical systems. The book series *Progress in Theoretical Chemistry and Physics* aims to report advances in methods and applications in this extended domain. It will comprise monographs as well as collections of papers on particular themes, which may arise from proceedings of symposia or invited papers on specific topics as well as initiatives from authors or translations.

The basic theories of physics – classical mechanics and electromagnetism, relativity theory, quantum mechanics, statistical mechanics, quantum electrodynamics – support the theoretical apparatus which is used in molecular sciences. Quantum mechanics plays a particular role in theoretical chemistry, providing the basis for the valence theories which allow to interpret the structure of molecules and for the spectroscopic models employed in the determination of structural information from spectral patterns. Indeed, Quantum Chemistry often appears synonymous with Theoretical Chemistry: it will, therefore, constitute a major part of this book series. However, the scope of the series will also include other areas of theoretical chemistry, such as mathematical chemistry (which involves the use of algebra and topology in the analysis of molecular structures and reactions), molecular mechanics, molecular dynamics and chemical thermodynamics, which play an important role in rationalizing the geometric and electronic structures of molecular assemblies and polymers, clusters and crystals, surface, interface, solvent and solid-state effects, excited-state dynamics, reactive collisions, and chemical reactions.

Recent decades have seen the emergence of a novel approach to scientific research, based on the exploitation of fast electronic digital computers. Computation provides a method of investigation which transcends the traditional division between theory and experiment. Computer-assisted simulation and design may afford a solution to complex problems which would otherwise be intractable to theoretical analysis, and may also provide a viable alternative to difficult or costly laboratory experiments. Though stemming from Theoretical Chemistry, Computational Chemistry is a field of research in its own right, which can help to test theoretical predictions and may also suggest improved theories.

The field of theoretical molecular sciences ranges from fundamental physical questions relevant to the molecular concept, through the statics and dynamics of isolated molecules, aggregates and materials, molecular properties and interactions, and the role of molecules in the biological sciences. Therefore, it involves the physical basis for geometric and electronic structure, states of aggregation, physical and chemical transformations, thermodynamic and kinetic properties, as well as unusual properties such as extreme flexibility or strong relativistic or quantum-field effects, extreme conditions such as intense radiation fields or interaction with the continuum, and the specificity of biochemical reactions.

Theoretical chemistry has an applied branch – a part of molecular engineering, which involves the investigation of structure–property relationships aimed at the design, synthesis, and application of molecules and materials endowed with specific functions, now in demand in such areas as molecular electronics, drug design, or genetic engineering. Relevant properties include conductivity (normal, semi- and supra-), magnetism (ferro- or ferri-), optoelectronic effects (involving nonlinear response), photochromism and photoreactivity, radiation and thermal resistance, molecular recognition and information processing, and biological and pharmaceutical activities, as well as properties favouring self-assembling mechanisms and combination properties needed in multifunctional systems.

Progress in Theoretical Chemistry and Physics is made at different rates in these various research fields. The aim of this book series is to provide timely and in-depth coverage of selected topics and broad-ranging yet detailed analysis of contemporary theories and their applications. The series will be of primary interest to those whose research is directly concerned with the development and application of theoretical approaches in the chemical sciences. It will provide up-to-date reports on theoretical methods for the chemist, thermodynamicist or spectroscopist, the atomic, molecular or cluster physicist, and the biochemist or molecular biologist who wish to employ techniques developed in theoretical, mathematical, or computational chemistry in their research programmes. It is also intended to provide the graduate student with a readily accessible documentation on various branches of theoretical chemistry, physical chemistry, and chemical physics.

Preface

These two volumes, which share the common title *Advances in the Theory of Atomic and Molecular Systems*, contain a representative selection of some of the most outstanding papers presented at the *Thirteenth International Workshop on Quantum Systems in Chemistry and Physics* (QSCP-XIII). The QSCP-XIII workshop, which ran from 6 to 12 July 2008, was held at the impressive site of the James B. Henry Center for Executive Development of Michigan State University, East Lansing, Michigan, USA, and was coordinated with the *Sixth Congress of the International Society for Theoretical Chemical Physics* (ISTCP-VI) that took place a week later in Vancouver, British Columbia, Canada. As a mark of the close collaboration which underpinned the scientific agendas of both meetings, three of the thirty three papers included in the present two volumes have been written by scientists who contributed in person at the ISTCP-VI Congress.

The QSCP-XIII workshop continued the series that was initiated by Roy McWeeny in April 1996 with a meeting held at San Miniato, near Pisa, Italy. Held every year, QSCP meetings bring together, in an informal atmosphere and with the aim of fostering collaboration, chemists and physicists who share interests in such areas as Novel Concepts and Methods in Quantum Chemistry, Molecular Structure and Spectroscopy, Atoms and Molecules in Electric and Magnetic Fields, Condensed Matter, Complexes and Clusters, Surfaces and Interfaces, Nano-Materials and Molecular Electronics, Reactive Collisions and Chemical Reactions, Computational Chemistry, Physics, and Biology, and Biological Modeling. The emphasis of the QSCP workshops is on broadly defined quantum-mechanical many-body methods, i.e., the development of innovative theory and its computational realization, along with their application to a broad range of scientific problems of relevance to chemistry, physics, biochemistry, and related fields.

The QSCP-XIII workshop, which was the first in the series held in North America, was truly international in nature. We welcomed more than 100 participants from 22 countries from North and South America, Europe, Africa, Asia, and Australia. The workshop was divided into 19 plenary sessions, during which a total of 64 scientific lectures were delivered in accordance with the usual QSCP "democratic" allocation of about 30 minutes for each lecture. These lectures were complemented by two 45-minute long keynote plenary talks which focused on science and relevant historical overviews, constituting a special session that was held on 6 July, as well as by a welcome presentation by Vice President for Research and Graduate Studies at Michigan State University, J. Ian Gray, given on 7 July, and 36 poster presentations divided into two poster sessions.

Following a tradition of QSCP workshops, the scientific programme of QSCP-XIII was accompanied by a social programme including a musical concert, a trip to the Henry Ford Museum in Dearborn, Michigan, and an award ceremony that took place at the banquet dinner. The musical concert, coordinated by Charles Ruggiero of Michigan State University, was performed by Danilo Mezzadri and Susan Ruggiero-Mezzadri, a flute and soprano duo who were accompanied by Judy Kabodian at the piano. The ceremony of the award honored five nominees and recipients of the *QSCP Promising Scientist Prize of the Centre de Mécanique Ondulatoire Appliquée* (CMOA). The first Prize was presented to Garnet K.-L. Chan from Cornell University, USA, and the second Prize was shared by David A. Mazziotti from the University of Chicago, USA, and Ágnes Szabados from Eötvös University, Hungary. T. Daniel Crawford from Virginia Tech, USA, and Stéphane Carniato from Université Pierre et Marie Curie, France, received certificates of nomination. For further details of the QSCP-XIII events and programme, including the abstracts of all lectures and poster presentations, we refer the reader to the workshop webpages at

www.chemistry.msu.edu/qscp13

The thirty three papers in the proceedings of QSCP-XIII are divided between the present two volumes in the following manner. The first volume, with the subtitle *Conceptual and Computational Advances in Quantum Chemistry*, contains twenty papers and is divided into six parts. The first part focuses on historical overviews of significance to the QSCP workshop series and quantum chemistry. The remaining five parts, entitled "High-Precision Quantum Chemistry," "Beyond Nonrelativistic Theory: Relativity and QED," "Advances in Wave Function Methods," "Advances in Density Functional Theory," and "Advances in Concepts and Models," address different aspects of quantum mechanics as applied to electronic structure theory and its foundations. The second volume, with the subtitle *Dynamics, Spectroscopy, Clusters, and Nanostructures*, contains the remaining thirteen papers and is divided into three parts: "Quantum Dynamics and Spectroscopy," "Complexes and Clusters," and "Nanostructures and Complex Systems."

We hope that together the present two volumes, with thirty three authoritative papers that are either of the advanced review type or original research articles, will provide readers with a good idea of the range of stimulating topics which made QSCP-XIII such a success. We thank both the contributors to these proceedings, who sent us their outstanding manuscripts, and the referees, who were willing to invest time in the unacknowledged effort of reviewing the work submitted. We greatly appreciate the help and advice they gave to authors and editors alike. We also thank Jeffrey R. Gour, Jesse J. Lutz, and Wei Li for their assistance with technical aspects of the manuscripts.

We are grateful to the participants of QSCP-XIII, not only for the high standard of the lectures and posters presented at the workshop, which is reflected in these proceedings, but also for the friendly and constructive atmosphere throughout the formal and informal sessions. Thanks to the participants, the QSCP workshops continue to provide a unique forum for the presentation and appraisal of new ideas in the broadly defined area of quantum systems in chemistry and physics.

We are grateful to the Honorary Chairs of QSCP-XIII for their support, encouragement, and advice. Specifically, we thank (in alphabetical order): Ernest R. Davidson of the University of Washington, USA, Zohra Ben Lakhdar of the University of Tunis, Tunisia, Raphael D. Levine of the Hebrew University of Jerusalem, Israel and the University of California, Los Angeles, USA, Rudolph A. Marcus of the California Institute of Technology, Pasadena, USA, and Roy McWeeny of the University of Pisa, Italy. We also express our gratitude to other members of the International Scientific Committee of QSCP-XIII, including Vincenzo Aquilanti of the University of Perugia, Italy, Erkki J. Brändas of the University of Uppsala, Sweden, Lorenz S. Cederbaum of the University of Heidelberg, Germany, Souad Lahmar of the University of Tunis, Tunisia, Aristides Mavridis of the National University of Athens, Greece, Hiroshi Nakatsuji of the Quantum Chemistry Research Institute, Kyoto, Japan, Josef Paldus of the University of Waterloo, Canada, Alia Tadjer of the University of Sofia, Bulgaria, Carmela Valdemoro of CSIC, Madrid, Spain, Oleg Vasyutinskii of the Ioffe Institute, St. Petersburg, Russia, and Y. Alexander Wang of the University of British Columbia, Vancouver, Canada, for their invaluable suggestions and collective wisdom.

We thank the Local Organizing Committee of QSCP-XIII, particularly several members of the Department of Chemistry at Michigan State University, including Thomas V. Atkinson, Thomas P. Carter, Jeffrey R. Gour, Janet K. Haun, and Paul A. Reed, for their dedication and hard work at all stages of the workshop preparation and organization, which resulted in the smooth running of the meeting.

Finally, no scientific meeting can be successful without sponsors, and QSCP-XIII was no different in this regard. We are grateful for the financial support provided to the organizers of QSCP-XIII by several offices and units at Michigan State University: the Office of the Vice President for Research and Graduate Studies, the Office of the Vice Provost for Libraries, Computing, and Technology, the AT&T Lectureships in Information Technology Endowment, the Colleges of Natural Science and of Engineering, the Office of International Studies and Programs, the Departments of Chemistry and of Biochemistry and Molecular Biology, and the Quantitative Biology Initiative, as well as corporate sponsors: SGI, JRT, and Dell.

It is the editors' hope that the present two volumes will not only convey the breadth, depth, and dynamism of the QSCP-XIII workshop itself, but also seed innovative ideas in the wider research community.

April 2009

Piotr Piecuch
Jean Maruani
Gerardo Delgado-Barrio
Stephen Wilson

Contents

Part IV   Advances in Wave Function Methods

Part V   Advances in Density Functional Theory

Part VI Advances in Concepts and Models

Part I
Historical Overviews

An Illustrated Overview of the Origins and Development of the *QSCP* Meetings

Jean Maruani

Abstract The origins and development of the *QSCP* meetings are recalled: from a congress organized in Paris in 1986 to honour Prof. Raymond Daudel, through Franco-Bulgarian cooperation between various teams then European contracts, in the frame of *COST* projects, involving a network of French, Spanish, British, Italian, Swedish, and Bulgarian scientists, till the holding of the first *QSCP* workshop near Pisa in 1996 to honour Prof. Stefan Christov. After that there was a meeting every year, always with proceedings published. This historical overview will be presented as an illustrated journey in picturesque cities of Western and Eastern Europe, North Africa, and North America, sprinkled with historical or philosophical anecdotes and insights.

Keywords: *CMOA, TMOE, COST, QSCP, PTCP*, Origins, Development, Overview

1 Introduction

For this thirteenth international workshop on Quantum Systems in Chemistry and Physics (*QSCP-XIII*), the second to take place out of Europe, Prof. Piotr Piecuch has asked me to present an illustrated overview of the origins and development of these now established yearly meetings. The presentation will necessarily be linked to my own personal memories, and no doubt that another member of our committee might have put the stress on different aspects of the story. However it may help the historians to see how processes in the scientific world may develop also along subjective lines. In order to make the presentation less tedious I have used a variety of pictures as well as historical or philosophical anecdotes or insights.

In the next section, the background will be recalled: the tradition of *CMOA du CNRS*, founded by Prof. Raymond Daudel, to establish scientific societies and to organize scientific meetings; the series of Franco-Bulgarian exchange conventions and the *COST* European projects that lead to the first *QSCP* workshop, organized by Prof. Roy McWeeny near Pisa in 1996 to honour Prof. Stefan Christov. In the third section, we will present an overview of the *QSCP* meetings from the 1996, Pisa workshop to the 2001, Sofia workshop, where the *'Promising Scientist Prize'* (*PSP*) of *CMOA*, announced in Uppsala in 2000, was first awarded.

Jean Maruani
Laboratoire de Chimie Physique - Matière et Rayonnement, CNRS and UPMC,
11 Rue Pierre et Marie Curie, 75005 Paris, France, e-mail: jean.maruani@upmc.fr

P. Piecuch et al. (eds.), *Advances in the Theory of Atomic and Molecular Systems*,
Progress in Theoretical Chemistry and Physics 19, DOI 10.1007/978-90-481-2596-8_1,
© Springer Science+Business Media B.V. 2009

In Section 4, the next three workshops, from Bratislava (where *QSCP's* were moved from April to September) to the famous conference center of *Les Houches* (where *QSCP's* acquired international recognition), and in Section 5, the meetings of Carthage, St. Petersburg, and Windsor, will be recalled. In Section 6, an overview will be given of the Lansing meeting, the proceedings of which make up the present volumes, and forthcoming workshops will be announced. Finally, we shall describe the ceremony of award of the *PSP* of *CMOA*, and conclude this overview.

2 The *CMOA* Background and *COST* Projects

In 1986, Professor Imre Czismadia (Toronto University) and I organized an interdisciplinary congress in Paris, in honour of Prof. Raymond Daudel, on the very general topics of 'molecular sciences'. There were about 180 participants, including eight Nobel laureates. The Proceedings, which gathered 4 volumes totalling 1800 pages, appeared as *'Molecules in Physics, Chemistry, and Biology'* in a bookseries that was founded at that time at Kluwer's (Fig. 1): *'Topics in Molecular Organization and Engineering' (TMOE)*.

Raymond Daudel (1920-2006) had been an assistant to Irène Joliot-Curie in Chemistry, Antoine Lacassagne in Medicine, and Louis de Broglie in Physics. In the 1950's he founded what was to become the *'Centre de Mécanique Ondulatoire Appliquée (CMOA) du CNRS'* in Paris. During its 30 years of existence, hundreds of scientists from all over the world (including some among the most famous) paid visits to the *CMOA*, and dozens worked there. The *CMOA* became known as an international place of gathering for quantum chemists and physicists involved in atomic, molecular, biochemical and solid-state sciences, including structure, dynamics, reactivity, and spectroscopy. A short history of the early years of this institute can be found on its web page: http://www.lcpmr.upmc.fr/CMOAhista.html.

Raymond Daudel was a great organizer of congresses and summer schools and founder of scientific institutions, as also were some other members of *CMOA*; e.g.:

- The *'International Academy of Quantum Molecular Sciences' (IAQMS)*, that he created, in 1967, together with Bernard Pullman, Per-Olov Löwdin, John Pople, Robert Parr, and other prominent scientists, including L. de Broglie, Fock, Fukui, Heitler, Herzberg, Hückel, Kotani, Longuet-Higgins, McWeeny, Mulliken, Pauling, Pauncz, Slater, Van Vleck, and others. This academy, seated at Menton (France), organizes an international congress once every three years and awards every year a prestigious medal, the first recipients of which were W. Kolos (Poland), R. Levine (Israel), A. Dalgarno (USA), R. Hoffmann (USA), E. Davidson (USA), J. Jortner (Israel), J. Cizek (Canada), etc.

- The *'European Academy of Arts, Sciences, and Humanities' (EAASH)*, that has come to act as a consulting body for such international organisations as *WHO*, *UNESCO*, etc. Since Daudel's passing in 2006, it is chaired by Prof. Guy de Thé (*Pasteur Institute*), its Secretary General remaining the painter Nicole D'Aggagio.

- Carl Moser, a former member of *CMOA*, founded the *'Centre Européen de Calcul Atomique et Moléculaire' (CECAM)*, which organizes regular workshops in France and other European countries.

- Imre Czismadia (a Hungarian-born Canadian), former regular visitor to the *CMOA* and member of the *European Academy*, founded the *'World Association of*

Theoretical Organic Chemists' (WATOC) - it changed its name afterwards - which organizes congresses alternating with those of *IAQMS* every three years.

- Janos Ladik (another Hungarian), former post-doc of the Pullman's, created the *'International Society for Theoretical Chemical Physics' (ISTCP)*, which holds congresses also every three years (there was one at Vancouver, British Columbia, Canada, in July 2008, following and coordinated with the *QSCP-XIII* meeting).

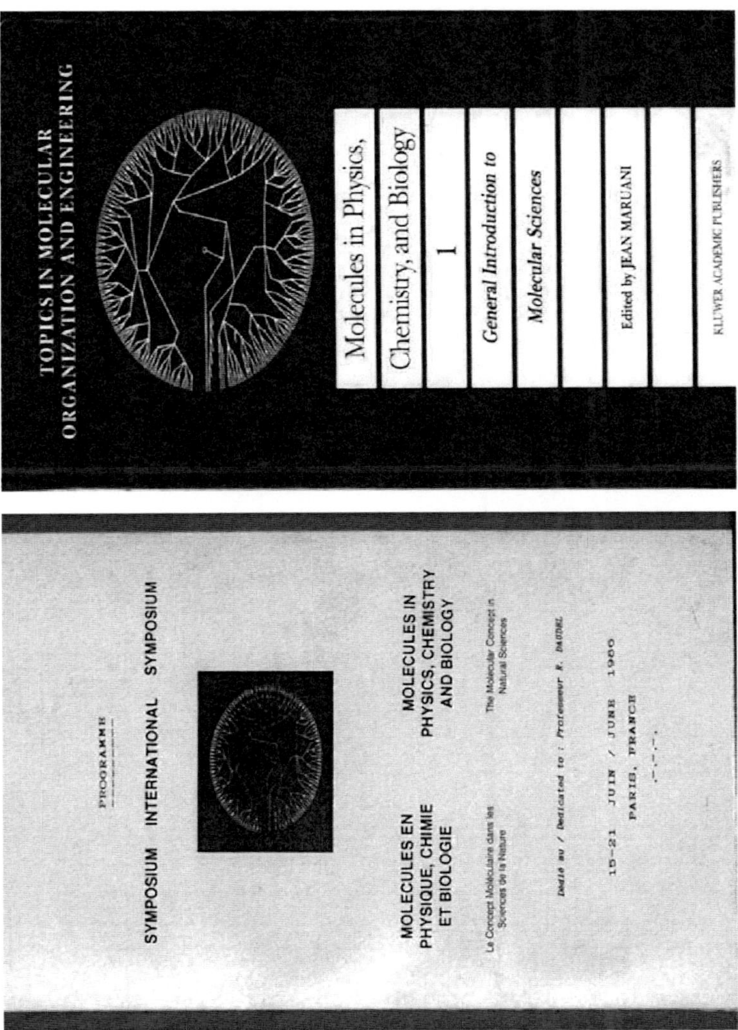

Fig. 1 The covers of the booklet of abstracts (lower) and of the first volume of proceedings (upper) of 'Daudel's Congress' (Paris, France, July 1986), which started the *TMOE* series (Kluwer, 1988) [3]

- Yves Smeyers (a Belgian-born Spaniard), former post-doc of Prof. Roland Lefebvre at *CMOA* and member of the *European Academy*, founded the *QUITEL*, which organizes Latin-language speaking meetings every two years, although their proceedings are nowadays usually published in English in *IJQC* ....

Raymond Daudel also published several volumes, including the first true textbook on *'Quantum Chemistry: Theory and Applications'*, with Roland Lefebvre and Carl Moser (Wiley, 1959). Earlier volumes dealt mostly with concepts (such as Pauling's in the 1930's or Coulson's in the 1950's), or specific topics (Eyring, Walter and Kimball in the 1930's, Wheland in the 1930's, Syrkin and Diatkina in the 1940's). Originally written in English, Daudel's textbook was translated into other languages, including Russian and Japanese.

After Raymond Daudel officially retired and the *CMOA du CNRS* broke up, in order to monitor the congress organized in 1986 to honour him, I created an association called *CMOA*, with a maximum number of 12 fellows, of which Pr. Daudel accepted to be the Honorary President. A detailed obituary of Raymond Daudel has appeared in vol. 16 of this bookseries.

The organization of the 1986 congress and the founding of the *TMOE* bookseries, with subsequent publication of the proceedings of the congress in the series, helped me improve my earlier experience in these matters [1, 2], and brought me in touch with a number of people who were to play a crucial role in the *QSCP* network: Bulgarian Academician Stefan Christov, Prof. Roy McWeeny from Pisa, a young, bright Polish fellow called Piotr Piecuch (who submitted a huge paper on molecular interactions) and, last but not least, a young, strange Bulgarian fellow called Rossen Pavlov, who stammered in a stressing manner but could make ladies faint by practising the old-fashioned hand-kissing, his long blond hair covering his bearded face while he was bowing his slender body down (Fig. 2).

Let me tell you how this started the chain of events that led to the *QSCP* network. Rossen Pavlov was a former scientific secretary of Julia Vassilieva Popova, a former director of an Institute of Biochemistry of the Bulgarian Academy of Sciences (and the wife of a former head of Bulgarian Secret Services). He was also a kind of guru for the daughter (and Minister of Culture) of former President Todor Zhivkov. Thanks to his friend Yavor Delchev, he later joined the Institute for Nuclear Research and Nuclear Energy. He was a person who could bring you in touch with anyone you wished. After I arranged for some European money to be sent to various Bulgarian Institutes and gave a few lectures (on exotic symmetries) at the University of Sofia, he and Prof. Mattey Mateev (a former Minister of Research) managed me to receive a *Doctorate Honoris Causa* from that university. He himself received his PhD in the late 1990's from the University of Paris-VI. He comes from that former bourgeois class who spoke French as a second language, and most of our common papers were therefore written in French.

We met again the following year in Sofia, where I was invited by ... Pr. Stefan Christov, at a *IUPAC* Congress. That was in 1987, when Western European policy was to draw Eastern Europe away from the Soviet Union by offering financial help through common projects. One of these projects was called *PECO* (later renamed *COST*), and Rossen Pavlov, stammering more than ever, insisted on involving me into a cooperation within this frame. But a single French fellow facing an army of Bulgarian scientists did not look serious enough for the European Commission, and it took four years before the first acceptable project could emerge.

The Bulgarians included the nuclear chemical physicists Rossen Pavlov and Yavor Delchev, the solid-state physicist Serguei Georgiev, the molecular spectroscopist Peter Raychev, the organic chemist Ivo Kanev, and a score of other people headed by Academician Stefan Christov. The project also involved colleagues from

Madrid (Yves Smeyers, who already knew Alia Tadjer in Sofia), from Pisa (Roy McWeeny, one of the founders of the reduced density matrix formalism, a keynote of our project), and from London (Stephen Wilson, whom Roy had known before but who was proposed by Sonia Rouve, a friend of Rossen Pavlov and the wife of a former 'attaché culturel' of Bulgaria in London).

Fig. 2 Lower left: Rossen Pavlov helping to prepare the Carthage workshop in Souad Lahmar's office. Upper left: voting at the Bulgarian consulate in *Parc du Belvedere* in Tunis. Lower right: Rossen sitting with daughter at a garden party in Zohra Lakhdar's home. Upper right: trying to find a drink in a sidewalk café in Tunis. This tenth *QSCP* meeting was the first held outside Europe

The founding meeting of our European project took place in an apartment lent by Anne Burchett, facing Sofia's Park Theatre, during a wild spring shower. Anne was a cousin of Yavor Delchev and the daughter of the nonconformist Australian journalist Wilfred Burchett, who was the first to take pictures at Hiroshima after the nuclear blast and to denounce the effects of radioactivity (Fig. 3). Later on he opposed the Vietnam War, married a Bulgarian woman (sister of Yavor's mother), and eventually joined Yavor's father as a dissident to the Soviet regime.

I was the only Western scientist facing a dozen Bulgarians, including the above named plus Stefan Christov and coworkers. But we had the agreement of Smeyers, McWeeny and Wilson. I drew up a project fulfilling the Bulgarians' requests, trying to include as many topics as possible in a hardly consistent patchwork (Fig. 4). Eventually, it appeared that the main objective of our Bulgarian fellows was to get some money to develop previous projects. We also, from Western Europe, became fond of these yearly ritual visits, with all our expenses paid by the European Commission: we could chat about our own work as well, while being treated as VIP's by our Bulgarian colleagues.

Fig. 3 Left: Wilfred Burchett denounced as anti-American for his reports on the Vietnam War. Right: His first report on the lethal effects of atomic irradiation after Hiroshima

We first received 80,000 'Ecus' to be shared between our teams, mainly for computer equipment (the Ecu was the ancestor of the Euro). For Bulgaria, where the average salary was about 50 dollars per month, that was a tremendous amount. The project then was renewed under a different framework (Fig. 4). All this lasted between 1992 and 2002 and brought us, altogether, half a million Ecus (including travel and living expenses between Bulgaria and Western Europe and sponsoring of *QSCP* meetings, Fig. 5).

Let me tell you an anecdote about the 'Ecu'. The name had been proposed by the former French President Giscard d'Estaing; it was the name of a past French currency while also being the acronym of an English expression: 'European Currency Unit'. Some believe it was for the very same reasons that Germans rejected it, although others claim it was because in German *Ein Ecu* would sound like *Eine Kühe* (a cow!). Therefore, when European currencies were unified the name *Euro* was adopted, with a symbol (€) reminding those of the USD ($) and of the BP (£).

In 1996, Yavor Delchev proposed to have a meeting to honour Academician Stefan Christov, who was at that time the official head of our project on the Bulgarian side (although the project was financially managed by Prof. Jordan Stamenov, head of the *INRNE*), while I was in charge on the Western side (Fig. 4). We agreed that Roy McWeeny would be the best person to organize that meeting, which took place in Italy, near Pisa. As we were sponsored by the European Commission, we had a preparatory meeting on the spot, with a few national representatives (Fig. 5).

3 The *QSCP* Meetings from Pisa through Sofia

It was in Pisa that we choose the generic name: *'Quantum Systems in Chemistry and Physics' (QSCP)*, in order to accommodate the wide variety of topics in which we were all supposed to be involved. The name was coined while we were sitting at *Hotel Di Stefano*, close to Pisa's leaning tower, wondering if our project would have the same stability! The workshop actually took place in a monastery close to San Miniato, a suburb of Galileo's city.

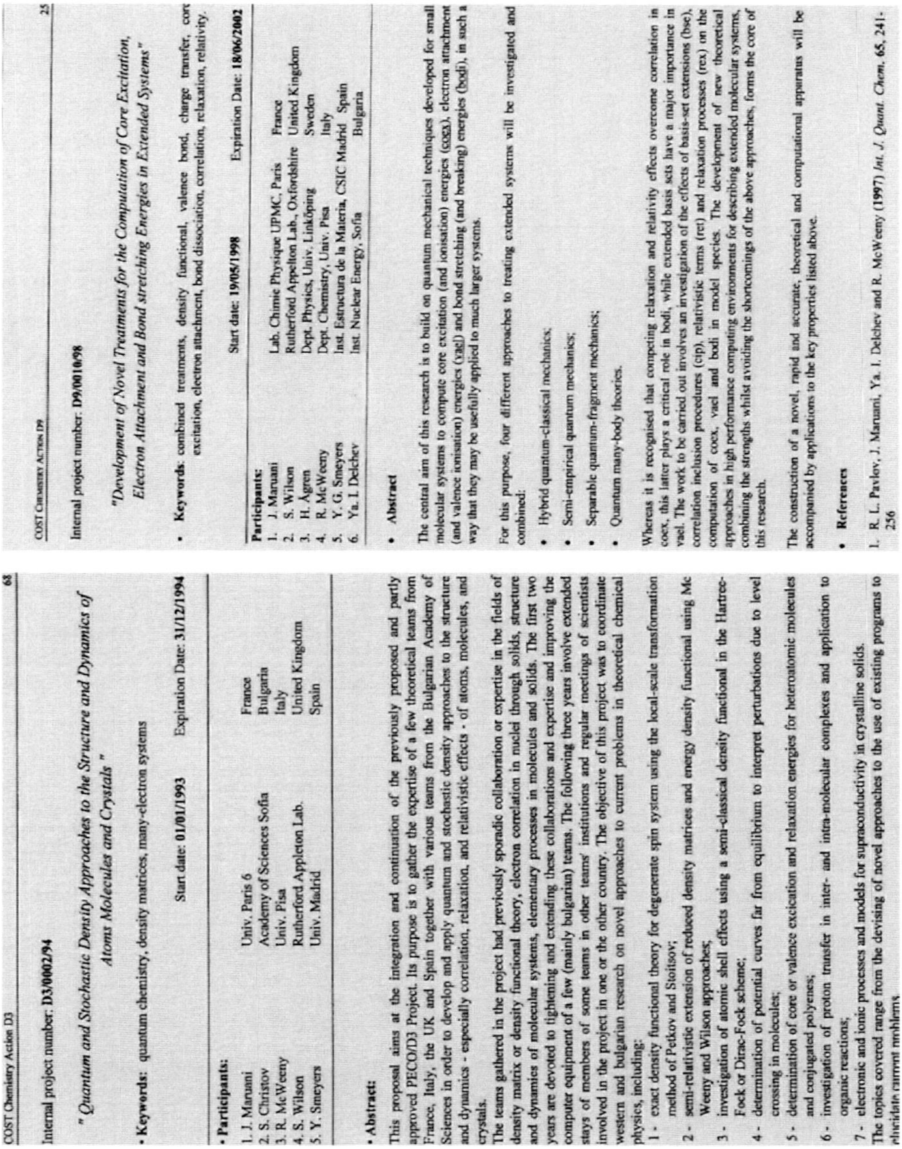

COST Chemistry Action D3 — 68

Internal project number: **D3/0002/94**

" Quantum and Stochastic Density Approaches to the Structure and Dynamics of Atoms Molecules and Crystals "

- **Keywords:** quantum chemistry, density matrices, many-electron systems

Start date: **01/01/1993**   Expiration Date: **31/12/1994**

- **Participants:**

1. J. Maruani	Univ. Paris 6	France
2. S. Christov	Academy of Sciences Sofia	Bulgaria
3. R. McWeeny	Univ. Pisa	Italy
4. S. Wilson	Rutherford Appleton Lab.	United Kingdom
5. Y. Smeyers	Univ. Madrid	Spain

- **Abstract:**

This proposal aims at the integration and continuation of the previously proposed and partly approved PECO/D3 Project. Its purpose is to gather the expertise of a few theoretical teams from France, Italy, the UK and Spain together with various teams from the Bulgarian Academy of Sciences in order to develop and apply quantum and stochastic density approaches to the structure and dynamics - especially correlation, relaxation, and relativistic effects - of atoms, molecules, and crystals.

The teams gathered in the project had previously sporadic collaboration or expertise in the fields of density matrix or density functional theory, electron correlation in nuclei through solids, structure and dynamics of molecular systems, elementary processes in molecules and solids. The first two years are devoted to tightening and extending these collaborations and expertise and improving the computer equipment of a few (mainly bulgarian) teams. The following three years involve extended stays of members of some teams in other teams' institutions and regular meetings of scientists involved in the project in one or the other country. The objective of this project was to coordinate western and bulgarian research on novel approaches to current problems in theoretical chemical physics, including:

1 - exact density functional theory for degenerate spin system using the local-scale transformation method of Petkov and Stoitsov;
2 - semi-relativistic extension of reduced density matrices and energy density functional using Mc Weeny and Wilson approaches;
3 - investigation of atomic shell effects using a semi-classical density functional in the Hartree-Fock or Dirac-Fock scheme;
4 - determination of potential curves far from equilibrium to interpret perturbations due to level crossing in molecules;
5 - determination of core or valence excication and relaxation energies for heteroatomic molecules and conjugated polyenes;
6 - investigation of proton transfer in inter- and intra-molecular complexes and application to organic reactions;
7 - electronic and ionic processes and models for supraconductivity in crystalline solids.

The topics covered range from the devising of novel approaches to the use of existing programs to elucidate current problems.

COST CHEMISTRY ACTION D9 — 25

Internal project number: **D9/0010/98**

"Development of Novel Treatments for the Computation of Core Excitation, Electron Attachment and Bond stretching Energies in Extended Systems"

- **Keywords:** combined treatments, density functional, valence bond, charge transfer, core excitation, electron attachment, bond dissociation, correlation, relaxation, relativity.

Start date: **19/05/1998**   Expiration Date: **18/06/2002**

Participants:

1. J. Maruani	Lab. Chimie Physique UPMC, Paris	France
2. S. Wilson	Rutherford Appelton Lab., Oxfordshire	United Kingdom
3. H. Ågren	Dept. Physics, Univ. Linköping	Sweden
4. R. McWeeny	Dept. Chemistry, Univ. Pisa	Italy
5. Y. G. Smeyers	Inst. Estructura de la Materia, CSIC Madrid	Spain
6. Ya. I. Delchev	Inst. Nuclear Energy, Sofia	Bulgaria

- **Abstract**

The central aim of this research is to build on quantum mechanical techniques developed for small molecular systems to compute core excitation (and ionisation) energies (coex), electron attachment (and valence ionisation) energies (vael) and bond stretching (and breaking) energies (bodi), in such a way that they may be usefully applied to much larger systems.

For this purpose, four different approaches to treating extended systems will be investigated and combined:

- Hybrid quantum-classical mechanics;
- Semi-empirical quantum mechanics;
- Separable quantum-fragment mechanics;
- Quantum many-body theories.

Whereas it is recognised that competing relaxation and relativity effects overcome correlation in coex, this latter plays a critical role in bodi, while extended basis sets have a major importance in vael. The work to be carried out involves an investigation of the effects of basis-set extensions (bse), correlation inclusion procedures (cip), relativistic terms (ret) and relaxation processes (rex) on the computation of coex, vael and bodi in model species. The development of new theoretical approaches in high performance computing environments for describing extended molecular systems, combining the strengths whilst avoiding the shortcomings of the above approaches, forms the core of this research.

The construction of a novel, rapid and accurate, theoretical and computational apparatus will be accompanied by applications to the key properties listed above.

- **References**

1. R. L. Pavlov, J. Maruani, Ya. I. Delchev and R. McWeeny (1997) *Int. J. Quant. Chem.* **65**, 241-256

Fig. 4 Our successive *COST* projects. Lower: COST-D3 (1993-1997), which replaced an earlier *PECO* project that started in 1992. Upper: COST-D9 (1998-2002)

It was also in Pisa that the pattern for following meetings was set up: 1 - the period would be mid-spring (later it became late summer); 2 - the duration would be 4-5 days (later it became 6-7 days), with a week-end in the middle to allow for cultural events and informal discussions; 3 - the venue would be a small (and every year different) European city with a high cultural background; 4 - all talks would have the same length (15 to 25 min, depending on the number of attendants, with 5 min for discussion), details being displayed on posters; 5 - registration fees would be all-inclusive: organisation expenses, room and board, social programme, etc.

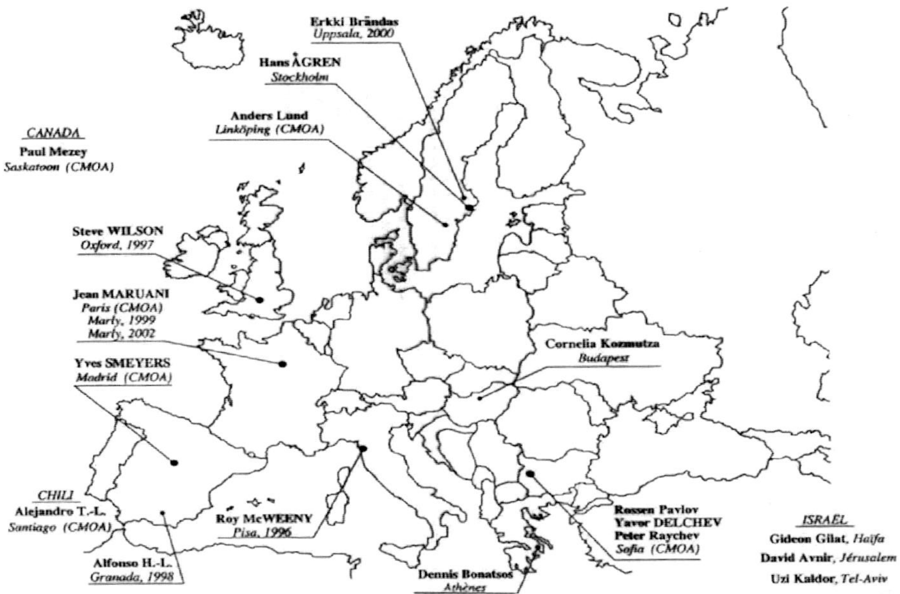

Fig. 5 The extended *QSCP* European network around the year 2000. The names of official national representatives are given in capital letters. There are also shown network contacts belonging to the Board of *CMOA* and / or having organized a *QSCP* meeting. Some other contacts in various countries are also listed

Also included in the 'registration fees' were the banquet dinner, a free copy of the proceedings, and a provision to help a few young fellows with limited financial means, by paying part or all of their local expenses, but never travel expenses. The social programme would include a city tour and a music concert. Later, other features were established, such as announcing the following meeting and awarding a prize at the banquet dinner.

Although it gathered only 65 people, this first *QSCP* workshop was a qualitative success. The European Commission delegate, Sylvie Benefice-Malouet, was so impressed that she decided to increase the financial allowance of our network. The proceedings of this workshop were published in *TMOE*, yielding the 16th and last volume of the bookseries (Fig. 6).

During the workshop, while touring Toscana in a bus, we managed to convince Stephen Wilson to organize another meeting, in England. He was then working at *Rutherford Appleton Laboratory* but he succeeded to involve Peter Grout from Oxford University. The second *QSCP* workshop took place at *Jesus College* (Fig. 7), and about 50 people attended the meeting. A poll showed that most participants expected to have their contribution published again. With the help of Erkki Brändas, two volumes appeared in *'Advances in Quantum Chemistry' (AQC)* [5].

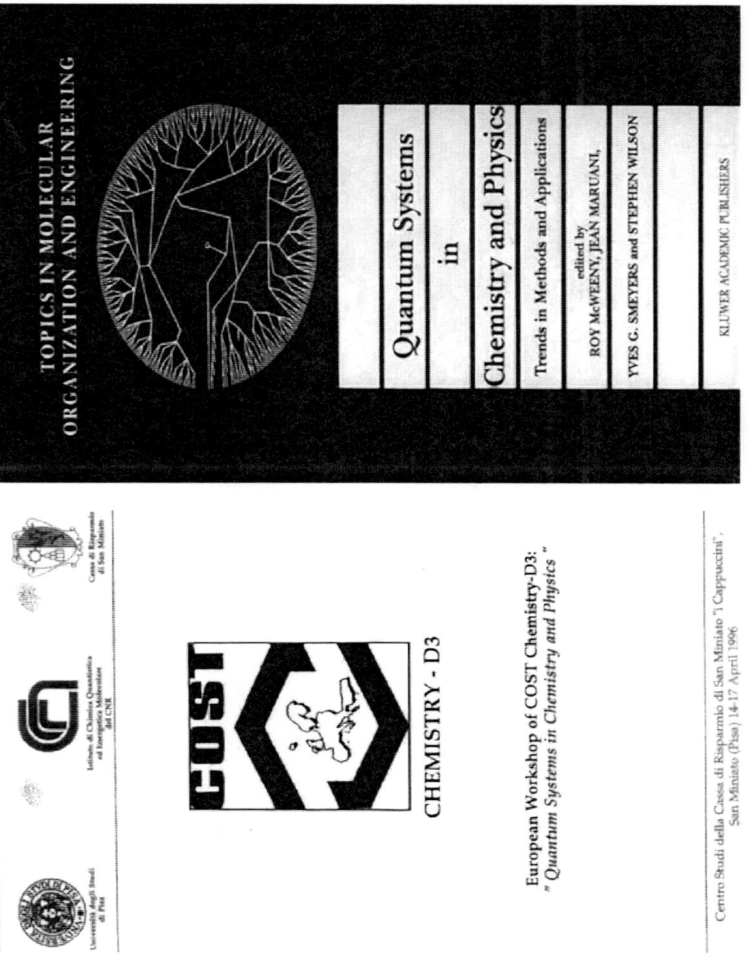

Fig. 6 The covers of the booklet of abstracts (lower) and of the book of proceedings (upper) of the Ist *QSCP* workshop (San Miniato, Italy, April 1996), which ended the bookseries (Kluwer, 1997) [4]

The following workshop was organized by a former student of Smeyers (and former post-doc of mine), Alfonso Hernandez-Laguna, at *Hotel Alixares* in Granada, Spain. I knew Alfonso had been disappointed by his proposal for a *QUITEL* symposium in Spain not being retained during an earlier meeting in Pucon, Chile. He was thus ready for a challenge: 'If you can't join them, beat them', I told him. The Granada workshop gathered 99 people, and its proceedings were published in two volumes, totalling over 800 pages, in the bookseries *'Progress in Theoretical Chemistry and Physics' (PTCP)*, that we had newly founded at Kluwer's [6].

Originally intended to be some kind of 'open encyclopaedia' of molecular sciences, that would update and extend our successful former book set *'Molecules in*

Physics, Chemistry, and Biology', this series was divided into two subsets: one (A) for monographs (yellowish cover) and the other one (B) for proceedings (greenish cover). The background pattern was an enlarged segment of the 'generative hyper-structure of anteriology relationships' of alkanes devised by Prof. Jacques-Emile Dubois (*ITODYS*, University Paris-VII), which illustrated the volume covers in the former bookseries (Figs 1 & 6).

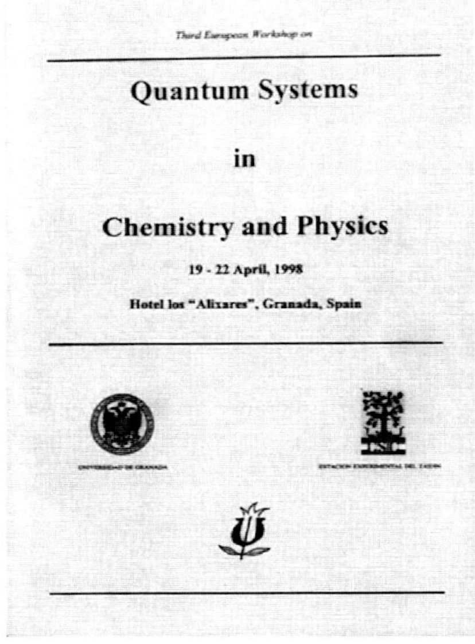

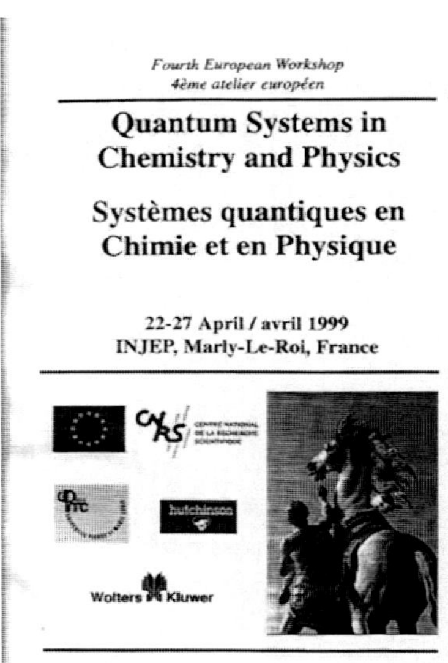

Fig. 7 From up and left to down and right: booklets of abstracts of the Oxford (England), Granada (Spain), Marly-le-Roi (France), and Uppsala (Sweden) *QSCP* workshops

The name of this bookseries was coined while Roy, Steve and I were having dinner at Granada, in front of a sculpture depicting Queen Isabella (who had then defeated the Moors and expelled the Jews) granting 'patent' to Columbus to search for India through the West (which led to the 'discovery' of America).

Then it was my turn to organize a *QSCP* meeting, near Paris (Fig. 8). By that time we had become used to select towns with special historical or cultural interest rather than big cities. The meeting took place at *INJEP* in Marly-le-Roi, the cradle of French royalty and summer residence of Louis-XIV. We had a memorable visit to the archaeological museum of Saint-Germain-en-Laye and a gorgeous banquet at the French *Senate House*. The meeting gathered 114 people, and its proceedings were published again as two volumes, totalling 750 pages, of *PTCP* [7].

Fig. 8 Left: Jean Maruani, Yves Smeyers, Stephen Wilson, and Roy McWeeny preparing *QSCP-IV* in Bureau Jean-Perrin at *LCP* in Paris. Upper right: the same with Christian Minot (left), Hans Ågren (middle right), and others in Maruani's home in Paris. Lower right: from the left, Sten Lunell, Erkki Brändas, Yves Smeyers, and others preparing *QSCP-V* in Lunell's office in Uppsala

A turning point occurred at the *QSCP-V* meeting, in 2000. In 1997, in order to have our European project renewed and also because I was working, at that time, on core excitations in molecules, we had added Hans Ågren (then at Linköping) to our network (Figs 4 & 5). But when his turn came to arrange for a meeting in Sweden, he was unable to do so because he had to take care of a newly born child and was busy moving to *KTH* in Stockholm. Now, I really wanted to have a meeting in Sweden, the country of the Nobel Prize, for the end of the millennium. Erkki Brändas then proposed to have one held in Uppsala, the seat of the oldest Scandinavian University, with the support of Sten Lunell (Fig. 8), provided he would be incorporated into our European network and the proceedings would be published in *AQC*, where he was an executive editor. It was done as he wanted, the meeting gathered 94 people, and yielded again two volumes in *AQC* [5].

In Figure 7 it can be noticed that the covers of the booklets of abstracts of the French and Swedish meetings present several similarities. The most striking one is that both covers bear, for the first time, a bilingual title. For the French this was required in order to get subsidies from French bodies. But why did the Swedes also have a bilingual title, and also in French? You may think that Erkki tried to return the favour I granted by complying with all his conditions. In addition, there were more attendants from France, sponsored by their government, than from any other country (except Sweden, of course). But then, why was it so?

I think this can be related to a Franco-Swedish connection that goes back far in history: first, with René Descartes and later, with Marshal Bernadotte (Fig. 9), but also with the numerous French Protestants who emigrated to escape religious persecutions after the Revocation of the Edict of Nantes. Jean-Louis Calais, a prominent member of Löwdin's group, was a descendant of these emigrants.

Fig. 9 Left: René Descartes (1596-1650), the founder of modern philosophy and mathematics, spent his last years teaching Queen Christina of Sweden. Accustomed to working in bed until noon, he suffered from the Queen's demands for early teaching and died of pneumonia. Right: Jean-Baptiste Bernadotte (1763-1844) was elected as King Carl XIV of Sweden and Norway after Finland was lost to Russia. He married Napoleon's ex-fiancée Eugénie Désirée Clary, with whom he begot Oskar I, who married the daughter of Napoleon's wife, Joséphine de Beauharnais. With the Bernadotte dynasty, the warlike Vikings entered a lasting culture of peace

It was at the banquet dinner of the Uppsala workshop that we announced the foundation of a *'Promising Scientist Prize'* of *CMOA* and called for nominations for the following meeting, which was to be held in Sofia, Bulgaria, in April 2001.

To have our sixth meeting in Bulgaria was a challenge: some members of our International Scientific Committee, although they had been received for years with kindness and efficiency, argued that our local organizers might not be ready to set up an efficient web site and organize a successful workshop. However, the meeting was a real success, with nearly 70 attendants, and yielded a thick special issue of

IJQC [5]. As one local organizer, Alia Tadjer, had political connections, we were accommodated at *Boyana Palace*, the 'White House' of the Bulgarian President, and had an interview on Bulgarian TV before our visit to Riila Monastery (Fig. 10). It was at the banquet of the Sofia workshop that the *PSP* of *CMOA* was awarded for the first time (Fig. 11).

Fig. 10 Left: A group visit to Riila Monastery, 80 km from Sofia, during *QSCP-VI*. Right: Roy Mc Weeny chatting with Alia Tadjer, one of the main organizers

Fig. 11 The first ceremony of award of the PSP of *CMOA*, held at *Boyana Palace* (Sofia, Bulgaria, April 2001). The three selected nominees (among seven who applied) stand beside their ceremony lady guide. From left to right: the Prize recipient, Edvardas Narevicius (Technion, Haifa, Israel), Eric Bittner (Houston, Texas, USA), and Wenjian Liu (Bochum, Germany)

4. The *QSCP* Meetings from Bratislava through *Les Houches*

Then we had a *QSCP* workshop at *Casta Paprinicka*, near Bratislava (Slovakia), where we were housed in a former 'Nomenclatura' Residence, with several pools and fitness rooms but no piano (surprisingly for a place so close to the Vienna and Prague of Mozart). There were hardly 40 attendants, mostly from Czechia or Slovakia, and the proceedings yielded a slim issue of *IJQC* [5]. The meeting suffered from several odds. First, I was then busy organizing the *ISTCP-IV* congress near Paris (Fig. 12), and Stephen Wilson was also busy holding a summer school at Oxford. Secondly, the main local organizer, Prof. Ivan Hubač, had insisted on having the workshop held in September while our participants were accustomed to having *QSCP* meetings in spring. Last but not least, we had stopped being sponsored by the European Commission, although (or maybe because) the fellow who had come in charge of *COST* Chemistry at Brussels was a Slovak. Nevertheless, there was a ceremony of award of the Prize of *CMOA*, with two nominees sharing 1,000 Euros: Marcin Hoffmann (Poznan, Poland) and Alexander Kuleff (Sofia, Bulgaria).

From this experience we reckoned that, if *ISTCP* and *QSCP* meetings were to take place again a few weeks apart on the same continent, it would be good to have them coordinated by the organizers. This was done very efficiently, in July 2008, between the *QSCP-XIII* (Lansing) workshop and *ISTCP-VI* (Vancouver) congress.

Starting with the Bratislava workshop, we held our meetings around August / September, rather than in April / May. There was no *CMOA* award at the meeting organized by Aristides Mavridis at Spetses, near Athens (Greece), in September 2003 (Fig. 13). But some nice memories remain, some of which are shown in pictures (Fig. 14). The meeting gathered about 70 people, and the proceedings were published in a special issue of *IJQC* [5].

Until the Bratislava workshop, our International Scientific Committee (ISC) had remained more or less as it was since the Uppsala meeting. Afterwards, Hans Ågren formally withdrew and was replaced by Erkki Brändas (earlier involved as editor-in-chief of *AQC* and *IJQC*), Yves Smeyers passed away and was replaced by his former student Gerardo Delgado-Barrio (then the President of the *Spanish Royal Society of Physics*), and Roy McWeeny shifted progressively from chemistry to sculpture. Besides, antispam regulations made it more difficult for Stephen Wilson to send announcements by mass e-mailing, and I took over this task while expanding its database.

Now I will tell you how we came to have the workshop following Spetses at *Les Houches*, near Grenoble (France). Following the *ISTCP* and *QSCP* meetings in 2002, Prof. Daudel had me invited by the *Tunisian Academy* to a *European Academy* meeting in Carthage, where I was to make a speech on the topic: 'Can cooperation in hard sciences help relieve tensions in the Mediterranean basin?' There I met Prof. Zohra Lakhdar, who was heading a small group of chemical physicists (she later received the prestigious *L'Oréal* Prize for women scientists), and asked her if she would be interested in organizing the next *QSCP* workshop. She agreed, and was invited to the Spetses meeting.

However, she did not show up at Spetses, and we urgently needed someone to announce the next meeting during the banquet. Now, in that island we had to go to buy drinkable water at a store; there, I met Prof. J.-P. Julien, who expressed his appreciation and asked where the next meeting would take place. I bent to him and whispered, as a secret: 'near Grenoble'. 'That's where I work!' he exclaimed; 'and who will be in charge?' I whispered again: 'You!' It worked .... After he consulted the Director of his Lab, Prof. Didier Mayou, the announcement could be made.

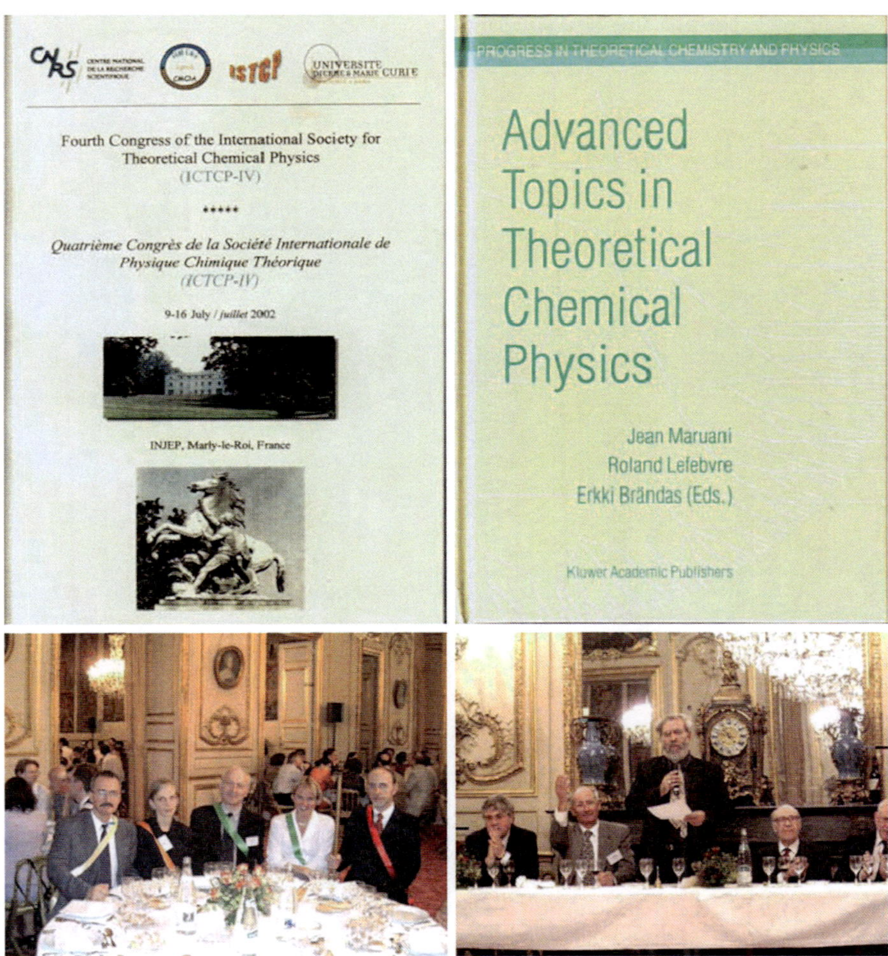

Fig. 12 Upper pictures: covers of the booklet of abstracts and of the book of proceedings in *PTCP* of the *ISTCP-IV* congress (Marly-le-Roi, France, July 2002); another volume of proceedings appeared in *IJQC* [8]. Lower pictures: the banquet dinner in the French *Senate House*; left: Paul Mezey, Agnes Vibok, Gerardo Delgado-Barrio, Angelina Hansman, and Osman Atabek holding office for the ceremony of award; right: Stephen Wilson, Roland Lefebvre, Jean Maruani, Raymond Daudel, and Janos Ladik sitting at the front table. There were three nominees to the Prize of *CMOA*, which was awarded to Agnes Nagy (Debrecen, Hungary)

The *QSCP-IX* workshop was incorporated into the summer programme of the famous *Ecole de Physique des Houches*, whose pattern had inspired *NATO Advanced Study Institutes*. It gathered about 70 people, and its proceedings appeared in *PTCP* [9]. Our banquet was held at restaurant *'La Calèche'* in Chamonix, close to the *Mont Blanc* (the tallest mountain in Europe). There were three selected nominees to the Prize of *CMOA*, which was awarded to Pr. Piotr Piecuch (Fig. 15). This time, Zohra Lakhdar was there, and announced the next meeting in Carthage.

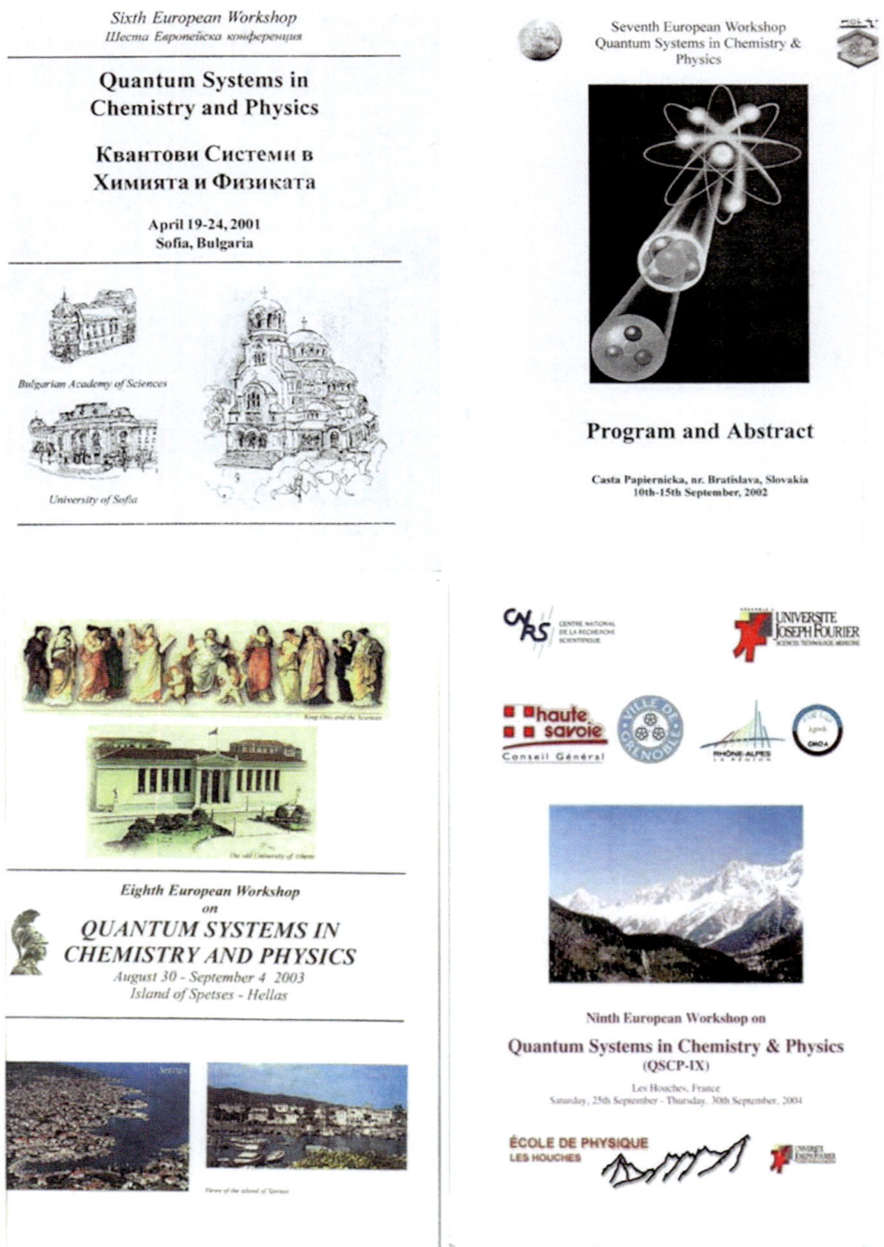

Fig. 13 From upper left to lower right: booklets of abstracts of the Sofia (Bulgaria), Bratislava (Slovakia), Spetses (Greece), and *Les Houches* (France) *QSCP* workshops

Fig. 14 Memories of the Greek meeting. Upper left: the conference room. Upper right: a view from a hotel room. Lower: Jean Maruani, Kate Wilson and Walter Kohn singing 'My Bonnie's gone over the Ocean' in the 2700-year old open-air theatre of Epidaur

5 The *QSCP* Meetings from Carthage through Windsor

As we were no longer sponsored by the European Commission since the Slovakian workshop, it became possible to have a meeting outside Europe. The initial suggestion was to organize it in Jerusalem, which is located in Asia but is historically one of the cradles of Western civilisation. However, none of our contacts in Israel (Fig. 5) were willing to take the risk, and we had to find another appropriate location for our tenth meeting. Although it had not worked at Spetses, we tried again Carthage, which had the privilege of presenting old symbolic links with the holy city.

Fig. 15 Upper: a group picture at *Les Houches*. Lower: Prof. R. Daudel handing his diploma to Piotr Piecuch during a *European Academy* meeting at *UNESCO* in Paris

It is known that, by looking hard enough, one can always find correlations between any two events. Titius-Bode's law, which relates the radii of planets' orbits to their rank from the Sun, is an example of such a correlation: even though it helped discover Allen's belt it is most likely fortuitous. Similarities between the conditions of assassination of Presidents Lincoln and Kennedy are also well documented: the most relevant is that both were fighting for civil rights and the weirdest one that, a week before they were shot, Lincoln was in Monroe, Maryland and Kennedy was … with Marilyn Monroe! Listed in the legend to Fig. 16 are a few historical similarities between Jerusalem and Carthage. Admittedly, neither of these two cities is, strictly speaking, in Europe. However, the very name of *Europe* comes from that of a Phoenician princess who was taken away to the *West* (for in Phoenician as in Hebrew, *'Erev* means *Sunset* or *Evening*).

Fig. 16 Left: Roman Ruins of Carthage. Right: Dome of the Rock in Jerusalem

1. About 950 BC, Hiram of Tyre spends 20 years helping King Solomon built his (the first) temple (and palace) in Jerusalem. About 815 BC, Phoenicians from Tyre build Carthage on the North-East Coast of North-Africa
2. In 146 BC, Romans led by Scipio, after a 3-year Punic war, reduce Carthage to ruins. In 70 AD, Romans led by Titus, after a 3-year Jewish war, reduce Jerusalem to ruins. A few centuries later, the King Genseric of the Vandals, sailing from Carthage, plunders Rome, while the Christian religion, expanding from Jerusalem, conquers Rome
3. According to Procopius of Caesarea, biographer of Justinian, the holy treasures from King Herod's (the second) temple, brought to Rome by Titus, were taken to Carthage by the Vandals, then to Byzance by Justinian, and finally back to monasteries in Jerusalem
4. About 690 AD, the fifth Umayyad caliph Abd el Malik builds the Dome of the Rock on the ruins of Herod's temple in Jerusalem. In 697, the same caliph resumes the conquest of North Africa, wins the Berbers to his side, and captures Carthage from the Byzantines

The *QSCP-X* workshop was held in the prestigious premises of the *Tunisian Academy*, a former Bey palace, the participants being housed at *Hotel Amilcar*. It gathered some 60 people (e.g., Fig. 17), and its proceedings were published as vol. 16 of *PTCP* [10]. The Prize of *CMOA* was shared between Richard Taïeb (France) and Majdi Hochlaf (Tunisia). Then came the next meeting, which was also to be held in an exceptional venue (Fig. 18).

Fig. 17 Left: Eugene Kryachko and Eduardo Ludeña gathered at the Carthage workshop. Right: Oleg Vasyutinskii chairing a talk given by Valentin Nesterenko

One of the cities that exerted fascination on us was St. Petersburg, which had recovered its imperial name from the former, soviet Leningrad. When I looked for a fellow to organize a meeting there, we thought of a few people who had come to earlier *QSCP* workshops (e.g., Fig. 17) or to the sixth *ISTCP* congress (Fig. 12): a name emerged: Oleg Vasyutinskii, earlier suggested by Hélène Lefebvre-Brion. It was not an easy task to convince him and I felt real great when, one evening, while working at home on computer, I received an e-mail from him telling me that, after consulting with his coworkers, he accepted to organize the 2007 *QSCP* workshop in St. Petersburg.

The meeting took place in August, due to weather conditions. We were housed at *Kochubey Palace* in Pushkin, a 'chic' suburb of St. Petersburg. The workshop gathered 96 scientists from 28 countries, including 32 from Russia. It was of great scientific value, and its proceedings yielded a large issue of *IJQC* [5]. It also left us with some unforgettable memories (Fig. 19 upper). Again, the Prize of *CMOA* was shared between two nominees (Fig. 19 lower). But, for the first time, the ceremony of award was not held in the banquet room. It was preceded by an obituary speech (accompanied by Vivaldi's *Nisi Dominus* aria) in the memory of Raymond Daudel, which appeared in vol. 15 of *PTCP*, then in print.

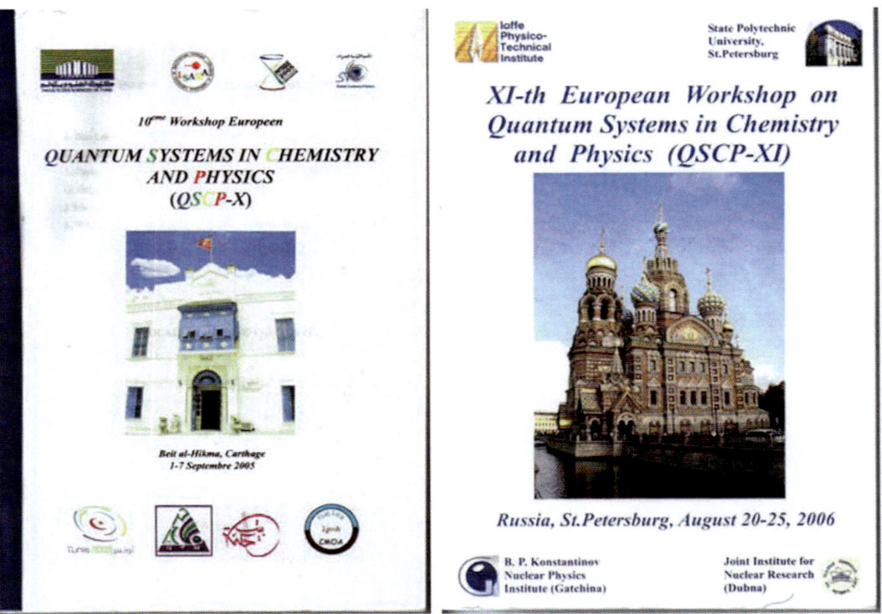

Fig. 18 Covers of the booklets of abstracts of the Carthage (Tunisia) and the St.-Petersburg (Russia) *QSCP* workshops

Twelve is a traditional 'magic' number: it is the approximate ratio of the Sun to the Moon cycles, has six dividers, and is used to express 'completeness' (hence the 12 constellations in the sky, tribes of Israel, apostles of Jesus, and even stars on the European flag - even though there are now 27 countries in the EU). This made me think again of Jerusalem (12 gates in *Revelation*) for our twelfth meeting. But eventually I turned to our old friend Stephen Wilson, on whom we could always rely. He proposed *Royal Holloway College* of the University of London, where his son was studying. This was a prestigious location indeed: *Royal Holloway* is very close to *Windsor Castle*, 'the largest, oldest inhabited castle in the world'. In addition, it had an unexpected connection with the holy city: this College was quoted in the *'Da Vinci Code'* as the place where the heroin, supposedly a descendant of Jesus through Merovingians (overthrown by Carolingians, oddly enough, the very same year when Umayyads were overthrown by Abbasids), studied Cryptology.

Fig. 19 Upper: a group picture at the entrance of *Kochubey Palace*. Lower: (left) the ballroom of the residence; (right) Erkki Brändas, Chair of the Selection Committee, discloses the names of the Prize recipients: Hiromi Nakai (Tokyo, Japan) and Luis Frutos-Gaite (Valencia, Spain)

Steve managed to organize a beautiful meeting (Fig. 20) under especially difficult conditions. It gathered some 70 people and, for the second time since Greece, we had Walter Kohn as a guest. The proceedings appeared as vol. 18 of *PTCP* [11] and some details about former *QSCP* meetings can be found in the Introduction in this volume. But due to the lack of eligible candidates there was no Prize award at this second English meeting.

Fig. 20 Upper: *Royal Holloway*'s main building, built on the model of *Château de Fontainebleau*. Lower left, a group picture at *Windsor Castle*, showing (back row) A. Kuleff, S. Wilson, J. Maruani, and (front row) C. Valdemoro, Maruani's, Wilson's and Karwowski's wives, P. Piecuch, and A. van der Avoird. Lower right, the cover of the booklet of abstracts

It was at this 'Windsor meeting' that the proposal, made at 'St Petersburg', to hold a *QSCP* meeting, for the first time, in the New World was confirmed.

6 From the *QSCP* Meeting of Lansing Onwards

The proposal came from Piotr Piecuch, whom I first met at 'Daudel's Congress' in 1986 (§ 2). We knew he would be as efficient as an organizer as he was a dynamic scientist and, moreover, would gather more money and select better nominees for the Prize of *CMOA*, which he had received earlier (see Fig. 15). But how would Lansing look after the cities of Galileo and Columbus, Queen Christina and Peter the Great, the blue billows of the homelands of Plato and Hannibal and the snowy peaks of Europe's Himalaya? It was not even one of the oldest settlements in the USA, as is St. Augustine, Florida, where many 'Sanibel' meetings had taken place.

Lansing, Michigan, would look great, indeed. When I walk over a bridge (e.g., Fig. 21 left) I am fascinated by the traffic. How could this unceasing, day and night flow of cars possibly last forever? Or why not, as part of what we call *sustainable development*, stop this traffic one day a week, say (as in Old Jerusalem), or even a month or a year, to help Nature partly recover from the assaults of Man? True, the Industrial Revolution started in England, then spread to the Continent. But its real take-off came from the rationalization of production in Ford's car factories, made notorious in a movie by Chaplin. The car industry is so linked to our modern way of life, for better or for worse, that the major move of western governments, after saving financial institutions, was to help the car industry. Now, the US car industry started to develop near Detroit, eighty miles south-east of Michigan State University, as recalled in the Ford Museum (Fig. 21 right). We visited the Ford Museum, as part of our social programme, in addition to our traditional music concert. Thus Lansing, the capital of the State of Michigan, was as great historically as a venue as the prestigious locations of our previous meetings.

Fig. 21 Left: Traffic under the bridge linking Figueroa to Bixel on W. 7th St, LA, CA. Right: announcing a 'car culture' show in the Ford Museum near Lansing, Michigan (the standing lady's late father had worked at Ford factories in the 1920's)

The Lansing *QSCP-XIII* workshop was not only the first one held in the New World, but also the first one held in July due to its coordination with the Vancouver *ISTCP-VI* congress. This made both meetings among the most successful of their kind. At Lansing: http://www2.chemistry.msu.edu/qscp13/, there were about 100 participants from 22 countries of all continents, more than half from the USA; and at Vancouver: http:// www2.bri.nrc.ca/ccb/istcp6/, there were close to 400 participants, the largest number ever reached. At Lansing, we were accommodated in the comfortable premises of their *James B. Henry Centre for Executive Development* (Fig. 22) of Michigan State University (*MSU*). There were five nominees selected for the Prize of *CMOA*: Pr. Garnet Chan (Cornell), who received a Prize of $1,000; Pr. David Mazziotti (Chicago) and Dr. Agnes Szabados (Budapest), who shared another Prize of $1,000; and Dr. Daniel Crawford (Virginia) and Pr. Stephane Carniato (Paris), who received a diploma and a gift from *CMOA* (Fig. 23).

Fig. 22 Upper: A group picture on the grounds of the *James B. Henry Centre*. Middle left: Registration desk, showing Janet Haun and Sharon Hammes-Schiffer on the left, Jeffrey Gour and Piotr Piecuch in the middle, and Marta Włoch and Jolanta Piecuch on the right. Middle right: Attending the music concert in the atrium, showing (in the front row) Marja and Jean Maruani, Erkki Brändas, and Anna and Jolanta Piecuch. Lower: A talk in the lecture room of the *James B. Henry Centre*

Listed in Table 1 are some key features of *QSCP* workshops held from 1996 (Pisa) to 2008 (Lansing), or expected in 2009 (El Escorial) and 2011 (Nakazawa).

At the banquet dinner of *QSCP-XIII*, the venue of the following (*QSCP-XIV*) workshop was disclosed: it will be organized by Pr. Gerardo Delgado-Barrio at *El Escorial*, near Madrid, Spain (Fig. 24), once again in September, 2009. One week later, at the banquet dinner of *ISTCP-VI*, the venue of a future (*QSCP-XVI*) workshop was also disclosed: it will be organized by Pr. Kiyoshi Nishikawa at Naka-zawa, Japan, in coordination with the *ISTCP-VII* congress organized by Pr. Hiromi Nakai at Waseda University, Tokyo, both in September 2011, again a week apart.

Fig. 23 Left: Pr. Piotr Piecuch opens the ceremony of award of the Prize of *CMOA*; to his right, Prs John McCracken (*MSU*) and Souad Lahmar (Carthage); to his left, Prs Jean Maruani (Paris) and Erkki Brändas (Uppsala). Right: Pr. Garnet Chan (Cornell) receives his Prize: Pr. Delgado-Barrio (Madrid) hands him the diploma and Mrs Janet Haun (*MSU*) the cheque

Fig. 23 (contd) Daniel Crawford (centre), who missed the banquet of *QSCP-XIII* at Lansing, receives his award from *CMOA* at the *ISTCP-VI* congress held a week later in Vancouver. To his left, Philip Hoggan, Eduardo Ludeña, M.A.C. Nascimento, Jose Alvarellos, Miguel Castro (was the organizer of *ISTCP-III* in Mexico in 1999), Alexander Kuleff (a *PSP* of *CMOA* in 2002), Marja Rantanen, Carmela Valdemoro; to his right, Piotr Piecuch (*PSP* in 2004), Agnes Nagy (*PSP* in 2002), Peter Surjan, Agnes Szabados (*PSP* in 2008), Erkki Brändas (Chair of *ISTCP*), Alexander Wang (the organizer of *ISTCP-VI*), and Hiromi Nakai (*PSP* in 2006, will be the organizer of *ISTCP-VII* in Tokyo in 2011). The picture was taken in the atrium of the conference building at the University of British Columbia

Fig. 24 Royal Palace of *El Escorial*, near Madrid, Spain, the location of *QSCP-XIV*

Table 1 *QSCP* workshops (~ means the workshop was held in the area of the city quoted)

Nb.	*Venue*	*Period*	*Main Organiser*	*Proceedings*
I	~ Pisa, Italy	April 1996	Roy McWeeny	*TMOE*, Kluwer
II	Oxford, England	April 1997	Stephen Wilson	*AQC*, Plenum
III	Granada, Spain	April 1998	A. Hernandez-Laguna	*PTCP*, Kluwer
IV	~ Paris, France	April 1999	Jean Maruani	*PTCP*, Kluwer
V	Uppsala, Sweden	April 2000	Erkki Brändas	*AQC*, Plenum
VI	Sofia, Bulgaria	April 2001	Alia Tadjer	*IJQC*, Wiley
VII	~ Bratislava, Slovakia	Sept. 2002	Ivan Hubač	*IJQC*, Wiley
VIII	~ Athens, Greece	Sept. 2003	Aristides Mavridis	*IJQC*, Wiley
IX	~ Grenoble, France	Sept. 2004	Jean-Pierre Julien	*PTCP*, Springer
X	Carthage, Tunisia	Sept. 2005	Souad Lahmar	*PTCP*, Springer
XI	~ St. Petersburg, Russia	Aug. 2006	Oleg Vasyutinskii	*IJQC*, Wiley
XII	~ Windsor, England	Sept. 2007	Stephen Wilson	*PTCP*, Springer
XIII	Lansing, Michigan, USA	July 2008	Piotr Piecuch	*PTCP*, Springer
XIV	El Escorial, Spain	Sept. 2009	G. Delgado-Barrio	*PTCP*, Springer
XV	To be announced	2010	undecided	undecided
XVI	Kanazawa, Japan	Sept. 2011	Kinoshi Nishikawa	undecided

7 The Promising Scientist Prize of *CMOA*

Announced at the banquet dinner of the Uppsala *QSCP* workshop, in April 2000, the Promising Scientist Prize (PSP) of *CMOA* was first awarded at the banquet of the Sofia *QSCP* workshop, in April 2001, then at the Paris *ISTCP* congress, in July 2002, and again at the Bratislava *QSCP* workshop, in September 2002, and afterwards at nearly all *QSCP* workshops, except those of Spetses, in April 2003, and Windsor, in April 2007. Up to the *MSU* meeting there were 21 selected nominees: 4 received the full Prize, 8 shared the Prize, and 9 received a certificate of nomination and a gift from *CMOA*.

Fig. 25 Outlook of the Diploma of *CMOA*

It was at Sofia that the rituals of the closing banquet and of the ceremony of award of the Prize of *CMOA* were set up, with scarce modifications at later workshops. These are the main features. 1. The banquet takes place on the eve of departure, but there are still a morning working session and an afternoon closing session on the next day. 2. A wide U-shaped front table is arranged for prominent members of the committees, with the Chair (main local organizer) at the centre and spouses on the wings, and two or more ranks of large round tables spaced by an alley leading from a podium to the front table. 3. Nominees would not know that they were selected before the ceremony starts, the surprise being part of the Prize! There is no Prize awarded *in abstentia*, a missing nominee being replaced by the next one on a list. 4. By the middle of the dinner, the Chair stands up and asks for silence: this is a sign for a person in charge of music to open the ceremony (usually with the opening of *Zarathustra* by Strauss), for a delegate of *CMOA* to head to the podium, for two selected men to pick up swords and go to stand face to face at the centre of the alley, and for a number of selected girls / boys to make their way to the tables of the nominees. They come from behind, gently knock their shoulder, and take them by the hand to the podium, where they meet the delegate of *CMOA*.

At this point (5) the President of the Selection Committee, sitting at the front table, stands up and starts acknowledging the number and quality of the nominees, congratulates all of them, and discloses the names of those who were selected, with a few words of *CV* for each of them. 6. Afterwards, for each nominee, the delegate of *CMOA* gives a signal for a specific piece of music to be played while he makes a short, formal speech, then awards the diplomas and gifts - first to the selected nominee (s) and then to the Prize winner (s). Each of them is led in turn by his / her escort, while the music is raised, to the main table, after passing between the men holding the swords, then is led back to his / her table by his / her escort, who stays standing behind him / her. As soon as all nominees are seated, the Chair asks for all participants (except the nominees) to stand up for hearty cheers.

At a preparatory meeting of the Sofia banquet Virginia McWeeny suggested that gifts should also be made to the main organizer, Alia Tadjer, for her extreme dedication. We thought it would be a great symbol if the gifts were offered by the nominees themselves, at the end of the ceremony, before the following workshop would be announced. Alia, who had a hard time trying to please everybody at the meeting, was close to tears when she saw the three nominees heading solemnly towards her with gifts in hands while the European hymn (Beethoven's *Ode to Joy*) was played. We maintained that part of the ceremony in later meetings.

Music is an essential ingredient in the ceremony of award (and the entire social programme) of *QSCP* meetings. The connection between Science and Music goes back far in history: from Pythagoras through Descartes, Euler, Helmholtz, Fourier, to our quantum chemists Löwdin and Longuet-Higgins [12]. Einstein played the violin, Prigogine the piano, Feynman the drums. The occurrence of discrete harmonic frequencies in standing acoustic waves was a clue that led Louis de Broglie to solving the wave-particle dualism by attaching a matter wave to every particle [13].

Displayed in Fig. 25 is a model of the diploma of the Prize of *CMOA*, which bears a *logo* and a *motto*. On the *logo* appears the sentence '*Fiat Lux*', from *Genesis* '*Iehi Or*' ('Let there be light'). This is to recall the role the elucidation of the properties of light played in the advent of the two great physical theories of the 20th century: *Relativity Theory*, which led to the mass-energy equivalence $E = mc^2$, and *Quantum Theory*, which stemmed from the frequency-energy equivalence $E = h\nu$, which combined together led de Broglie to the matter-wave relationship: $\lambda = h / p$ [13}, also recalled on the *logo*. The *motto* is a sentence attributed to King Solomon by physician and writer François Rabelais: '*Science sans conscience n'est que ruine de l'âme*' ('Knowledge without wisdom means ruin to the soul'). The nominees are awarded their diploma for 'scientific and human endeavor and achievement'.

For details on the conditions of eligibility, names of nominees, and pictures of ceremonies, we refer to the web site: http://www.lcpmr.upmc.fr/prize.html.

8 Conclusion

In this paper we have recalled the origins and development of the *QSCP* network, from European contracts in the frame of *COST* projects to the organization of annual workshops and the foundation of the *PTCP* bookseries. As the scientific contents of the *QSCP* meetings have already been documented in the 18 volumes of proceedings (the list of which is detailed in Ref. 10), this historical overview was conceived rather as an illustrated journey sprinkled with various anecdotes and insights. However, one may also like to think that, behind these circumstances, there may be some kind of underlying leading thread.

Already in classical mechanics or in classical optics one can use, to compute the trajectory followed by a matter particle or by a light ray, either a *deterministic, derivative* formulation (Hamilton's equations for position and momentum, Descartes' laws of reflection and refraction) or a *finalistic, integral* approach (Maupertuis' principle of least action integral, Fermat's principle of stationary optical path). The latter formulations have been understood, through quantum mechanics, as resulting from interference between waves associated with matter particles or light rays, constructive along the effective, real trajectory and destructive along all other, virtual paths. The generalized, corresponding formulation of this concept is embedded in the probability amplitude expression of Feynman's quantification principle [14].

In the biological sciences, only deterministic approaches to microevolution (as those involving Darwinian mechanisms of natural and sexual selection) have been well rationalized. There is no true understanding of the macroevolution process in which we are embedded, and not just external observers as in the physical sciences. This has lead to such speculations as 'intelligent design'. However, the existence of *selected trajectories for phylum evolution* cannot be discarded: in a way, chance and necessity play a role similar in the Darwinian theory of biological evolution as in the Copenhagen interpretation of quantum mechanics. If biological phenomena are indeed, as Schrödinger had felt it [15], a manifestation of quantum (not statistical) laws at a macroscopic level, then something like *constructive and destructive interference* may operate also among living systems, and hence within social structures, such as ... our own *QSCP* network!

The vitality of the *QSCP* network, which has reached international status, has survived the termination of European sponsorship. We expect bright future ahead, perhaps with the extended scope: *'Quantum Systems in Physics, Chemistry ... and Biology'*.

Acknowledgements I wish to thank all the persons, famous or anonymous, who took part in the adventures of *CMOA, TMOE, COST, QSCP*, and / or *PTCP* in France, Bulgaria, Italy, England, Spain, Sweden, Slovakia, Greece, Tunisia, Russia, the USA and other countries, and I apologize to those whom I could not mention in this historical overview. I am most grateful to Prof. Piotr Piecuch for asking me to present this general record in the proceedings of the great meeting that he organized at Lansing, Michigan, USA, in July 2008. Professor Roland Lefebvre (Paris) and Jeff Gour (*MSU*) are gratefully acknowledged for a critical reading of the manuscript.

References

Hereafter (except for Refs 12 to 15) are quoted only volumes (co) edited by the author of this overview preliminary to the founding of the *QSCP* network, or as proceedings of *QSCP* or related meetings published in this bookseries. For a detailed list of *QSCP* workshops and proceedings see the Introduction in Ref. 11.

1. J. Maruani & J. Serre (eds.), *Symmetries and Properties of Nonrigid Molecules: A Comprehensive Survey*, as Stud. Phys. Theor. Chem. **23** (Elsevier, 1983), 536 pp.

2. R. Daudel, J.-P. Korb, J.-P. Lemaistre, J. Maruani (eds.), *Structure and Dynamics of Molecular Systems*, vol. **1** (Reidel, 1985), 290 pp.; vol. **2** (Reidel, 1986), 316 pp.
3. J. Maruani (ed.), *Molecules in Physics, Chemistry, and Biology*, as TMOE **1-4**
    Vol. 1: *General Introduction to Molecular Sciences* (Kluwer, 1988), 266 pp.
    Vol. 2: *Physical Aspects of Molecular Systems* (Kluwer, 1988), 720 pp.
    Vol. 3: *Electronic Structure and Chemical Reactivity* (Kluwer, 1989), 452 pp.
    Vol. 4: *Molecular Phenomena in Biological Sciences* (Kluwer, 1989), 378 pp.
4. R. McWeeny, J. Maruani, Y.G. Smeyers, S. Wilson (eds.), *Quantum Systems in Chemistry and Physics: Trends in Methods and Applications* (1997), as TMOE **16** (Kluwer, 1997), 406 pp.
5. For the exact reference to proceedings of *QSCP* meetings not published in *PTCP*, see the Introduction in Ref. 11.
6. A. Hernandez-Laguna, J. Maruani, R. McWeeny, S. Wilson (eds.), *Quantum Systems in Chemistry and Physics*, as PTCP **2-3**
    Vol. 1: *Basic Problems and Model Systems* (Kluwer, 2000), 418 pp.
    Vol. 2: *Advanced Problems and Complex Systems* (Kluwer, 2000), 400 pp.
7. J. Maruani, C. Minot, R. McWeeny, Y.G. Smeyers, S. Wilson (eds.), *New Trends in Quantum Systems in Chemistry and Physics*, as PTCP **6-7**
    Vol. 1: *Basic Problems and Model Systems* (Kluwer, 2001), 434 pp.
    Vol. 2: *Advanced Problems and Complex Systems* (Kluwer, 2001), 322 pp.
8. J. Maruani, R. Lefebvre, E. Brändas (eds.), *Advanced Topics in Theoretical Chemical Physics -* Proceedings ICTCP-IV (Advanced Reviews), as PTCP **12** (Kluwer, 2003), 540 p.
    J. Maruani, R. Lefebvre, E. Brändas (eds.), Proceedings ICTCP-IV (Research Articles), as IJQC **99** / 4-5 (Wiley, 2004), 700 p.
9. J.-P. Julien, J. Maruani, D. Mayou, S. Wilson, G. Delgado-Barrio, *Recent Advances in the Theory of Chemical and Physical Systems*, as PTCP **15** (Springer, 2006), 600 p.
10. S. Lahmar, J. Maruani, S. Wilson, G. Delgado-Barrio, *Topics in the Theory of Chemical and Physical Systems*, as PTCP **16** (Springer, 2007), 310 p.
11. S. Wilson, P.J. Grout, G. Delgado-Barrio, J. Maruani, P. Piecuch, *Frontiers in Quantum Systems in Chemistry and Physics*, as PTCP **18** (Springer, 2008), 600 p.
12. J. Maruani, R. Lefebvre, M. Rantanen, 'Science and Music: from the music of the depths to the music of the spheres', in PTCP **12** (Kluwer, 2003), pp. 479-514.
13. L. de Broglie, 'Recherches sur la Théorie des Quanta' (Thesis, Paris Sorbonne, 1924), and Ann. Phys. **10** III, 22 (1925).
14. R. Feynman, Rev. Mod. Phys. **20**, 367 (1948).
15. E. Schrödinger, *What is Life?* (first edition, 1944). Combined publication with *Mind and Matter* (Cambridge University Press, 1967), 180 pp.

Methylene: A Personal Perspective

James F. Harrison

Abstract Efforts to unravel the electronic and geometric structure of the methylene molecule are presented from a personal perspective.

Keywords: Methylene · Singlet-triplet separation · Geometry

1 Introduction

When I was asked to say a few words at the welcoming dinner for this conference Professor Piecuch suggested that I say something about methylene, a molecule which has been important in my career and which has played a crucial role in the development of computational quantum chemistry. I initially demurred, thinking that everyone knows the methylene story but upon reflection I realized that while I still remember much of the methylene story as if it happened yesterday, it did not. In fact it began in 1960, almost a half century ago, and ended in 1986, almost a quarter of a century ago. Most of this audience was not born when it began and many were still in middle school during the final chapter. Coincidentally I had reason to revisit this story recently in connection with a commemorative issue of the *Canadian Journal of Chemistry* [1] honoring Gerhard Herzberg, perhaps the central figure in the methylene story. Many of the observations I will share with you today are detailed in this chapter. There are several excellent reviews [2–4] addressing the history of methylene, and this talk is not intended to replace these. I will instead present a more informal picture of the highlights of the methylene story, at least as I remember them!

2 The Bond Angle in the Ground State of CH$_2$

Let me remind you of the methylene molecule. It is an 8-electron system consisting of one carbon atom and two H atoms. Two of the electrons are cloistered in the $1s$ orbital on carbon and two are in each of the CH bonds leaving two nonbonding

J.F. Harrison (✉)
Michigan State University, East Lansing, Michigan 48824, USA,
e-mail: harrison@chemistry.msu.edu

P. Piecuch et al. (eds.), *Advances in the Theory of Atomic and Molecular Systems*,
Progress in Theoretical Chemistry and Physics 19, DOI 10.1007/978-90-481-2596-8_2,
© Springer Science+Business Media B.V. 2009

electrons. If one imagines the two H atoms bonding to the two singly occupied $2p$ orbitals on carbon one would expect a highly bent singlet state with the two nonbonding electrons in the carbon $2s$ orbital. One can then imagine electrons flowing from the H to the carbon, generating a positive charge on the hydrogen atoms which would then repel one another causing the bond angle to increase toward $100°$. Simultaneously the $2s$ electrons would hybridize and form a lone pair in the plane of the molecule and to the rear of the CH bonds with the resulting state being of $^1A_1$ symmetry [1]. This gedanken experiment leaves an empty $2p$ orbital on carbon, perpendicular to the molecular plane. One can imagine an excited state being formed by exciting one of the lone pair electrons into this empty $p\pi$ orbital forming either a singlet or a triplet of B_1 symmetry [2].

This is a rather accurate representation of the situation as we understand it today. At small angles the $^1A_1$ symmetry is the ground state with the $^3B_1$ being the first excited state. However in the linear molecule one can imagine the two CH bonds being formed from sp hybridized orbitals on carbon with the two $2p$ electrons in $p\pi$ orbitals perpendicular to the HCH line. With this scenario the $^3\Sigma_g^-$ is the ground state and the singlet is of $^1\Delta_g$ symmetry. The question becomes what happens when one bends the molecule away from linearity where the triplet is the ground state or when one opens the HCH angle from a $90°$ angle where the singlet is the ground state. In the linear configuration the CH bonds are essentially sp hybrids and as one bends the molecule these bonds loose s character and become weaker. Simultaneously the in plane $p\pi$ orbital hosting an unpaired electron acquires s character and drops in energy. So we have one stabilizing and one destabilizing feature and in my opinion it is impossible to predict qualitatively what the outcome will be. It really must be settled by experiment or computation.

The first mention of the methylene molecule in the traditional scientific literature seems to be by Mulliken [5] in a 1932 Physical Review article titled "Electronic Structures of Polyatomic Molecules and Valence. II. Quantum Theory of the Chemical Bond." Mulliken was interested in the nature of double bonds and in particular the double bond in ethylene, which he analyzed in terms of the constituent CH_2 fragments. In the course of this analysis he proposed that the ground state of CH_2 was of $^1A_1$ symmetry with an angle of about $110°$ and that there was a low-lying $^3B_1$ state. This was a remarkable illustration of Mulliken's legendary insight. It is humbling to note that quantum mechanics as we know it was only 6 years old.

Many qualitative and semiempirical studies followed Mulliken's analysis [6], but it was not until 1960 that the first ab initio calculation on CH_2 was published [7] by J. M. Foster and S. F. Boys in the famous April 1960 issue of the *Reviews of Modern Physics* which contained the papers from a conference on Molecular Quantum Mechanics that was held at the University of Colorado in June of 1959. This is the same Boys who in 1950 introduced the idea of using oscillator functions or as we know them now, gaussian functions as basis functions for calculating the electronic structure of polyatomic molecules.

The methylene calculation of Foster and Boys was done in a Slater orbital basis consisting of eight single exponentials, six on carbon and one on each hydrogen atom. It is interesting to note that the carbon basis had two functions representing the carbon $1s$ orbital and one each for the $2s$ and the three components of the $2p$. This emphasis on the $1s$ shell reflects the lack of experience in the field. Now one would certainly try to represent the valence orbitals more accurately than the inner shell core orbital. The basis was orthogonalized and the resulting orbitals used to generate 128 determinants of $^3B_1$ and $^1B_1$ symmetry and a comparable number for the $^1A_1$ state. The calculations were carried out on the EDSAC2 computer which was a vacuum tube machine that had 1024 40-bit words and could execute an add and subtract instruction in 100–170 μs and a multiplication in 210–340 μs. Input was via paper tape!

Compare that speed (10^4–10^5 instructions/s) to the recently unveiled IBM super-computer, Roadrunner, at the Los Alamos National Laboratory which can execute 10^{15} instructions/s.

Foster and Boys predicted that the ground state of CH_2 had $^3B_1$ symmetry with a bond angle of 129° and that the $^1A_1$ state had a bond angle of 90° and was 39 mH or 25 kcal/mol above the triplet. There were no experimental data to compare with so these results stood as predictions.

The Foster and Boys calculation took place at a pivotal point in the evolution of computational chemistry. At this time many were not sure how or whether the discipline would develop. There was a significant philosophical gap between theoreticians who preferred to use empirical calculations to understand chemical systems qualitatively and those who wanted to use computers to perform accurate calculations. Coulson [8] called the first group a posteriori-ists and the second ab initio-ists. A few excerpts from Coulson's comments [8] given in an after dinner talk at the Boulder conference where Foster and Boys presented their work on CH_2 are representative.

> Anyone who attended all of the sessions this week could be in little doubt but that the first and second halves were quit different – almost alien to each other. In its simplest form this difference was associated with the use of large-scale use of electronic computers, though, as I shall say latter, I think there is a deeper aspect of it than just this.
>
> ... the speeding up of calculations, and the design of even faster machines, should enable us to extend the range of effectively exact solutions. I am inclined to think that perhaps the range 6–20 electrons belongs to this picture
>
> I see little chance-and even less desirability-in dealing in this accurate manner with systems containing more than 20 electrons
>
> It looks as if somewhere around 20 electrons there is an upper limit to the size of a molecule for which accurate calculations are ever likely to become practicable. This range of 1–20 includes many interesting questions (e.g., the dissociation of F_2, the shapes of CH_2 and CH_3 in their ground and excited states, the reaction $H+H_2$ going to H_2+H, and much else), but there is a great deal that it leaves out!

Boys was one of the more visible of the ab initio-ists, and to a large extent the viability of the field rested on the accuracy of his results for CH_2.

It was not long afterwards that Herzberg's 1961 Bakerian lecture [9] on the structure of CH_3 and CH_2 was published in the *Proceedings of the Royal Society*. Herzberg had observed the vacuum ultraviolet spectrum of CH_2, CD_2, CHD, and $^{13}CH_2$ and concluded that CH_2 was a linear triplet in the ground state and that the first excited state was a highly bent $^1A_1$. Herzberg was arguably the most prominent spectroscopist in the world and his opinion carried enormous weight. This disagreement between the first ab initio calculation of a polyatomic molecule and experiment had a demoralizing effect on the electronic structure community. On the basis of his analysis of the electronic spectrum of CD_2, Herzberg concluded that CH_2 in its triplet ground state is "linear or nearly linear." On the comparison with the calculations of Foster and Boys he says: "Our experimental values are distinctly different from these predictions but not excessively so, when the approximate nature of the calculations is considered."

It should be noted that in the Bakerian lecture where Herzberg reported the CH_2 results he allows that the molecule might be non linear if there was pre-dissociation in the excited triplet state. Indeed he estimated that a bent triplet would have an angle of $140°$!

As one might imagine from the comments of Coulson there was a question about the long-term role of computer calculations in chemistry and the tone of the day is well represented in a comment by H. C. Longuet-Higgins who published [10] an article in Molecular Physics in 1962 arguing on the basis of semiempirical calculations that CH_2 was linear. He commented, "It may be that future theoretical progress will require elaborate variational calculations such as those of Foster and Boys on CH_2 but until the results of such machine experiments can be interpreted physically, there would seem to be a place for more empirical theories such as we now describe." The significance of this comment is intensified when one remembers that Longuet-Higgins was the Professor of Theoretical Chemistry at Cambridge where Boys held a lesser position.

The second ab initio calculation on CH_2 was by me and Lee Allen and was published [11] in 1969. I entered the graduate program in chemistry at Princeton in 1962 and as was the custom spent a year and a half taking various courses in anticipation of the second year comprehensive examinations. One did not spend a lot of time on research until these were completed and in the spring of 1964 I began to do research seriously. At this time Jerry Whitten came to Lee's group as a postdoc and was developing the idea of a gaussian lobe basis [12]. In a gaussian lobe basis one represents atomic s functions as we do nowadays but higher angular momentum functions required particular linear combinations of the $1s$ functions offset from the nucleus whose atomic orbitals they were to represent. This basis allowed one to do the integrals required for calculations on polyatomic molecules with impunity as all integrals were essentially between $1s$ gaussians. This approach was also being pursued by Preuss in Germany. In spite of the ability to study polyatomic systems I began a series of calculations, using these functions, on the ground and excited states of the diatomic radicals BH and then NH. I calculated many excited states and potential curves as well as many spectroscopic observables. Sometime in 1965 Lee told me I needed one more large study to graduate so I began to look for an interesting problem. It was not long before I noted that CH_2 was isoelectronic with the triplet radical NH and I was off and running. I had no idea of the significance of CH_2 until I was well into the calculation.

These calculations predicted that the ground state of CH_2 had $^3B_1$ symmetry with an angle of $138°$ with the $^1A_1(108°)$ symmetry 50 mh higher, confirming the results of Boys and Foster and disagreeing with Herzberg. Although written in my thesis in 1966 they where not published until 1969 because life intervened. I went to Indiana University for 2 years with Harrison Shull and Lee went on an extended sabbatical with Coulson. As the end of my post doc drew near I contacted Lee and we assembled the manuscript, which was sent to the *Journal of the American Chemical Society* in September of 1968, my first quarter at MSU and published in February of 1969.

Things were quiet for a while at least on the surface. It was the lull before the storm of 1970. There was however activity in the form of preprints. In May of 1970 R. A. Bernheim, H. W. Bernard, P. S. Wang, L. S. Wood, and P. S. Skell, submitted [13] a paper to the *Journal of Chemical Physics* titled "EPR of Triplet CH_2." These authors were the first to observe the ESR spectrum of CH_2 in a Xenon matrix at 4.2 K. The ESR spectrum of a triplet molecule is characterized by a spin Hamiltonian containing two parameters, called D and E. D measures the spin distribution along the principal axis, say z, and E measures the anisotropy in the spin distribution in the x and y directions. If E is zero the molecule is linear (at least for a triatomic). Bernheim et al. measured $E = 0.003$ cm^{-1}, small but not zero. They said that CH_2 is slightly bent in the Xenon environment but did not claim that free CH_2 was bent. Indeed they did not reference either of the two ab initio papers that predicted a bent molecule.

The following month, on June 3, 1970, Bender and Schaefer submitted a note [14] to the *Journal of the American Chemical Society* titled "New Theoretical Evidence for the Nonlinearity of the Triplet Ground State of Methylene" in which they reported CI calculations on the $^3B_1$ state. The basis on carbon was Dunning's as yet unpublished $4s2p$ (double zeta) contraction of Huzinaga's $9s5p$ set and the H basis was a single $4s$ contracted function for a total of 12 basis functions. The calculation included 408 configurations and was one of the largest to date. These authors formally challenged Herzberg, saying, "Nevertheless, on the basis of the present and previous ab initio calculations and the stated numerical uncertainties, we conclude that the CH_2 ground state is nonlinear with a geometry close to r $= 1.096$ Å, $\Theta = 135.1°$." The previous calculations referenced were those of Foster and Boys [7], and Harrison and Allen [11].

Five weeks later in early July, 1970, I submitted a manuscript [15] to the *Journal of Chemical Physics* titled "An Ab Initio Study of the Zero Field Splitting Parameters of $^3B_1$ Methylene" in which I computed a CH_2 angle of $132.5°$ and in addition calculated the D and E EPR parameters. I knew that besides the Penn State group the Bell Lab group were trying to measure the ESR spectrum of CH_2 and I thought that an ab initio calculation of the ESR parameters D and E would bolster the case for a bent CH_2 radical. Assuming of course that they agreed with experiment! This was a difficult calculation as I needed to evaluate matrix elements of the spin-dipole, spin-dipole interaction, and these depend on $1/r_{12}^3$. Fortunately in a gaussian lobe basis these could be reduced to the usual integrations over error functions. This was the first ab initio calculation of the D and E parameters for a polyatomic molecule.

The calculated D parameter was in good agreement with the recently published experiments of Bernheim et al. (exp $= 0.69$ cm^{-1}, calc $= 0.71$ cm^{-1}) but the E

parameter differed considerably (exp $= 0.003\,\text{cm}^{-1}$, calc $= 0.05\,\text{cm}^{-1}$). Things did not look good! However after considerable thought I concluded that something was wrong with the interpretation of the Herzberg and Bernheim experiments and in this paper I wrote [15]

> The small value of E found experimentally suggests a nearly linear triplet in accord with Herzberg's interpretation of the electronic spectrum of CH_2. This agreement is perplexing since every ab-initio study of CH_2 predicts a highly bent triplet (130–140°) with a low-energy difference between the bent and linear forms. The most extensive calculation to date estimates this difference at 6.7 kcal/mol. While these calculations are certainly capable of improvement one does not expect the predicted geometry to change dramatically. It seems that this persistent discrepancy between theory and experiment warrants a critical evaluation of the experimental data.

This was the second formal challenge to Herzberg's interpretation.

August passed uneventfully and in September the Bell Lab group consisting of Wasserman, Yager, and Kuck submitted a manuscript [16] to *Chemical Physics Letters* titled "EPR of CH_2: A Substantially Bent and Partially Rotating Ground State Triplet." This group observed two EPR spectra of methylene (same D, different Es) which they associated with two different sites in the Xe matrix. They concluded that the CH_2 molecule was rotating about the long axis (the b2 axis), which would not change D very much but would result in a small rotationally averaged E. After an elaborate analysis of possible barriers to rotation they suggested that CH_2 has an angle of 136° in good agreement with the ab initio calculations of Harrison and Allen [11], Bender and Schaefer [14], and Harrison [15].

In October the Berkley group, O'Neil, Schaefer, and Bender submitted a manuscript [17] to the *Journal of Chemical Physics* titled "C_{2v} Potential Energy Surfaces for Seven Low-Lying States of CH_2" in which they extend their earlier note on $^3B_1$ to other symmetries. All of these calculations used the Dunning $4s2p$ contractions of Huzinaga's $9s5p$ set for carbon and a $2s$ contraction of Huzinaga's $4s$ set on H.

In October E. Wasserman, V. J. Kuck, R. S. Hutton, and W. A. Wagner from Bell Labs and Rutgers, submitted a paper [18] titled "EPR of CH_2 and CHD; Isotope Effects, Motion and Geometry of Methylene" to the *Journal of the American Chemical Society* and concluded again that free CH_2 has an angle of 136°.

In November I submitted a manuscript [19] to the *Journal of the American Chemical Society* titled "Electronic Structure of Carbenes I, CH_2, CHF, and CF_2" in which I calculated a CH_2 angle of 132.5°. I was certain of the CH_2 angle and began to study the effect of substituents on the geometry and singlet–triplet splitting in carbenes.

In December Herzberg and Johns submitted a note [20] to the *Journal of Chemical Physics* titled "On the Structure of CH_2 in its Triplet Ground State." Herzberg and Johns opened the note with the sentence: "Recent electron-spin resonance work on CH_2 in solid matrices by Bernheim, Bernard, Wang, Wood and Skell and Wasserman, Yager, and Kuck, as well as theoretical calculations by Harrison and Allen and Bender and Schaefer, suggest strongly that CH_2 is bent in its triplet ground state."... They continued

In view of the diverging results it is perhaps worth pointing out that there is a possibility of reinterpreting the vacuum ultraviolet spectrum in terms of a bent form of the radical

The considerations given here point strongly toward the bent structure of the triplet ground ($^3B_1$) of CH_2, as first suggested by the electron-spin resonance work and the ab initio calculations.

We are much indebted to Dr. Bernheim, Dr. Wasserman, Dr. Bender, and Dr. Harrison for sending us preprints of their papers.

As I mentioned earlier when Herzberg reported the CH_2 results in his Bakerian Lecture he allows that the molecule might be non linear if there was pre-dissociation in the excited triplet state. Indeed he estimated that a bent triplet would have an angle of 140°!

What is fascinating about this line is the primary role of the ESR experiments. While not discounting the ab initio calculations it is clear from discussions I had with Professor Herzberg when he gave the Renaud Lectures at MSU in 1974 that it would have taken him longer to come around to the idea that CH_2 was linear but for the ESR results.

This is another interesting aspect of the belief that CH_2 was linear in its triplet ground state. In the Bakerian lecture Herzberg allows that the molecule might be non linear if there was pre-dissociation in the excited triplet state. Indeed he estimated that a bent triplet would have an angle of 140°. So the Foster and Boys prediction of 129, the Harrison and Allen prediction of 138, and the Bender and Schaeffer prediction of 136, were not sufficiently convincing to suggest that the pre-dissociation scenario might be viable.

There were three EPR papers submitted after the Herzberg retraction appeared in print. In January of 1971 Bernheim et al. submitted [21] a paper to *Journal of Chemical Physics* titled "^{13}C Hyperfine Interactions in CD_2," in which they deduce an angle of 137.7 from the hyperfine splitting.

In February 1971 the Bell labs and Rutgers group followed [22] with the manuscript "^{13}C Hyperfine Interactions and Geometry of Methylene." They estimated the %s character from ^{13}C hyperfine interaction and used %s versus angle data from Harrison [15] to fix angle at 137°.

The third was submitted [23] in June 1974 by Wasserman et al., titled "Zero-Field Parameters of Free CH_2; Spin-Orbit Contributions in Xenon," in which the authors estimate the D and E parameters of CH_2, free from any environmental (matrix) effects and the results are in good agreement with my calculations [15].

While both theory and experiment agreed that CH_2 was bent, the Vacuum UV data placed the angle between 128 and 140° and the EPR data suggested 137, but these were not direct observations and relied on various models. The issue was settled in 1983 by Bunker and Jensen [24] who constructed a potential surface from the available spectroscopic data and concluded that the angle is 133.8± 0.1°.

Twenty-three years after the first ab initio calculation of the angle (129) the issue was closed. Foster and Boys had missed the angle by 5°!

3 Singlet–Triplet Separation

The next issue was the singlet–triplet separation, and this story, like the story of the CH_2 angle, involves a disagreement between theory and experiment. But first let us take an overview of the situation. Between 1967 and 1972 there were five experimental determinations [25–29] of the singlet–triplet splitting and these are listed in Table 1. Two of these [25, 27] measured 1–3 kcal/mol and three [26, 28, 29] deduced 8–9 kcal/mol. I say deduced because backing these numbers out of the experimental data required considerable effort and skill.

The theory, up to 1972 placed this number around 1 eV or 23 kcal/mol and the results of a few representative calculations are listed in Table 2. There were two reasons why calculations of this era predicted this separation. First the $^1A_1$ state requires a two-configuration representation for an SCF function which is of comparable quality to a single determinant SCF on the $^3B_1$. Secondly one needs a d orbital in the carbon basis to provide a balanced description of the two states. The two-configuration nature of the problem was understood and discussed in my CH_2 paper with Lee Allen and to the best of my knowledge this is the first reference to the concept of a state requiring two configuration for an adequate description. The need for a d orbital was slow in coming and it was not until the 1972 paper of Hay, Hunt, and Goddard [30] who reported a (GVB+CI) using a double zeta plus a single d on carbon and Bender et al. [31] who did a two-configuration SCF

Table 1 Experimental singlet–triplet separation in CH_2

Year	Authors	Method	T^0 (kcal/mol)
1967	Halberstadt and McNesby	RRKM(Ph)	~ 2.5
1968	Rowland, McKnight, and Lee	RRKM(Ph)	9
1970	Carr, Eder, and Topor	Eq(Ph)	1–2
1971	Hase, Phillips, and Simons	Th(Ph)	~ 9
1972	Frey	Eq(Ph)	~ 8

Table 2 Early theoretical estimates of the singlet–triplet separation in CH_2

Author	Year	$(^1A_1-^3B_1)$ kcal/mol	Comment
Foster and Boys	1960	24.4	sp-CI
(Meyer)	1968	27.0	sp-SCF
Harrison and Allen	1969	31.9	sp-VB
Harrison	1971	24.4	sp-CI
O'Niel, Schaefer, and Bender	1971	22.2	sp-CI
Del Bene	1971	33.8	sp-SCF
Lathan, Hehre, Curtis, and Pople	1971	37.0	sp-SCF
Chu, Siu, and Hayes	1972	19.9	sp-CI
Hay, Hunt, and Goddard	1972	11.5	spd-GVB-CI
Bender, Schaefer, Franceschetti, and Allen	1972	11.0	spd-CI
Staemmler	1973	9.2	spd-CI
Harrison	1974	9.7	Analysis

with a triple zeta set on carbon augmented with a d on carbon and a p on hydrogen that the splitting came into reasonable agreement with experiment. The Cal Tech group calculated a separation of 11.5 kcal/mol while the Berkley–Princeton group calculated 11 ± 2 kcal/mol.

In my 1974 Accounts of Chemical Research [32] article I pointed out that a combination of spectroscopic experiments and theory supported a singlet–triplet separation of 9.7 kcal/mol. Because of the similarity in the electron correlation in both states the $^3B_1$–$^1B_1$ separation should be much more reliable than the $^1A_1$–$^3B_1$ separation and most calculations of this former separation gave a value around 42.5 kcal/mol. This, combined with the modified experimental [33] value for the $^1A_1$–$^1B_1$ splitting results in a singlet–triplet splitting of 9.7 kcal/mol. Things looked good and in my 1974 Accounts article I concluded, "While theory and experiment concur on many of the characteristics of CH_2, a few loose ends remain." In particular I noted "Also, while an analysis of the spectrum arising from the $^3A_2 \leftarrow {}^3B_1$ transition fixes the angle of the $^3B_1$ state at 136°, it also demands that the $^3A_2$ state be strongly bent with a bond angle of 125°. The nature of this $^3A_2$ state has not been characterized theoretically." Soon thereafter, David Wernette and I published [34], "The $^3A_2$ and $^3B_2$ states of CH_2" where we showed that indeed there were three $^3A_2$ states in the required energy range with bond angles of 127, 120, and 113°. It is unfortunate that this study has fallen through the cracks and even in the most recent study of the $^3A_2$ state [35] (published 24 years later) it is not mentioned.

The agreement between experiment and theory as to the singlet–triplet separation in CH_2 changed dramatically in 1976 when the Lineberger group published [36] the first direct observation of this splitting. They generated CH_2^- in the $^2B_1$ state and ionized this negative ion to produce CH_2 in various states, but most significantly in the $^1A_1$ and $^3B_1$ states. By measuring the kinetic energy of the ejected electrons one can determine, directly, the singlet–triplet separation. Most disconcertingly the experiment put this number at 19.5 ± 0.7 kcal/mol, nearly twice the value predicted by the largest calculations and in remarkable agreement with the earlier (pre d orbital) calculations.

The year following the Lineberger experiment the Berkeley [37] (R. R. Luccese and H. F. Schaefer III) and Cal Tech [38] (Larry Harding and Bill Goddard) groups expanded their computational effort by enlarging the basis set and doing large CI. The Cal Tech group used a double zeta sp basis plus $2ds$ on C and a set of p functions on H, plus a set of diffuse s and p functions on carbon and the GVB+POLCI method, generating 12084 determinants for the $^1A_1$ and 7916 for the $^3B_1$. The resulting energy separation was 0.48 eV or 11 kcal/mol. The Berkeley group did a similar calculation and obtained 13.5 kcal/mol. Many other computational groups, for example, B. O. Roos and P. M. Siegbahn [39], and then C. W. Bauschlicher and I. Shavitt [40] entered the fray and all agreed that 19.5 was not possible.

Harding and Goddard in the above calculation also calculated the CH_2^- energy and vibrational frequencies. They noted that if the CH_2^- was being ionized in an excited vibrational level and not, as Lineberger had assumed, in the ground vibrational level, the resulting singlet–triplet separation would be too large. By reinterpreting Lineberger's data Harding and Goddard predicted that the correct separation was 9 kcal/mol.

A conference was organized by Lineberger at Boulder in 1978 to discuss the situation. Once again computational chemistry was being challenged by experimental results obtained by one of the most prestigious experimental groups in the world. I remember the conference well. It was all about reliability, and this time there was no waffling on the part of the theoreticians. The hot band theory seemed to hold the day but the experimentalists were not really sure. They spent the next 6 years constructing another instrument that would eliminate the possibility of hot bands in CH_2^-, and in 1984 they announced [41] that the hot band interpretation was correct and the singlet–triplet separation was 9 kcal/mol. Theory had survived another challenge.

As noted above there were three reviews of the CH_2 story written in the mid-1980s. The first was by I. Shavitt and titled "Geometry and Singlet–Triplet Energy Gap in Methylene: A Critical Review of Experimental and Theoretical Determinations", and two excerpts follow:

> In both experimental and theoretical work, the principal uncertainty is often not in the raw data collected, but in their interpretation. Experimental measurements are frequently interpreted and adjusted on the basis of simplified theoretical models.
>
> From the theoretical point of view, it is seen that for problems such as the singlet–triplet gap in methylene there is no alternative to sophisticated and laborious calculations. Even if simple models sometimes produce correct answers, there is no independent way, in general, to recognize when this has happened.

The second review was Goddard's 1985 Science article, "Theoretical Chemistry Comes Alive: Full Partner with Experiment" in which he says "During the last decade advances in computational techniques and in the extraction of chemically useful concepts from electronic wave functions have put theorists into the mainstream of chemistry" and noted the significance of the theoretical determination of the geometry and singlet–triplet gap in CH_2.

The third major review dealing with methylene was Schaefer's 1986 Science article titled "Methylene: A Paradigm for Computational Quantum Chemistry" in which he says "The year 1970 has been suggested as a starting date for the 'third age of quantum chemistry,' in which theory takes on not only qualitative but also quantitative value. In fact, each of the years 1960, 1970, 1972, and 1977 is of historical value in unraveling the structure and energetics of the CH_2 molecule, methylene." I certainly agree especially in view of the number of submissions in 1970 related to CH_2.

In this talk I have focused on aspects of the methylene story that I am most familiar with. It is a remarkable story about a simple molecule the understanding of which was instrumental in establishing the credibility of computational chemistry as we know and enjoy it today. The story has many authors, too many for me to mention this evening but the paper [1] "CH_2 revisited" has a more complete list of contributors.

References

1. A. Kalemos, T. H. Dunning, Jr., A. Mavridis, and J. F. Harrison, Can. J. Chem. **82**, 684 (2004)
2. I. Shavitt, Tetrahedron **41**, 1531 (1985)
3. W. A. Goddard III, Science **227**, 917 (1985)
4. H. F. Schaefer III, Science **231**, 1100 (1986)
5. R. S. Mulliken, Phys. Rev. **41**, 751 (1932)
6. J. F. Harrison, *Carbene Chemistry*, ed. by W. Kirmse (Academic Press, Inc., New York and London, 1971)
7. J. M. Foster and S. F. Boys, Rev. Mod. Phys. **32**, 305 (1960)
8. C. A. Coulson, Rev. Mod. Phys. **32**, 170 (1960)
9. G. Herzberg, Proc. R. Soc. London, Ser. A **262**, 291 (1961)
10. P. C. H. Jordan and H. C. Longuet-Higgins, Mol. Phys. **5**, 121 (1962)
11. J. F. Harrison and L. C. Allen, J. Am. Chem. Soc. **91**, 807 (1969)
12. J. L. Whitten, J. Chem. Phys. **44**, 359 (1966)
13. R. A. Bernheim, H. W. Bernard, P. S. Wang, L. S. Wood, and P. S. Skell, J. Chem. Phys. **53**, 1280 (1970)
14. C. F. Bender and H. F. Schaefer III, J. Am. Chem. Soc. **92**, 4984 (1970)
15. J. F. Harrison, J. Chem. Phys. **54**, 5415 (1971)
16. E. Wasserman, W. A. Yager, and V. J. Kuck, Chem. Phys. Lett. **7**, 409 (1970)
17. S. V. O'Neil, H. F. Schaefer III and C. F. Bender, J. Chem. Phys. **53**, 162 (1971)
18. E. Wasserman, V. J. Kuck, R. S. Hutton, and W. A. Wagner, J. Am. Chem. Soc. **92**, 7491 (1970)
19. J. F. Harrison, J. Am. Chem. Soc. **93**, 4112 (1971)
20. G. Herzberg and J. W. C. Johns, J. Chem. Phys. **54**, 2276 (1971)
21. R. A. Bernheim, H. W. Bernard, P. S. Wang, L. S. Wood, and P. S. Skell, J. Chem. Phys. **54**, 3223 (1971)
22. E. Wasserman, V. J. Kuck, R. S. Hutton, E. D. Anderson and W. A. Yager, J. Chem. Phys. **54**, 4120 (1971)
23. E. Wasserman, R. S. Hutton, V. J. Kuck and W. A. Yager, J. Chem. Phys. **55**, 2593 (1971)
24. P. R. Buenker and P. Jensen, J. Chem. Phys. **79**, 1224 (1983)
25. M. L. Haberstadt and J. R. McNesby, J. Am. Chem. Soc. **89**, 3417 (1967)
26. P. S. Rowland, C. McKnight and E. K. C. Lee, Ber. Bunsenges. Phys. Chem. **72**, 236 (1968)
27. R. W. Carr Jr., T. W. Eder, and M. G. Topor, J. Chem. Phys. **53**, 4716 (1970)
28. W. L. Hase, R. J. Phillips, and J. W. Simons, Chem. Phys. Lett. **12**, 161 (1971)
29. H. M. Frey, J. Chem. Soc. Commun., **18**, 1024 (1972)
30. P. J. Hay, W. J. Hunt, and W. A. Goddard III, Chem. Phys. Lett. **13**, 30 (1972)
31. C. F. Bender, H. F. Schaefer III, D. R. Franceschetti, and L. C. Allen, J. Am. Chem. Soc. **94**, 6888 (1972)
32. J. F. Harrison, Acc. Chem. Res. **7**, 378 (1974)
33. G. Herzberg and J. W. C. Johns, Proc. Roy. Soc. (London) A **295**, 107 (1966)
34. J. F. Harrison and D. A. Wernette, J. Chem. Phys. **62**, 2918 (1975)
35. Y. Yamaguchi and H. F. Schaefer III, J. Chem. Phys. **106**, 8753 (1997)
36. P. F. Zittel, G. B. Ellison, S. V. O'Neil, E. Herbst, W. C. Lineberger, and W. P. Reinhardt, J. Am. Chem. Soc. **98**, 3731 (1976)
37. R. R. Lucchese and H. F. Schaefer III, J. Am. Chem. Soc. **99**, 6765 (1977)
38. L. B. Harding and W. A. Goddard III, J. Chem. Phys. **67**, 1777 (1977)
39. B. O. Roos, P. M. Siegbahn, J. Am. Chem. Soc. **99**, 7716 (1977)
40. C. W. Bauschlicher, Jr. and I. Shavitt, J. Am. Chem. Soc. **100**, 739 (1978)
41. D. G. Leopold, K. K. Murray, and W. C. Lineberger, J. Chem. Phys. **81**, 1048 (1984)

Part II
High-Precision Quantum Chemistry

Free Complement Method for Solving the Schrödinger Equation: How Accurately Can We Solve the Schrödinger Equation

H. Nakatsuji and H. Nakashima

Abstract Free complement (FC) method provides a general and systematic method of solving the Schrödinger equation. In this method, the Hamiltonian of the system modified for the singularity of the potential is used to generate the FC functions that span the exact wave function of the system. Thus, by applying the variation principle to the sum of the complement functions, which we call FC wave function, we can calculate the essentially exact wave function and energy for the ground and excited states of the system. We here show that the Schrödinger equation can be solved to an arbitrary accuracy with the FC method by examining the upper and lower bounds of the energy, local energy, H-square error, cusp condition, and so on, for the helium atom.

Keywords: Solving the Schrödinger equation · Free complement method · Cusp condition · Upper and lower bounds

1 Introduction

This chapter summarizes briefly the lecture of Nakatsuji given on July 10, 2008, at the QSCP-13 workshop at Lansing organized by Prof. Piotr Piecuch of the Michigan State University. Let us first celebrate our exciting memories of this workshop for its high-quality science and good performance, and nice organization, all of which were due to the careful coordination and organization of the workshop by Profs. P. Piecuch and J. Maruani. So, let us first deeply thank Profs. Piotr Piecuch and Jean Maruani for all of this.

The Schrödinger equation has long been believed to be insoluble for over 80 years, since it was discovered by Prof. Erwin Schrödinger in 1926 [1], though it was believed to govern all of chemistry and most of physics [2]. For this reason, all we could have done in quantum science was to formulate "approximate" theories to "understand" or "interpret" the main features of chemical phenomena [2]. Thus,

H. Nakatsuji (✉) and H. Nakashima
Quantum Chemistry Research Institute, JST, CREST, Nishikyo-ku, Kyoto 615-8245, Japan,
e-mail: h.nakatsuji@qcri.or.jp

P. Piecuch et al. (eds.), *Advances in the Theory of Atomic and Molecular Systems*,
Progress in Theoretical Chemistry and Physics 19, DOI 10.1007/978-90-481-2596-8_3,

quantum chemistry has long been characterized as an approximate science that can never predict phenomena in full accuracy. However, recently, we have found simple, general, and accurate methods of solving the Schrödinger equation [3–13]. We referred to them as iterative complement (configuration) interaction (ICI) method [3, 4] and the free ICI [6] or free complement (FC) method [12], the latter two being the same.

The FC method is completely different from the conventional quantum chemistry. In the state-of-the-art quantum chemistry, one first defines Hartree–Fock orbitals based on the initially chosen basis set and then expands many-electron correlated wave functions by means of the Hartree–Fock orbitals. In this approach, any theory lies between the Hartree–Fock and the full CI and so, the full CI is a goal of this type of the theory. However, the full CI cannot be the exact solution of the Schrödinger equation because of the incompleteness of the basis set first introduced. When we use numerical Hartree–Fock that is free from the basis set, the full CI becomes infinite expansion that cannot be handled in principle.

Explicitly correlated wave function theory [14] is another important approach in quantum chemistry. One introduces inter-electron distances together with the nuclear–electron distances and set up some presumably accurate wave function and applies the variation principle. The Hylleraas wave function reported in 1929 [15] was the first of this theory and gave accurate results for the helium atom. Many important studies have been published since then even when we limit ourselves to the helium atom [16–28]. They clarified the natures and important aspects of very accurate wave functions. However, the explicitly correlated wave function theory has not been very popularly used in the studies of chemical problems in comparison with the Hartree–Fock and electron correlation approach. One reason was that it was generally difficult to formulate very accurate wave functions of general molecules with intuitions alone and another reason was that this approach was rather computationally demanding.

Thus, quantum chemistry has long been a science mainly for understanding and interpretation. It was difficult for quantum chemists to become truly confident on the calculated results. One reason was the approximate nature of the theory and another reason was an incompleteness of the basis set. For example, many people might have experienced the feeling of "maybe, my basis set was not good enough." In the author's opinion, quantitative reliability is a key of the theory. Otherwise, one cannot do "confident prediction." For getting truly quantitative reliability in theoretical quantum science, there is no other way than solving the Schrödinger equation and the Dirac–Coulomb equation accurately.

2 Free Complement Method

In 1999, one of the authors got an inspiration that the Schrödinger equation might be able to be solved. He clarified the structure of the exact wave function and showed a method of obtaining the exact wave function by introducing the ICI method and its variants [3, 4]. However, there still existed a big obstacle, called singularity problem [6]. Namely, the integrals involved in the formulation diverge to infinity when the Hamiltonian involves Coulomb potential, as it does for atoms and molecules. However, a simple idea came. Instead of solving the original Schrödinger equation,

$$(H - E)\psi = 0, \tag{1}$$

one may solve an equivalent equation, called scaled Schrödinger equation [6].

$$g(H - E)\psi = 0. \tag{2}$$

The factor g is called scaling function. It is always positive but can become zero only at the singular points. Even there, the g function must satisfy

$$\lim_{r \to 0} gH \neq 0 < \infty, \tag{3}$$

for not to erase the information of the Hamiltonian at the singular regions. Then, we can formulate the simplest ICI (SICI) method based on the scaled Schrödinger equation as

$$\psi_{n+1} = [1 + C_n g(H - E_n)]\,\psi_n, \tag{4}$$

where E_n is defined by $\langle \psi_n | g(H - E_n) | \psi_n \rangle = 0$. This SICI was also proved to become exact at convergence, and for the existence of the g-function, we do not encounter the singularity problem in the course of the iterative calculations.

When we do the SICI calculations to n-th iteration, the right-hand side of Eq. (4) becomes a sum of the analytical functions multiplied with the coefficients C_i. Now, we reformulate it as follows. We take all the independent analytical functions from there and group them as $\{\phi_i\}$, which we refer to as complement functions, and using them, we expand again our wave function as

$$\psi_{(n+1)} = \sum_i^{M_n} c_i \phi_i. \tag{5}$$

We referred to this wave function as the free ICI wave function. It converges faster to the exact wave function than the original SICI one, because of the increased freedom. In the SICI scheme, the $(n + 1)$-th result, ψ_{n+1}, depends on all the former results, ψ_m and $C_m (m = 0 \ldots n)$, but in the free ICI method, all the coefficients c_i are reoptimized at each n, and therefore, this method is not an iterative method. Then, the naming, the free ICI method may be confusing. So, hereafter we use the new name "free complement (FC)" method instead of the free ICI method. We refer to n of the FC method as an *order*, instead of an iteration number. Thus, the FC method gives a general method of solving the SE in an analytical expansion form.

The FC formalism for the exact wave function may be summarized as follows.

1. The Hamiltonian defines the system.
2. The Hamiltonian paves the way toward its exact wave functions in the analytical expansion form starting from a given initial function ψ_0: Eq. (4) in the SICI case or Eq. (5) in the FC formalism.
3. This formalism is applicable for any system when its Hamiltonian is defined unambiguously.
4. We have no basis set nightmare: the complement functions, which may correspond to the basis set, are generated by the Hamiltonian of the system and so should be a best possible functions for the system.

A general method for calculating the unknown coefficients in the FC wave function given by Eq. (5) is the variation principle. Applying the variation principle to the FC wave function, we obtain the secular equation

$$(\mathbf{H} - E\mathbf{S})\mathbf{C} = 0, \tag{6}$$

where the Hamiltonian and overlap matrices are defined by

$$\mathbf{H} = \begin{pmatrix} \ddots & & \\ \cdot & \int \phi_i H \phi_j \, d\tau & \cdot \\ & & \ddots \end{pmatrix}, \mathbf{S} = \begin{pmatrix} \ddots & & \\ \cdot & \int \phi_i \phi_j \, d\tau & \cdot \\ & & \ddots \end{pmatrix}. \tag{7}$$

For simple few-electron atoms and molecules, these matrix elements are easily calculated. We apply here our FC formalism only to such systems. Then, starting from the initial wave function ψ_0 and using some appropriate scaling function g, we can calculate the solution of the Schrödinger equation in an analytical expansion form. The accuracy of the calculated results would depend on the choices of ψ_0, g, and the expansion order n. We show here that, in principle, we can get the solution of the Schrödinger equation to any desired accuracy in this formalism.

3 Super-Accurate FC Calculation of Helium Atom

Helium atom is the simplest case for which the Schrödinger equation cannot be solved in a closed form. There have been many attempts to solve the Schrödinger equation of the helium atom accurately, starting from the famous study by Hylleraas [15–28]. These studies have produced a lot of important insights about the nature of the accurate wave functions of atoms and molecules. We applied the FC method described above to the helium atom immediately after this method was discovered [6]. It gave a strong support that the FC method was correct and useful. We have given more extended accurate calculations [9, 10] and examined the accuracy of the calculated wave functions by studying several properties that are the stringent test of the exactness of the wave functions [12, 13]. We have further studied the effect of nuclear motion [29] and the excited states with and without considering the effect of nuclear motion [30].

Here we overview our applications to the helium atom ground state. In the Hylleraas coordinate defined by

$$s = r_1 + r_2, \quad t = r_1 - r_2, \quad u = r_{12}, \tag{8}$$

the Hamiltonian in the fixed nucleus approximation is given by

$$
\begin{aligned}
H = &-\left(\frac{\partial^2}{\partial s^2} + \frac{\partial^2}{\partial t^2} + \frac{\partial^2}{\partial u^2}\right) - 2\frac{s(u^2 - t^2)}{u(s^2 - t^2)}\frac{\partial^2}{\partial s \partial u} - 2\frac{t(s^2 - u^2)}{u(s^2 - t^2)}\frac{\partial^2}{\partial u \partial t}, \\
&- \frac{4s}{s^2 - t^2}\frac{\partial}{\partial s} - \frac{2}{u}\frac{\partial}{\partial u} + \frac{4t}{s^2 - t^2}\frac{\partial}{\partial t} - \frac{4sZ}{s^2 - t^2} + \frac{1}{u},
\end{aligned}
\tag{9}
$$

where the last two terms represent the nuclear–electron attraction potential (Z is nuclear charge) and the electron–electron repulsion potential. The other terms originate from the kinetic operator. Using these potentials, we chose the g-function as

$$g = \frac{1}{V_{Ne}} + \frac{1}{V_{ee}}. \tag{10}$$

The initial function ψ_0 was chosen as

$$\psi_0 = [1 + \ln(s + u)]\exp(-\alpha s), \tag{11}$$

where the exponent α was dealt with as a variation parameter. The logarithmic dependence on s and u was introduced to describe well the three-particle coalescence region [16, 18, 20]. Then, the FC calculations are automatic and its wave function is guaranteed to become essentially exact at convergence. The FC wave function in this case is written as

$$\psi = \sum_i c_i \, s^{l_i} t^{m_i} u^{n_i} [\ln(s + u)]^{j_i} \exp(-\alpha s), \tag{12}$$

where l_i runs both positive and negative [9, 19] integers, $\{m_i, n_i\}$ run non-negative integers (m_i is even integers) and j_i is 0 or 1.

Table 1 shows the convergence of the variational energy [9]. The bold face implies that the figure is confidently reliable. A landmark calculation of the helium atom with the explicitly correlated wave function approach was done by Schwartz [28], who obtained the energy correct to 37 digits by applying the variation principle to his intuitively generated trial wave function. This was a surprising result. In the FC method, all we have to do is to fix ψ_0 and g function. Then, the FC formalism automatically generates a series of analytical functions in the form of Eq. (5). It is generated by the successive applications of the Hamiltonian and the g-function of the system to the starting wave function ψ_0 as expressed by Eq. (4). So, no severe intuition is necessary. Because this FC algorithm is automatic, we could continue the calculations up to the order n of 27 and obtained the energy correct to 41 digits.

Table 1 Ground-state energies of the helium atom calculated with the g function given by Eq. (10) and the initial function ψ_0 given by Eq. (11)[a]

n[a]	M_n^b	Optimal α	Energy (a.u.)[c]
0	2	1.827	**−2.865** 370 819 026 71
1	10	1.475	**−2.903** 536 812 281 53
2	34	1.627	**−2.903 724** 007 321 45
3	77	1.679	**−2.903 724 375** 094 16
4	146	1.683	**−2.903 724 377** 022 34
5	247	1.679	**−2.903 724 377 034** 05
6	386	1.693	**−2.903 724 377 034 119** 011 25
7	569	1.704	**−2.903 724 377 034 119 592** 84
8	802	1.707	**−2.903 724 377 034 119 598** 24
9	1091	1.713	**−2.903 724 377 034 119 598 309** 973 48
10	1442	1.724	**−2.903 724 377 034 119 598 311 136** 32
11	1861	1.738	**−2.903 724 377 034 119 598 311 158** 76
12	2354	1.757	**−2.903 724 377 034 119 598 311 159 23**
13	2927	1.779	**−2.903 724 377 034 119 598 311 159 244** 938 53
14	3586	1.806	**−2.903 724 377 034 119 598 311 159 245 187** 71
15	4337	1.837	**−2.903 724 377 034 119 598 311 159 245 194** 18
16	5186	1.866	**−2.903 724 377 034 119 598 311 159 245 194** 39
17	6139	1.899	**−2.903 724 377 034 119 598 311 159 245 194 403** 526 60
18	7202	(1.93)	**−2.903 724 377 034 119 598 311 159 245 194 404** 346 36
19	8381	(1.96)	**−2.903 724 377 034 119 598 311 159 245 194 404** 433 80
20	9682	(1.99)	**−2.903 724 377 034 119 598 311 159 245 194 404 444** 83
21	11111	(2.02)	**−2.903 724 377 034 119 598 311 159 245 194 404 446** 40
22	12674	(2.05)	**−2.903 724 377 034 119 598 311 159 245 194 404 446 646** 839 61
23	14377	(2.08)	**−2.903 724 377 034 119 598 311 159 245 194 404 446 687** 685 92
24	16226	(2.11)	**−2.903 724 377 034 119 598 311 159 245 194 404 446 695** 101 79
25	18227	(2.14)	**−2.903 724 377 034 119 598 311 159 245 194 404 446 696** 542 44
26	20386	(2.17)	**−2.903 724 377 034 119 598 311 159 245 194 404 446 696 840** 21
27	22709	(2.20)	**−2.903 724 377 034 119 598 311 159 245 194 404 446 696 905** 37
Ref. 28	10259		−2.903 724 377 034 119 598 311 159 245 194 404 440 049 5

[a] Order of the FC wave function.
[b] Number of complement functions at order n.
[c] Surely correct digit is shown by the bold face.

The 37-digit accuracy was attained at order 20. There, the number of the complement analytical functions was 9682, which is a bit smaller than 10,256, the number of the analytical functions used by Schwartz.

More recently, we have found that the exponential integral (Ei) function describes the three-particle coalescence region better than the logarithmic function [10]. So, starting with the Ei function, we could obtain a better energy at the order $n = 27$ with smaller number of variables; the energy was correct up to 43 digits.

We have applied the same method as above to the helium iso-electronic ions with Z from 1 to 10. The resultant FC wave functions had exactly the same form as Eq. (12) with only one difference in the exponents α. We performed the calculations up to the order 20 and obtained the lowest variational energies ever obtained [9]. The calculations consisted of three steps: (1) complement function generation step using MAPLE [31], (2) integral evaluation step, (3) diagonalization step in arbitrary accuracy. It took about 3 hours for the step (1), 2 days for the step (2), and 1.5 days for the step (3), with a single Intel(R) Core2 Duo 2.8 GHz workstation. We used MAPLE also in the second step, which means that this step can be substantially accelerated. Anyway, roughly 4 days were enough to get the world best energies and the analytical wave functions of the helium iso-electronic ions.

4 Properties Suitable for Checking the Exactness of the Calculated Wave Functions

To verify the exactness of the calculated results, the calculated energy alone is insufficient. We examine here several quantities that offer stringent test about the exactness of the calculated energy and wave function. Most of the properties shown here are useful only for the wave functions near the exact limit, otherwise, they show quite arbitrary numbers.

The Schrödinger equation is a local equation that must be satisfied at any local coordinate r. It is written as

$$\frac{H\psi(r)}{\psi(r)} = E(const.) \quad (^{\forall}r), \tag{13}$$

where $\psi(r)$ is the wave function at a coordinate, r. The left-hand side of Eq. (13) is called local energy, $E_L(r)$, as

$$E_L(r) = \frac{H\psi(r)}{\psi(r)}. \tag{14}$$

If ψ is not an exact wave function, then $E_L(r)$ may depend on r. If $E_L(r)$ is a constant at any point r, then Eq. (14) becomes Eq. (13), which is the Schrödinger equation. Therefore, the constancy of the local energy over the coordinate r is a straightforward test of how well the wave function ψ satisfies the Schrödinger equation.

In the formulation of the structure of the exact wave function, we introduced the H-square equation [3, 4],

$$\langle\psi|(H - E)^2|\psi\rangle = 0 \tag{15}$$

as the equation that is equivalent to the Schrödinger equation. When we define the left-hand side of Eq. (15) as

$$\sigma^2 = \langle\psi|(H - E)^2|\psi\rangle \tag{16}$$

for the normalized wave function and call it as H-square error, it is also the quantity that is very sensible to the exactness of the calculated wave function, because it is an integral sum of the positive quantities over all the coordinates. $\sigma^2 = 0$ means that the corresponding wave function is exact. The H-square error is also related to the local energy by

$$\sigma^2 = \langle E_L^2 \rangle_{\psi^2} - \langle E_L \rangle_{\psi^2}^2 , \tag{17}$$

where $\langle Q \rangle_{\psi^2}$ represents the expectation value of Q over the weight function $|\psi|^2$. Thus, σ^2 is the variance of the local energy weighted by $|\psi|^2$.

When we use the variation principle, the calculated energy is an upper bound to the exact energy, but as far as we do not know the exact energy, we cannot say how close the calculated energy is to the exact energy. A good theoretical way is to calculate the lower bound to the exact energy at the same time. If we can calculate both upper and lower bounds to the exact energy in high accuracy, we can predict the energy of the system with the error bars. The utility of such method lies entirely in the smallness of the error bars. As far as we use the variation principle, the upper bound of the exact energy is calculated twice more accurately than the accuracy of the wave function itself.

For the lower energy bound, Weinstein formulated the following expression [32],

$$E_{lower}^W = \langle \psi | H | \psi \rangle - \sqrt{\sigma^2}. \tag{18}$$

The Weinstein's lower bound is calculated for any state when its σ^2 and energy expectation value are known. However, a problem of this method is that the quality (accuracy) of this lower bound is not good enough: it is usually too low to be useful. Another method was proposed by Temple [33] for the ground state as

$$E_{lower}^T = \langle \psi | H | \psi \rangle - \frac{\sigma^2}{E_1 - \langle \psi | H | \psi \rangle}, \tag{19}$$

which requires the energy expectation value, σ^2 and, furthermore, the exact energy E_1 of the first excited state having the same symmetry as the ground state. In general, the exact energy E_1 is not known and so we have to modify Eq. (19). If one replaces E_1 with its lower bound energy, then one obtains the energy that is lower than the Temple's lower bound energy given by Eq. (19). We used the Weinstein's formula, Eq. (18), for calculating the lower bound to the first excited state, $E_1^W = \langle \psi_1 | H | \psi_1 \rangle - \sqrt{\sigma_1^2}$, where ψ_1 and σ_1^2 are the quantities for the first excited state. Then, $E_1 \geq E_1^W$. When this further satisfies $E_1^W > \langle \psi | H | \psi \rangle$, then we can define the modified Temple's lower bound energy by

$$E_{lower}^{T'} \equiv \langle \psi | H | \psi \rangle - \frac{\sigma^2}{E_1^W - \langle \psi | H | \psi \rangle}, \tag{20}$$

which satisfies

$$E_{exact} \geq E_{lower}^T \geq E_{lower}^{T'}. \tag{21}$$

The modified Temple's lower bound energy can be calculated only with the available theoretical quantities. Combining Eq. (21) with the result of the variation calculation, E_{upper}, we obtain

$$E_{upper} \geq E_{exact} \geq E_{lower}^{T'}. \tag{22}$$

When we calculate both the upper and lower bound of Eq. (22), we can say that the exact energy should lie in a definite region of the energy.

The cusp values of the wave function are also the necessary conditions of the exact wave function. Kato [34] rigorously derived the cusp conditions for many-electron systems as

$$\left. \frac{\partial \bar{\psi}}{\partial r} \right|_{r=0} = \lambda \psi(r = 0), \tag{23}$$

where $\bar{\psi}$ represents the spherically averaged wave function around the inter-particle coalescence region, $r = 0$. The value λ should be $-Z$ (nuclear charge) for the electron–nucleus coalescence and 1/2 for the electron–electron singlet-pair coalescence. We examine here the cusp values for the helium atom. The electron–nucleus and electron–electron cusp values for the helium wave function ψ are expressed, similarly to Eq. (23), as

$$Cusp(\mathbf{r}') = \frac{1}{\psi(r = 0)} \cdot \left. \frac{\partial \psi}{\partial r} \right|_{r=0}, \tag{24}$$

where $r = |\mathbf{r}_1 - \mathbf{r}_2|$ [13]. The difference from Eq. (23) is that the cusp value of Eq. (23) depends on the other coordinate $\mathbf{r}' = \mathbf{r}_3 - \mathbf{r}_1$ [35]. Without any spherical average, if the particles 1 and 2 approach each other perpendicularly to $\mathbf{r}'$, i.e., $\mathbf{r} \cdot \mathbf{r}' = 0$, then the cusp value still depends on the distance r' ($Cusp(r')$), but at any r', it should be $-Z$ for the electron–nucleus case and 1/2 for the electron–electron singlet-pair coalescence case.

5 Exactness Check of the Calculated Wave Functions

We examine the exactness of the helium wave function calculated by the FC method by calculating the quantities summarized in the above section.

Figure 1 shows the plots of the local energy at different orders up to $n = 27$ [12]. The helium nucleus is at the origin, one electron is located at $z = 0.5$ a.u. on the z-axis and the other electron moves along the z-axis from $z = -1.0$ to $+1.0$, experiencing the nuclear singularity at the origin and the electron singularity at $z = 0.5$. The vertical axis shows the relative value of the local energy, E_Y, which is scaled by the factor, ε, shown on the top of the vertical axis of each graph. Therefore, the local energy at each point, E_L, is calculated from the energy, E, shown on each

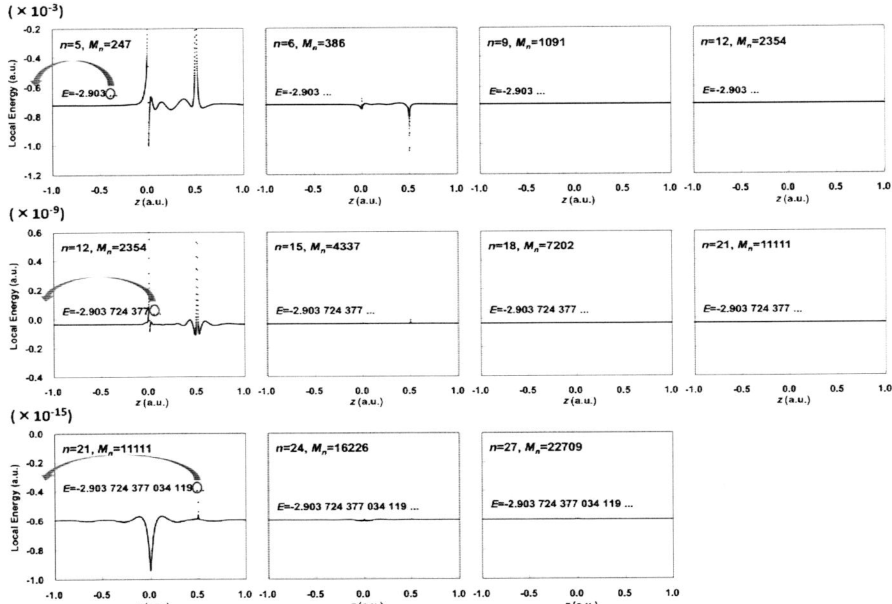

Fig. 1 Local energy plots of the FC wave functions for the orders n from 5 to 27. The *arrows* and *circles* in the *left-hand figures* show the digits of the total energy, in which the local energy is shown on the vertical axis changes

graph and the value of E_Y from $E_L = E + \varepsilon E_Y$. At the order $n = 5$, the local energy oscillates near the nucleus and another electron in the order of 10^{-3} a.u. However, at $n = 6$, the local energy becomes almost constant, except for the regions very close to the singularities. At $n = 9$ and $n = 12$, the local energy appears to be constant in the scale of 10^{-3} a.u. However, when we use a microscope and enlarge the figure by a scale of 10^6, we again see the fluctuations near the nuclear and electron singularities. Again, as we increase the order n from 12 to 15, 18, and 21, these fluctuations disappear and the local energy becomes completely flat. The same is true again in the last three figures in a finer scale.

In Fig. 2, we showed a very fine-detailed behavior of the local energy near the nuclear singularity and the electron singularity. Though there are fluctuations there, their half widths are very narrow, of the order of 10^{-5} a.u. and the heights of $0.5 - 1 \times 10^{-16}$ a.u. In all other regions of the space, the local energy is highly constant.

We next show in Table 2 the H-square error σ^2 and the energy lower bound calculated by the modified Temple equation. As the order n of the FC wave function increases, the H-square error gradually decreases and converges towards zero, the exact value. It is as small as 1.29×10^{-32} at $n = 27$. When the H-square error becomes zero, it means that the wave function becomes exact. So, this table means that, as the order n increases, the FC wave function approaches the exact wave

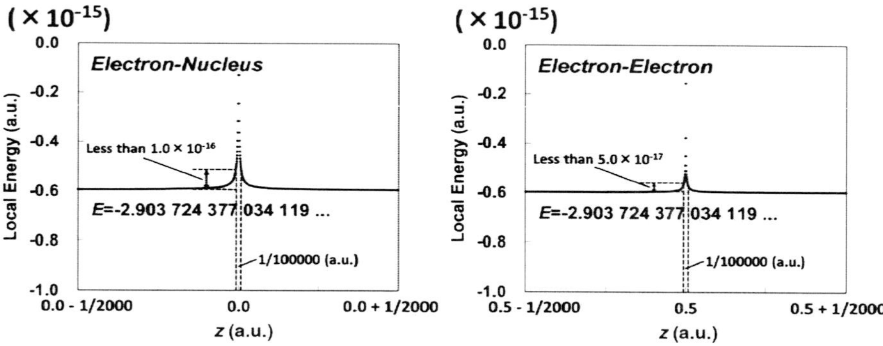

Fig. 2 Local energy plots at $n = 27$ ($M_n = 22709$) around the regions very close to the electron–nucleus ($z = 0$) and electron–electron ($z = 0.5$) singularities, where the local energy singularities are within 1.0×10^{-16} and 5.0×10^{-17} a.u., respectively, for an interval of $1/100000$ a.u.

Table 2 Convergence of the H-square error, σ^2, and the modified Temple's energy lower bound with increasing order n of the FC wave function

Order, n	M_n^a	H-square error, σ^2	Energy lower bound[b]
5	247	2.934869×10^{-9}	**−2.903 724 380** 97
6	386	4.782529×10^{-10}	**−2.903 724 377** 674
9	1091	1.095586×10^{-15}	**−2.903 724 377 034** 121 066
12	2354	5.007353×10^{-21}	**−2.903 724 377 034 119 598 317** 869
15	4337	1.835489×10^{-24}	**−2.903 724 377 034 119 598 311 161** 704
18	7202	5.372350×10^{-27}	**−2.903 724 377 034 119 598 311 159 252** 393
21	11111	4.000913×10^{-29}	**−2.903 724 377 034 119 598 311 159 245** 248
24	16226	5.665577×10^{-31}	**−2.903 724 377 034 119 598 311 159 245 195** 163
27	22709	1.293955×10^{-32}	**−2.903 724 377 034 119 598 311 159 245 194** 421 785

[a] Number of the complement functions for order n.
[b] Correct figure is expressed in bold face.

function, as shown clearly by the theoretical formulation [3, 4, 6]. This table confirms this numerically and further shows that the convergence speed is good.

As the order n of the FC wave function increases, the accuracy of the energy lower bound also increases. It approaches the exact value from below. This is in contrast to the variational energy shown in Table 1, which approaches the exact value from above. Using these lower and upper bounds to the exact energy, we can confidently predict that the exact energy should lie between the two bound energies, that is,

$$-2.903\ 724\ 377\ 034\ 119\ 598\ 311\ 159\ 245\ 194\ 421\ 785 < E_{\text{exact}} <$$
$$-2.903\ 724\ 377\ 034\ 119\ 598\ 311\ 159\ 245\ 194\ 404\ 446\ 696\ 905\ 37, \qquad (25)$$

where the bold-face digits show that this number is confidently correct. Thus, we can predict in confidence that the exact non-relativistic energy of the helium atom in the fixed nucleus approximation is $-\mathbf{2.903\ 724\ 377\ 034\ 119\ 598\ 311\ 159\ 245\ 194\ 4}$ a.u., which is correct to 32 digits. From the experience of the variational calculations, we can estimate the correct digits of the exact energy in higher accuracy, as shown in the upper bound of the exact energy shown in the above formula (25). Though we are confident about the correctness of the bold-face digits, the most a priori estimate of the exact energy is due to the accurate calculations of both of the upper and lower bounds of the exact energy, as shown in the above formula (25).

Finally, we examine the cusp values for the helium atom. Table 3 shows the nuclear–electron and electron–electron cusp values at the distance $r' = 1.0$ a.u., which was explained below Eq. (24). Both nuclear–electron and electron–electron cusp values approach the exact values of -2.0 and 0.5, respectively, as the order n of the FC calculation increases. At $n = 27$, the cusp values are correct to 22 digits, which is about a half of the correct digits of the variational energy, 41 digits, given in Table 1. This result is natural from a theoretical point of view.

Table 3 Electron–nucleus and electron–electron cusp values of the FC wave functions of helium atom[a]

Order, n	M_n^b	$Cusp_{(n)}^{N-e}(r', \theta' = \pi/2)$ (a.u.) $r' = 1.0$	$Cusp_{(n)}^{e-e}(r', \theta' = \pi/2)$ (a.u.) $r' = 1.0$
0	2	−1.71469176186748896865309	0.11230823813251103311346903
1	10	−**2.0**3769530273805345107342	**0.5**388261133165970364082007
2	34	−**2.00**250405223447850022359	**0.49**3191384355565889412867
3	77	−**2.0000**95337681210167679009	**0.499**313304921485831193970
4	146	−**1.99996**0751001985481002999	**0.499926**4052122836846890241
5	247	−**2.000004**0349019463261245	**0.4999888**12723545347905886
6	386	−**2.0000002**52006472327502965	**0.49999934**81518041394710234
7	569	−**1.999999990**139523025500075	**0.49999986**78901444684668622
8	802	−**2.0000000001**732245168503244	**0.499999984**3747191143208148
9	1091	−**1.99999999999**54741231937599	**0.4999999991**854449195682342
10	1442	−**2.0000000000**028992139510927	**0.49999999999**520485964408054
11	1861	−**1.9999999999999**4506214502899	**0.499999999999**68561359042406
12	2354	−**2.0000000000000**770317391674	**0.5000000000000**789589242552
13	2927	−**1.9999999999999**18885202492	**0.5000000000000**750757147237
14	3586	−**2.00000000000000**7641057691	**0.5000000000000**128229900478
15	4337	−**1.99999999999999**8856501484	**0.50000000000000**08582572593
16	5186	−**2.0000000000000000**172175707	**0.49999999999999992**93406817
17	6139	−**1.99999999999999999**68175846	**0.4999999999999999**9740766894
18	7202	−**2.00000000000000000**06492680	**0.4999999999999999999**57258457
19	8381	−**1.999999999999999999**8818275	**0.4999999999999999999**6645368
20	9682	−**2.00000000000000000000**112782	**0.50000000000000000000**0124903
21	11111	−**1.9999999999999999999**73697	**0.49999999999999999999**850833
22	12674	−**2.0000000000000000000**16585	**0.50000000000000000000**087634
23	14377	−**1.99999999999999999999**97746	**0.4999999999999999999**989519
24	16226	−**1.99999999999999999999**98202	**0.4999999999999999999**995815
25	18227	−**2.00000000000000000000000**911	**0.500000000000000000000**05676
26	20386	−**1.99999999999999999999999**660	**0.499999999999999999999**97157
27	22709	−**2.0000000000000000000000**175	**0.50000000000000000000000**691
Exact		**−2.0**	**0.5**

[a] Correct figure is expressed in bold face.

[b] Number of the complement functions at order n.

6 Concluding Remarks

We have shown here that the FC method for solving the Schrödinger equation gives a series of analytical complement functions that span the exact wave function. By increasing the number of the complement functions with increasing the order n, one can calculate the solutions of the Schrödinger equation as accurately as one desires. This was shown for the helium atom as an example. Not only the variational energy, which is an upper bound of the exact energy, but also the other properties like local energy, H-square error, lower energy bound, and nuclear–electron and electron–electron cusp values all approached the exact values as the order n of the FC method increased. Theoretically, the variational energy is always more accurate than the wave function itself and other properties. The present results constitute a numerical proof that with the FC method one can calculate the solution of the Schrödinger equation as accurately as one desires.

We could not show here the results of solving the relativistic Dirac–Coulomb equation. The FC method can be extended to the case of the Dirac–Coulomb equation with only a small modification [36]. It is important to use the inverse Dirac–Coulomb equation to circumvent the variational collapse problem which often appears in the relativistic calculations [37].

For complex atoms and molecules, the analytical integrations involved in Eqs. (6) or (7) are difficult to perform. For such cases, we have proposed the local Schrödinger equation (LSE) method. It is based on the potential exactness of the FC wave function given by Eq. (5) for large n. For more details, we refer to Ref. [11]. Using the LSE method, we can calculate the analytic wave function of atoms and molecules without doing the analytical integrations. This method is very general, since the integrations in Eq. (7) are difficult to perform for the complement functions of general atoms and molecules.

The programs for the variational calculations of the helium atom and its iso-electronic ions and the diagonalization program in arbitrary accuracy used for obtaining the data shown in this chapter can be obtained with charges. For details, please refer to the web site of our QCRI (www.qcri.or.jp).

References

1. E. Schrödinger, Phys. Rev. **28**, 1049 (1926)
2. P. A. M. Dirac, Proc. Roy. Soc. A **123**, 714 (1929)
3. H. Nakatsuji, J. Chem. Phys. **113**, 2949 (2000)
4. H. Nakatsuji, E. R. Davidson, J. Chem. Phys. **115**, 2000 (2001)
5. H. Nakatsuji, M. Ehara, J. Chem. Phys. **117**, 9 (2002); ibid. **122**, 194108 (2005)
6. H. Nakatsuji, Phys. Rev. Lett. **93**, 030403 (2004)
7. H. Nakatsuji, Phys. Rev. A **72**, 062110 (2005)
8. H. Nakatsuji, Bull. Chem. Soc. Jpn. **78**, 1705 (2005)
9. H. Nakashima, H. Nakatsuji, J. Chem. Phys. **127**, 224104 (2007)
10. Y. I. Kurokawa, H. Nakashima, H. Nakatsuji, Phys. Chem. Chem. Phys. **10**, 4486 (2008)
11. H. Nakatsuji, H. Nakashima, Y. Kurokawa, A. Ishikawa, Phys. Rev. Lett. **99**, 240402 (2007)

12. H. Nakashima, H. Nakatsuji, Phys. Rev. Lett. **101**, 240406 (2008)
13. H. Nakatsuji, H. Nakashima, Int. J. Quantum Chem., **109**, 2248 (2009)
14. *Explicitly Correlated Wave Functions in Chemistry and Physics – Theory and Applications*, ed. by J. Rychlewski (Kluwer, Dordrecht, 2003)
15. E. A. Hylleraas, Z. Phys. **54**, 347 (1929)
16. J. H. Bartlett, Jr., Phys. Rev. **51**, 661 (1937)
17. T. H. Gronwall, Phys. Rev. **51**, 655 (1937)
18. V. A. Fock, Izv. Akad. Nauk. SSSR, Ser. Fiz. **18**, 161 (1954)
19. T. Kinoshita, Phys. Rev. **105**, 1490 (1957)
20. K. Frankowski, C. L. Pekeris, Phys. Rev. **146**, 46 (1966)
21. A. J. Thakkar, T. Koga, Phys. Rev. A **50**, 854 (1994)
22. S. P. Goldman, Phys. Rev. A **57**, R677 (1998)
23. G. W. F. Drake, Phys. Scr. **T83**, 83 (1999)
24. G. W. F. Drake, M. M. Cassar, R. A. Nistor, Phys. Rev. A **65**, 054501 (2002)
25. J. S. Sims, S. A. Hagstrom, Int. J. Quantum Chem. **90**, 1600 (2002)
26. V. I. Korobov, Phys. Rev. A **66**, 024501 (2002)
27. A. J. Thakkar, T. Koga, Theor. Chem. Acc. **109**, 36 (2003)
28. C. Schwartz, Int. J. Mod. Phys. E **15**, 877 (2006)
29. H. Nakashima, H. Nakatsuji, J. Chem. Phys. **128**, 154107 (2008)
30. H. Nakashima, Y. Hijikata, H. Nakatsuji, J. Chem. Phys. **128**, 154108 (2008)
31. Computer code MAPLE, Waterloo Maple Inc., Waterloo, Ontario, Canada; see http://www.maplesoft.com
32. D. H. Weinstein, Proc. Natl. Acad. Sci. USA **20**, 529 (1934)
33. G. Temple, Proc. Roy. Soc. A **119**, 276 (1928)
34. T. Kato, Commun. Pure Appl. Math. **10**, 151 (1957)
35. V. A. Rassolov, D. M.Chipman, J. Chem. Phys. **104**, 9908 (1996)
36. H. Nakatsuji, H. Nakashima, Phys. Rev. Lett. **95**, 050407 (2005)
37. R. N. Hill, C. Krauthauser, Phys. Rev. Lett. **58**, 83 (1987)

Energy Computation for Exponentially Correlated Four-Body Wavefunctions

Frank E. Harris

Abstract Formulas are presented for efficient computation of the energy of four-body quantum-mechanical Coulomb systems with wavefunctions consisting of fully correlated exponentials premultiplied by arbitrary integer powers of the interparticle distances. Using the interparticle distances as coordinates, the potential energy is easily expressed in terms of basic integrals involving these wavefunctions. All the contributions to the kinetic energy are also expressible using the same basic integrals, but it is useful to organize the computations in ways that take advantage of the relations between integrals and that illustrate the underlying symmetry of the formulation. The utility of the formulation presented here is illustrated by an "ultra-compact" computation of the ground state of the Li atom.

Keywords: Few-body problems · Correlated wavefunctions · Li atom

1 Introduction

Bases of fully exponentially correlated wavefunctions [1, 2] provide more rapid convergence as a function of expansion length than any other type of basis thus far employed for quantum mechanical computations on Coulomb systems consisting of four particles or less. This feature makes it attractive to use such bases to construct "ultra-compact" expansions which exhibit reasonable accuracy while maintaining a practical capability to visualize the salient features of the wavefunction. For this purpose, exponentially correlated functions have advantages over related expansions of Hylleraas type [3], in which the individual-term explicit correlation is limited to pre-exponential powers of various interparticle distances (generically denoted r_{ij}). The general features of the exponentially correlated expansions are well illustrated for three-body systems by our work on He and its isoelectronic ions, for

F.E. Harris (✉)
Department of Physics, University of Utah, Salt Lake City, Utah 84112, and Quantum Theory Project, University of Florida, Gainesville, Florida 32611,
e-mail: harris@qtp.ufl.edu

P. Piecuch et al. (eds.), *Advances in the Theory of Atomic and Molecular Systems*,
Progress in Theoretical Chemistry and Physics 19, DOI 10.1007/978-90-481-2596-8_4,

which exponentially correlated four-term wavefunctions (with no pre-exponential r_{ij}) yield energies within 40 microhartrees of the fully converged limit, a range of correlation-dependent properties with three- to five-digit accuracy and demonstrable regularities in the wavefunction parameters as a function of the ionic charge [4–6].

There have now been several applications reported for fully exponentially correlated four-body wavefunctions [7–9], also limited to bases without pre-exponential r_{ij}. While it was found that pre-exponential r_{ij} are relatively unimportant for three-body systems, they can be expected to contribute in a major way to the efficiency of expansions for three-electron systems such as the Li atom and its isoelectronic ions, as is obvious from the fact that the zero-order description of the ground states of such systems has electron configuration $1s^2 2s$.

A practical reason that pre-exponential r_{ij} have not been used with exponentially correlated four-body wavefunctions has been the difficulty of managing analytical formulas for the integrals that thereby result; that difficulty has now been reduced in importance by the author's recent presentation[10] of a recursive procedure for the integral generation.

This communication outlines formulas that can be used when the energy is described (for S states) entirely in terms of the interparticle coordinates r_{ij} and extends earlier work [11, 12] that shows how the combinations of integrals that describe the kinetic energy can be related to the overlap and Coulomb interaction integrals that enter the evaluation of the electrostatic potential energy.

Use of the formulation is illustrated by a single-configuration computation of the Li atom ground state.

2 Problem Formulation

We consider here the kinetic- and potential-energy matrix elements for four-body Coulomb systems that are described by wavefunctions consisting of terms each of the generic form

$$\psi = r_{12}^{n_{12}} r_{13}^{n_{13}} r_{14}^{n_{14}} r_{23}^{n_{23}} r_{24}^{n_{24}} r_{34}^{n_{34}} e^{-\alpha_{12} r_{12} - \alpha_{13} r_{13} - \alpha_{14} r_{14} - \alpha_{23} r_{23} - \alpha_{24} r_{24} - \alpha_{34} r_{34}}. \tag{1}$$

Here r_{ij} is the distance between particles i and j, the n_{ij} are non-negative integers, and the α_{ij} are parameters that are assigned values such that the energy matrix elements involving ψ converge. For effective management of what otherwise would become cumbersome notationally, we identify the wavefunction of Eq. (26) as $\psi(\boldsymbol{\alpha}, \mathbf{n})$, where $\boldsymbol{\alpha}$ and $\mathbf{n}$ are, respectively, shorthand for the sets $\{\alpha_{12}, \alpha_{13}, \ldots, \alpha_{34}\}$ and $\{n_{12}, n_{13}, \ldots, n_{34}\}$; two wavefunctions with different pre-exponential powers and exponential parameters can therefore be described by notations such as $\psi(\boldsymbol{\alpha}, \mathbf{n})$ and $\psi(\boldsymbol{\beta}, \mathbf{m})$.

All the matrix elements needed for energy computations using wavefunctions of the type $\psi(\boldsymbol{\alpha}, \mathbf{n})$ can be shown to reduce to integrals of the generic type

$$F(\boldsymbol{\gamma}, \mathbf{p}) \equiv \int d\mathbf{r}_{12}\, d\mathbf{r}_{13}\, d\mathbf{r}_{14} \prod_{1 \le i < j \le 4} \left(r_{ij}^{p_{ij}} e^{-\gamma_{ij} r_{ij}} \right), \tag{2}$$

where the integral is over the entire nine-dimensional space of the $\mathbf{r}_{1i}$ coordinates. Analytical formulas for $F(\boldsymbol{\gamma}, \mathbf{p})$ of general $\boldsymbol{\gamma}$ and $\mathbf{p}$ are known [13, 14] (see also the Appendix to this chapter), and their use has recently been made less cumbersome by the development of recursive formulas [10] connecting integrals with contiguous sets of pre-exponential powers $\mathbf{p}$.

The matrix elements of interest here are of the forms $\langle \psi(\boldsymbol{\alpha}, \mathbf{n}) | \psi(\boldsymbol{\beta}, \mathbf{m}) \rangle$, $\langle \psi(\boldsymbol{\alpha}, \mathbf{n}) | V | \psi(\boldsymbol{\beta}, \mathbf{m}) \rangle$, and $\langle \psi(\boldsymbol{\alpha}, \mathbf{n}) | T | \psi(\boldsymbol{\beta}, \mathbf{m}) \rangle$, where V and T are, respectively, the Coulomb and kinetic-energy operators. For a four-body system, with respective charges $q_1, \ldots, q_4$ and masses $m_1, \ldots, m_4$, with all quantities expressed in Hartree atomic units, these matrix elements can be written entirely in terms of the interparticle coordinates [12], with the potential-energy matrix elements given as

$$\langle \psi(\boldsymbol{\alpha}, \mathbf{n}) | V | \psi(\boldsymbol{\beta}, \mathbf{m}) \rangle = \sum_{1 \leq i < j < \leq 4} \left\langle \psi(\boldsymbol{\alpha}, \mathbf{n}) \left| \frac{q_i q_j}{r_{ij}} \right| \psi(\boldsymbol{\beta}, \mathbf{m}) \right\rangle, \tag{3}$$

while the kinetic-energy matrix element can be written in either of the two equivalent forms

$$\langle \psi(\boldsymbol{\alpha}, \mathbf{n}) | T | \psi(\boldsymbol{\beta}, \mathbf{m}) \rangle = \sum_i \frac{1}{2m_i} \sum_{j(\neq i)} \left[\left\langle \frac{\partial \psi(\boldsymbol{\alpha}, \mathbf{n})}{\partial r_{ij}} \frac{\partial \psi(\boldsymbol{\beta}, \mathbf{m})}{\partial r_{ij}} \right\rangle \right.$$

$$\left. + \sum_{k(\neq ij)} \left\langle \cos \Theta_{ijk} \frac{\partial \psi(\boldsymbol{\alpha}, \mathbf{n})}{\partial r_{ik}} \frac{\partial \psi(\boldsymbol{\beta}, \mathbf{m})}{\partial r_{ij}} \right\rangle \right] \tag{4}$$

and

$$\langle \psi(\boldsymbol{\alpha}, \mathbf{n}) | T | \psi(\boldsymbol{\beta}, \mathbf{m}) \rangle$$

$$= -\sum_i \frac{1}{2m_i} \sum_{j(\neq i)} \left[\left\langle \psi(\boldsymbol{\alpha}, \mathbf{n}) \left(\frac{\partial^2 \psi(\boldsymbol{\beta}, \mathbf{m})}{\partial r_{ij}^2} + \frac{2}{r_{ij}} \frac{\partial \psi(\boldsymbol{\beta}, \mathbf{m})}{\partial r_{ij}} \right) \right\rangle \right.$$

$$\left. + \sum_{k(\neq ij)} \left\langle \psi(\boldsymbol{\alpha}, \mathbf{n}) \cos \Theta_{ijk} \frac{\partial^2 \psi(\boldsymbol{\beta}, \mathbf{m})}{\partial r_{ij} \partial r_{ik}} \right\rangle \right]. \tag{5}$$

In the above equations $1 \leq i, j, k \leq 4$, $\cos \Theta_{ijk}$ stands for

$$\cos \Theta_{ijk} \equiv (\hat{\mathbf{r}}_{ij} \cdot \hat{\mathbf{r}}_{ik}) = \frac{r_{ij}^2 + r_{ik}^2 - r_{jk}^2}{2 r_{ij} r_{ik}}, \tag{6}$$

and the $\langle \ldots \rangle$ brackets denote integration over the same range as that of Eq. (2). The task presently at hand is to obtain the matrix elements of overlap, V, and T as convenient combinations of the $F(\boldsymbol{\gamma}, \mathbf{p})$.

3 Evaluation of Matrix Elements

The overlap matrix elements are trivial cases of Eq. (2):

$$\langle \psi(\boldsymbol{\alpha}, \mathbf{n}) | \psi(\boldsymbol{\beta}, \mathbf{m}) \rangle = F(\boldsymbol{\alpha} + \boldsymbol{\beta}, \mathbf{n} + \mathbf{m}) \,, \tag{7}$$

where $\boldsymbol{\alpha} + \boldsymbol{\beta}$ denotes the parameter set $\{\alpha_{ij} + \beta_{ij}\}$ and $\mathbf{n} + \mathbf{m}$ describes the set $\{n_{ij} + m_{ij}\}$. Because V is a multiplicative operator, its matrix elements also reduce immediately in a similar fashion. Defining $\mathbf{1}_{ij}$ to stand for the index set whose ij element is unity, with all other elements zero, Eq. (3) assumes the form

$$\langle \psi(\boldsymbol{\alpha}, \mathbf{n}) | V | \psi(\boldsymbol{\beta}, \mathbf{m}) \rangle = \sum_{1 \le i < j < \le 4} q_i \, q_j \, F(\boldsymbol{\alpha} + \boldsymbol{\beta}, \mathbf{n} + \mathbf{m} - \mathbf{1}_{ij}) \,. \tag{8}$$

To proceed to the kinetic energy, we use the basic differentiation formula $\partial \psi(\boldsymbol{\alpha}, \mathbf{n}) / \partial r_{ij} = n_{ij} \psi(\boldsymbol{\alpha}, \mathbf{n} - \mathbf{1}_{ij}) - \alpha_{ij} \psi(\boldsymbol{\alpha}, \mathbf{n})$, thereby reducing Eq. (4) to the form

$$\begin{aligned}
\langle \psi(\boldsymbol{\alpha}, \mathbf{n}) | T | \psi(\boldsymbol{\beta}, \mathbf{m}) \rangle = &\sum_{1 \le i < j \le 4} \frac{1}{2\mu_{ij}} \Big[\alpha_{ij} \beta_{ij} F(\boldsymbol{\alpha} + \boldsymbol{\beta}, \mathbf{n} + \mathbf{m}) \\
&\quad - (\alpha_{ij} m_{ij} + \beta_{ij} n_{ij}) F(\boldsymbol{\alpha} + \boldsymbol{\beta}, \mathbf{n} + \mathbf{m} - \mathbf{1}_{ij}) \\
&\quad + n_{ij} m_{ij} F(\boldsymbol{\alpha} + \boldsymbol{\beta}, \mathbf{n} + \mathbf{m} - \mathbf{2}_{ij}) \Big] \\
&+ \sum_{i=1}^{4} \frac{1}{2m_i} \Bigg(\sum_{\substack{j < k \\ j,k \ne i}} \Big[(\alpha_{ij} \beta_{ik} + \alpha_{ik} \beta_{ij}) G_{ijk}(\boldsymbol{\alpha} + \boldsymbol{\beta}, \mathbf{n} + \mathbf{m}) \\
&\quad + (n_{ij} m_{ik} + n_{ik} m_{ij}) G_{ijk}(\boldsymbol{\alpha} + \boldsymbol{\beta}, \mathbf{n} + \mathbf{m} - \mathbf{1}_{ij} - \mathbf{1}_{ik}) \Big] \\
&\quad - \sum_{\substack{jk \\ j,k \ne i}} (\alpha_{ij} m_{ik} + \beta_{ij} n_{ik}) G_{ijk}(\boldsymbol{\alpha} + \boldsymbol{\beta}, \mathbf{n} + \mathbf{m} - \mathbf{1}_{ik}) \Bigg) \,.
\end{aligned} \tag{9}$$

Here and henceforth, $\alpha_{ji} = \alpha_{ij}$, $\beta_{ji} = \beta_{ij}$, $n_{ji} = n_{ij}$, $m_{ji} = m_{ij}$, $\mu_{ij} = m_i m_j / (m_i + m_j)$, and $\mathbf{2}_{ij}$ stands for $2(\mathbf{1}_{ij})$. The new quantities G_{ijk}, whose computation is discussed in the next section, have definition

$$G_{ijk}(\boldsymbol{\gamma}, \mathbf{p}) \equiv \int d\mathbf{r}_{12} \, d\mathbf{r}_{13} \, d\mathbf{r}_{14} \, \cos \Theta_{ijk} \prod_{1 \le \mu < \nu \le 4} \left(r_{\mu\nu}^{p_{\mu\nu}} e^{-\gamma_{\mu\nu} r_{\mu\nu}} \right) . \tag{10}$$

It is apparent from Eq. (6) that $G_{ikj}(\boldsymbol{\gamma}, \mathbf{p}) = G_{ijk}(\boldsymbol{\gamma}, \mathbf{p})$, and that fact has been used in the derivation of Eq. (8). Note, however, that $G_{ijk}(\boldsymbol{\gamma}, \mathbf{p})$ is not equivalent to $G_{jik}(\boldsymbol{\gamma}, \mathbf{p})$.

4 Evaluation of G_{ijk}

The G_{ijk} can be reduced to sums of the integrals $F(\gamma, \mathbf{p})$ by direct application of Eq. (3), but doing so introduces integrals F whose occurrence can be avoided by developing suitable relations that give G_{ijk} entirely in terms of overlap and potential-energy matrix elements that are already needed for other parts of the computation. This fact was first recognized by Rebane [11], and a straightforward way of obtaining these relations (in the absence of pre-exponential r_{ij}) has previously been described by the present author [12]. A starting point for the extension needed here is provided by exploiting the equivalence of the two forms of the kinetic energy matrix element given in Eqs. (4) and (5); equating these for the case $\mathbf{n} = \mathbf{m} = \mathbf{0}$, where $\mathbf{0}$ is the zero index set, we then found (cf. [12]) for each i, written in the present notation,

$$\sum_j \beta_{ij} \left[(\alpha_{ij} + \beta_{ij}) F(\boldsymbol{\alpha} + \boldsymbol{\beta}, \mathbf{0}) - 2\, F(\boldsymbol{\alpha} + \boldsymbol{\beta}, -\mathbf{1}_{ij}) \right.$$
$$\left. + \sum_{k(\neq i,j)} (\alpha_{ik} + \beta_{ik}) G_{ijk}(\boldsymbol{\alpha} + \boldsymbol{\beta}, \mathbf{0}) \right] = 0 \,. \tag{11}$$

Since the quantities of Eq. (10) within square brackets are functions only of $\boldsymbol{\alpha}+\boldsymbol{\beta}$, we can keep these quantities constant while making independent variations of the β_{ij} (doing so by making compensating changes in the α_{ij}), so that the satisfaction of Eq. (10) can only be maintained if the square-bracketed quantities individually vanish. In this way, we obtain an identity for each distinct ij ordered pair, which we write with $\boldsymbol{\alpha} + \boldsymbol{\beta}$ replaced by $\boldsymbol{\gamma}$:

$$K_{ij} \equiv \gamma_{ij} F(\boldsymbol{\gamma}, \mathbf{0}) - 2\, F(\boldsymbol{\gamma}, -\mathbf{1}_{ij}) + \sum_{k(\neq i,j)} \gamma_{ik} G_{ijk}(\boldsymbol{\gamma}, \mathbf{0}) = 0 \,. \tag{12}$$

Finally, we differentiate K_{ij} as indicated here, using Leibniz's rule to evaluate the multiple derivatives:

$$\prod_{1 \leq \mu < \nu \leq 4} \left(-\frac{\partial}{\partial \gamma_{\mu\nu}} \right)^{p_{\mu\nu}} K_{ij} = \gamma_{ij} F(\boldsymbol{\gamma}, \mathbf{p}) - (p_{ij} + 2) F(\boldsymbol{\gamma}, \mathbf{p} - \mathbf{1}_{ij})$$
$$+ \sum_{k(\neq i,j)} \left[\gamma_{ik} G_{ijk}(\boldsymbol{\gamma}, \mathbf{p}) - p_{ik} G_{ijk}(\boldsymbol{\gamma}, \mathbf{p} - \mathbf{1}_{ik}) \right] = 0 \,. \tag{13}$$

For each i, there will be three instances of Eq. (13), and we regard these as equations to be solved for $G_{ijk}(\boldsymbol{\gamma}, \mathbf{p})$, $G_{ijl}(\boldsymbol{\gamma}, \mathbf{p})$, and $G_{ikl}(\boldsymbol{\gamma}, \mathbf{p})$ (where j, k, and l are the members of 1,2,3,4 other than i). These equations take the form

$$\gamma_{ik} G_{ijk}(\boldsymbol{\gamma}, \mathbf{p}) + \gamma_{il} G_{ijl}(\boldsymbol{\gamma}, \mathbf{p}) = s_{ij} , \tag{14}$$

$$\gamma_{ij} G_{ijk}(\boldsymbol{\gamma}, \mathbf{p}) + \gamma_{il} G_{ikl}(\boldsymbol{\gamma}, \mathbf{p}) = s_{ik} , \tag{15}$$

$$\gamma_{ij} G_{ijl}(\boldsymbol{\gamma}, \mathbf{p}) + \gamma_{ik} G_{ikl}(\boldsymbol{\gamma}, \mathbf{p}) = s_{il} , \tag{16}$$

where

$$s_{ij} \equiv (p_{ij} + 2) F(\boldsymbol{\gamma}, \mathbf{p} - \mathbf{1}_{ij}) - \gamma_{ij} F(\boldsymbol{\gamma}, \mathbf{p}) + \sum_{k(\neq i, j)} p_{ik} G_{ijk}(\boldsymbol{\gamma}, \mathbf{p} - \mathbf{1}_{ik}) . \tag{17}$$

The solution for $G_{ijk}(\boldsymbol{\gamma}, \mathbf{p})$ is

$$G_{ijk}(\boldsymbol{\gamma}, \mathbf{p}) = \frac{\gamma_{ij} s_{ij} + \gamma_{ik} s_{ik} - \gamma_{il} s_{il}}{2 \gamma_{ij} \gamma_{ik}} . \tag{18}$$

Inserting explicit expressions for the s_{ij}, we reach

$$\begin{aligned}
G_{ijk}(\boldsymbol{\gamma}, \mathbf{p}) = \frac{1}{\gamma_{ij} \gamma_{ik}} \Bigg[& (\gamma_{il}^2 - \gamma_{ij}^2 - \gamma_{ik}^2) F(\boldsymbol{\gamma}, \mathbf{p}) + \gamma_{ij}(p_{ij} + 2) F(\boldsymbol{\gamma}, \mathbf{p} - \mathbf{1}_{ij}) \\
& + \gamma_{ik}(p_{ik} + 2) F(\boldsymbol{\gamma}, \mathbf{p} - \mathbf{1}_{ik}) - \gamma_{il}(p_{il} + 2) F(\boldsymbol{\gamma}, \mathbf{p} - \mathbf{1}_{il}) \\
& + \gamma_{ij} \sum_{\mu=k,l} p_{i\mu} G_{ij\mu}(\boldsymbol{\gamma}, \mathbf{p} - \mathbf{1}_{i\mu}) + \gamma_{ik} \sum_{\mu=j,l} p_{i\mu} G_{ik\mu}(\boldsymbol{\gamma}, \mathbf{p} - \mathbf{1}_{i\mu}) \\
& - \gamma_{il} \sum_{\mu=j,k} p_{i\mu} G_{il\mu}(\boldsymbol{\gamma}, \mathbf{p} - \mathbf{1}_{i\mu}) \Bigg] ;
\end{aligned} \tag{19}$$

the formulas for $G_{ijl}(\boldsymbol{\gamma}, \mathbf{p})$ and $G_{ikl}(\boldsymbol{\gamma}, \mathbf{p})$ can be obtained by permutation of the indices in Eq. (14).

Equation (14) can be regarded as a self-starting recursive formulation, as the $G_{ijk}(\boldsymbol{\gamma}, \mathbf{p})$ for $\mathbf{p} = \mathbf{0}$ require only values of the same F that are needed for the overlap and potential-energy matrix elements, while those for non-zero $\mathbf{p}$ require in addition only $G(\boldsymbol{\gamma}, \mathbf{p}')$ for $\mathbf{p}'$ whose index sum is one less than that of $\mathbf{p}$.

Formulas such as Eq. (13) lead to many relations among the integrals F and G. For use in this chapter, we note that in the special case that $i = 1$, $j = 4$, and $\mathbf{p} = \mathbf{1}_{14}$, Eq. (13) assumes the form

$$\gamma_{12} G_{142}(\boldsymbol{\gamma}, \mathbf{1}_{14}) + \gamma_{13} G_{143}(\boldsymbol{\gamma}, \mathbf{1}_{14}) = 3 F(\boldsymbol{\gamma}, \mathbf{0}) - \gamma_{14} F(\boldsymbol{\gamma}, \mathbf{1}_{14}) ; \tag{20}$$

also, for $i = 4$, $j = 1$,

$$\gamma_{24} G_{412}(\boldsymbol{\gamma}, \mathbf{1}_{14}) + \gamma_{34} G_{413}(\boldsymbol{\gamma}, \mathbf{1}_{14}) = 3 F(\boldsymbol{\gamma}, \mathbf{0}) - \gamma_{14} F(\boldsymbol{\gamma}, \mathbf{1}_{14}) . \tag{21}$$

5 Application to Li

We present now an illustrative calculation for the ^2S electronic ground state of the Li atom, using a single-term spatial wavefunction that can be characterized as an exponentially correlated $1s1s'2s$ Slater-type orbital (STO) product. For the purpose of this illustration, we restrict the spin state (the three-electron doublet space has dimension two) to that in which the $1s$ and $1s'$ are singlet-coupled. Thus, the space-spin wavefunction has the form

$$\Psi = (1 - P_{23} - P_{24} - P_{34} + P_{342} + P_{423})\psi(\boldsymbol{\alpha}, 1_{14})\chi \ , \tag{22}$$

where in ψ the nucleus is Particle 1, the $1s$ and $1s'$ electrons are Particles 2 and 3, and the $2s$ STO electron is Particle 4. The symbols P_{ij} denote interchange of Particles i and j, while P_{ijk} represents the permutation of Particles $2, 3, 4$ into i, j, k. The quantity χ is a spin state described by

$$\chi = 2^{-1/2}\left[\alpha(2)\beta(3)\alpha(4) - \beta(2)\alpha(3)\alpha(4)\right] \ . \tag{23}$$

Thus, α_{12} and α_{13} represent the $1s$ and $1s'$ screening parameters, α_{14} is the $2s$ screening parameter, and $\alpha_{23}, \alpha_{24}, \alpha_{34}$ produce electron–electron correlation.

We seek to minimize $E = \langle\Psi|T + V|\Psi\rangle/\langle\Psi|\Psi\rangle$ with respect to $\boldsymbol{\alpha}$; the parameters in T and V are $m_1 = \infty$, $m_2 = m_3 = m_4 = 1$, $q_1 = +3$, and $q_2 = q_3 = q_4 = -1$. The only terms in the evaluation of E that require further comment are those arising from T; these can be permuted to one of the two generic forms $\langle\psi(\boldsymbol{\alpha}, 1_{14})|T|\psi(\boldsymbol{\beta}, 1_{14})\rangle$ or $\langle\psi(\boldsymbol{\alpha}, 1_{13})|T|\psi(\boldsymbol{\beta}, 1_{14})\rangle$, where $\boldsymbol{\beta}$ is a permutation of $\boldsymbol{\alpha}$. Starting with Eq. (8), a straight-forward specialization to the present case leads to

$$
\begin{aligned}
\langle\psi(\boldsymbol{\alpha}, 1_{14})|T|\psi(\boldsymbol{\beta}, 1_{14})\rangle =& \sum_{1\leq i<j\leq 4} \frac{\alpha_{ij}\beta_{ij}}{2\mu_{ij}} F(\boldsymbol{\alpha} + \boldsymbol{\beta}, 2_{14}) \\
&+ \frac{1}{2\mu_{14}}\left[F(\boldsymbol{\alpha} + \boldsymbol{\beta}, 0) - (\alpha_{14} + \beta_{14})F(\boldsymbol{\alpha} + \boldsymbol{\beta}, 1_{14})\right] \\
&+ \sum_{i=1}^{4} \frac{1}{2m_i} \sum_{\substack{j<k \\ j,k\neq i}} (\alpha_{ij}\beta_{ik} + \alpha_{ik}\beta_{ij})G_{ijk}(\boldsymbol{\alpha} + \boldsymbol{\beta}, 2_{14}) \\
&- \frac{1}{2m_1} \sum_{j=2,3} (\alpha_{1j} + \beta_{1j})G_{1j4}(\boldsymbol{\alpha} + \boldsymbol{\beta}, 1_{14}) \\
&- \frac{1}{2m_4} \sum_{j=2,3} (\alpha_{4j} + \beta_{4j})G_{41j}(\boldsymbol{\alpha} + \boldsymbol{\beta}, 1_{14}) \ .
\end{aligned}
\tag{24}
$$

This equation can be drastically simplified; its fourth and fifth lines can be recognized as instances of Eqs. (20) and (21), after which they combine with the second line of Eq. (22) to yield the far more compact final result:

$$\langle \psi(\boldsymbol{\alpha}, \mathbf{1}_{14}) | T | \psi(\boldsymbol{\beta}, \mathbf{1}_{14}) \rangle = \sum_{1 \le i < j \le 4} \frac{\alpha_{ij}\beta_{ij}}{2\mu_{ij}} F(\boldsymbol{\alpha} + \boldsymbol{\beta}, \mathbf{2}_{14}) - \frac{1}{\mu_{14}} F(\boldsymbol{\alpha} + \boldsymbol{\beta}, \mathbf{0})$$

$$+ \sum_{i=1}^{4} \frac{1}{2m_i} \sum_{\substack{j < k \\ j,k \neq i}} (\alpha_{ij}\beta_{ik} + \alpha_{ik}\beta_{ij}) G_{ijk}(\boldsymbol{\alpha} + \boldsymbol{\beta}, \mathbf{2}_{14}) . \tag{25}$$

The other generic kinetic-energy matrix element needed here is of the form $\langle \psi(\boldsymbol{\alpha}, \mathbf{1}_{13}) | T | \psi(\boldsymbol{\beta}, \mathbf{1}_{14}) \rangle$. In the present instance, it specializes to

$$\langle \psi(\boldsymbol{\alpha}, \mathbf{1}_{13}) | T | \psi(\boldsymbol{\beta}, \mathbf{1}_{14}) \rangle = \sum_{1 \le i < j \le 4} \frac{1}{2\mu_{ij}} \alpha_{ij}\beta_{ij} F(\boldsymbol{\alpha} + \boldsymbol{\beta}, \mathbf{1}_{13} + \mathbf{1}_{14})$$

$$- \frac{\alpha_{14}}{2\mu_{14}} F(\boldsymbol{\alpha} + \boldsymbol{\beta}, \mathbf{1}_{13}) - \frac{\beta_{13}}{2\mu_{13}} F(\boldsymbol{\alpha} + \boldsymbol{\beta}, \mathbf{1}_{14}) + \frac{1}{2m_1} G_{134}(\boldsymbol{\alpha} + \boldsymbol{\beta}, \mathbf{0})$$

$$+ \sum_{i=1}^{4} \frac{1}{2m_i} \sum_{\substack{j < k \\ j,k \neq i}} (\alpha_{ij}\beta_{ik} + \alpha_{ik}\beta_{ij}) G_{ijk}(\boldsymbol{\alpha} + \boldsymbol{\beta}, \mathbf{1}_{13} + \mathbf{1}_{14})$$

$$- \sum_{j=2,3} \left[\frac{\alpha_{1j}}{2m_1} G_{1j4}(\boldsymbol{\alpha} + \boldsymbol{\beta}, \mathbf{1}_{13}) + \frac{\alpha_{4j}}{2m_4} G_{41j}(\boldsymbol{\alpha} + \boldsymbol{\beta}, \mathbf{1}_{13}) \right]$$

$$- \sum_{j=2,4} \left[\frac{\beta_{1j}}{2m_1} G_{1j3}(\boldsymbol{\alpha} + \boldsymbol{\beta}, \mathbf{1}_{14}) + \frac{\beta_{3j}}{2m_3} G_{31j}(\boldsymbol{\alpha} + \boldsymbol{\beta}, \mathbf{1}_{14}) \right] . \tag{26}$$

There is no obvious simplification for Eq. (A1).

A check on the foregoing formulas is provided by a comparison with computations that have been laboriously (and carefully) carried out by six-dimensional numerical integration (A.V. Turbiner, private communication). The analytical and numerical methods are in agreement within the precision of the latter.

Table 1 Computed energies of Li ground state (non-relativistic, Coulomb interaction only, infinite-mass nucleus) for various wavefunctions, in Hartree atomic units. "This research" is for a correlated exponential premultiplied by the electron–nuclear distance for the $2s$ electron, with the parameters given in Table 33

Restricted Hartree–Fock[a]	-7.432726
Hylleraas[b] (1 term)	-7.417907
(4 terms)	-7.444700
(5 terms)	-7.472382
Pure correlated exponential (1 term)	-7.454694
This research (1 term)	-7.471281
Exact[c]	-7.478060

[a] Koga, Tatawaki, and Thakkar [18].
[b] Larsson [17].
[c] Puchalski and Pachucki [19].

Table 2 Parameters for the spatial wavefunction $\psi(\boldsymbol{\alpha}, \mathbf{1}_{14})$ for the energy-optimized ground-state Li wavefunction Ψ of the form given in Eq. (20)

α_{12}	3.299363
α_{13}	2.355741
α_{14}	0.701786
α_{23}	−0.217170
α_{24}	−0.022287
α_{34}	−0.038647

The $\{\alpha_{ij}\}$ leading to the minimum value of E can now be found by a conjugate gradient algorithm; the result is shown in Table 1, with the parameter set given in Table 33. It is apparent that the one-term exponentially correlated wavefunction yields a better energy than either a well converged restricted Hartree–Fock computation or a single-term Hylleraas function; in fact, the current result is comparable in quality with the 5-term Hylleraas function from the detailed study by Larsson [17] and is also considerably better than an optimum single-term exponentially correlated function without a pre-exponential factor.

The wavefunction described with the optimum $\{\alpha_{ij}\}$ corresponds well with a somewhat perturbed $1s1s'2s$ configuration. Two of the electrons depend on the electron–nuclear distance with α values (screening parameters) that correspond to a partially screened interaction with the +3-charged Li nucleus in a split-shell electron distribution. The third electron (that with pre-multiplying r) has an α value somewhat larger than for a hydrogenic $2s$ orbital, indicative of the fact that the inner-shell electrons do not completely shield the Li nucleus. The electron–electron α values all reflect the existence of electron–electron repulsion, with the effect most pronounced for the $1s–1s'$ interaction. All these observations are consistent with the notion that the exponentially correlated wavefunction gives an excellent zero-order description of the electronic structure of Li.

Acknowledgement Supported in part by the US National Science Foundation, Grant PHY-0601758. The author also thanks Professor A. V. Turbiner and Dr. N. L. Guevara for communicating the results of their numerically based Li computations and for stimulating discussions.

Appendix: Evaluation of $F(\boldsymbol{\gamma}, \mathbf{p})$

In principle, the integral for $F(\boldsymbol{\gamma}, \mathbf{p})$ appearing in Eq. (2) can be reduced to one over the six r_{ij} dimensions, but that seems not to provide a route to its analytic evaluation, though it has been a starting point for evaluating $F(\boldsymbol{\gamma}, \mathbf{p})$ numerically (A.V. Turbiner, private communication). Fromm and Hill [13], using Fourier representation theory and ingenious methods of contour integration, produced in 1987 a formula for the "generating integral" (that with all the $p_{ij} = -1$) from which all instances of Eq. (2) could formally be obtained by differentiation with respect to the γ_{ij}. The Fromm–Hill formula is cumbersome, containing 19 terms, each featuring a special function (the dilogarithm) with a different complicated argument, and with troublesome termwise singularities within the interior of the parameter space for which $F(\gamma, \mathbf{p})$ is defined. It is helpful to use evaluation methods that provide a formal singularity cancellation [14] and to use recursive methods to avoid the escalating complexity of the multiple differentiations [10].

A check on the correctness of the Fromm–Hill formula is available because a simpler formula, due to Remiddi [15], applies for wavefunctions of the Hylleraas type (for which the three γ_{ij} corresponding in atomic problems to electron–electron correlation are set to zero). Recurrence formulas are also available for these Hylleraas functions [16]. For further detail relative to these issues, the reader is referred to Refs. [10, 12–16].

References

1. L. M. Delves, T. Kalotas, Aust. J. Phys. **21**, 1 (1968)
2. A. J. Thakkar, V. H. Smith, Jr., Phys. Rev. A **15**, 1 (1977)
3. E. A. Hylleraas, Z. Phys. **54**, 347 (1929)
4. F. E. Harris, V. H. Smith, Jr., J. Phys. Chem. A **109**, 11413 (2005)
5. F. E. Harris, V. H. Smith, Jr., Adv. Quantum Chem. **48**, 407 (2005)
6. F. E. Harris, V. H. Smith, Jr., in *Symmetry, Spectroscopy and SCHUR*, ed. by R. C. King, M. Bylicki, J. Karwowski (Nicolaus Copernicus University Press, Torun, 2006), pp. 127–137
7. T. K. Rebane, V. S. Zotev, O. N. Yusupov, Zh. Eksp. Teor. Fiz. **110**, 55 (1996) [JETP **83**, 28 (1996)]
8. V. S. Zotev, T. K. Rebane, Opt. Spekt. **85**, 935 (1998) [Opt. Spect. **85**, 856 (1998)]
9. F. E. Harris, A. M. Frolov, V. H. Smith, Jr., Int. J. Quantum Chem. **100**, 1086 (2004)
10. F. E. Harris, Phys. Rev. A **79**, 032517 (2009)
11. T. K. Rebane, Opt. Spekt. **75**, 945 (1993) [Opt. Spect. **75**, 557 (1993)]
12. F. E. Harris, A. M. Frolov, V. H. Smith, Jr., J. Chem. Phys. **119**, 8833 (2003)
13. D. M. Fromm, R. N. Hill, Phys. Rev. A **36**, 1013 (1987)
14. F. E. Harris, Phys. Rev. A **55**, 1820 (1997)
15. E. Remiddi, Phys. Rev. A **44**, 5492 (1991)
16. K. Pachucki, M. Puchalski, E. Remiddi, Phys. Rev. A **70**, 032502 (2004)
17. S. Larsson, Phys. Rev. **169**, 49 (1968)
18. T. Koga, H. Tatewaki, A. J. Thakkar, Phys. Rev. A **47**, 4510 (1993)
19. M. Puchalski, K. Pachucki, Phys. Rev. A **73**, 022503 (2006)

Part III

Beyond Nonrelativistic Theory: Relativity and QED

The Equivalence Principle from a Quantum Mechanical Perspective

E.J. Brändas

Abstract In previous studies, we have attempted to unify quantum mechanics with the theory of special and general relativity based on analytic extensions of quantum mechanics by the use of an elementary complex symmetric ansatz. We will here present a formal re-derivation of the extended dynamics essentially starting from the Maxwell's equations. The formulation displays Einstein's law of light deflection in a gravitational field significantly as a quantum mechanical effect and illustrates further the materialization of a Schwarzschild-like singularity connected with the emergence of an Jordan block with a Segrè characteristic larger than unity. The analysis throws additional light on the particle-wave behaviour near the singularity as well as resolves one of the main problems afflicting the principle of equivalence.

Keywords: Klein–Gordon equation · Maxwell's equation · Complex symmetry · Jordan blocks · Special and general relativity · Electromagnetic and gravitational fields · Schwarzschild radius

1 Introduction

We have recently attempted to integrate quantum mechanics with the theory of special and general relativity. In addition to provide an explanation of the Einstein relativity laws based on the quantum mechanical super-position principle, we have also analysed the boundary condition problem associated with a Schwarzschild-type singularity as well as the quantum analogue of a black hole concurrent with a gravitational collapse. In this development there exists, among other things, several questions and disagreements undermining the general understanding, one of them being the conflict between the relativity of acceleration and the Principle of Equivalence [1, 2].

E.J. Brändas (✉)
Department of Quantum Chemistry, Uppsala University, SE-751 20 Uppsala, Sweden,
e-mail: Erkki.Brandas@kvac.uu.se

P. Piecuch et al. (eds.), *Advances in the Theory of Atomic and Molecular Systems*,
Progress in Theoretical Chemistry and Physics 19, DOI 10.1007/978-90-481-2596-8_5,

It is well known that the Principle of Equivalence has received discerning criticism over the years all from tidal dysfunctions and the Ehrenfest paradox to the application of the principle to the propagation of light. Although most explanations rest on the analysis of phenomena, which profess to discriminate between a flat and a curved world therefore providing a guide for a proper interpretation of the principle, cf. the discussions related to the weak and the strong equivalence principle, the most visible violation stands out, i.e. that the bending of a light ray in an accelerated box will be half as large as the bending in a box at rest in a gravitational field.

In this article, we will re-derive and recover some of our previous results based on analytic extensions of quantum mechanics by the use of an elementary complex symmetric ansatz. The idea rests on a reformulation of standard partial differential operators, e.g. the Maxwell and/or the Dirac and Klein–Gordon equations [3–5], posed as an operator-secular equation with a non-positive definite metric [6–8]. After briefly summarizing the results previously obtained, we will redevelop the theory starting from Maxwell's equations akin to the existence of general vector fields. In this representation, one finds that Einstein's laws of relativity are construed from superposition principles associated with general wave propagation. Also Schwarzschild-type singularities occur in the formalism as emerging Jordan blocks of the operator matrix. The occurrence of black holes disguised as strongly correlated coherent dissipative structures [9] (cf. the concept of Off-Diagonal Long-Range Order, ODLRO [10]) implies an explanation of the origin of gravitational inferences as well as offers an explicit justification of the Einstein law of the gravitational deviation of light.

Finally the results obtained will provide a resolution of the above-mentioned inconsistency between the Einstein equivalence principle and the relativity of acceleration. In conclusion, we will also bring up the non-participation of gravitational waves in our picture.

2 The Complex Symmetric Model of the Theory of Special Relativity

We will first provide a review of our model. Although it is fundamentally of quantum origin, there is an important generalization built in, i.e. the provision to analytically extend quantum mechanical quantities such as resolvents, Green's functions, S-matrices and general spectral properties, when appropriate, into the complex plane [11–13]. As we will see, this provides the means to include dynamical characteristics such as time, length and temperature scales into the theory [6]. In addition, the formulation contributes a wider set of broken symmetry solutions leading up to the emergence of singular structures like Jordan blocks. Even if this looks like a complication, it turns out to be a "blessing in disguise." Nevertheless, proper invariance laws, such as gauge invariance and covariance, will be appropriately embedded when required by the situation.

The characteristics developed here emanate from the so-called complex scaling method [11] of atomic and molecular physics [14, 15]. The observance to incorporate complex symmetric interactions in effect provides for transitions from specific quantum non-local representations to classical locality.

To summarize the discussion in previous reports [6–9], we will employ a simple complex symmetric ansatz obtaining a Klein–Gordon-like equation with precise restrictions and constraints. Although not explicitly written out here it is easy to identify the present construction with a standard arrangement in terms of a non-positive definite metric. Specifically, we write

$$H = \begin{pmatrix} m & -iv \\ -iv & -m \end{pmatrix},$$

(1)

where, using mass units, the diagonal matrix elements are energies (operators) associated with particle antiparticle states (if fermions, the Dirac equation analogue would be required [8]), respectively, and $-iv$, $v = p/c$, is the complex symmetric interaction; the minus sign is by convention only. Here, c is the velocity of light and p the momentum of the particle. The quantum particle (with mass m) is described by the state vector $|m\rangle$ and the associated antiparticle (assigned with a negative energy $-m$) by the state vector $|\bar{m}\rangle$. In the interaction-free case, the diagonal elements become $\pm m_0$, i.e. the equivalent of the particle rest mass. The associated vectors $|m_0\rangle$ and $|\bar{m}_0\rangle$ can, without loss of generality, be chosen orthonormal, while $|m\rangle$ and $|\bar{m}\rangle$ in general become bi-orthogonal.

Transforming the matrix, Eq. (1), to classical canonical form assuming that it can be diagonalized yields for the roots as $\lambda_\pm = \pm m_0$

$$\lambda^2 = m_0^2 = m^2 - v^2,$$

(2)

finding the well-known expression

$$m^2 c^4 = m_0^2 c^4 + p^2 c^2.$$

(3)

The quantities defined in Eq. (3) must in general be identified as operators; for example p, which is conventionally a self-adjoint operator, looses this property in its extended form. Nevertheless, it will be consistent with the relationship $p = mv$ (v is the velocity relative a system in rest, wherever the rest mass of the particle is m_0) with appropriate modifications made for a particle in an electromagnetic or other field. We also note that p should be a vector operator although we treat it here temporarily as scalar operator, the square root of p^2. Our operator-secular equation generates as expected a Klein–Gordon-type equation, with the resulting eigensolutions.

$$\begin{aligned} |m_0\rangle &= c_1|m\rangle + c_2|\bar{m}\rangle; \lambda_+ = m_0 \\ |\bar{m}_0\rangle &= -c_2|m\rangle + c_1|\bar{m}\rangle; \lambda_- = -m_0 \end{aligned}$$

(4)

or

$$|m\rangle = c_1|m_0\rangle - c_2|\bar{m}_0\rangle;$$
$$|\bar{m}\rangle = c_2|m_0\rangle + c_1|\bar{m}_0\rangle \tag{5}$$

with

$$c_1 = \sqrt{\frac{1+X}{2X}} \qquad m = \frac{m_0}{X}; \qquad c_1^2 + c_2^2 = 1. \tag{6}$$
$$c_2 = -i\sqrt{\frac{1-X}{2X}}$$

In Eq. (6), $X = \sqrt{1-\beta^2}$; $\beta = p/mc = $ ("classical particles") $= v/c$. As already mentioned, the vectors in Eq. (5) are bi-orthogonal and hence the absolute value squared of c_1 and c_2 do not sum up to unity. Note also that $v/c \to 1$ implies

$$|c_1|^2/|c_2|^2 \to 1. \tag{7}$$

We also emphasize that Eqs. (4) and (5) in general are operator relations, i.e. where the functional forms contain operators. Although this is usually referred to as "formal matters," it is not self-evident that this should work since we are dealing with non-self-adjoint operator extensions. Nevertheless, there are several fine points to be made here regarding the development of quantum mechanics and we refer to [9] for some recent considerations on this issue. We will not distinguish between the functional and the operator elements unless the situation itself will make the interpretation ambiguous.

Since the present differential equation, Eq. (3), derives from a complex symmetric structure, there are two main consequences, namely, (i) under appropriate perturbations there appears generic complex resonance energies and (ii) it may not be possible to bring the matrix to a diagonal form. As an example of a mathematically rigorous development, we mention the theory of dilation analytic operators [11] of current use in atomic and molecular physics. Hence, the theory outlined here should apply to atomic and molecular systems and their antiparticle partners. Considering initially the first point (the second point will be handled in the following sections), we make the apposite replacements

$$m_0c^2 \to m_0c^2 - i\frac{\Gamma_0}{2}; \qquad \tau_0 = \frac{\hbar}{\Gamma_0}$$
$$mc^2 \to mc^2 - i\frac{\Gamma}{2}; \qquad \tau = \frac{\hbar}{\Gamma}, \tag{8}$$

where Γ, τ and Γ_0, τ_0 are, respectively, the half width and lifetime of the state and $\hbar$ is Planck's constant divided by 2π. Inserting the definitions (8) into the secular equation, separating the real and imaginary parts, we get the following contractions

$$\Gamma_0 = \Gamma\sqrt{1-\beta^2}; \quad \tau = \tau_0\sqrt{1-\beta^2}. \tag{9}$$

It is now trivial to compare times in the two scales, i.e.

$$t = \frac{t_0}{\sqrt{1 - \beta^2}}. \tag{10}$$

Enforcing Lorentz invariance, the length scales can now be brought in. Alternatively, one can define analogous equations for conjugate variables, see Section 5.

$$l = \frac{l_0}{\sqrt{1 - \beta^2}}; \ t = \frac{t_0}{\sqrt{1 - \beta^2}}; \ m = \frac{m_0}{\sqrt{1 - \beta^2}}. \tag{11}$$

Summarizing the situation, we have derived well-known relations of the special theory of relativity as a consequence of the quantum mechanical superposition principle, involving matter–antimatter quantum states. In passing, we notice that one can include appropriate electromagnetic fields [6] through

$$(E_{op} - eA_0)^2 = m_0^2 c^4 + (p - \frac{e}{c} \vec{A})^2 c^2, \tag{12}$$

where in Eq. (12) $(A_0, \vec{A})$ are the usual vector and scalar potentials.

In the following section, we will merge the gravitational field to the present formalism [7] before we return to an alternative formulation based on Maxwell's equations.

3 The Gravitational Field and General Relativity

It is actually a straightforward procedure to include gravitational interactions within the present framework. Although gravity is associated with a tensor field, we will here initiate the formulation by augmenting the present complex symmetric model, in the basis $|m, \bar{m}\rangle$, with the "scalar" interaction (the word scalar is placed in quotation marks since the potential will be built into an appropriate matrix formalism, see also a more detailed formulation in the next section):

$$m\kappa(r) = m\mu/r; \ \mu = \frac{G \cdot M}{c^2}, \tag{13}$$

ending up with the modified Hamiltonian matrix

$$\mathbf{H} = \begin{pmatrix} m - m\kappa(r) & -iv \\ -iv & -(m - m\kappa(r)) \end{pmatrix} \tag{14}$$

with μ the gravitational radius, G the gravitational constant, M a "classical non-rotating mass" (which does not change sign when $m \to -m$) and $v = p/c$ as before. We will deal with the fundamental nature of M and the emergence of black hole-like objects in the Appendix, see also [7]. At this point, it is sufficient to say that our model concerns a quantum particle (for a discussion on spin, see [8]) in a gravitational field created by a black hole object M, the latter founded on a multidimensional quantum formulation by the use of ODLRO [7, 10].

Note that the operator, $\kappa(r) \geq 0$, depends on the coordinate r of the particle m, with the origin at the centre of mass of M. It is important to distinguish between the coordinates r (and t) of a flat Euclidean space and the scales defining the curved space–time geometry defined by the (operator)-secular problem Eq. (14). We get directly

$$
\begin{aligned}
H &= \begin{pmatrix} m(1 - \kappa(r)) & -i\upsilon \\ -i\upsilon & -m(1 - \kappa(r)) \end{pmatrix}; \\
\lambda^2 &= m^2(1 - \kappa(r))^2 - p^2/c^2 \\
\lambda &= m_0(1 - \kappa(r)); \upsilon - p/c
\end{aligned}
\tag{15}
$$

with the eigenvalues $\lambda_\pm$ (properly scaled for convenience), i.e.

$$
\begin{aligned}
m_0^2 &= m^2 - p^2/(1 - \kappa(r))^2 c^2 \\
\lambda_\pm/(1 - \kappa(r)) &= \pm m_0 = \pm\sqrt{m^2 - p^2/(1 - \kappa(r))^2 c^2} \\
m &= m_0/\sqrt{1 - \beta'^2}; \ \beta' \leq 1; \ 1 > \kappa(r) \\
\beta' &= p/mc(1 - \kappa(r)) = \upsilon/c(1 - \kappa(r)).
\end{aligned}
\tag{16}
$$

Eqs. (15) and (16) can be suitably written in the representation

$$
H = \begin{pmatrix} m & -ip'/c \\ -ip'/c & -m \end{pmatrix}; \ p' = p/(1 - \kappa(r)).
\tag{17}
$$

We will now make some remarks concerning the limitations incurred by complex symmetry. This relates to the magnitude of the kinematical interaction and of the subsequent discontinuous variation of eigenvalues as mass increases with momentum. This will direct us to point (ii), see the comment under Eq. (7) in the previous section, leading up the examination and conception of the emergence of so-called Jordan blocks. Electromagnetic and/or gravitational fields can easily generate such effects. The first remark relates to the relativistic Kepler problem for a particle of non-zero rest mass. Assuming a central force, see e.g. discussions in [7, 8], analogous to the classical case, one finds that the quantity mD, where D is the well-known area velocity, is a constant of motion, i.e. $mD = m_0 A$, which defines the constant A. For a circular orbit with a constant angular velocity, this coincides with the angular momentum $\hat{m}\upsilon r$, where in the quantum situation $\hat{m}$ is an operator. Determining the value of A from the conserved angular momentum, i.e. evaluating the latter for the limiting velocity c assumed at the limiting distance μ, the gravitational radius ($m_0 \neq 0$, φ is the standard polar angle), we obtain

$$(\mathbf{r} \times \mathbf{p}) \cdot \mathbf{n} = mD = m_0 A = m_0 \upsilon r = m_0 \mu c$$

$$D = r^2 \frac{d\varphi}{dt} \tag{18}$$

where $\mathbf{n}$ is the unit vector in the direction of the angular momentum. Since the m operator commutes with the operator $\mathbf{H}$ in Eq. (15), they have simultaneous eigenvalues. This holds in the extended picture as well leading to the simple condition

$$\upsilon = \kappa(r)c = \mu c / r \tag{19}$$

For the non-zero mass particle one finds, cf. again the remark related to point (ii) above, that a Jordan block (degeneracy with Segrè characteristic 2) occurs at a certain radius R_{LS} (provided the mass M is entirely localized inside the sphere with radius R_{LS}), i.e.

$$\frac{m}{2} = m\upsilon / c = m\kappa(r); \ r = R_{LS} = 2\mu. \tag{20}$$

Here R_{LS} is the famous Laplace–Schwarzschild radius of the Schwarzschild solution of the theory of general relativity. The singular behaviour of Eqs. (15), (16) and (17) follows directly by observing that $\beta' \to 1$, as $r \to R_{LS} = 2\mu$. By the use of Eq. (19), the matrix in Eq. (15) may be written as

$$\mathbf{H} = \begin{pmatrix} m(1 - \kappa(r)) & -im\kappa(r) \\ -im\kappa(r) & -m(1 - \kappa(r)) \end{pmatrix} \tag{21}$$

At $r = R_{LS}$, the matrix becomes (note that m is not uniquely defined as $\mathbf{H}$ becomes singular, see more below)

$$\mathbf{H}_{\text{deg}} = \frac{1}{2} \begin{pmatrix} m & -im \\ -im & -m \end{pmatrix} \to \mathbf{H}_{\text{deg}} = \begin{pmatrix} 0 & m \\ 0 & 0 \end{pmatrix}$$

$$|0\rangle = \frac{1}{\sqrt{2}}|m\rangle - i\frac{1}{\sqrt{2}}|\bar{m}\rangle \tag{22}$$

$$|\bar{0}\rangle = \frac{1}{\sqrt{2}}|m\rangle + i\frac{1}{\sqrt{2}}|\bar{m}\rangle,$$

displaying the Jordan block structure as well as the associated similarity (also unitary) transformation. From Eqs. (15), (16) and (19), we obtain the relation

$$m = \frac{m_0(1 - \kappa(r))}{\sqrt{1 - 2\kappa(r)}}, \tag{23}$$

explicitly showing a singularity at $r = R_{LS}$. In this representation, we observe that either $m \to \infty$ adiabatically with m_0 finite or $m_0 \to 0$ adiabatically with m finite. Written differently, m/m_0 is singular at $r = R_{LS}$.

What about "zero rest-mass particles?" As is seen below, we must look at this problem separately and denote the proper gravitational interaction by $\kappa_0(r)$ for particles with $m_0 = 0$. Solving the secular equations, Eqs. (15) and (16), for a zero eigenvalue one finds

$$\lambda^2 = m^2(1 - \kappa_0(r))^2 - p^2/c^2 = 0, \tag{24}$$

where $\kappa_0(r)$, as mention above, is indexed to indicate the law for particles with zero rest mass, i.e. $m_0 = 0$,

$$H = \begin{pmatrix} m(1 - \kappa_0(r)) & -ip/c \\ -ip/c & -m(1 - \kappa_0(r)) \end{pmatrix};$$
$$\lambda^2 = m^2(1 - \kappa_0(r))^2 - p^2/c^2 = 0. \tag{25}$$

To be consistent with the degeneracy condition, see Eqs. (20) and (22), we must ensure that photons will be trapped or confined inside the black hole object M. Evaluating expectation values on both sides of Eq. (24), demanding the boundary condition

$$\langle p \rangle = 0; \ \langle r \rangle = R_{LS},$$

one obtains the following requirement at the Schwarzschild boundary:

$$\kappa_0(r) = 2\mu/r = 2\kappa(r) = \frac{2G \cdot M}{c^2}. \tag{26}$$

Equation (26) displays a gravitational law that is commensurate with the effect of light deflection in a gravitational field. Rewriting Eq. (25), we obtain by analogy with Eq. (22)

$$H_{\text{deg}} = \begin{pmatrix} p/c & -ip/c \\ -ip/c & -p/c \end{pmatrix} \rightarrow H_{\text{deg}} = \frac{1}{c}\begin{pmatrix} 0 & 2p \\ 0 & 0 \end{pmatrix}$$
$$|0\rangle = \frac{1}{\sqrt{2}}|m\rangle - i\frac{1}{\sqrt{2}}|\bar{m}\rangle \tag{27}$$
$$|\bar{0}\rangle = \frac{1}{\sqrt{2}}|m\rangle + i\frac{1}{\sqrt{2}}|\bar{m}\rangle$$

Note the difference between Eqs. (22) and (27). In the former, space coordinates appear to have been lost, while in the latter only the space-dependent operator remains in the off-diagonal position. The unitary property of the associated transformation implies that bi-orthogonality can be made to coincide with "ordinary" orthogonality for the "vacuum" states, simply denoted as o and $\bar{o}$.

To study this situation further, we will assume an empty space outside a spherically symmetric black hole object. It is easy to see that Eq. (25) combined with

(23) is equivalent with the Schwarzschild gauge in the minimal two component metric or

$$ds^2 = (1 - \kappa_0(r))c^2 dt^2 - (1 - \kappa_0(r))^{-1} dr^2 - r^2 d\Omega^2 \tag{28}$$

where as usual Ω contains the colatitude and longitude angles. From Eq. (25) follows the "semi-classical" relations

$$0 = (1 - \kappa_0(r))^2 c^2 dt^2 - dx^2 - dy^2 - dz^2 \tag{29}$$

or

$$0 = (1 - \kappa_0(r))c^2 dt^2 - (1 - \kappa_0(r))^{-1} dr^2 \tag{29'}$$

To prove the equivalence with (28), we will combine this result for the "zero length element" with a rewrite of Eq. (23), i.e.

$$m = \frac{m_0(1 - \kappa(r))}{\sqrt{1 - 2\kappa(r)}} = \frac{\lambda_0}{\sqrt{1 - 2\kappa(r)}}, \tag{30}$$

where λ_0 is the positive (real part) of the secular equation associated with the matrix

$$\boldsymbol{H} = m \begin{pmatrix} 1 - \kappa(r) & -i\kappa(r) \\ -i\kappa(r) & -(1 - \kappa(r)) \end{pmatrix} \tag{31}$$

Since eigenvalues in this setting may come out complex, i.e.

$$m = m_r - i\Gamma; \; \lambda_0 = \lambda_r - i\Gamma_0, \tag{32}$$

we obtain by projection of the real and imaginary parts of (30) and (31)

$$\tau = \tau_0\sqrt{1 - 2\kappa(r)}; \; \Gamma = \frac{\hbar}{\tau}; \; \Gamma_0 = \frac{\hbar}{\tau_0} \tag{33}$$

or for differential "times"

$$d\tau^2 = d\tau_0^2(1 - 2\kappa(r)). \tag{34}$$

Equations (29') and (34) prove the equivalence sought for. In the next section, we will reformulate the present operator equations from the viewpoint of the classical Maxwell's equations.

4 Maxwell's Equations and Gravitation

We have up to this point presented a generalized description, which goes beyond classical features as it give reasons for features like the contraction of scales and the integration of some general dynamical features of general gravitational interactions from a quantum mechanical perspective. In this section, we will reconsider the situation starting from Maxwell's equations for free space

$$\begin{aligned}
\nabla \cdot \mathbf{E} &= 0 \\
\nabla \times \mathbf{E} &= -\frac{\partial \mathbf{B}}{\partial t} \\
\nabla \cdot \mathbf{B} &= 0 \\
\nabla \times \mathbf{B} &= \frac{1}{c^2}\frac{\partial \mathbf{B}}{\partial t}
\end{aligned} \tag{35}$$

where as usual $\mathbf{E}$ and $\mathbf{B}$ are the vectors associated with the electric and the magnetic fields, respectively, and with Δ the nabla operator. In addition to the trivial solution, one obtains the wave equation for the displacement ζ (c is the speed of light in free space)

$$\begin{aligned}
\Box^2 \zeta &= 0 \\
\Box^2 &= \nabla^2 - \frac{1}{c^2}\frac{\partial^2}{\partial t^2}
\end{aligned}$$

from which the electric and magnetic fields can be obtained ($\hat{\mathbf{n}}$ is the unit vector in the direction of propagation, using a unit k-vector in the dispersion relation)

$$\begin{aligned}
\mathbf{E} &= \mathbf{E}_0 \zeta(\hat{\mathbf{n}}\mathbf{r} - ct)) \\
\mathbf{B} &= \frac{1}{c}\hat{\mathbf{n}} \times \mathbf{E}.
\end{aligned} \tag{36}$$

Eqs. (1), (2) and (3) will thus be analogous to

$$\begin{vmatrix} i\hbar\dfrac{\partial}{\partial t} & -\hbar c\nabla \\ -\hbar c\nabla & -i\hbar\dfrac{\partial}{\partial t} \end{vmatrix} = \hbar^2\frac{\partial^2}{\partial t^2} - \hbar^2 c^2\nabla^2 = -\hbar^2 c^2\Box^2, \tag{37}$$

with the eigenvalue relation, cf. the Klein–Gordon equation, becoming

$$-\hbar^2\Box^2\varphi = m_0^2 c^2\varphi, \tag{38}$$

where the zero rest mass case is identical to the Maxwell free space condition mentioned above. However, in order to solve for the eigenvalues, Eqs. (4), (5) and (6), we need to consider the incompatibility between the diagonal elements and the off-diagonal vector components occurring in the matrix in Eq. (37). This can be resolved either by inserting the positive square root of p^2 or by introducing an appropriate field in analogy with Eq. (36). Such a field may have a more general dependence on the space–time variables than that indicated in Eq. (36). We may in the general case realize the variables of the Dirac ket explicitly.

In order to continue, we will as a consequence introduce and generalize the following familiar formalism

$$\langle \mathbf{r} | \mathbf{p} \rangle = (2\pi \hbar)^{-3/2} e^{i/\hbar \tilde{\mathbf{r}} \cdot \mathbf{p}}, \tag{39}$$

where

$$\mathbf{r} = \begin{pmatrix} x \\ y \\ z \end{pmatrix} ; \quad \mathbf{p} = \begin{pmatrix} p_x \\ p_y \\ p_z \end{pmatrix} \tag{40}$$

to include all four dimensions. In a conventional bra-ket notation, we obtain

$$\left\langle \mathbf{r}, -ict \;\middle|\; \mathbf{p}, \frac{iE}{c} \right\rangle = (2\pi \hbar)^{-2} e^{i/\hbar (\tilde{\mathbf{r}} \cdot \mathbf{p} - Et)}. \tag{41}$$

bearing in mind the complex conjugate in the bra-position, subscribing to a complex symmetric construction

$$\langle \mathbf{x}^* \;|\; \mathbf{\Pi} \rangle = (2\pi \hbar)^{-2} e^{i/\hbar (\tilde{\mathbf{x}} \cdot \mathbf{\Pi})}, \tag{42}$$

with the obvious definitions

$$\mathbf{x} = \begin{pmatrix} x \\ y \\ z \\ ict \end{pmatrix} ; \mathbf{\Pi} = \begin{pmatrix} p_x \\ p_y \\ p_z \\ iE/c \end{pmatrix} \tag{43}$$

Rewriting our equations displaying the standard operator identifications, i.e.

$$\vec{\mathbf{p}} = -i\hbar \vec{\nabla} = -i\hbar \left(\underline{e}_x \frac{\partial}{\partial x} + \underline{e}_y \frac{\partial}{\partial y} + \underline{e}_z \frac{\partial}{\partial z} \right) = -i\hbar \underline{e} \cdot \begin{pmatrix} \frac{\partial}{\partial x} \\ \frac{\partial}{\partial y} \\ \frac{\partial}{\partial z} \end{pmatrix}$$

$$E = i\hbar \frac{\partial}{\partial t}; \quad \mathbf{\Pi} = -i\hbar \left(\nabla, i/c \frac{\partial}{\partial t} \right) \tag{44}$$

$$\vec{\mathbf{\Pi}} = -i\hbar \left(\underline{e}_x \frac{\partial}{\partial x} + \underline{e}_y \frac{\partial}{\partial y} + \underline{e}_z \frac{\partial}{\partial z} + i/c\, \underline{e}_t \frac{\partial}{\partial t} \right) = -i\hbar \underline{\varepsilon} \cdot \begin{pmatrix} \frac{\partial}{\partial x} \\ \frac{\partial}{\partial y} \\ \frac{\partial}{\partial z} \\ i/c \frac{\partial}{\partial t} \end{pmatrix} ; \; \vec{\mathbf{x}} = \underline{\varepsilon} \cdot \begin{pmatrix} x \\ y \\ z \\ ict \end{pmatrix},$$

the operator-secular equation yields (identity operator suppressed)

$$\lambda^2 = \left(E^2 - p^2 c^2\right) = -c^2 \, \vec{\Pi} \cdot \vec{\Pi} = -c^2 \varepsilon \, \widetilde{\Pi} \cdot \Pi \, \tilde{\varepsilon} = -c^2 \Pi^2 = \hbar^2 c^2 \Box^2 = m_0^2 c^4. \tag{45}$$

In short form, we find that

$$\Pi^2 = -\hbar^2 \left(\triangle - \frac{1}{c^2} \frac{\partial^2}{\partial t^2} \right) \tag{46}$$

and furthermore

$$\langle \mathbf{x}^* \left| -\Pi^2 \right| \mathbf{\Pi} \rangle = m_0^2 c^2 \langle \mathbf{x}^* \mid \mathbf{\Pi} \rangle, \tag{47}$$

the latter in principle defining the rest mass uniquely, provided we know the appropriate space–time boundary conditions.

To understand this statement as well as attempting to resolve one of the main inconsistencies between the relativity of acceleration and the Principle of Equivalence, we use Eq. (28) to introduce the gravitational interaction as follows (in the present case $m_0 \to 0$, static, spherically symmetric case), i.e. let

$$\Pi^2 \to \Pi_{\mathrm{grav}}^2 = \left(1 - \frac{2Gm}{c^2 r} \right)^{-1} p_r^2 - \left(1 - \frac{2Gm}{c^2 r} \right) \frac{E^2}{c^2} \tag{48}$$

or employing its appropriate symmetrized forms

$$\Pi^2 \to \Pi_{\mathrm{grav}}^2 = \left(1 - \frac{2Gm}{c^2 r} \right)^{-1/2} p_r^2 \left(1 - \frac{2Gm}{c^2 r} \right)^{-1/2} - \left(1 - \frac{2Gm}{c^2 r} \right)^{1/2}$$
$$\frac{E^2}{c^2} \left(1 - \frac{2Gm}{c^2 r} \right)^{1/2} \tag{49}$$
$$\Pi^2 \to \Pi_{\mathrm{grav}}^2 = p_r \left(1 - \frac{2Gm}{c^2 r} \right)^{-1} p_r - \frac{E}{c} \left(1 - \frac{2Gm}{c^2 r} \right) \frac{E}{c},$$

where we have made a change of the coordinate and the reciprocal coordinate system: $\mathbf{x}' = \alpha^{-1} \mathbf{x}$, $\mathbf{\Pi}' = \alpha \mathbf{\Pi}$, where α in general is a 4×4 similarity transformation, but here restricted to two dimensions on the basis (r, ict) and $(p_r, iE/c)$ with corresponding modifications in Eqs. (39), (40), (41), (42), (43) and (44)

$$\alpha = \begin{pmatrix} \left(1 - \frac{2Gm}{c^2 r} \right)^{-1/2} & 0 \\ 0 & \left(1 - \frac{2Gm}{c^2 r} \right)^{1/2} \end{pmatrix}. \tag{50}$$

Note that we have utilized a general complex symmetric notation to analyse the consequences of the transformation instead of traditional covariant formalisms. Nevertheless, it is significant to point out that coordinate relations, formulated in connection with Maxwell type equations, refer to, and include, a scaled, curved space, cf. the relations obtained from the operator-secular equations (1–12). Another way to appreciate this feature is to posit the adjoint equations based on the matrix

$$\begin{pmatrix} -i\hbar\frac{\partial}{\partial E} & \hbar\nabla_p \\ \hbar\nabla_p & -i\hbar\frac{\partial}{\partial E} \end{pmatrix}$$

from which analogous secular equations, cf. Eqs. (1), (2), (3), (4) and (5), yield direct space–time relations in accordance with Lorentz invariance and appropriately also general covariance. From Eq. (49), we find the formal identity

$$\langle (r, ict)^* \left| \Pi^2 \right| p_r, iE/c \rangle \rightarrow \langle (r', ict')^* | \Pi'^2 | p'_r, iE'/c \rangle = \tag{51}$$
$$\langle (r', ict')^* \left| \Pi^2_{\text{grav}} \right| p'_r, iE'/c \rangle = \langle (r, ict)^* \left| \Pi^2_{\text{grav}} \right| p_r, iE/c \rangle,$$

As can be seen from Eq. (51), the interaction has been built into the formulation in such away that the surrounding gravitational field materializes as a result of the geometry characterizing the gravitational source. Hence, the coordinate transformation α defines the appropriate boundary conditions for the quantum formulation for a spherically symmetric static vacuum.

In a sense, the formulation mimics the classical formulation. Nevertheless, it is interesting to note that the Schwarzschild-like singularity depends on the restrictions imposed by the complex symmetric ansatz and yet is essentially "classical" or rather quasi-classical outside the domain boundary described in this representation by the Schwarzschild radius. The main contrast of this interpretation stems from a precise foundation on the quantum mechanical superposition principle.

Except from the emergence of so-called Jordan blocks, the equations here appear mostly to agree with the classical picture. One the other hand, structures can be formulated inside the singularity, a condition that is not satisfactorily resolved by all classical theories. For more details on the specific degeneracy condition, see Eq. (22), where all interactions/correlations condense or unify according to Yang's ODLRO [10], we refer to the appendix, see also Refs. [7, 9]. In the final section, we will explain how the present results remove some inadequacies of the illustrious Principle of Equivalence as well as promote the non-participation of gravitational waves.

5 The Principle of Equivalence

Equivalence principles have always been and remain a hot topic. In typical disputes, one usually considers the so-called Einstein Equivalence Principle essentially stating that free fall and inertial motion are physically equivalent. In a more detailed argument, one distinguishes between weak, Einstein and strong equivalence principles, incorporating all effects from free fall to universal features of the theory of general relativity. Although a massive collection of inconsistencies has been pointed out and examined, see e.g. [1, 2], even Einstein himself made mistakes in his derivations, the general consensus appears to be that Einstein was to all intents and purposes correct. As already stated in the introduction, most rationalizations are sited upon the investigation of phenomena, which attempt to discriminate between a flat and a curved world providing manuals, in this way, for an appropriate interpretation of the principle. Yet the most flagrant violation notably sticks out, i.e. the bending of a light ray in an accelerated box will be half as large as the bending in a box at rest in a gravitational field.

In order to discuss this problem as well as connecting the present results with the (microscopic) Equivalence Principle, we need to support the discussion in terms of quantum mechanics, therefore the reference to the word "microscopic." Hence the bending of a light ray should be properly analysed as a scattering experiment, where the differential cross section displays a maximum at the appropriate direction. The microscopic equivalence principle, based on quantum mechanics, does not need or even allow the complication of an observer. Before going on we will make a brief account of the results, presented here, appropriate for the present interpretation.

As already emphasized, the present structure envisages various physical laws of relativity theory with the remarkable feature, contrary to accepted classical beliefs, that the laws follow as a direct consequence of the (extended) quantum mechanical super-position principle. This understanding projects a generalized quantum description which transcends classical features like the contraction of scales in the special theory and the emergence of a Schwarzschild-like domain in the general theory of relativity and, most importantly, preserves the Einstein gravitational law of light deflection. We have integrated, without contradiction, the electromagnetic field in the special theory and contrived the Schwarzschild geometry in connection with Jebsen–Birkoff theorem [16, 17].

The generation of Jordan blocks by a gravitational field has been explicitly examined. As stated, an electromagnetic fluctuation could also produce "triangular structures," which might further lead to important relationships and conjectures with respect to the properties of the vacuum as well as the mass generation puzzle. We have noted that the restriction $p/m \leq c$ (in the gravitational free case) is obtained from the singular behaviour of the matrix eigenvalues as well as the quotient m/m_0. For instance considering the matrix, Eq. (1), for $v = m$ (m finite), e.g. produced by an electromagnetic fluctuation, we get

$$H_{\text{deg}}^{\text{field}} = \begin{pmatrix} m & -im \\ -im & -m \end{pmatrix} \tag{52}$$

By a unitary transformation, see Eqs. (22) and (27), it follows that $H_{\text{deg}}^{\text{field}}$ becomes

$$\overline{H}_{\text{deg}}^{\text{field}} = \begin{pmatrix} 0 & 2m \\ 0 & 0 \end{pmatrix} \tag{53}$$

or taking the complex conjugate of Eq. (52) (time reversal in this simple picture) or the Hermitian adjoint of (53) yields under the same transformation

$$\overline{\boldsymbol{H}}^{+\,\mathrm{field}_{\mathrm{deg}}} = \begin{pmatrix} 0 & 0 \\ 2m & 0 \end{pmatrix} \tag{54}$$

Analogous formulas are obtained for $m \to -m$. It is appealing to consider the present Jordan form, cf. Eq. (22), as a description of the vacuum as a "particle antiparticle superposition," since there appear no particle states in the diagonal of Eqs. (53) and (54). The energy is hidden, i.e. assigned to a transition between the states o and $\bar{o}$. In the Appendix, we have showed, see also Ref. [7], that a strongly correlated many-body theory (including ODLRO [10]), based on the particle–antiparticle states denoted by o and $\bar{o}$, leads to massive energy stabilization on the condition that a *fundamental interaction* operates inside a principal Schwarzschild-like domain. The model, in this study founded on the Klein–Gordon formalism, has been extended to a Dirac 4×4 ansatz, including modifications of the small component and allowing particle structures inside the Schwarzschild radius, see Ref. [8] for more details.

We are now ready to return to the main problem of this revision, i.e. the conflict between observable gravitation and the relativity of acceleration. As stated in various critical examinations of the Principle of Equivalence, a physicist will measure that the bending of light in an accelerating box will only be half as large as the bending in a box at rest in a gravitational field. The common answer, often given by relativists, is that the "second half" of the bending comes from the amount of space curvature, notwithstanding the fact that this space curvature should be a consequence of the Equivalence Principle.

In the present version of the microscopic principle of equivalence, we have been compelled to distinguish the gravitational interaction between particles of zero and non-zero rest mass, respectively, with a classical mass M, see Appendix for details. The photon experiences an interaction, which is twice the magnitude of that expected from a particle with $m_0 \neq 0$. It seems like the photon is "aware" of the fact that the classical mass M, if reduced to an apposite black-hole-like entity based on geminal pairs of the states, represented by o and $\bar{o}$, also contain an equal amount of "hidden" mass from associated antiparticles. At any rate, applying the microscopic version of the equivalence principle yields that a physicist, *forbidden by the laws of quantum mechanics to be enclosed in the box*, expects the outcome of the light scattering experiment in an accelerating frame to be consistent with the same experiment in a gravitational field. It is important to remember that in this set up there are no classical mass objects, with non-zero rest mass, permitted in the "laboratory."

One might speculate whether the present picture would predict or even allow gravitational waves. It is evident that the present microscopic model will bring up the non-participation of gravitational waves. This is not at all surprising, since it is supported by Birkoff's theorem [17] that states that any spherically symmetric solution of the vacuum field equations must be stationary and asymptotically flat and given by the Schwarzschild metric with the concomitant conclusion that any spherically pulsating object does not emit gravitational waves.

Acknowledgement These results have been presented in parts at QSCP XIII held at Michigan State University, East Lansing, Michigan, USA, and ISTCP VI held at the University of British Columbia, Vancouver, Canada, July 6–12 and 19–24, 2008, respectively. The author thanks the organizers of QSCP XIII, Prof. Piotr Pieuch and ISTCP VI, Prof. Yan Alexander Wang for an excellent programme and organization as well as generous hospitality. This research has been supported in part by the Swedish Foundation for Strategic Research.

6 Appendix

We will here review some aspects of Yang's famous concept of ODLRO [10]. The formulation has been adapted to several problems of complex phenomena in condensed matter [9, 15]. In this development, we have employed Coleman's notion of an extreme state [18] as a precursor for the onset of the strongly correlated (condensed) phase and/or of a coherent dissipative state, for more details see Refs. [9, 19] and references therein. We will first supply the necessary machinery before proceeding to the conversion of a black hole type entity into the organization of ODLRO.

The formulation is based on a hierarchy of reduced density matrices, which under specific conditions experience anomalous behaviour. The fundamental quantity is the many-particle fermion density matrix. The N particle (and its p-reduced companions) representable density matrix $\Gamma^{(p)}$ is defined as follows [20] in the Löwdin normalization (x_i is a combined space–spin coordinate)

$$\Gamma^{(p)}(x_1 \ldots x_p | x_1' \ldots x_p') = \binom{N}{p} \int \Psi^*(x_1 \ldots x_p, x_{p+1} \ldots x_N)$$
$$\Psi(x_1' \ldots x_p', x_{p+1} \ldots x_N) dx_{p+1} \ldots dx_N, \qquad (55)$$

where the wave function $\Psi(x_1 \ldots x_N)$ represents fermionic many-body quantum mechanical systems. We will specifically focus on $p = 2$. The fundamental concept of ODLRO was introduced by Yang [10] in association with his famous proof of the largest bound for $\Gamma^{(2)}$. He showed that the manifestation of a macroscopically large eigenvalue in the second-order-reduced (fermion) density matrix leads to a new physical order, cf. the theories of superconductivity and superfluidity. Without going into too many details and proofs, we will summarize the development in the most compact form.

We will employ a set of n localized pair functions $\boldsymbol{h} = (h_1, h_2, \ldots h_n)$ or so-called geminals, obtained from appropriate pairings of the fermionic basis functions spanning the appropriate spin orbital space under consideration. In our specific application to relativity theory, the selection will be the aforementioned particle–antiparticle basis functions. The following transformations, which are not part of the original ODLRO development, will be found to be valuable not only for the ensuing compactness of the formulation but also for reasons beyond this exposition, see e.g. Ref. [15] fore more details, i.e.

$$
\mathbf{B} = \frac{1}{\sqrt{n}}
\begin{pmatrix}
1 & \omega & \omega^2 & \cdot & \omega^{n-1} \\
1 & \omega^3 & \omega^6 & \cdot & \omega^{3(n-1)} \\
\cdot & \cdot & \cdot & \cdot & \cdot \\
\cdot & \cdot & \cdot & \cdot & \cdot \\
1 & \omega^{2n-1} & \omega^{2(2n-1)} & \cdot & \omega^{(n-1)(2n-1)}
\end{pmatrix} ; \quad \omega = \mathrm{e}^{\frac{i\pi}{n}}
\tag{56}
$$

The unitary transformation in (56) connects $\mathbf{h}$ with a coherent and a correlated basis g and $\mathbf{f}$, respectively, through

$$
\begin{aligned}
|\mathbf{h}\rangle \mathbf{B} &= |\mathbf{g}\rangle = |g_1, g_2, \ldots g_n\rangle \\
|\mathbf{h}\rangle \mathbf{B}^{-1} &= |\mathbf{f}\rangle = |f_1, f_2, \ldots f_n\rangle.
\end{aligned}
\tag{57}
$$

In passing we note that the functions in the set g are completely delocalized over the region of sites defined by the "localized" particle–antiparticle basis $\mathbf{h}$, while the $\mathbf{f}$-basis contains all possible phase-shifted contributions from each site in accordance with Eqs. (56) and (57) above. Some interrelationships can be recognized here. The first connection concerns Coleman's so-called extreme state [18], cf. the theories of superconductivity and superfluidity based on ODLRO. The second observation relates to the identification of the present finite dimensional representation as a precursor for possible condensations, developing correlations and coherences that may extend over macroscopic dimensions. If $\mathbf{h}$ is a set of two-particle determinants and the N-particle fermionic wave function is constructed from an AGP, antisymmetrized geminal power, based on g_1, see Eq. (57), then the reduced density matrix can be represented as

$$
\Gamma^{(2)} = \Gamma_{\mathrm{L}}^{(2)} + \Gamma_{\mathrm{S}}^{(2)} = \lambda_{\mathrm{L}} |g_1\rangle\langle g_1| + \lambda_{\mathrm{S}} \sum_{k=2}^{n} |g_k\rangle\langle g_k| + \text{the tail.}
\tag{58}
$$

For the exact prerequisites with respect to the existence of the so-called extreme state giving rise to the representation (58) of the two-particle reduced density matrix including the "tail contribution" (not explicitly displayed here and resulting from the remaining "unpaired" pair configurations), we refer to Coleman's classical treatise [18]. Note also that a completely different statistical argument for the present representation has been given within a more general quantum logical frame, see for instance [9, 21, 22]. In strongly correlated situations, the tail will play no explicit role.

From the theoretical analysis follows that the density matrix, (58), exhibits a large eigenvalue λ_{L} and a degenerate small one λ_{S} given by ($2n \geq N$)

$$
\begin{aligned}
\lambda_{\mathrm{L}} &= \frac{N}{2} - (n-1)\lambda_{\mathrm{S}} \\
\lambda_{\mathrm{S}} &= \frac{N(N-2)}{4n(n-1)}
\end{aligned}
$$

Since we have n basis pair functions, or geminals, and $2n$ spin orbitals (here particle- and antiparticle functions), the number of fermion pairings and pair configurations is

$$\binom{N}{2}, \binom{2n}{2},$$

respectively. The dimension of the "box contribution" defined in Eq. (58) is n and consequently the dimension of the left out "tail contribution" must be $2n(n-1)$. Coleman's theorem leads to the result that eigenvalues associated with the missing "tail" are identical to λ_S [18]. For the extreme state one thus attains a "large" eigenvalue λ_L and a $(n-1)(2n+1)$ set of degenerate eigenvalues λ_S. Under specific conditions one finds that λ_L may drastically increase to a macroscopic size, i.e. will approach the number of physical pairs, $N/2$ (in the Löwdin normalization), developing the new phase according to Yang's ODLRO. A key element in this particular transition is the extreme state (precursor) level and the significance of the full basis g.

As mentioned above, analogous equations can be derived in a statistical framework. This works both for the case of localized fermions in a specific pairing mode and/or bosons subject to a quantum transport environment. An additional interconnection regards the relevance of the basis $\mathbf{f}$, which is related to the surprising fact, that a transformation of the form (58), unites canonical Jordan blocks to a convenient complex symmetric form. We will not discuss this possibility here except call attention to a potential relationship between temperature scales and Jordan block formation by thermal correlations, see e.g. [9, 15, 22] for more details.

Consider now a finite number of fermion- or particle–antiparticle pairs in a vacuum (or particle-like environment, cf. Cooper pairs in a superconductor), i.e.

$$|0\rangle \wedge |\bar{0}\rangle = i\,|m_0\rangle \wedge |\overline{m}_0\rangle \tag{59}$$

using Eq. (22) or (27). To properly understand (59), we briefly refer to Eq. (21). The transformation linking o and $\bar{o}$ with the particle–antiparticle pair presumes a gravitational interaction in (21) of such a magnitude that a Jordan block singularity emerges. From the knowledge of the second order reduced density matrix, it is straightforward to obtain the energy W as

$$W = \mathrm{Tr}\left\{ H_{12} \Gamma^{(2)} \right\} \tag{60}$$

where H_{12} is a general two-body potential between the constituents, i.e. the particle–antiparticle combinations based on the pairing $|0\rangle \wedge |\bar{0}\rangle$. It is easy to demonstrate the following simplifications. From (60), one obtains

$$W = \lambda_L \langle g_1 | H_{12} | g_1 \rangle + \lambda_S \sum_{k=2}^{n} \langle g_k | H_{12} | g_k \rangle. \tag{61}$$

For a "sufficiently" localized basis $\boldsymbol{h}$, i.e.

$$\langle h_k | H_{12} | h_l \rangle = \langle h_k | H_{12} | h_k \rangle \delta_{kl} \tag{62}$$

one obtains for large n

$$W \approx \frac{N}{2} w$$
$$w = \frac{1}{n} \sum_{k=1}^{n} \langle h_k | H_{12} | h_k \rangle \tag{63}$$

Assuming on the other hand that the pairing is based on one fundamental interaction with all matrix elements $\langle h_k | H_{12} | h_l \rangle = w_{LS}$, being constant, provided that the localization centres are all inside a fundamental radius (cf. the discussion related to Eq. (21) and the appearance of Schwarzschild radius). A first observation is that the second term in (61) vanishes rigorously. The result is that we obtain remarkable energy stabilization, cf. Eqs. (61) and (62) (for $n \gg N/2$)

$$W = \lambda_L \langle g_1 | H_{12} | g_1 \rangle = \lambda_L n w_{LS} = \frac{N}{2} \bar{n} w_{LS}; \ (n-1) < \bar{n} < n \tag{64}$$

In passing we note that $n = N/2$ corresponds to the celebrated independent particle model or the Hartree–Fock equation of great importance, e.g. in the development of atomic and molecular physics. We also find that the fundamental interaction mentioned above is unitarily connected with the electromagnetic interactions between the particle m_0 and the antiparticle $-m_0$. Since we are not making any explicit distinction between the Klein–Gordon and the Dirac equation, see Ref. [8] for some fine points, we are not able here to integrate the electro-weak theory although in principle this should be possible.

The macroscopically large energy stabilization, obtained as a consequence of a unique fundamental interaction, implies the possible existence of a quantum mechanical version of a (fermionic-quasibosonic) black hole. Here we will not speculate in any detail on the character of condensation of neutrons and/or nuclei immersed in a gas of electrons, as well as the limiting size of the system for the gravitational collapse to occur.

The emergence of a black hole as an ODLRO coherent system is thus to some extent analogous to a Bose condensate. We note that the stabilization energy W is proportional to the product of the number of the particles of the system, $N/2$ and the number of available quantum states n, with the proportionality factor given by the matrix element w_{LS}, the latter depending on the fundamental interaction H_{12} and conventionally on the local properties (cusps, etc.) of h_k at the site k. The main problem of this analogy rests on the definition of the site k. In usual descriptions of ODLRO, we are treating either light carriers moving in a nuclear skeleton or nuclei moving in a field of defect or excess electrons or in other words the former corresponding to condensation in coordinate space and the latter in momentum space. In the present case, these space–time elements are consumed by the singularity. Nevertheless we note (i) the association of the energy W with a macroscopically large system of mass M interacting with N particles of mass m and $-m$, respectively, and (ii) that the condensate will contain an equal amount of particles and antiparticles, cf. the explanation behind the Einstein law of gravitational reflection.

References

1. A. Eddington, *The Mathematical Theory of Relativity*, 2nd edition (Cambridge University Press, Cambridge, 1924)
2. J. L. Synge, *Relativity, The General Theory* (North-Holland Publishing Company, Amsterdam, 1960)
3. W. Gordon, Z. Physik **40**, 117 (1926)
4. O. Klein, Z. Physik **41**, 407 (1927)
5. P. A. M. Dirac, Proc. Roy. Soc. (London) **A117**, 610 (1928); ibid. **A118**, 351 (1928); ibid. **A126**, 360 (1930)
6. E. Brändas, Int. J. Quantum Chem. **106**, 2836 (2006)
7. E. Brändas, Adv. Quantum Chem. **54**, 115 (2008)
8. E. J. Brändas, in *Progress in Theoretical Chemistry and Physics*, vol. 18, *Frontiers in Quantum Systems in Chemistry and Physics*, ed. by S. Wilson, P. J. Grout, J. Maruani, G. Delgado-Barrio, P. Piecuch (Springer, Berlin, 2008), pp. 239–255
9. E. Brändas, in *Self-Organization of Molecular Systems: From Molecules and Clusters to Nanotubes and Proteins,* ed. by N. Russo, V. Ya. Antonchenko, E. Kryachko (NATO Science for Peace and Security Series A: Chemistry and Biology, Springer Science+Business Media B.V., Munich, Heidelberg), 2009, pp 49–87
10. C. N. Yang, Rev. Mod. Phys. **34**, 694 (1962)
11. E. Balslev, J. M. Combes, Commun. Math. Phys. **22**, 280 (1971)
12. C. E. Reid, E. Brändas, in *Lecture Notes in Physics*, vol. 325, *Resonances – The Unifying Route Towards the Formulation of Dynamical Processes Foundations and Applications in Nuclear, Atomic and Molecular Physics*, ed. by E. Brändas, N. Elander (Springer, Berlin, 1989), pp. 475–483
13. E. Brändas, M. Rittby, N. Elander, J. Math. Phys. **26**, 2648 (1985); E. Engdahl, E. Brändas, M. Rittby, N. Elander, Phys. Rev. **A37**, 3777 (1988)
14. N. Moiseyev, Phys. Rep. **302**, 211 (1998)
15. *Lecture Notes in Physics*, vol. 325, *Resonances – The Unifying Route Towards the Formulation of Dynamical Processes – Foundations and Applications in Nuclear, Atomic and Molecular Physics*, ed. By E. Brändas, N. Elander (Springer, Berlin, 1989)
16. J. T. Jebsen, Ark. Mat. Ast. Fys. **15**, 1 (1921)
17. G. D. Birkoff, *Relativity and Modern Physics* (Cambridge University Press, Cambridge, 1921)
18. A. J. Coleman, Rev. Mod. Phys. **35**, 668 (1963)
19. E. Brändas, Adv. Quantum Chem. **41**, 121 (2002)
20. P.-O. Löwdin, Phys. Rev. **97**, 1474 (1955)
21. E. J. Brändas, in *Dynamics during Spectroscopic Transitions*, ed. by E. Lippert and J. D. Macomber (Springer Verlag, Berlin 1995), pp 148–241
22. E. Brändas, C. H. Chatzidimitriou-Dreismann, Int. J. Quantum Chem. **40**, 649 (1991)

Relativistically Covariant Many-Body Perturbation Procedure

Ingvar Lindgren, Sten Salomonson, and Daniel Hedendahl

Abstract A covariant evolution operator (CEO) can be constructed, representing the time evolution of the relativistic wave function or state vector. Like the nonrelativistic version, it contains (quasi-)singularities. The regular part is referred to as the Green's operator (GO), which is the operator analogue of the Green's function (GF). This operator, which is a field-theoretical concept, is closely related to the many-body wave operator and effective Hamiltonian, and it is the basic tool for our unified theory. The GO leads, when the perturbation is carried to all orders, to the Bethe–Salpeter equation (BSE) in the equal-time or effective-potential approximation. When relaxing the equal-time restriction, the procedure is fully compatible with the exact BSE. The calculations are performed in the photonic Fock space, where the number of photons is no longer constant. The procedure has been applied to helium-like ions, and the results agree well with S-matrix results in cases when comparison can be performed. In addition, evaluation of higher-order quantum-electrodynamical (QED) correlational effects has been performed, and the effects are found to be quite significant for light and medium-heavy ions.

Keywords: Covariance · Many-body perturbation · Quantum-electrodynamical effects · Electron correlation · Fine structure · Heliumlike ions

1 Introduction

Relativistic covariance is an important concept in a relativistic theory. Well-known examples of covariant theories are Maxwell's theory of electromagnetism and Einstein's special theory of relativity. Many-body perturbation theories available today,

I. Lindgren (✉)
Physics Department, University of Gothenburg, Sweden, e-mail: ingvar.lindgren@physics.gu.se

S. Salomonson
Physics Department, University of Gothenburg, Sweden, e-mail: sten.salomonson@physics.gu.se

D. Hedendahl
Physics Department, University of Gothenburg, Sweden, e-mail: daniel.hedendahl@physics.gu.se

P. Piecuch et al. (eds.), *Advances in the Theory of Atomic and Molecular Systems,*
Progress in Theoretical Chemistry and Physics 19, DOI 10.1007/978-90-481-2596-8_6,
© Springer Science+Business Media B.V. 2009

on the other hand, are NOT relativistically covariant. A covariant many-body theory would, in principle, include electron correlation as well as quantum-electrodynamical (QED) effects to arbitrary order. In this chapter, such a procedure will be outlined. The first question is to what extent this is an important problem, and where effects beyond the present procedures are expected to appear.

One important example is the study of *highly charged ions*, which are suitable objects for testing QED at very strong fields. With improved accelerators, particularly the new FAIR facility – *Facility for Antiproton and Ion Research* – now under construction at GSI in Darmstadt (see Fig. 1), new possibilities will appear to study transition energies (fine and hyperfine structure, Lamb shift), g-factors, etc. with higher accuracy than was previously possible. In order to match the new experimental situation, it is important that theoretical procedures also are being developed. In the future, *combined effects of QED and relativistic electron correlation* will be increasingly important – effects that presently cannot be evaluated.

Another example is the precision studies of light atoms and ions, such as the fine structure of the helium atom and helium-like ions. Very accurate fine-structure separations have been determined for the helium atom and the Li^+ ion, but accurate results have also been achieved for somewhat heavier ions, such as F^{+7} and Si^{+12} [1, 2]. For the helium atom, there is a significant discrepancy between the experimental and theoretical results (see Fig. 2), the reason of which is presently unknown. The most accurate experimental results are obtained by Gabrielse [3] and Inguscio et al. [4]. The theoretical calculations have been performed by Drake and

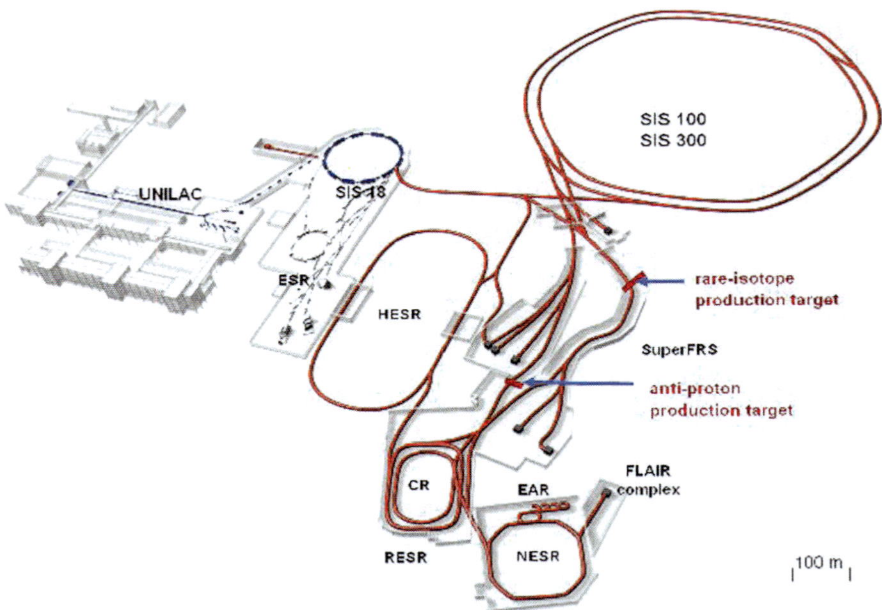

Fig. 1 FAIR project at GSI

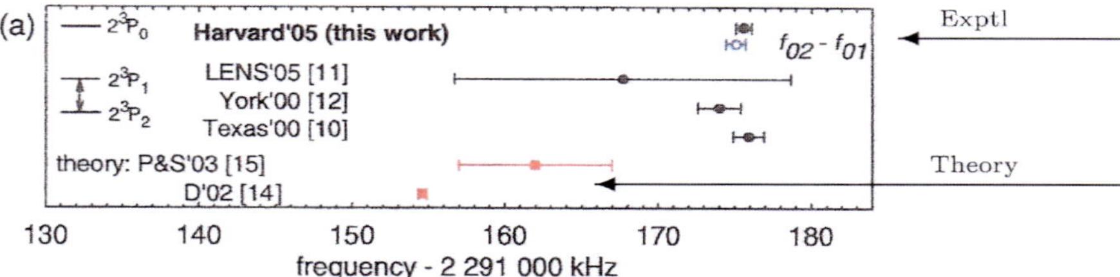

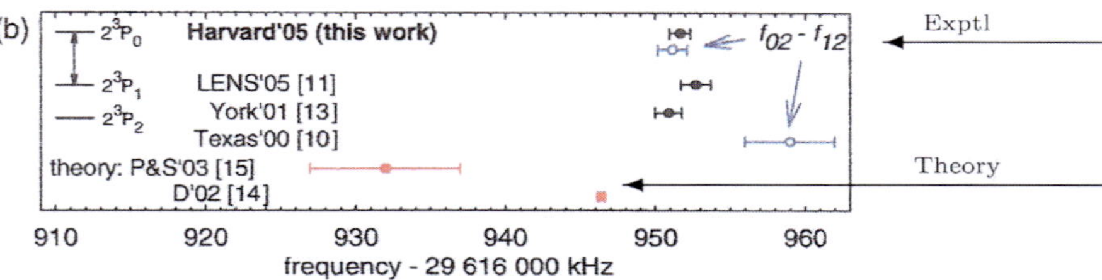

Fig. 2 Experimental and theoretical results for the fine-structure separations of the lowest P-state of the helium atom, $2^3 P_1 - 2^3 P_2$ *top*, $2^3 P_0 - 2^3 P_1$ *bottom* (picture taken from Ref. [3])

coworkers [5] as well as by Pachucki and Sapirstein [6]. The theoretical calculations are based upon non-relativistic wave functions of Hylleraas type with built-in electron correlation, while relativistic and QED effects are treated analytically in an α, $Z\alpha$ power expansion. Our aim is to develop a *numerical* procedure for calculating the *combined* relativistic, QED and correlation effects.

The many-body perturbative procedures are now well developed non-relativistically as well as relativistically [7]. Here, electron correlation can be treated essentially to all orders by methods of coupled-cluster type, while QED effects are at most included to first order. For pure QED calculations, several methods have been developed. Most frequently used is the S-matrix formulation [8], which has been successfully applied particularly to highly charged ions. More recently, two other methods have been developed, the *two-times Green's function* (GF) *technique* by Shabaev et al. [9] and the *covariant evolution operator* (CEO) *technique* developed by the Göteborg group [10]. The latter two methods have the advantage over the S-matrix method that they can be applied to *quasi-degenerate states*, like fine-structure separations. As an illustration, we consider the calculations performed by the Göteborg group a few years ago on some light helium-like ions, compared with experimental data and the calculation by Drake et al. [11] (see Table 1). Our theoretical results agree with the experimental results within the assigned uncertienties, while some results of Drake fall outside the limits of error.

The methods for QED calculations presently available can for practical reasons only be applied to second order (two-photon exchange), which particularly for light systems yields an insufficient description of the electron correlation.

The CEO method has the advantage, compared to other techniques for QED calculations, that it has a structure quite similar to that of MBPT, which opens up the

Table 1 Fine-structure separations for some helium-like ions (from Ref. [10])

Z	$^3P_1 - {}^3P_0$	$^3P_2 - {}^3P_0$	$^3P_2 - {}^3P_1$	
9	701(10) μH		4364,517(6)	Expt'l
	680	5050	4362(5)	Drake
	690	5050	4364	Göteborg
10	1361(6)	8455(6)	265880	Drake
	1370	8460	265880	Göteborg
18		124960(30)		Expt'l
		124810(60)		Drake
		124940		Göteborg

possibility of merging the two effects, as has been described in our recent publications [12, 13]. This will make it possible to develop for the first time a *relativistic MBPT procedure that is fully covariant*. Before going into this problem, we shall briefly summarize the standard time-independent and time-dependent perturbation procedures.

2 Time-Independent Perturbation Procedure

2.1 Bloch Equation

We assume that we have a set of *target states*, satisfying the non-relativistic Schrödinger equation

$$H\Psi^\alpha = E^\alpha \Psi^\alpha \quad (\alpha = 1, 2, \ldots .d) \tag{1}$$

H is the Hamiltonian of the system[1]

$$H = \sum_{i=1}^{N} h_S(i) + \sum_{i<j}^{N} \frac{e^2}{4\pi r_{ij}}, \tag{2}$$

where the first term is a sum of single-electron Schrödinger Hamiltonians and the second term represents the electrostatic interaction between the electrons. For each target state, we assume that there exists a *model state*, which in the *intermediate normalization* (IN) is the projection of the target state onto the *model space*

$$\Psi_0^\alpha = P\Psi^\alpha. \tag{3}$$

Assuming the model states to be linearly independent, a *wave operator* transforms the model states back to the target states,

---

[1] We use here relativistic units, $c = \hbar = m_e = \epsilon_0 = 1$, $e^2 = 4\pi\alpha$, α being the fine-structure constant.

$$\Psi^\alpha = \Omega \Psi_0^\alpha. \tag{4}$$

An *effective Hamiltonian* can be defined so that, operating entirely within the model space, it generates the exact energies of all the target states

$$H_{\mathrm{eff}}\Psi_0^\alpha = E^\alpha \Psi_0^\alpha. \tag{5}$$

In IN, we have

$$H_{\mathrm{eff}} = PH\Omega P. \tag{6}$$

By partitioning the Hamiltonian into a model Hamiltonian (H_0) and a perturbation (V),

$$H = H_0 + V; \tag{7}$$

the wave operator satisfies the *generalized Bloch equation* [14, 7]

$$\left[\Omega, H_0\right]P = \left(V\Omega - \Omega V_{\mathrm{eff}}\right)P. \tag{8}$$

Here,

$$V_{\mathrm{eff}} = PV\Omega P \tag{9}$$

is the *effective interaction* and the effective Hamiltonian becomes

$$H_{\mathrm{eff}} = PH_0P + V_{\mathrm{eff}}. \tag{10}$$

The Bloch equation (8) can be used to generate a perturbation expansion of Rayleigh–Schrödinger type also in the case of quasi-degeneracy by means of an extended model space. If the model space is *complete*, i.e. contains all configurations that can be formed by the valence electrons, then it can be shown that the expansion can be represented graphically by linked diagrams only, known as the *linked diagram* or *linked cluster theorem*,[2]

$$\left[\Omega, H_0\right]P = \left(V\Omega - \Omega V_{\mathrm{eff}}\right)_{\mathrm{linked}}P. \tag{11}$$

By means of the *exponential ansatz* $\Omega = \exp(T)$, the Bloch equation leads directly to the *coupled-cluster approach*. For general open-shell systems, it is often convenient to use the *normal-ordered exponential* [15]

$$\Omega = \{\exp(T)\}, \tag{12}$$

---

[2] A *linked* diagram can consist of disconnected pieces, as long as they are all *open* in the sense that they do not operate entirely within the model space.

which eliminates unwanted contractions between the cluster operators. For a complete model space, it then follows that the graphical representation of the expansion consists of *connected* diagrams only

$$\left[T, H_0\right]P = \left(V\Omega - \Omega V_{\text{eff}}\right)_{\text{conn}} P. \tag{13}$$

Mukherjee has recently modified the normal-ordered exponential so that certain wanted contractions are maintained [16]

$$\Omega = \{\{\exp(T)\}\}, \tag{14}$$

which improves the convergence in certain cases.

2.2 Perturbation Expansion

As mentioned, the Bloch equation (8) is valid for arbitrary quasi-degenerate model spaces. For simplicity, though, here we shall illustrate how the expansion is performed for a degenerate model space (with energy E_0). We can then express the Bloch equation in the form

$$\Omega P = \Gamma_Q(E_0)\left(V\Omega - \Omega V_{\text{eff}}\right)P; \qquad \Gamma_Q(E_0) = \frac{Q}{E_0 - H_0}, \tag{15}$$

which is essentially the original form of the equation, given by Bloch [17, 18]. In first order, we have

$$\Omega^{(1)}P = \Gamma_Q(E_0)VP \tag{16}$$

and in second order

$$\Omega^{(2)}P = \Gamma_Q(E_0)\left(V\Omega^{(1)} - \Omega^{(1)}V_{\text{eff}}^{(1)}\right)P, \tag{17}$$

where $V_{\text{eff}}^{(1)} = PVP$ is the first-order effective interaction. The second term is a so-called *folded term*, because it is traditionally drawn in a folded way (see Fig. 3),

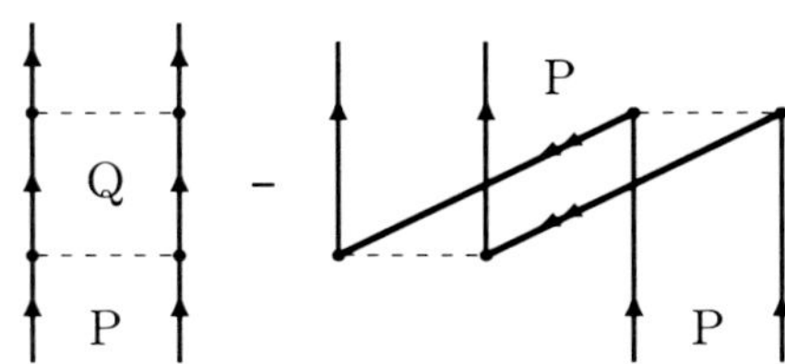

Fig. 3 Graphical representation of the second-order wave operator (17). The *solid lines* represent single-electron orbitals and the *dashed lines* instantaneous Coulomb interactions. The second folded diagram represents the part with the intermediate state in the model space (P)

where the two parts can be evaluated independently. We shall see that this kind of effect plays an important role in the unified theory we are developing.

2.3 Versions of MBPT/CCA

What is indicated here is a *multi-reference* approach, in which a multiple of states are treated simultaneously. This is particularly advantageous in calculating transitions energies. The *valence universal* version, valid for different stages of ionization, is particularly useful in evaluating ionization energies or electron affinities. A serious disadvantage with the multi-reference approach is that it often leads to so-called *intruder states*, i.e. states that do not belong to the group of target states under study but penetrate into the energy range of target states of the same symmetry when the perturbation is turned on. When this happens, the perturbation expansion no longer converges.

The effects of intruder states are generally more severe for molecules than for atoms, due to more dense energy levels. Therefore, even if there are ways of avoiding – or at least reducing – the effect of intruder states in the multi-reference approach, it is when the interest lies entirely in one or a few particular states, more advantageous to study one state at a time in a *state-specific approach.*

It is outside the scope of this chapter to deal further with the various approaches of MBPT/CCA, which are well documented in the literature. Our main goal is to combine many-body calculations with QED, and here it is irrelevant exactly which many-body approach that is used.

2.4 Standard Relativistic MBPT: QED Effects

The standard relativistic MBPT procedures are based upon the *projected Dirac–Coulomb–Breit approximation* [19]

$$H = \Lambda_+ \left[\sum_{i=1}^{N} h_D(i) + \sum_{i<j}^{N} \frac{e^2}{4\pi r_{ij}} + H_B \right] \Lambda_+, \tag{18}$$

where h_D is the single-electron Dirac Hamiltonian and H_B is the instantaneous Breit interaction.

$$H_B = -\frac{e^2}{8\pi} \sum_{i<j} \left[\frac{\alpha_i \cdot \alpha_j}{r_{ij}} + \frac{(\alpha_i \cdot r_{ij})(\alpha_j \cdot r_{ij})}{r_{ij}^3} \right], \tag{19}$$

where Λ_+ is a projection operator that eliminates the negative-energy solutions of the Dirac equation. This approximation is also known as the *no-virtual-pair approximation* (NVPA).

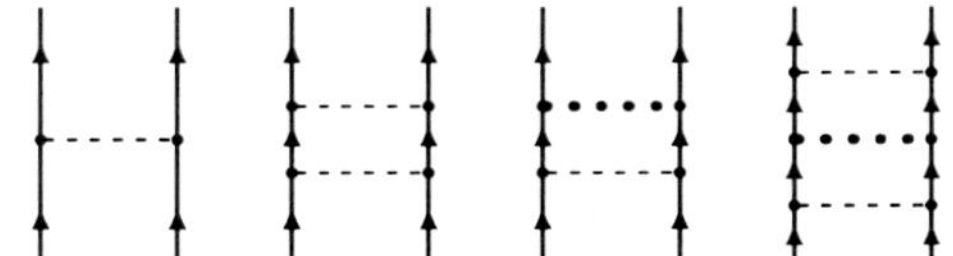

Fig. 4 Graphical representation of the NVPA for helium-like systems. The *dashed line* represents, as before, the Coulomb interaction and the *dotted line* the instantaneous Breit interaction

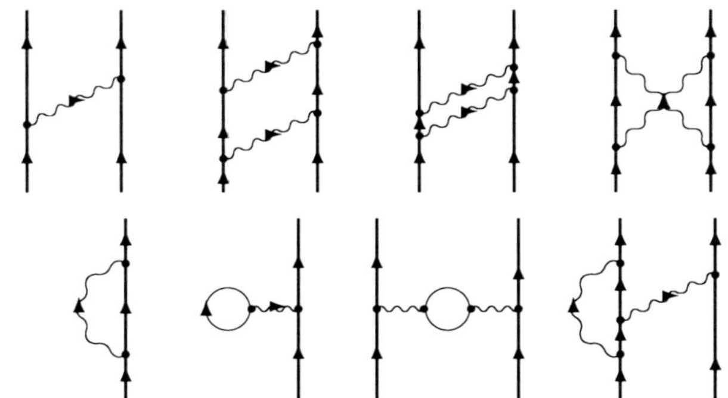

Fig. 5 Some low-order non-radiative (*upper line*) and radiative (*lower line*) "QED effects." The *wavy lines* represent the covariant photon exchange. These diagrams are *Feynman diagrams*, where the *orbital lines* can represent particle as well as hole or anti-particle states

The diagrammatic representation of the NVPA for a two-electron system is exhibited in Fig. 4. The effects beyond the NVPA are traditionally referred to as *QED effects*, some of which are shown graphically in Fig. 5.

3 Time-Dependent Perturbation Theory

3.1 Evolution Operator

The time-dependent Schrödinger state vector has in the Schrödinger picture (SP) the time dependence

$$|\chi(t)\rangle = e^{-iH(t-t_0)} |\chi(t_0)\rangle. \tag{20}$$

In the *interaction picture* (IP), the SP state vectors and operators are transformed according to

$$|\chi_I(t)\rangle = e^{iH_0 t} |\chi_S(t)\rangle; \qquad V_I(t) = e^{iH_0 t} V e^{-iH_0 t}. \tag{21}$$

This leads to the Schrödinger equation in IP

$$i\frac{\partial}{\partial t} |\chi_I(t)\rangle = V_I(t) |\chi_I(t)\rangle. \tag{22}$$

The *time-evolution operator* in IP, $U(t, t_0)$, is defined by[3]

$$|\chi(t)\rangle = U(t, t_0)\,|\chi(t_0)\rangle \qquad (t > t_0), \tag{23}$$

and it satisfies the differential equation

$$i\frac{\partial}{\partial t} U(t, t_0) = V(t)\,U(t, t_0). \tag{24}$$

We assume that an *adiabatic damping* is applied

$$V(t) \to V(t)\,\mathrm{e}^{-\gamma|t|}, \tag{25}$$

where γ is a small, positive number that eventually tends to zero. This leads to the expansion [20]

$$U_\gamma(t, t_0) = \sum_{n=0}^{\infty} \frac{(-\mathrm{i})^n}{n!} \int_{t_0}^{t} \mathrm{d}t_1 \ldots \int_{t_0}^{t} \mathrm{d}t_n\, T\big[V(t_1)\ldots V(t_n)\big]\,\mathrm{e}^{-\gamma(|t_1|+|t_2|\ldots+|t_n|)}, \tag{26}$$

where T is the *time-ordering operator*. The perturbation is represented by the interaction between an electron and the radiation fields

$$V(t) = \int \mathrm{d}^3x\, \mathcal{H}(t, \boldsymbol{x}) \tag{27}$$

with

$$\mathcal{H}(x) = -e\hat{\psi}(x)^{\dagger}\alpha^{\mu}A_{\mu}(x)\hat{\psi}(x), \tag{28}$$

where $x = (t, \boldsymbol{x})$ is the four-dimensional space–time coordinate and $\hat{\psi}(x)$, $\hat{\psi}(x)^{\dagger}$ and A_{μ} are the electron-field and the photon-field operators, respectively. This perturbation operates in the *extended photonic Fock space*, where the number of photons is no longer constant.[4] The expansion (26) then becomes

$$U(t, t_0) = \sum_{n=0}^{\infty} \frac{(-\mathrm{i})^n}{n!} \int_{t_0}^{t} \mathrm{d}x_1^4 \ldots \int_{t_0}^{t} \mathrm{d}x_n^4\, T\big[\mathcal{H}(x_1)\ldots\mathcal{H}(x_n)\big]\,\mathrm{e}^{-\gamma(|t_1|+|t_2|\ldots+|t_n|)}, \tag{29}$$

where the integrations are performed over all space and over time as indicated. The exchange of a single photon is represented by TWO perturbations of this kind.

---

[3] In the following we shall work mainly in the interaction picture and leave out the subscript "I".

[4] Also the Fock space is a form of Hilbert space, and therefore we shall refer to the Hilbert space with a constant number of photons as the *restricted (Hilbert) space* and the space with a variable number of photons as the *extended or photonic Fock space*.

3.2 Gell-Mann–Low Theorem

Gell-Mann and Low [21] have shown that for a closed-shell system the state vector

$$|\chi(0)\rangle = |\Psi\rangle = \lim_{\gamma \to 0} \frac{U_\gamma(0, -\infty)|\Phi\rangle}{\langle \Phi|U_\gamma(0, -\infty)|\Phi\rangle} \tag{30}$$

satisfies the time-independent Schrödinger equation

$$(H_0 + V(0))|\Psi\rangle = E|\Psi\rangle. \tag{31}$$

Here,

$$|\Phi\rangle = \lim_{t \to -\infty} |\chi(t)\rangle \tag{32}$$

is the *parent state*, equal to the limit of the time-dependent target function as the perturbation is adiabatically turned off. In the single-reference case, this is identical to the model state (3). The Gell-Mann–Low (GML) theorem can be extended to a general open-shell system [10]

$$\left| \chi^\alpha(0)\rangle = |\Psi^\alpha\rangle = \lim_{\gamma \to 0} \frac{N^\alpha U_\gamma(0, -\infty)|\Phi^\alpha\rangle}{\langle \Phi^\alpha|U_\gamma(0, -\infty)|\Phi^\alpha\rangle}. \right. \tag{33}$$

In this case, the parent state $|\Phi^\alpha\rangle$ is not necessarily identical to the model state, $|\Psi_0^\alpha\rangle = P|\Psi^\alpha\rangle$, which is the reason for the appearance of the normalization factor N^α. In this multi-reference case, the wave functions satisfies similar equations

$$(H_0 + V(0))|\Psi^\alpha\rangle = E^\alpha|\Psi^\alpha\rangle \tag{34}$$

It should be observed that a condition for the GML relations to hold is that the *perturbation is time-independent in the SP*, apart from the adiabatic damping. With the Fock-space perturbation (27), this condition is fulfilled, but it is NOT true for any time-dependent perturbation, acting in the restricted space. Therefore, in the present formalism, which is based upon the GML theorem, *we have to work in the photonic Fock space*.

We can define a wave operator in the photonic space in the same way as before (4)

$$|\Psi^\alpha\rangle = \boldsymbol{\Omega}|\Psi_0^\alpha\rangle \tag{35}$$

(using bold-face symbol to distinguish it from the standard wave operator). From Eq. (34), we can also define a corresponding effective Hamiltonian

$$P\left(H_0 + V_{\mathrm{eff}}\right)|\Psi_0^\alpha\rangle = E^\alpha|\Psi_0^\alpha\rangle \tag{36}$$

and

$$V_{\mathrm{eff}} = P V(0)\, \Omega\, P \tag{37}$$

Of course, the effective Hamiltonian/interaction lies in the model space, which is a part of the restricted Hilbert space with constant number of photons, while the wave operator now acts in the extended space.

4 Covariant Evolution Operator and the Green's Operator

4.1 Definitions

The evolution operator (23) is a non-relativistic concept, since time evolves only in the positive direction. In relativistic applications, we must allow time to run also *backwards* in the negative direction, which represents the propagation of *hole or antiparticle states* with negative energy. This leads to the CEO, introduced by Lindgren, Salomonson and coworkers [10].

Here, we shall define the CEO by means of the GF, using the Feynman kernel, which leads to relativistically covariance. The field-theoretical single-particle GF can be defined [22]:

$$G(x, x_0) = \frac{\langle 0_{\mathrm{H}} | T[\hat{\psi}_{\mathrm{H}}(x)\hat{\psi}_{\mathrm{H}}^{\dagger}(x_0)] | 0_{\mathrm{H}} \rangle}{\langle 0_{\mathrm{H}} | 0_{\mathrm{H}} \rangle}, \tag{38}$$

where T is the *Wick time-ordering* operator and $\hat{\psi}_{\mathrm{H}}$, $\hat{\psi}_{\mathrm{H}}^{\dagger}$ are the electron field operators in the Heisenberg representation. The state $|0_{\mathrm{H}}\rangle$ is the *Heisenberg vacuum*. In the vacuum expectation, all normal-ordered products vanish. Therefore, in transforming the time-ordered product to normal ordering by means of Wick's theorem, only fully contracted terms will remain.

By transforming to the interaction picture, the vacuum expectation above can be expanded in analogy with the time evolution operator

$$\langle 0_{\mathrm{H}} | T[\hat{\psi}_{\mathrm{H}}(x)\hat{\psi}_{\mathrm{H}}^{\dagger}(x_0)] | 0_{\mathrm{H}} \rangle = \sum_{n=0}^{\infty} \frac{(-i)^n}{n!} \int \cdots \int \mathrm{d}t_1 \cdots \mathrm{d}t_n$$

$$\times \langle 0 | T[V(t_1) \cdots V(t_n)\, \hat{\psi}(x)\hat{\psi}^{\dagger}(x_0)] | 0 \rangle \, \mathrm{e}^{-\gamma(|t_1|+|t_2|\cdots)} \tag{39}$$

with *integrations over all times*. We now define the single-particle CEO[5]

____

[5] In this definition, we shall allow photons operators to remain uncontracted, for reasons that will be apparent later.

$$U^1_{\text{Cov}}(t, t_0) = \iint d^3x\, d^3x_0\, \hat{\psi}^\dagger(x)\langle 0_{\text{H}}|T[\hat{\psi}_{\text{H}}(x)\hat{\psi}^\dagger_{\text{H}}(x_0)]|0_{\text{H}}\rangle\,\hat{\psi}(x_0) \qquad (40)$$

with integration over the space coordinates of x and x_0. This is obviously relativistically covariant.

The CEO is an *operator* in contrast to the GF, which is a *function*. In Fig. 6, we compare diagrams for single-photon exchange for the standard evolution operator, the GF and the CEO.

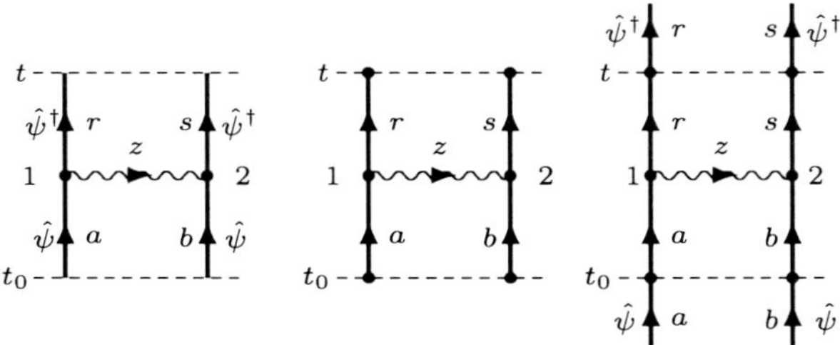

Fig. 6 Comparison between the standard evolution operator, the Green's function and the covariant evolution operator for single-photon exchange in the equal-time approximation. The *solid lines* between *heavy dots* represent electron propagators and the *free lines* electron creation and absorption operators

The exchange of a single retarded photon is represented by *two* contracted perturbations of the type (27). The corresponding single-photon CEO can be shown to be [10][6]

$$\langle rs|U_{\text{sp}}(t)|ab\rangle = \frac{e^{-it(E_0-E_{\text{out}})}}{E_0 - E_{\text{out}}}\,\langle rs|V_{\text{sp}}(E_0)|ab\rangle, \qquad (41)$$

where V_{sp} is the effective single-photon potential

$$\langle rs|V_{\text{sp}}(E_0)|ab\rangle =$$
$$\left\langle rs\left|\int_0^\infty dk\, f(k)\left[\frac{1}{E_0 - \varepsilon_r - \varepsilon_b - (k - i\gamma)_r} + \frac{1}{E_0 - \varepsilon_s - \varepsilon_a - (k - i\gamma)_s}\right]\right|ab\right\rangle.(42)$$

Here, ε_x represent the orbital energies and $E_0 = \varepsilon_a + \varepsilon_b$ and $E_{\text{out}} = \varepsilon_r + \varepsilon_s$ are the initial and final energies of the system, respectively. $(x)_r$ represents an expression with the sign of the orbital r, and $f(k)$ is a gauge-dependent function of the photon momentum. Note that this potential depends on the initial energy, E_0.

---

[6] When operating on unperturbed states with the adiabatic damping, the initial time is $t_0 = -\infty$, which we normally leave out.

4.2 Connection to MBPT

The vacuum expectation (39) contains singularities, which are eliminated by the denominator in the definition of the GF (38). For the CEO, which is an operator, the situation is more complex. We shall refer to the regular part of the CEO as the *Green's operator* (GO), which we separate into open and closed parts

$$\mathcal{G}(t, t_0) = 1 + \mathcal{G}_{\mathrm{op}}(t, t_0) + \mathcal{G}_{\mathrm{cl}}(t, t_0). \tag{43}$$

The open and closed parts of this operator are together identical to the previously introduced *reduced CEO*, $\widetilde{U}(t, t_0)$ [23, 10]

$$\widetilde{U}(t, t_0) = \mathcal{G}_{\mathrm{op}}(t, t_0) + \mathcal{G}_{\mathrm{cl}}(t, t_0). \tag{44}$$

The parts of the GO are defined by

$$\begin{cases} QU(t, t_0)P = \mathcal{G}_{\mathrm{op}}(t, t_0) \cdot PU(0, t_0)P \\ PU(t, t_0)P = P + \mathcal{G}_{\mathrm{cl}}(t, t_0) \cdot PU(0, t_0)P, \end{cases} \tag{45}$$

where P is the projection operator for the model space and $Q = 1 - P$ for the complementary space. The *heavy dot* implies that the two parts are evaluated separately in the same way as in folded diagrams (Fig. 3).

It is easy to show that for $t = 0$

$$U(0, t_0)P = \left(1 + \mathcal{G}_{\mathrm{op}}(0, t_0)\right) \cdot PU(0, t_0)P, \tag{46}$$

known as the *factorization theorem*. Inserting this into the GML formula (33) yields

$$|\Psi^{\alpha}\rangle_{\mathrm{Rel}} = \left(1 + \mathcal{G}_{\mathrm{op}}(0, -\infty)\right) \cdot P \frac{N^{\alpha}\hat{U}_{\mathrm{Cov}}(0, -\infty)|\Phi^{\alpha}_{\mathrm{Rel}}\rangle}{\langle\Phi^{\alpha}_{\mathrm{Rel}}|\hat{U}_{\mathrm{Cov}}(0, -\infty)|\Phi^{\alpha}_{\mathrm{Rel}}\rangle}. \tag{47}$$

The expression to the right of the dot is the model state $P|\Psi^{\alpha}\rangle_{\mathrm{Rel}} = |\Psi^{\alpha}_0\rangle_{\mathrm{Rel}}$, which implies that the expression to the left is the *relativistically covariant wave operator* (also a Fock-space operator)

$$\boxed{\Omega_{\mathrm{Cov}} = 1 + \mathcal{G}_{\mathrm{op}}(0, -\infty).} \tag{48}$$

The *relativistically covariant effective interaction* can be shown to be [10]

$$\boxed{V_{\mathrm{eff}}^{\mathrm{Cov}} = P\left(i\frac{\partial}{\partial t}\mathcal{G}_{\mathrm{cl}}(t, -\infty)\right)_{t=0}P.} \tag{49}$$

4.3 Model-Space Contributions

In the definition of the GF (38), the singularities of the vacuum expectation appear only in the form of disconnected diagrams. For the GO, on the other hand, (quasi-) singularities can appear also for connected diagrams, when an intermediate state lies in the model space. We shall consider here two-electron systems, and we shall see how these singularities can be eliminated.

From the definitions (45) of the GO, it follows that the open part can be expanded as

$$\mathcal{G}_{\mathrm{op}}(t)P = U_{\mathrm{op}}(t)P - \mathcal{G}_{\mathrm{op}}(t) \cdot \mathcal{G}_{\mathrm{cl}}(0)P - \mathcal{G}_{\mathrm{op}}(t) \cdot \mathcal{G}_{\mathrm{cl}}(0) \cdot \mathcal{G}_{\mathrm{cl}}(0)P - \cdots \tag{50}$$

The negative terms are referred to as *counterterms*, which eliminate the (quasi-)singularities of the CEO, U_{op}.

We consider a ladder diagram with two photons, shown in Fig. 7 (*left*), and we assume that we operate on a model-space state of energy $\mathcal{E}$. (We recall that the perturbation in this formalism is the interaction (27), which implies that each retarded photon exchange is a second-order perturbation.) For the single photon we have for $t = 0$

$$\mathcal{G}^{(1)}(\mathcal{E})P_{\mathcal{E}} = U_{\mathrm{sp}}^{(1)}(\mathcal{E})P_{\mathcal{E}} = \Gamma_{\mathcal{E}}\, V_{\mathrm{sp}}(\mathcal{E})P_{\mathcal{E}} = \frac{1}{\mathcal{E} - H_0}\, V_{\mathrm{sp}}(\mathcal{E})P_{\mathcal{E}}, \tag{51}$$

where $V_{\mathrm{sp}}(\mathcal{E})$ is the energy-dependent single-photon potential (42). (Also the evolution operator and the GO depend on the energy to the right, which we have indicated, leaving out the time.) The two-photon CEO can be expressed as a product of two single-photon CEOs, both with the same energy parameter $\mathcal{E}$,

$$U_{\mathrm{sp}}^{(2)}(\mathcal{E})_{\mathrm{Ladd}}P_{\mathcal{E}} = U_{\mathrm{sp}}^{(1)}(\mathcal{E})\, U_{\mathrm{sp}}^{(1)}(\mathcal{E})\, P_{\mathcal{E}} = \frac{1}{\mathcal{E} - H_0}\, V_{\mathrm{sp}}(\mathcal{E})\frac{1}{\mathcal{E} - H_0}\, V_{\mathrm{sp}}(\mathcal{E})P_{\mathcal{E}}. \tag{52}$$

We assume now that we have an intermediate model-space state of energy $\mathcal{E}' \approx \mathcal{E}$, which makes the expression quasi-singular,

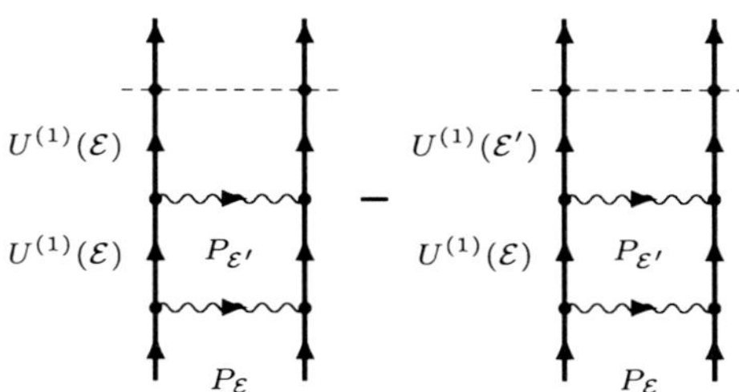

Fig. 7 Elimination of singularity of the second-order evolution operator by means of a counterterm (*second diagram*)

$$U_{\text{sp}}^{(2)}(\mathcal{E})_{\text{Ladd}} P_{\mathcal{E}} = U_{\text{sp}}^{(1)}(\mathcal{E}) \, P_{\mathcal{E}'} \, U_{\text{sp}}^{(1)}(\mathcal{E}) \, P_{\mathcal{E}} = \frac{1}{\mathcal{E} - H_0} \, V_{\text{sp}}(\mathcal{E}) \frac{P_{\mathcal{E}'}}{\mathcal{E} - H_0} \, V_{\text{sp}}(\mathcal{E}) P_{\mathcal{E}}$$

$$= U_{\text{sp}}^{(1)}(\mathcal{E}) \, \frac{1}{\mathcal{E} - \mathcal{E}'} \, P_{\mathcal{E}'} V_{\text{sp}}(\mathcal{E}) P_{\mathcal{E}}. \tag{53}$$

Here, $P_{\mathcal{E}'} V_{\text{sp}}(\mathcal{E}) P_{\mathcal{E}}$ is the single-photon effective interaction, which is identical to the second-order effective interaction (37) with the Fock-space wave operator.

The counterterm looks similar, but the second factor has the energy parameter $\mathcal{E}'$ (it does not operate beyond the heavy dot)

$$U_{\text{sp}}^{(2)}(\mathcal{E})_{\text{counter}} P_{\mathcal{E}} = -U_{\text{sp}}^{(1)}(\mathcal{E}') P_{\mathcal{E}'} \cdot P_{\mathcal{E}'} U_{\text{sp}}^{(1)}(\mathcal{E}) \, P_{\mathcal{E}}$$

$$= -U_{\text{sp}}^{(1)}(\mathcal{E}') \, \frac{P_{\mathcal{E}'}}{\mathcal{E} - \mathcal{E}'} \, V_{\text{sp}}(\mathcal{E}) P_{\mathcal{E}}. \tag{54}$$

This eliminates the (quasi-)singularity, but there is a finite remainder,

$$\frac{U_{\text{sp}}^{(1)}(\mathcal{E}) - U_{\text{sp}}^{(1)}(\mathcal{E}')}{\mathcal{E} - \mathcal{E}'} \, P_{\mathcal{E}'} V_{\text{sp}}(\mathcal{E}) P_{\mathcal{E}} = \frac{\delta U_{\text{sp}}^{(1)}(\mathcal{E})}{\delta \mathcal{E}} \, V_{\text{eff}}^{(1)} \Rightarrow \frac{\partial U_{\text{sp}}^{(1)}(\mathcal{E})}{\partial \mathcal{E}} \, V_{\text{eff}}^{(1)}. \tag{55}$$

Differentiating the single-photon CEO, we find that the two-photon GO becomes

$$\mathcal{G}(\mathcal{E})^{(2)} P_{\mathcal{E}} = \Gamma_{\mathcal{E}} \left(V_{\text{sp}}(\mathcal{E}) \, \Omega^{(1)} - \Omega^{(1)} V_{\text{eff}}^{(1)} + \frac{\delta V_{\text{sp}}(\mathcal{E})}{\delta \mathcal{E}} \, V_{\text{eff}}^{(1)} \right) P_{\mathcal{E}}, \tag{56}$$

where $\Gamma_{\mathcal{E}}$ is defined in Eq. (51). The last two terms are due to the intermediate model-space state, and we refer to them as the *model-space contribution* (MSC). This is quite analogous to the folded term in Eq. (17) in standard MBPT, the only difference being that we now have an additional term, due to the energy dependence of the perturbation.

We have so far only considered multiple single-photon exchange, but this can be generalized to the full exchange of irreducible interactions, $\mathcal{V}(\mathcal{E})$, shown in Fig. 8, yielding

$$\mathcal{G}(\mathcal{E})^{(2)} P_{\mathcal{E}} = \Gamma_{\mathcal{E}} \left(\mathcal{V}(\mathcal{E}) \, \Omega^{(1)} - \Omega^{(1)} V_{\text{eff}}^{(1)} + \frac{\delta \mathcal{V}(\mathcal{E})}{\delta \mathcal{E}} \, V_{\text{eff}}^{(1)} \right) P_{\mathcal{E}}, \tag{57}$$

where we now have

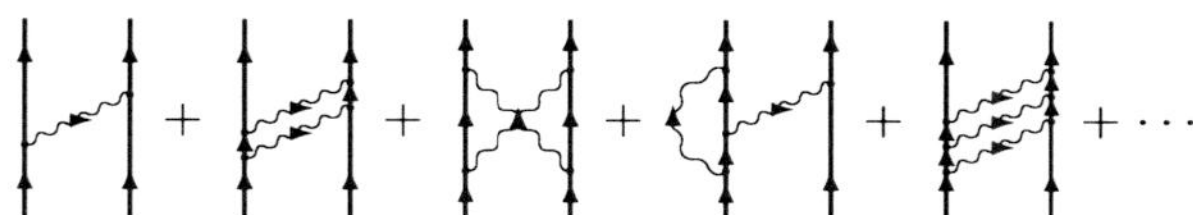

Fig. 8 Irreducible potential interactions, acting in the restricted space

$$V_{\text{eff}} = P V(\mathcal{E}) \Omega P \tag{58}$$

with the operators acting in the restricted space.

5 Connection to Bethe–Salpeter Equation

5.1 Equal-Time Approximation

When the procedure of the previous section is continued, one finds that the GO can be expended as [12, 13]

$$\mathcal{G}(\mathcal{E}) = \mathcal{G}_0(\mathcal{E}) + \sum_{n=1}^{\infty} \frac{\delta^n \mathcal{G}_0(\mathcal{E})}{\delta \mathcal{E}^n} \left(V_{\text{eff}} \right)^n, \tag{59}$$

where $\mathcal{G}_0(\mathcal{E})$ represents the GO without any intermediate model-space states (no folds). It then follows that the second term of the expansion represents the entire MSC.

When we operate with the expansion (59) on the model state $|\Phi\rangle$ with energy E_0, the result is

$$\mathcal{G}(E_0)|\Phi\rangle = \left[\mathcal{G}_0(\mathcal{E}) + \sum_{n=1}^{\infty} \frac{\delta^n \mathcal{G}_0(\mathcal{E})}{\delta \mathcal{E}^n} \left(\Delta E \right)^n \right]_{\mathcal{E}=E_0} |\Phi\rangle. \tag{60}$$

The effective interaction (58) with the full irreducible potential and with the wave operator in the restricted Hilbert space is also equal to $V_{\text{eff}} = P V(0) \Omega P$ with the interaction (27) and the Fock-space wave operator. Therefore, according to Eq. (36),

$$V_{\text{eff}}|\Phi\rangle = (E - E_0)|\Phi\rangle = \Delta E|\Phi\rangle. \tag{61}$$

This is a Taylor series, and the result can be expressed as

$$\mathcal{G}(E_0)|\Phi\rangle = \mathcal{G}_0(E)|\Phi\rangle. \tag{62}$$

This implies that *the effect of the model-space contributions is to shift the energy parameter from the model energy E_0 to the target energy E.*

From

$$\mathcal{G}_0(E_0) = 1 + \left[\frac{1}{E_0 - H_0} \mathcal{V}(E_0) + \frac{1}{E_0 - H_0} \mathcal{V}(E_0) \frac{Q}{E_0 - H_0} \mathcal{V}(E_0) + \cdots \right], \tag{63}$$

we then find that the GO *with* MSC becomes

$$\mathcal{G}(E_0) = \mathcal{G}_0(E) = 1 + \left[\frac{1}{E - H_0} \mathcal{V}(E) + \frac{1}{E - H_0} \mathcal{V}(E) \frac{Q}{E - H_0} \mathcal{V}(E) + \cdots \right]. \tag{64}$$

The open part of the GO represents the open part of the wave operator (48), i.e.

$$Q\Psi = \left[\frac{Q}{E - H_0}\,\mathcal{V}(E) + \frac{Q}{E - H_0}\,\mathcal{V}(E)\frac{Q}{E - H_0}\,\mathcal{V}(E) + \cdots\right]|\Phi\rangle \tag{65}$$

or

$$Q(E - H_0)\Psi = Q\mathcal{V}(E)\Psi \tag{66}$$

From Eq. (61), we have

$$P(E - H_0)|\Psi\rangle = V_{\text{eff}}(E)|\Phi\rangle = P\mathcal{V}(E)\Psi, \tag{67}$$

which yields

$$\boxed{(E - H_0)|\Psi\rangle = \mathcal{V}(E)|\Psi\rangle.} \tag{68}$$

This is the *Bethe–Salpeter equation (BSE) in the effective-potential form.* We can regard this equation as the projection of the Fock-space equation (34) onto the restricted space.

We have here assumed that the CEO represents the time evolution of the relativistic wave function (23), which has the consequence that it depends only on a single initial and a single final time, the same for all particles. In the next section, we shall relax this restriction and let the times be independent for the individual particles. Then we will retrieve the exact BSE. This leads to a manifestly covariant concept, although it is not in accord with standard quantum mechanics.

5.2 The Full Bethe–Salpeter Equation

The Dyson equation for the two-particle GF, illustrated in Fig. 9, can be expressed as

$$G(x, x'; x_0, x_0') = G_0(x, x'; x_0, x_0') + \iiiint \mathrm{d}^4x_1\mathrm{d}^4x_2\mathrm{d}^4x_1'\mathrm{d}^4x_2'$$
$$\times\, G_0(x, x'; x_2, x_2')\,(-\mathrm{i})\Sigma^*(x_2, x_2'; x_1, x_1')\,G(x_1, x_1'; x_0, x_0'), \tag{69}$$

where Σ^* is the *proper* or *irreducible* two-particle self-energy (that cannot be separated into two or more self-energy parts). G_0 is the zeroth-order two-particle GF, which can also be "dressed" with single-particle self-energy insertions.

Bethe and Salpeter [24] as well as Gell-Mann and Low [21] argue that a similar equation can be set up for a two-particle wave function. We assume we have a single-reference situation and let the Dyson equation act on the unperturbed wave function of model function $\Phi(x_0, x_0')$ (with $t_0 = t_0' = -\infty$). With

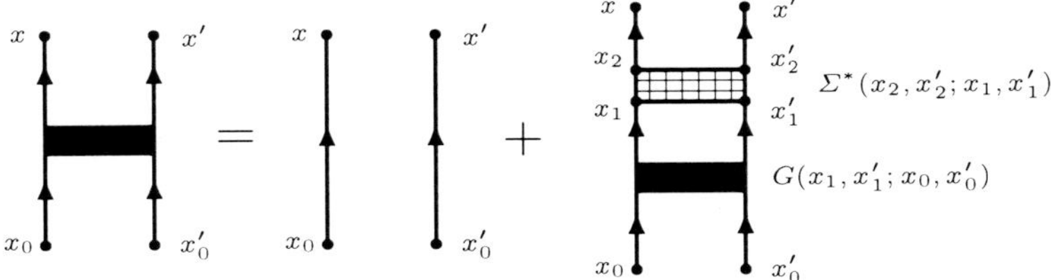

Fig. 9 Graphical representation of the Dyson equation for the two-particle Green's function. The *crossed box* represents the proper or irreducible two-particle self energy

$$\Psi(x, x') = \iint \mathrm{d}^3 x_0 \mathrm{d}^3 x_0' \, G(x, x'; x_0, x_0') \, \Phi(x_0, x_0') \tag{70}$$

and

$$\Phi(x, x') = \iint \mathrm{d}^3 x_0 \mathrm{d}^3 x_0' \, G_0(x, x'; x_0, x_0') \, \Phi(x_0, x_0'), \tag{71}$$

we have

$$\Psi(x, x') = \Phi(x, x') + \iiiint \mathrm{d}^4 x_1 \mathrm{d}^4 x_2 \mathrm{d}^4 x_1' \mathrm{d}^4 x_2'$$
$$\times \, G_0(x, x'; x_2, x_2') \, (-\mathrm{i}) \Sigma^*(x_2, x_2'; x_1, x_1') \, \Psi(x_1, x_1'). \tag{72}$$

This is the famous *Bethe–Salpeter equation*, which is illustrated graphically in Fig. 10. In the treatment of Bethe–Salpeter and Gell-Mann–Low free-electron propagators are used, and then the first inhomogeneous term cannot contribute to a bound state. Here, we shall work in the Ferry picture with bound-state propagators, and then this term should remain.

By means of the two-times GO, Eq. (70) can be expressed as an operator relation

$$\boxed{|\Psi(t, t')\rangle = \mathcal{G}(t, t'; -\infty) |\Phi\rangle,} \tag{73}$$

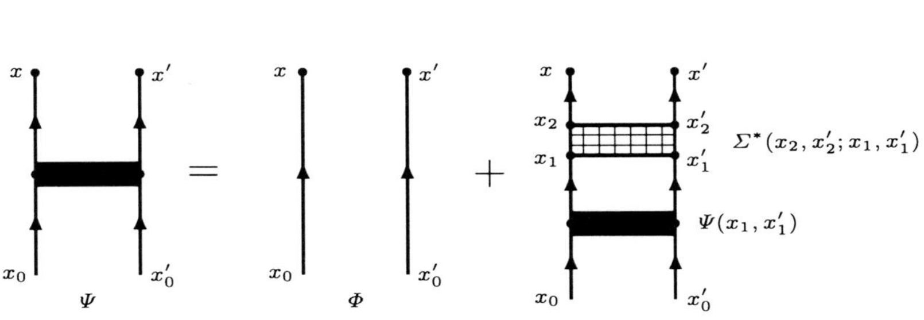

Fig. 10 Graphical representation of the *inhomogeneous* Bethe–Salpeter equation (72). This is similar to Fig. 9 but operates now on the unperturbed wave function

which implies that *the two-times GO* essentially represents wave operator of the Bethe–Salpeter state vector.

6 Implementation

In order to implement the procedure developed above, it is convenient to work in the photonic Fock space, where the number of photons is no longer constant.

We consider for simplicity the single-reference case and start with the Fock-space relation (34) and the corresponding Fock-space Bloch equation

$$\left[\boldsymbol{\Omega}, H_0\right] P = \left(V(0)\boldsymbol{\Omega} - \boldsymbol{\Omega} V_{\mathrm{eff}}\right) P. \tag{74}$$

We use here the Coulomb gauge, where the interaction can be separated in an instantaneous Coulomb part and a Breit interaction that can be retarded. The Breit part is represented by two interactions of the type (27) with the $f(k)$ function in Eq. (42) given by

$$f_{\mathrm{C}}(k) = \boldsymbol{\alpha}_1 \cdot \boldsymbol{\alpha}_2 \frac{\sin(kr_{12})}{\pi r_{12}} - (\boldsymbol{\alpha}_1 \cdot \mathbf{V}_1)(\boldsymbol{\alpha}_2 \cdot \mathbf{V}_2) \frac{\sin(kr_{12})}{\pi k^2 r_{12}}, \tag{75}$$

where the nabla operators do not operate beyond the factor shown. The terms here represent the Gaunt and scaler-retardation parts of the interaction, respectively.

The function $f_{\mathrm{C}}(k)$ can be expanded in partial waves

$$f_{\mathrm{C}}(k) = \sum_{l=0}^{\infty} \left[V_G^l(kr_1) \cdot V_G^l(kr_2) - V_{\mathrm{sr}}^l(kr_1) \cdot V_{\mathrm{sr}}^l(kr_2) \right]. \tag{76}$$

In the photonic Fock space, the perturbation is then of the form

$$V = V_{\mathrm{C}} + V_G^l(kr) + V_{\mathrm{sr}}^l(kr) \tag{77}$$

which is time independent. This is illustrated in Fig. 11. Applying this simple perturbation in the Fock space is equivalent to applying the complicated perturbation in Fig. 8 in the restricted space.

The photonic-Fock-space Bloch equation now becomes

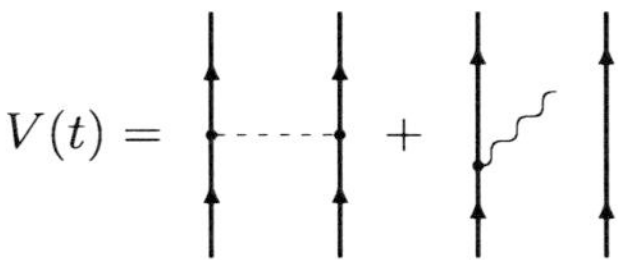

Fig. 11 Graphical representation of the perturbation (77), acting in the photonic Fock space

$$\left[\boldsymbol{\Omega}, H_0\right] P = \left(V_{\mathrm{C}} + V_1^l + V_2^l\right) \boldsymbol{\Omega} P - \boldsymbol{\Omega} V_{\mathrm{eff}}, \tag{78}$$

letting V^l represent the Gaunt term as well as the scalar retardation. Applying, for instance, first a series of Coulomb interactions, then a perturbation V^l, then a new series of Coulomb interactions, a second V^l perturbation and finally a new series of Coulomb interactions leads to the result shown in Fig. 12. This represents a single time-dependent photon with crossing Coulomb interactions, evaluated with correlated wave function. In addition, folded diagrams have to be included, which also represent the energy derivatives (56). By closing the photon on the same electron line, corresponding self-energy and vertex correction effects are obtained (of course, after proper renormalization).

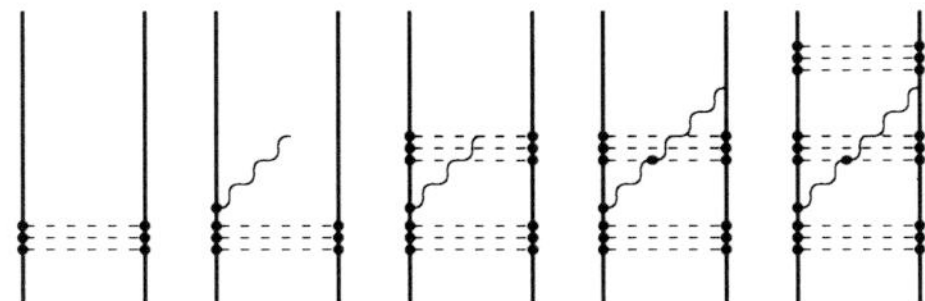

Fig. 12 Graphical representation of the perturbative solution of the Fock-space Bloch equation (78)

The procedure presented here has been applied to a number of light and medium-heavy helium-like ions, and the results agree well with standard S-matrix results in cases where comparison can be made. In addition, effects of single retarded photon with correlation – with and without crossing Coulomb interactions – have been evaluated and found to yield effects that are quite significant and more important than second-order QED effects for light elements. More details about the implementation procedure and numerical results will appear in a forthcoming publication.

7 Summary and Conclusions

We have presented a relativistically covariant many-body perturbation procedure, based upon the CEO and the GO. This represents a unification of the many-body perturbation theory and quantum electrodynamics. Applied to all orders, the procedure leads in the equal-time approximation to the BSE in the effective-potential form. By relaxing this restriction, the procedure is consistent with the full BSE. The new procedure will be of importance in cases where QED effects beyond first order in combination with high-order electron correlation are significant.

References

1. E. G. Myers, H. S. Margolis, J. K. Thompson, M. A. Farmer, J. D. Silver, M. R. Tarbutt, Phys. Rev. Lett. **82**, 4200 (1999)
2. T. R. DeVore, D. N. Crosby, E. G. Myers, Phys. Rev. Lett. **100**, 243001 (2008)
3. T. Zelevinsky, D. Farkas, G. Gabrielse, Phys. Rev. Lett. **95**, 203001 (2005)
4. G. Giusfredi, P. C. Pastor, P. DeNatale, D. Mazzotti, C. deMauro, L. Fallani, G. Hagel, V. Krachmalnicoff, M. Ingusio, Can. J. Phys. **83**, 301 (2005)
5. G. W. F. Drake, Can. J. Phys. **80**, 1195 (2002)
6. K. Pachucki, J. Sapirstein, J. Phys. B **33**, 5297 (2000)
7. I. Lindgren, J. Morrison, *Atomic Many-Body Theory* (Second edition, Springer-Verlag, Berlin, 1986, reprinted 2009)
8. P. J. Mohr, G. Plunien, G. Soff, Phys. Rep. **293**, 227 (1998)
9. V. M. Shabaev, Phys. Rep. **356**, 119 (2002)
10. I. Lindgren, S. Salomonson, B. Åsén, Phys. Rep. **389**, 161 (2004)
11. G. W. F. Drake, Can. J. Phys. **66**, 586 (1988)
12. I. Lindgren, S. Salomonson, D. Hedendahl, Can. J. Phys. **83**, 183 (2005)
13. I. Lindgren, S. Salomonson, D. Hedendahl, Phys. Rev. A **73**, 062502 (2006)
14. I. Lindgren, J. Phys. B **7**, 2441 (1974).
15. I. Lindgren, Int. J. Quantum Chem. **S12**, 33 (1978)
16. D. Jena, D. Datta, D. Mukherjee, Chem. Phys. **329**, 290 (2006)
17. C. Bloch, Nucl. Phys. **6**, 329 (1958)
18. C. Bloch, Nucl. Phys. **7**, 451 (1958)
19. J. Sucher, Phys. Rev. A **22**, 348 (1980)
20. A. L. Fetter, J. D. Walecka, *The Quantum Mechanics of Many-Body Systems* (McGraw-Hill, NY, 1971)
21. M. Gell-Mann, F. Low, Phys. Rev. **84**, 350 (1951)
22. C. Itzykson, J. B. Zuber, *Quantum Field Theory* (McGraw-Hill, NY, 1980)
23. I. Lindgren, B. Åsén, S. Salomonson, A.-M. Mårtensson-Pendrill, Phys. Rev. A **64**, 062505 (2001)
24. E. E. Salpeter, H. A. Bethe, Phys. Rev. **84**, 1232 (1951)

Relativistic Variational Calculations for Complex Atoms

Charlotte Froese Fischer

Abstract Variational methods can determine a wide range of atomic properties for bound states of simple as well as complex atomic systems. Even for relatively light atoms, relativistic effects may be important. In this chapter we review systematic, large-scale variational procedures that include relativistic effects through either the Breit–Pauli Hamiltonian or the Dirac–Coulomb–Breit Hamiltonian but where correlation is the main source of uncertainty. Correlation is included in a series of calculations of increasing size for which results can be monitored and accuracy estimated. Examples are presented and further developments mentioned.

Keywords: Complex atoms · Correlation · Multiconfiguration Hartree-Fock · Relativistic effects · Variational methods

1 Introduction

Large amounts of atomic data (wavelengths, transition probabilities, hyperfine structure constants, isotope shifts) are needed, for example, in astrophysical applications, fusion diagnostics, and plasma modeling. In astrophysics, iron and it ions are observed in a variety of astronomical objects. Because of the complexity of iron spectra, many lines are observed for which data are needed. The investigation of atmospheres of chemically peculiar stars has led to a growing need of data for heavy elements including the lanthanides [1] where spectra contain thousands of lines and only a few have been identified. In the lighting industry, cerium is now being used as an additive in a number of high-intensity discharge lamps in order to obtain dramatic improvements in luminous efficacy while still maintaining good color-rendering. The excellent performance of Ce is attributable to the approximately 20,000 emission lines across its visible spectrum. Optimization by computer modeling of lamps containing Ce is desirable, but very little data exist. The large

C.F. Fischer (✉)
Vanderbilt University, Nashville, TN 37235; National Institute of Standards and Technology, Gaithersburg, MD 20899-8422, USA, e-mail: charlotte.fischer@nist.gov

P. Piecuch et al. (eds.), *Advances in the Theory of Atomic and Molecular Systems,*
Progress in Theoretical Chemistry and Physics 19, DOI 10.1007/978-90-481-2596-8_7,
© Springer Science+Business Media B.V. 2009

number of lines effectively precludes experimental determination of a sufficient number of transition probabilities leaving calculation as the only realistic option. Generally when large amounts of data are required the needed accuracy is reduced. At the other extreme, the accurate prediction of a single property that requires all available computer resources may also be complex. An example is the calculation of the magnetic dipole hyperfine constant for the $5d^{10}6s$ $^2S_{1/2}$ ground state of atomic gold where calculations need to include relativistic effects and triple excitations have been shown to be important [2].

Variational methods that optimize the orbital basis for wave function expansions are the methods of choice for complex atoms. Two quite general numerical methods have been implemented and used extensively. In the first, a radial orbital basis that depends only on nl quantum numbers, is determined by the application of the variational procedure to an energy expression derived using the nonrelativistic Hamiltonian, followed by a configuration interaction (CI) calculation using a Breit–Pauli Hamiltonian that includes relativistic corrections to the energy of order α^2, where α is the fine-structure constant. This procedure is appropriate for light atoms where correlation is more important than relativistic effects and the size of the nucleus can be ignored. In the second, a radial orbital basis that depends on nlj quantum numbers is determined using the Dirac–Coulomb (DC) Hamiltonian, followed by a CI calculation in which the Hamiltonian may include other corrections of which the Breit and sometimes the self-energy corrections are the most important. These broad approaches have been implemented in the Atomic Structure Package ATSP2K [3] and the General Relativistic Atomic Structure Package GRASP2K [4]. Both packages have some limitations. In the former, the configurations associated with radial orbitals may include an arbitrary number of f electrons but the occupation of subshells with $l > 3$ is restricted to at most two. In GRASP2K, subshells with $j > 7/2$ are restricted to at most two electrons. Thus there may be an arbitrary number of g_- ($j = 7/2$) electrons but at most two g_+ ($j = 9/2$) electrons. The largest atom for which this code can determine the Dirac–Hartree–Fock (DHF) energy has the atomic number $Z = 122$, assuming the ground state is $5g^26d^{10}7p^68s^2$ [5]. Both these codes rely on very general angular momentum methods for the evaluation of matrix elements for a variety of operators. The MCDFGME code implemented by Indelicato and Desclaux [6] relies on expansions in terms of determinants and consequently does not suffer from the above angular limitation. Furthermore, the Breit correction can be included at the variational stage rather than as a perturbative correction. DHF results (referred to as DF in their paper) that include Breit and quantum electrodynamic (QED) corrections have been published for isoelectronic sequences with 2–105 electrons and nuclear charges up to $Z = 118$ [7]. Studies of QED and relativistic corrections in superheavy elements have also been reported [8].

The present chapter reviews the factors that need to be considered when variational procedures are developed for complex atoms or ions where missing correlation is the main source of uncertainty in the accuracy of a calculation. The goal is to have a procedure for bound state calculations that can be applied to any element of the periodic table. Systematic methods for assessing accuracy and controlling the size of a calculation will be discussed.

2 Configuration States and the Hartree–Fock Approximation

In non-relativistic theory, a configuration state function (CSF) $\Phi(\gamma L S \pi)$ is an anti-symmetric N-electron function that is an eigenfunction of the total angular momentum operators L^2, L_z, S^2 and S_z, where L and S represent the coupling of angular l_i and spin s_i momenta, respectively, for $i = 1, \ldots N$ and π represents the parity of the state. In this notation, γ specifies the configuration expressed in terms of subshells and their occupation, $(n_j l_j)^{q_j}$ where $\sum q_j = N$, and any additional quantum numbers needed to uniquely define the CSF. For some partially filled d-subshells a seniority number is sufficient but a quasispin quantum number has been shown to have theoretical advantages [9]. For partially filled f-subshells, both schemes require additional quantum numbers but associated tables are much smaller for the quasispin representation [10]. The ATSP2K package uses the latter. When more than one open shell is present, additional quantum numbers are also needed to represent the coupling of the open shells as in $2s2p(^3P)3s\ ^2P$. Since the parity of a CSF is related to γ, the parity quantum number is often omitted. The energy of the CSF is the matrix element $\mathcal{E}(\gamma L S) = \langle \Phi(\gamma L S) | \mathcal{H} | \Phi(\gamma L S) \rangle$. Upon integration over all angular and spin variables, the result is an expression in terms of radial integrals (see [11] for examples). The angular momentum theory used for evaluating matrix elements assumes a single orthonormal basis for which the radial orbitals, $P(nl; r)$, depend only on nl quantum numbers. These functions are solutions of the Hartree–Fock (HF) equations that minimize the energy defined by the energy expression subject to normalization and orthogonality constraints between orbitals with the same orbital angular momentum l.

In the Breit–Pauli approximation, CSFs are expressed in intermediate coupling as $\Phi(\gamma L S J)$ where J is the quantum number for the coupling of the total angular and spin momenta. The Breit–Pauli energy of the CSF is $\langle \Phi(\gamma L S J) | \mathcal{H}_{BP} | \Phi(\gamma L S J) \rangle$, where $\mathcal{H}_{BP}$ is the Breit–Pauli Hamiltonian [11]. The latter is not diagonal in LS, and a Breit–Pauli calculation for a configuration normally includes an expansion over all LS terms of the configuration, a topic that will be discussed in the next section.

In a DHF approximation, the orbital and spin angular momenta of the individual electrons are coupled to yield momenta j_i which are then coupled to form the total J, a coupling referred to as jj-coupling. Configuration states defined in terms of LSJ coupling may be transformed to one or more CSFs in jj-coupling. In atomic structure calculations it is useful to preserve this progression from LS (non-relativistic), to LSJ (Breit–Pauli), to J (DC or Dirac–Coulomb–Breit), where the HF approximation is a single CSF. For this reason, it is customary to describe non-relativistic and fully relativistic calculations in non-relativistic terminology with codes transforming CSFs to the jj-coupling. Table 1 shows this progression for the $2p^2\ ^3P_0$ ground state of carbon. Note the near degeneracy of the two CSFs in jj-coupling.

It generally is believed that relativistic effects are small for light atoms but there are exceptions. The HF energies of $p^5\ell\ ^3L$ and ^1L are degenerate when $L = \ell$. These terms only exist for $\ell > 0$. Consequently, there is a spin-orbit interaction between these two CSFs when $J = \ell$ which results in a wave function with highly

Table 1 Energy in Hartrees for $2p^2\ {}^3P_0$ ground state of carbon for different Hamiltonians

Hamiltonian	Energy	CSF expansion	
Non-relativistic	−37.688619	$1.0000\ 2p^2\ {}^3P$	
+ Rel. Shift	−37.7022911	$1.0000\ 2p^2\ {}^3P$	
Breit–Pauli	−37.702329	$1.0000\ 2p^2\ {}^3P_0$	$0.00144\ 2p^2\ {}^1S_0$
Dirac–Coulomb	−37.7051309	$0.8175\ 2p^2_{1/2}\ (J = 0)$	$-0.5760\ 2p^2_{3/2}\ (J = 0)$
+ Breit	−37.702395	$0.8173\ 2p^2_{1/2}\ (J = 0)$	$-0.5762\ 2p^2_{3/2}\ (J = 0)$

mixed singlet/triplet composition. The lightest element for which this can occur is neon ($Z = 10$).

3 Wave Function Expansions and Correlation

In CI methods, an atomic state wave function (ASF) labelled $\Psi(\gamma LSJ\pi)$ is defined as a linear combination of CSFs,

$$\Psi(\gamma LSJ\pi) = \sum_i c_i \Phi(\gamma_i L_i S_i J\pi). \tag{1}$$

In this notation, γLSJ is now a *label*, usually representing the dominant component of the wave function: only J and π are strict quantum numbers.

Associated with this expansion is a set of radial orbitals as well as expansion coefficients. In a fully variational calculation the energy must be stationary with respect to the variation of all free variables. The radial orbitals together with the expansion coefficients are determined through an iterative process until results are self-consistent and an energy is stationary. Many variations are available.

1. Some radial orbitals may already have been determined and are to remain unchanged. When the theoretical model of a computational process assumes a common core, these orbitals can be determined first and kept fixed in subsequent calculations. This procedure was used in transition probability studies for many lines of a spectrum where energy differences are more important than a minimum total energy [12, 13]. Such calculations are referred to as "spectrum" calculations and ensure a balance in the energies from different variational calculations. Relaxation effects can be treated as a correlation correction.

2. The optimization is applied not to the energy expression for a single ASF but to a weighted average of energy expressions. In the GRASP2K program this is referred to as an extended optimized level (EOL) calculation. When the radial orbitals for a multiplet do not depend significantly on the total J or parity, it is convenient to optimize them for all levels at once. The variation of the DHF orbitals with respect to J can be treated as a correlation correction in subsequent calculations. For example, in Ce the $4f5d^2 6s$ configuration has associated with it 162 CSFs with $J = 0 - 8$. The orbital dependence on the energy level for the low-lying levels is small and it is convenient to assume the orbitals are the

same for all J and optimize on all levels in a given energy region simultaneously. In Breit–Pauli calculations, the ATSP2K code requires the use of the same orthonormal radial orbitals for the expansion over LS terms. This limitation can be satisfied by optimizing the orbitals for the non-relativistic multiconfiguration HF (MCHF) expansions simultaneously for the different LS terms, a process referred to as "simultaneous" optimization.

The expansion coefficients are eigenvectors of the interaction matrix. Sparse matrix methods are used since, as the size of the expansion increases, more and more matrix elements are zero. An implementation of the Davidson method [14] is used for large cases. Since it is based on the multiplication of the interaction matrix by a vector, the method can readily be parallelized [15].

Although variational methods are quite general and any CSF with the required quantum numbers may be included in an expansion, it is extremely helpful to follow a computational model for the inclusion of correlation in an efficient manner such as single (S) and double (D) substitutions from a multireference set of CSFs. These "substitutions" are often referred to as "excitations" but when the state under consideration is an excited state, some substitutions may be to CSFs with a lower energy and the term is misleading. An example would be a calculation for the $3p^3\ 2P^o$ ASF of aluminum. The substitution $3p^2 \rightarrow 3s^2$ is a de-excitation and ensures that the excited state is orthogonal to a lower state of the same symmetry. For some properties, such as hyperfine constants [2], it may be necessary to include selected triple (T) and quadruple (Q) substitutions. Such expansions increase in size rapidly with the size of the orbital set.

In order to control the size of expansions, the subshells in the atomic system may be classified as either inactive, core (C), or valence (V) subshells. Substitutions from the inactive core are omitted. The SD substitutions from the remaining active electrons represent three different types of correlation: VV, CV, and CC correlation according to the type of orbitals that enter the double substitution process. All substitutions are from occupied subshells to unfilled subshells and not only to unoccupied subshells. Depending on the atomic property, not all these types of correlation are important. For example, with a neon-like core, CC correlation is negligible, after a few stages of ionization, for transition energies between low-lying levels of the Al-like sequence that involve only outer electrons. The contribution to the energy from CV correlation also decreases with degree of ionization but may be important for transition matrix elements, particularly for transitions involving excited states with open shells. VV correlation always needs to be considered since it accounts for near degeneracy effects. In variational methods CC correlation affects the potential for the valence electrons and is most important for neutral atoms [16]. Though wave function expansions can be defined by specifying the multireference set and the SD(TQ) substitutions, a simple but concise configuration notation can sometimes be used. Thus $1s^2 2s^2 2p^6 \{3, 4\}\{3, 4, 5, 6, 7\}^2$ represents the set of CSFs with an inactive Ne-like core, one nl orbital with either $n = 3$ or $n = 4$ and two $nln'l'$ orbitals for which $3 \leq n, n' \leq 7$. All possible CSFs are included with the given parity and total angular momentum. An expansion may be over a union of multiple sets. The addition of $1s^2 2s^2 2p^5 \{3\}^2 \{3, 4\}^2$ to the above expansion would include a limited amount of CV correlation. Such a notation is convenient when not all TQ substitutions are included.

Usually a core consists of filled shells or at least filled subshells but for Ce and possibly other lanthanides, it is convenient to include the open f-shell as part of the core. Figure 1 shows the mean radii of orbitals for the $[\mathrm{Kr}]4d^{10}5s^2 5p^6 4f 5d^2 6s$ configuration with $J = 0 - 8$ from a DHF calculation. This notation for the configuration follows the convention of listing the subshells in decreasing order of electron energy. But correlation between subshells is largest when the two shells overlap in a region of space and their orbitals have similar mean radii and it is helpful to list the orbitals in order of increasing mean radii. Though the $4f$ orbitals have an energy less than that of $5s$ and $5p$, the mean radius of the $4f$ orbital is between that of the filled $4d$ and $5s, 5p$ subshells. Thus the open f-subshell is in the middle of filled subshells considerably removed from the outer $5d^2 6s$ electrons. Atomic properties such as the $6s - 6p$ excitation energy and transition probability will not be significantly affected by this correlation inside the atom, since the energy of all states with the same core will be affected to a similar degree and the contribution to the energy will roughly cancel in the difference. In Ce, the assumption that $[\mathrm{Kr}]4d^{10}4f 5s^2 5p^6$ $^2F_J^o$ is an inactive core and that the valence electrons are $5d^2 6s$, leads to reasonable results, but needs to be tested further. The fact that the core has quantum numbers of $J = 5/2$ and $7/2$ with a small splitting means that there are many closely spaced levels for the atom when the core is coupled to the outer three electrons for a final total quantum number J.

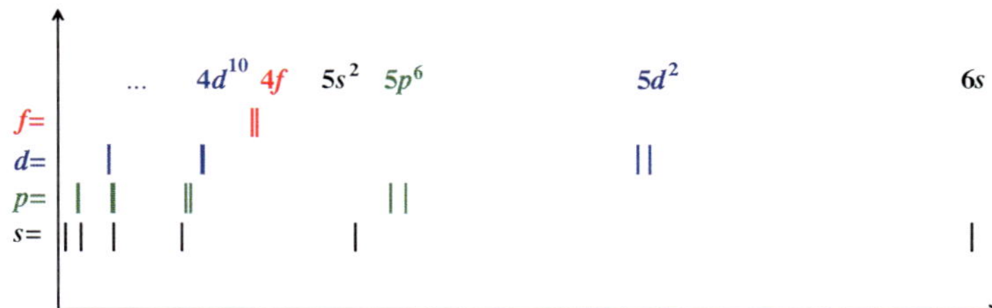

Fig. 1 Mean radii (arbitrary units) of orbitals in the $[\mathrm{Kr}]4d^{10}5s^2 5p^6 4f 5d^2 6s$ configuration of Ce

4 Systematic Methods

Orbital optimization in variational methods is achieved through an iterative process that requires initial estimates. Many difficulties can be avoided by adopting a systematic procedure where, instead of one large calculation with many orthogonality constraints, a series of smaller calculations are performed with larger and larger orbital sets and the results from one are part of the input for the next. In atomic calculations, it is convenient to classify a calculation by the maximum principal quantum number of the orbital set or a closely related quantity. Associated with this process is the notion of convergence which can be used to monitor the calculation. At the same time, other checks related to the property of interest can be applied as well.

The concepts are illustrated clearly in Table 2 for a simple calculation of the line strength (S) and the weighted oscillator strength (gf) for the $2s^2\ {}^1S_0 - 2s3p\ {}^1P_1^o$ transition in N^{3+} [17]. All wave function expansions were obtained by substitutions from the reference CSF – $2s^2\ {}^1S_0$ for the ground state and $2s3p\ {}^1P_1^o$ for the excited state. Orbitals were reoptimized for each correlation study. For a transition probability calculation there are two tests of accuracy. The first is the observed transition energy which is the difference of the total energies of the two states (expressed in units of cm^{-1} in Table 2). The second is the agreement in the oscillator strength calculated in the length and velocity form. For an exact wave function the length value gf_ℓ should be the same as the velocity value gf_v. Table 2 shows that VV correlation is not significantly affecting the transition energy and that length and velocity gf-values are not converging to the same value. Thus VV correlation (SD substitutions) is inadequate for this problem. The inclusion of CV correlation (SDT substitutions) greatly improves the agreement in the two forms of the oscillator strength. The notation "$8i$," for example, indicates that orbital angular quantum numbers were restricted to i-orbitals, or $l \leq 6$. The agreement in length and velocity forms of the oscillator strength is now much better but the transition energy is not in close agreement with observation [18]. The full correlation calculation included a complete active space (CAS) (SDTQ substitutions) expansion for orbitals with $n \leq 5$ and SD expansions for larger n. In order to compare with observed transition energies, it is necessary to include both relativistic and finite mass corrections. The

Table 2 MCHF results from systematic calculations of transition data for the $2s^2\ {}^1S_0 - 2s3p\ {}^1P_1^o$ transition in N^{3+}

n	$\Delta E(cm^{-1})$	S_ℓ	gf_ℓ	gf_v
Valence correlation (VV)				
3	60219	0.105682	0.019331	0.021702
4	60151	0.076661	0.014007	0.016349
5	60141	0.071330	0.013031	0.015418
6	60151	0.069579	0.012711	0.015245
Core-valence (CV) and Valence correlation (VV)				
7	60241	0.051224	0.009373	0.009483
8i	60237	0.050524	0.009245	0.009175
9i	60236	0.050196	0.009185	0.009141
10i	60235	0.050097	0.009166	0.009143
Full correlation				
7	60200.5	0.052610	0.009620	0.009480
8i	60181.3	0.050634	0.009256	0.009093
9i	60181.2	0.050029	0.009145	0.008974
10i	60181.6	0.049286	0.009009	0.008883
$\vdots$				
Extrap. (rm)	60189.8	0.048671	0.008878	0.008821
Obs. [18]	60187.34			
Estimated gf			0.00885(5)	

latter is important since nitrogen is a light atom. From these results extrapolated values of S and gf were obtained along with an estimated value of gf and its uncertainty. The extrapolated energy is in good agreement with observation.

Computational procedures such as those illustrated in Table 2 are easy to apply to simple atomic systems but need to be modified for complex calculations when expansions "explode" as the orbital set increases. The current GRASP2K code encounters convergence problems when two correlation orbitals of the same symmetry are varied simultaneously. A practical and stable procedure in this case is to introduce only one new orbital of each symmetry and optimize only the new orbitals. These orbitals are sometimes referred to as a "layer" [2]. In this way, a basis of relativistic orbitals is built for a final calculation that includes other effects such as Breit and QED corrections. Such a procedure was used in the determination of transitions probabilities for Fe^{3+} [19].

The ground configuration of Fe^{3+} is $3d^5$ with 37 levels ($1/2 \leq J \leq 13/2$) of which the lowest level is $^6S_{5/2}$. Earlier Breit–Pauli calculations had established extensive admixtures of different LSJ terms in the wave function for some atomic state functions indicating important relativistic effects [20]. To confirm these results, MCDHF calculations were undertaken with wavefunction expansions as shown in Table 3. The expansions at the $n = 3$ stage take into account the near degeneracy effects within the $3s, 3p, 3d$ orbitals but only the $3s \rightarrow 3d$ substitutions from $3s$ were allowed. All orbitals were optimized, including the core orbitals. To these expansions were added new CSFs obtained from SD substitutions with $3s^2$ inactive, a $3p^6$ core, and $3d^5$ as the valence electrons to the $n = 4$ orbital set $\{4s, 4p, 4d, 4f\}$, restricted to CV and VV correlation. Note that the size of the expansion (for all J) increased by a factor of about 100. The same substitution procedure was used for the $n = 5$ orbital set but the expansions for $n = 6$ orbitals were restricted to valence correlation only. At each stage, the new CSFs were added to the expansions from the smaller orbital sets. In this way, once the orbital sets were at their largest, the new CSFs were targeted on the most important correlation.

Table 3 Expansions for the different layers of a calculation for Fe^{3+}

n	Inactive	Core	Valence	SD correl.	Size
3	[Ne]	$3s^2$	$3p^6 3d^5$	CV+VV	378
4	[Mg]	$3p^6$	$3d^5$	CV+VV	37 486
5	[Mg]	$3p^6$	$3d^5$	CV+VV	213 037
6g	[Mg]	$3p^6$	$3d^5$	+ VV	243 370

5 Variational Results

The total energy that is the basis for a variational calculation is rarely of interest. Instead, what is of interest is the energy difference between two or more states. When this difference is small, it is important to maintain a balance, particularly in core correlation that typically is much larger than the energy difference.

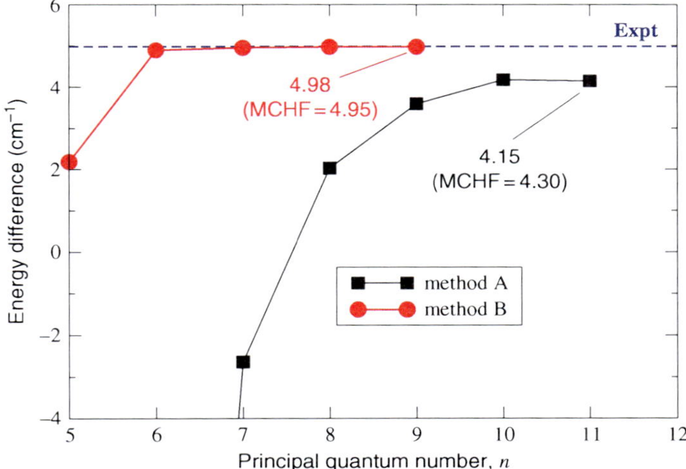

Fig. 2 Convergence of $4f - 4d$ energy difference for independent optimization (Method A) and for limited core-correlation from simultaneous optimization followed by independent optimization for core-valence correlation (method B)

For example, in lithium [21], the experimental $4d - 4f$ transition energy of (4.988 ± 0.003) cm^{-1} [22] agrees well with the all-order perturbation theory value of $\Delta E(SD) = 4.99$ cm^{-1}, $\Delta E(SDT) = 4.92$ cm^{-1}, and also a balanced MCHF calculation for which $\Delta E(MCHF) = 4.98$ cm^{-1}. Figure 2 shows the convergence of two variational procedures using the MCHF method for orbitals and energy and the non-relativistic energy corrected using the Breit-Pauli Hamiltonian. In Method A, the two states were optimized separately with triple substitutions up to $n = 8$ and some restrictions thereafter [21]. Because there is less CV correlation for the $4f$ state than the $4d$ state, the variational process produced orbitals that placed a greater emphasis on CC correlation for $4f$. Initial energy differences were negative and only for higher n did the difference become positive. The high n limit may have been affected by the expansion restrictions. In this example, CC correlation affects the potential for the outer electrons and should be the same for both states. On the other hand, Method B first determined the $n = 3$ natural orbitals for the $1s^2$ ^1S core and, with these orbitals fixed, obtained additional orbitals from expansions of the form $\{1, 2, 3\}\{1, 2, 3, \ldots, n\}^2$. A common set of core orbitals also ensured that the one-body relativistic effects from the core are the same (except for relaxation effects) for the two states. The final values, with and without relativistic effects, are shown in the figure: (4.15, 4.30) cm^{-1} for Method A and (4.98, 4.95) cm^{-1} for Method B. Note the smaller relativistic correction in method B.

The "spectrum" calculations for iso-electronic sequences [12, 13] are an example where only energy differences are important since these determine the wavelengths in observed spectra. It is often more difficult to determine the energy of an excited state relative to the ground state than to another excited state. For example, it was easy to determine the $4d - 4f$ transition energy in lithium with method B since both $4d$ and $4f$ are well outside the core, but a calculation of the energy of $4f$ relative to the $2s$ ground state is expected to be more difficult.

A calculation for a heavy atom requires a proper treatment of both relativistic and correlation effects. Few such calculations have been performed for atoms with an atomic number $Z > 100$. As a test of the capability of GRASP2K, calculations were undertaken for Lr ($Z = 103$) to determine the $[\mathrm{Rn}]5f^{14}7s^26d\ ^2D_{3/2} - [\mathrm{Rn}]5f^{14}7s^27p\ ^2P^o_{1/2}$ energy separation [24]. Coupled cluster calculations had been performed by Eliav et al. [23] using an analytic G-spinor basis of the same size as for the homologous Lu ($Z = 71$) atom. Table 4 shows how energy differences change as CV and CC substitutions from more and more shells of the atom are included in the expansion. For each correlation model, orbitals up to $n = 8$ (for Lu) and $n = 9$ for Lr were included in a systematic way. g-Orbitals were found to be important although h-orbitals had a negligible effect and were omitted. The mean radii of the occupied orbitals are such that $\langle 4d \rangle < \langle 4f \rangle < \langle 5s \rangle$ and so, in retrospect, it is not surprising that CV with an inner f subshell in addition to valence correlation among the outer three electrons is not a stable energy difference, one that will not change dramatically when more correlation is added. Only when all SD substitutions from the $5s^25p^64f^{14}6s^2nl$ shells were included (Model III for Lu) is the difference stable, improving somewhat when $4d^{10}$ is also included and the orbital set expanded to $n = 8$. The DHF Breit and QED corrections were found to be small. The result from Model III) for Lu agrees closely with the relativistic coupled cluster separation with the variational value form Model IV) in slightly better agreement with experiment and with the $6s^25d\ ^2D_{3/2}$ being the lower in energy. Similar calculations were performed for Lr but now $7s^27p$ has the lower energy. The agreement with the relativistic coupled-cluster result is not as good as in Lu, possibly because the same number of analytic basis functions were used for both atoms. However, both calculations clearly identify $7s^27p\ ^2P^o_{1/2}$ as the lower state, in contradiction with

Table 4 Expansions and transition energies (ΔE in cm^{-1}) for the different computational models for $6s^25d\ ^2D_{3/2} - 6s^26p\ ^2P^o_{1/2}$ in Lu ($Z = 71$) and $7s^26d\ ^2D_{3/2} - 7s^27p\ ^2P^o_{1/2}$ in Lr ($Z = 103$)

Model	Inactive	Core	SD correl.	Size ($^2D_{3/2}$ / $^2P^o_{1/2}$)	ΔE
Lutetium					
I)	[Xe]	$4f^{14}$	CV+VV	4354 / 2071	3989
II)	[Cd]	$5p^64f^{14}$	CV+VV	5600 / 2764	8004
III)	[Kr]$4d^{10}$	$5s^25p^64f^{14}$	CC+CV+VV	128 763 / 36 974	3857
IV)	[Kr]	$4d^{10}5s^25p^64f^{14}$	CC+CV+VV	305 717 / 87 251	4186
RCC [23]					3828
Obs.					4136
Lawrencium					
I)	[Xe]	$5f^{14}$	CV+VV	3659 / 1842	−1298
II)	[Cd]	$6p^65f^{14}$	CV+VV	4708 / 2495	1339
III)	[Kr]$5d^{10}$	$6s^26p^65f^{14}$	CC+CV+VV	125 325 / 37 333	−1953
IV)	[Kr]	$5d^{10}6s^26p^65f^{14}$	CC+CV+VV	330 252 / 95 969	−1127
RCC [23]					−1388

the designation in many websites for periodic tables [25] but confirming the ground state listed at the National Institute of Standards and Technology [26] website.

Similar studies have been performed by Fritzsche [27] for nobelium ($Z = 102$) and the homologous ytterbium ($Z = 70$) with emphasis on the rapidly increasing QED effects with increasing Z. He found that the self-energy correction (part of QED) was determined more reliably by the methods implemented in RELCI [28] than in GRASP.

6 Further Developments

All the methods mentioned here relied upon an orthonormal orbital basis. This requirement poses problems, for example, for the calculation of the dipole hyperfine constant A of the $5d^{10}6s$ $^1S_0$ ground state in gold. Correlation from all shells of the atom contribute to the hyperfine constant and eight s, p, d, f, g and three h correlation orbitals were used in addition to the occupied orbitals in a recent fully relativistic calculation [2]. With n orbitals of a given symmetry there are $n(n - 1)/2$ orthogonality constraints, thus they increase rapidly with n and the GRASP2K numerical grid needed to be considerably refined in order to retain accuracy in multiconfiguration DHF theory. Figure 3 shows the slow convergence as a function of the size of the expansion not to mention the large expansions that were needed. Both convergence and computational efficiency could be improved through the use of separate non-orthogonal orbitals for corrections to the different shells of the atom. These could then be combined in a single non-orthogonal CI problem leading to a generalized eigenvalue problem. The matrix can be generated using determinants as in the non-orthogonal Breit–Pauli CI program [31].

These ideas can be described simply for the $1s^2 2s^2$ $^1S_0$ ground state of a four-electron system. There is a strong interaction with $1s^2 2p^2$ $^1S_0$ and radial orbitals

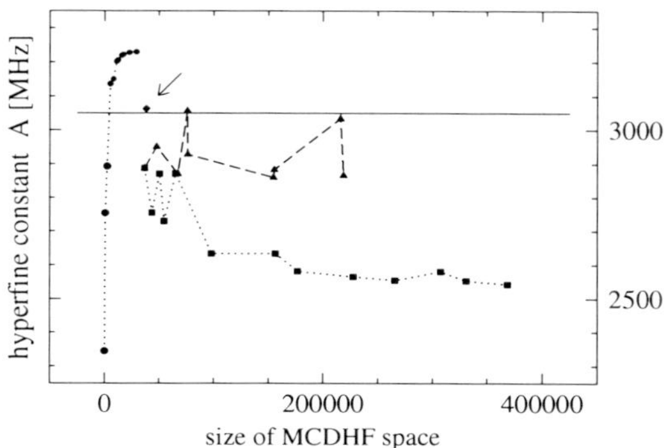

Fig. 3 Magnetic dipole hyperfine constant A for the $^2S_{1/2}$ ground state of gold as a function of the size of the MCDHF expansion: *circles* – CV, *squares* – SD, *triangles* – SDT expansions. The arrow refers to a value from Ref. [29] and the horizontal line indicates the experimental value [30]

can be obtained that minimize the energy of the $n = 2$ CAS expansion. Let this wave function be Ψ_0. First-order corrections consist of single (S) and double (D) substitutions. Of these there are seven types of D substitutions:

$$1s^2(\,^1S) \bullet nln'l(\,^1S)\ 1s2s(\,^{1,3}S) \bullet nln'l(\,^{1,3}S)\ 1s2p(\,^{1,3}P) \bullet nln'l'(\,^{1,3}P)$$
$$2s^2(\,^1S) \bullet nln'l'(\,^1S)\ 2p^2(\,^1S) \bullet nln'l'(\,^1S)$$

with $n, n' \geq 3$. In Ref. [32, 33] these are referred to as "symbolic" states since neither n, n' nor l, l' are specified. For each type, the linear combination over n, n' defines a partial wave consisting of a "subconfiguration" CSF, $\gamma_c L_c S_c \pi_c$ and a coupled two-electron CSF with quantum numbers L_v, S_v, π_v, coupled to form $^1S_0$ for which the parity π is even where $\bullet$ represents a coupling of the two components. Radial orbitals for the expansion

$$\Psi = \Psi_0 + \Psi(\gamma_c L_c S_c \pi_c, L_v S_v, \pi_v; LS\pi) \tag{2}$$

can readily be evaluated since the interaction matrix elements between the first-order components are simply two-electron Coulomb matrix elements for which well-known formulas are available [32] and expressions for interactions with Ψ_0 are known [32, 33]. Variational methods have the best convergence when a solution is unique. For sums over $nsn's\ ^1S$ it is customary to use the natural orbital expansion. Similar rapidly convergent expansions exist for other two-electron symmetry adapted partial waves but may require more than one orthonormal set of radial basis functions for a given orbital symmetry [34, 11]. For example,

$$\Psi(1s2p\ ^1P^o) = a_1|1s2p_1\ ^1P^o\rangle + a_2|2s3p_1\ ^1P^o\rangle + \ldots \tag{3}$$
$$b_1|2p_23d_1\ ^1P^o\rangle + b_2|3p_24d_1\ ^1P^o\rangle + \ldots \tag{4}$$
$$\ldots \tag{5}$$

where the orbital subscript is a set indicator and all orbitals within a set are orthonormal. These ideas are being investigated.

Other limitations are the use of finite difference methods for solving the integro-differential equations of eigenvalue type as two-point boundary-value problems and the self-consistent field procedure that improves one orbital at a time. This is particularly slow when orbitals of the same symmetry, constrained through orthogonality, need to rotate in orbital space. B-spline methods replace the differential equations by generalized eigenvalue problems and orbitals can easily be improved simultaneously [35]. For the Dirac equation, stable procedures have been determined without the introduction of spurious states [36]. The effect of a finite nucleus can be included simply through a modification of the grid. By taking advantage of the fact that the set of eigenvectors of a generalized eigenvalue problem define an effectively complete orbital basis, the variational method could be supplemented by many-body perturbation theory for capturing the remaining small effects. It is not clear that non-orthogonal orbitals are needed. The study of angular data that is needed for SD expansions [32, 33] suggests that relatively small amounts of angular data would be needed when expansions are expressed as subcomponents and coupled pairs of electrons coupled to the final LS (or J) and symmetry. The data structure of the interaction matrix is shown in Fig. 4 for a VV and CV calculation for beryllium. The angular data needed for generating this matrix is independent of the number of orbitals used for SD expansions. Further studies are needed.

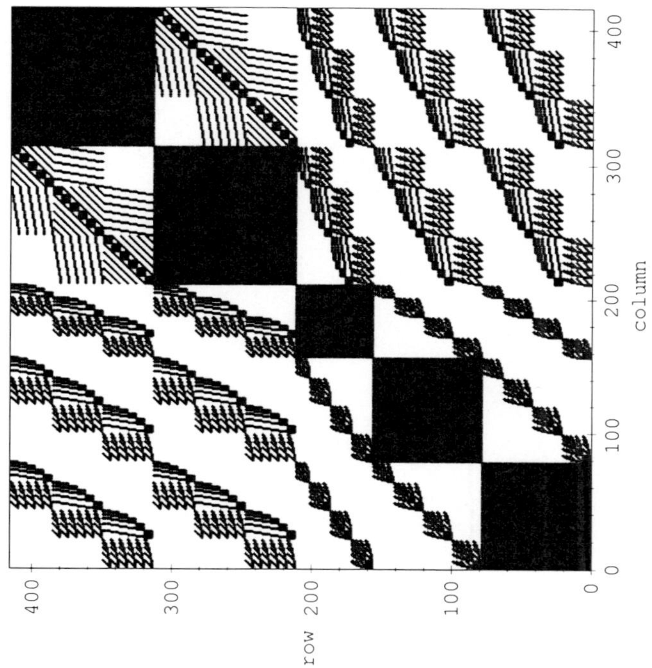

Fig. 4 Structure of the configuration interaction matrix for an SD expansion for the ground state of Be including VV and CV correlation. White regions represent regions where matrix elements are zero

Acknowledgement Stimulating discussions and collaborations with J. Bieroń, D. Ellis, G. Gaigalas, M. R. Godefroid, P. Jönsson, Y. Ralchenko, and O. Zatsarinny are gratefully acknowledged.

References

1. A. Jorissen, Phys. Scr. **T112**, 73–80 (2004)
2. J. Bieroń, C. Froese Fischer, P. Jönsson, P. Pyykkö, J. Phys. B: At. Mol. Phys. **41**, 115002 (2008)
3. C. Froese Fischer, G. Tachiev, G. Gaigalas, M. R. Godefroid, Comput. Phys. Commun. **176**, 559–579 (2007)
4. P. Jönsson, X. He, C. Froese Fischer, I. P. Grant, Comput. Phys. Commun. **176**, 597–692 (2007)

5. *Wikipedia Extended Periodic Table*, en.wikipedia.org/wiki/Periodic_table_(extended)
6. P. Indelicato, J.-P. Desclaux, MCDFGME: a MultiConfiguration Dirac Fock and General Matrix Elements Program (released 2005), http://dirac.spectro.jussieu.fr/mcdf
7. G. C. Rodrígues, P. Indelicato, J. P. Santos, P. Patté, F. Parente, At. Data Nucl. Data Tables **86**, 117–233 (2004)
8. P. Indelicato, J. P. Santos, S. Boucard, J.-P. Desclaux, Eur. Phys. J. D **45**, 155–170 (2007)
9. Z. Rudzikas, *Theoretical Atomic Spectroscopy* (Cambridge Univ. Press, Cambridge, 1997)
10. G. Gaigalas, Z. Rudzikas, C. Froese Fischer, At. Data Nuc. Data Tables **70**, 1–30 (1998)
11. C. Froese Fischer, T. Brage, P. Jönsson, *Computational Atomic Structure* (Institute of Physics Pub., Bristol, 1997)
12. C. Froese Fischer, G. Tachiev, At. Data Nucl. Data Tables **87**, 1–184 (2004)
13. C. Froese Fischer, G. Tachiev, A. Irimia, At. Data Nucl. Data Tables **92**, 607–812 (2006)
14. A. Stathopoulos, C. Froese Fischer, Comput. Phys. Commun. **79**, 268–290 (1994)
15. A. Stathopoulos, A. B. Ynnerman, C. Froese Fischer, Int'l J. Supercomputer Appl. and High Performance Computing **10**, 41–61 (1996)
16. T. Brage, C. Froese Fischer, Phys. Scr. **T47**, 18–28 (1993)
17. C. Froese Fischer, M. Godefroid, J. Olsen, J. Phys. B: Atom. Mol. Opt. Phys. **30**, 1163–1172 (1997)
18. Yu. Ralchenko, A. E. Kramida, J. Reader, NIST ASD Team (2008). *NIST Atomic Spectra Database* (version 3.1.5), [Online]. Available: http://physics.nist.gov/asd3 [2008, June 20]. National Institute of Standards and Technology, Gaithersburg, MD
19. C. Froese Fischer, R. H. Rubin, M. Rodríguez, Mon. Not. Royal Astron. Soc. **391**, 1828–1837 (2008)
20. C. Froese Fischer, R. H. Rubin, Mon. Not. Royal Astron. Soc. **355**, 461–474 (2004); Erratum *ibid* 355, 400 (2004)
21. M. S. Safronova, C. Froese Fischer, Y. Ralchenko, Phys. Rev. A **76**, 054502 (2007)
22. L. J. Radziemski, R. Engleman, Jr., J. W. Brault, Phys. Rev. A. **52**, 4462 (1995)
23. E. Eliav, U. Kaldor, Y. Ishikawa, Phys. Rev. A **52**, 291–296 (1995)
24. Y. Zou, C. Froese Fischer, Phys. Rev. Lett. **88**, 183001 (2002)
25. *Periodic Tables*: (http://en.wikipedia.org/wiki/Lawrencium and http://education.jlab.org/itselemental)
26. *Ground Levels and Ionization Energies for Neutral Atoms*, National Institute of Standards and Technoligy (NIST) (http://physics.nist.gov/PhysRefData/IonEnergy/tblNew.html)
27. S. Fritzsche, Eur. Phys. J. D **33**, 15–21 (2005)
28. S. Fritzsche, C. Froese Fischer, G. Gaigalas, Comput. Phys. Commun. 148, 103–123 (2002)
29. S. Q. Song, G. F. Wang, A. P. Ye, G. Jiang, J. Phys. B: At. Mol. Opt. Phys. **40**, 475–484 (2007)
30. H. Dahmen, S. Penselin, Z. Phys. **200**, 456–466 (1967)
31. O. Zatsarinny, C. Froese Fischer, Comput. Phys. Commun. **124**, 247–289 (2000)
32. C. Froese Fischer, D. Ellis, Lithuanian J. Phys. **44**, 121–134 (2004)
33. R. Matulioniene, D. Ellis, C. Froese Fischer, Lithuanian J. Phys. **48**, 35–48 (2008)
34. C. Froese Fischer, J. Comput. Phys. **13**, 502–521 (1973)
35. C. Froese Fischer, in *Advances in Atomic and Molecular Physics*, Vol. 55, ed. by E. Arimondo, P. R. Berman, C.C. Lin, pp. 235–291 (2007)
36. C. Froese Fischer, O. Zatsarinny, Comput. Phys. Commun. **180**, 879–886 (2009)

Part IV
Advances in Wave Function Methods

Linear Scaling Local Correlation Extensions of the Standard and Renormalized Coupled-Cluster Methods

Wei Li, Piotr Piecuch, and Jeffrey R. Gour

Abstract This chapter reviews our recent effort toward extension of the local correlation approach, termed "cluster-in-molecule" and abbreviated as CIM, to the coupled-cluster (CC) theory with singles and doubles (CCSD) and CC methods with singles, doubles, and non-iterative triples, including CCSD(T) and the completely renormalized CR-CC(2,3) approach. The resulting CIM-CCSD, CIM-CCSD(T), and CIM-CR-CC(2,3) algorithms are characterized by the linear scaling of the CPU time with the system size, the use of orthonormal orbitals in the CC subsystem calculations, the explicit and natural parallelism, the high efficiency of the CIM-CC system and CC subsystem calculations, and the purely non-iterative treatment of local triples corrections. By comparing the results of the canonical and CIM CC calculations for normal alkanes and water clusters, it is demonstrated that the CIM-CCSD, CIM-CCSD(T), and CIM-CR-CC(2,3) approaches accurately reproduce the corresponding canonical CC correlation and relative energies, while offering savings in the computer effort by orders of magnitude. The difficulties with achieving similarly high accuracies in local CC calculations that employ basis sets with diffuse functions are addressed by suggesting a new variant of the CIM-CC theory whose performance is illustrated by the CIM-CCSD calculations for water clusters, as described by the basis sets as large as 6-311++G(d, p).

Keywords: Coupled-cluster theory · Local correlation methods · Cluster-in-molecule formalism · Linear scaling algorithms · Single-reference coupled-cluster methods · CCSD approach · CCSD(T) approach · Completely renormalized coupled-cluster approaches · CR-CC(2,3) approach · Large molecular systems · Bond breaking · Normal alkanes · Water clusters

P. Piecuch (✉)
Department of Chemistry and Department of Physics and Astronomy, Michigan State University, East Lansing, Michigan 48824, USA, e-mail: piecuch@chemistry.msu.edu

W. Li
Department of Chemistry, Michigan State University, East Lansing, Michigan 48824, USA
e-mail: weili@msu.edu

J.R. Gour
Department of Chemistry, Michigan State University, East Lansing, Michigan 48824, USA
e-mail: gourjeff@chemistry.msu.edu

P. Piecuch et al. (eds.), *Advances in the Theory of Atomic and Molecular Systems*, Progress in Theoretical Chemistry and Physics 19, DOI 10.1007/978-90-481-2596-8 8,
© Springer Science+Business Media B.V. 2009

1 Introduction

The coupled-cluster (CC) theory, employing the exponential wave function ansatz [1–5], has become the *de facto* standard for high accuracy calculations for atomic and molecular systems (cf., e.g., Refs. [6–12] for selected reviews). The basic single-reference CC method with singly and doubly excited clusters (CCSD), in its spin-orbital [13, 14] and spin-adapted [15, 16] formulations, and the non-iterative CC approaches that account for the effects of connected triply excited clusters [11, 12, 17–26], particularly the widely used CCSD(T) method [18] and its increasingly popular completely renormalized extension termed CR-CC(2,3) [24–26], which improves the CCSD(T) results in the bond breaking and biradical regions of molecular potential energy surfaces, offer an excellent compromise between high accuracy in describing the many-electron correlation effects and relatively low computer costs. In view of these successes, there is a growing need for advancing CC algorithms and methodology, so that one could apply them to the increasingly large and complex molecular problems at the high level of correlation treatment the CC approaches offer. Indeed, the computer costs of the CCSD calculations are defined by the iterative steps that scale as $n_o^2 n_u^4$, where n_o and n_u are the numbers of occupied and unoccupied orbitals, respectively, that are included in the post-Hartree–Fock (post-HF) or other post-self-consistent-field (post-SCF) calculations, or as N^6, where N provides a measure of the molecular size. In the case of the non-iterative corrections due to triply excited clusters defining the CCSD(T) and CR-CC(2,3) methods, one must invoke the additional $n_o^3 n_u^4$ (or N^7) steps. With the computational steps of these types, one can apply the CCSD, CCSD(T), and CR-CC(2,3) approaches to systems with up to about 100 correlated electrons and a few hundred basis functions, but it is extremely difficult to go to larger systems with the CPU steps that scale as N^6 or N^7, in spite of the extraordinary advances in computer architectures in recent years.

One may attempt to extend the applicability of CC methods to somewhat larger molecular problems through code parallelization, and impressive advances have been made in this area (cf., e.g., Refs. [27–37]), but parallelization of CC equations alone is not sufficient to move the area of CC calculations to systems that are much larger than what one can handle today. For example, the intrinsic N^7 scaling of CCSD(T) or CR-CC(2,3) means that one would have to dedicate 128 processors just to double the size of the problems one can handle today with these kinds of methods without increasing the CPU time. Even this is often difficult to accomplish, since it is hard to achieve a perfect scalability through parallelization of CC equations and, eventually, one runs out of the available memory and disk due to the huge amounts of data that characterize the conventional CC calculations. Thus, in addition to code parallelization and relying on computer technology, one must find ways of reducing the intrinsic scaling laws that define the dependence of the costs of CC calculations on the system size without sacrificing the accuracies these methods offer.

The primary source of the N^6–N^7 scaling of the CPU operation count that characterizes the conventional, i.e., canonical CCSD, CCSD(T), and CR-CC(2,3) calculations is the use of the highly delocalized HF molecular orbitals (MOs). It is well established though that electron correlation in nonmetallic systems is a local property of electrons, so that it is natural to expect that by using localized MOs (LMOs) or atomic orbitals (AOs) one should be able to reduce the intrinsic power-scaling laws that characterize the CC and other post-SCF calculations. This chapter summarizes our recent effort aimed at the development, efficient computer implementation, testing, and applications of the linear scaling local correlation CCSD, CCSD(T), and CR-CC(2,3) approaches utilizing LMOs and based on the theoretical framework termed "cluster-in-molecule", abbreviated as CIM [38–40].

The idea of using LMOs in correlated electronic structure calculations is quite old and dates back to the early 1960s [41, 42]. However, the exploitation of LMOs in the actual correlated calculations had to wait until about two decades later. Among the earliest uses of localized orbitals in the context of molecular CCD (CC doubles) [3, 4, 43, 44] and CCSD calculations are the works of Laidig et al. [45–47], Cullen and Zerner [14], Förner et al. [48, 49], and Takahashi and Paldus [50]. In particular, the studies reported by Förner et al. [48, 49] and Takahashi and Paldus [50] demonstrated that a small fraction of all cluster amplitudes is sufficient to recover bulk of the CC correlation energy when one uses local orbital bases. Following these pioneering efforts, significant progress has occurred in the area of reducing the large costs of canonical CC calculations with the help of the suitably chosen LMOs or AOs by directly attacking the intrinsic power-scaling laws that characterize the conventional CC schemes employing delocalized orbitals [38–40, 51–74] (the analogous low-order scaling algorithms developed in the context of other kinds of correlated ab initio calculations, including, for example, the promising work on local methods of the multi-reference configuration interaction (CI) type by Carter et al. [75–78], are not discussed here; for more comprehensive recent reviews of the local CC as well as non-CC work and the relevant references, we refer the reader to the introductions to Refs. [40, 63, 67, 78]). In addition to the intrinsically local CC methods that aim at the description of the many-electron correlation effects by computing the entire system in the LMO or AO bases, there exist a number of the so-called energy-based fragmentation schemes, in which the system is divided into smaller fragments and the final energy is assembled from the energies of fragments in an appropriate manner. The implementations and applications of such schemes at the CC level were described in Refs. [79–81].

Of all local CC methods to date, the approaches that have received the most attention are those of Hampel and Werner [55] and Schütz and Werner [56–60]. These authors reported the fully operational, low-order [55] or linear scaling implementations of the local CCSD [56, 60], CCSD(T) [57, 58], and CCSDT-1 [59] methods. Their implementations of local CC methods exploit the local correlation formalism of Pulay [82] and Pulay and Saebø [83–87], in which one solves the CC equations in a basis of orthonormal occupied LMOs obtained with one of the conventional MO localization schemes [88–90] and non-orthogonal unoccupied orbitals constructed from the projected AOs (PAOs), while dividing the large system of interest into orbital domains to which excitations defining the CC ansatz are restricted.

An alternative local CC approach that utilizes the non-orthogonal AO basis rather than orthonormal occupied LMOs and non-orthogonal unoccupied PAOs and that results, at least in theory, in a linear scaling algorithm was described in Ref. [54]. Yet another type of local CC approach that makes use of the redundant sets of occupied and unoccupied PAOs was considered in Ref. [61], and further development of the idea of exploiting PAOs to represent the occupied and unoccupied orbital spaces was discussed in Ref. [68]. The interesting idea of a dynamically screened local CC method using the minimum basis set of atom-centered orbitals enveloping the occupied space and AOs to represent the unoccupied space was presented in Ref. [67]. A promising new type of local CC approach, which leads to a near linear-scaling algorithm that produces, at the same time, smooth potential energy surfaces and in which one uses orthonormal occupied and unoccupied LMOs and a suitably defined "bump function" that multiplies the standard CC equations to produce a sparse equation system, was recently developed by Head-Gordon and coworkers [62–64].

In spite of the above and other advances, which have certainly contributed to our understanding of the challenges facing the CC calculations for large systems, the use of local CC methods in actual applications remains rather sporadic. To respond to this unsatisfactory situation, we have initiated a new research program aimed at the development and highly efficient implementation of the linear scaling local CCSD, CCSD(T), and CR-CC(2,3) approaches that can be traced to the early work by Förner et al. [48, 49] and that exploits the general CIM formalism laid down by Li and coworkers [38–40] which is, in turn, close in overall philosophy to the more recent natural linear scaling (NLS) CC method of Ref. [66] (cf., also, Ref. [69] for the analogous considerations at the semiempirical CCSD level). There also are some similarities between the CIM-CC methods pursued in this work and the divide-and-conquer (DC) CC technique of Kobayashi and Nakai [91]. The CIM-CC methodology relies on the idea of orbital domains, referred to as subsystems, similar to that exploited in the local correlation CC methods of Refs. [56–60], but, unlike in the local CC work described in Refs. [56–60], the unoccupied PAOs that are assigned to each CIM subsystem are orthogonalized or, depending on the specific implementation, remain orthogonal to a very good approximation, so that one can make use of the conventional CC algorithms for orthonormal bases. Moreover, unlike in the local correlation CC methods of Refs. [56–60], the CIM-CC approaches that we pursue lead to very transparent and intuitive algorithms in which, beginning with the integral transformation and ending up with the target CC work, the CC calculation for a large system is split into a number of independent and relatively inexpensive subsystem calculations, which are subsequently used to determine the final correlation energy for the entire large system out of the correlation energy contributions defining the individual CIM subsystems. Since the individual subsystems exploited in the CIM-CC considerations are more or less independent of the system size, the CPU time of the CIM-CC calculations scales nearly linearly with the size of the system. Furthermore, the memory requirements of the largest CC calculation that one has to carry out within the CIM framework are practically independent of the system size.

The key to the understanding of the CIM-CC ideas is the realization of the fact that the total correlation energy of a large system is as a sum of contributions from the occupied orthonormal LMOs and their respective occupied and unoccupied orbital domains as long as each of those domains provides an accurate representation of the most essential contributions to the correlation energy associated with a given occupied LMO. As explained in Ref. [38], where the initial justification for the CIM formalism was provided, by solving CC equations for individual subsystems, which consist of occupied LMOs and their corresponding unoccupied orbital domains, one picks up significant cluster amplitudes that are responsible for the most essential contributions to the total correlation energy. The main difference between the CIM-CC methods and the NLS-CC approaches of Ref. [66], which all rely on splitting the CC correlation energy into contributions from the individual occupied orbitals and the associated orbital domains, is in the use of the orthonormal occupied LMOs, obtained with the conventional Boys localization technique [88], and the properly orthogonalized unoccupied PAOs employed in the former approach to define the relevant subsystems rather than the localized natural bond orbitals (NBOs) [92] exploited in the latter method to define the subsystems or subunits. Another difference between the CIM-CC and the NLS-CC techniques is in the fact that in the CIM-CC methods one splits the total correlation energy into individual subsystem contributions while calculating the reference HF energy for the entire system prior to the localization of occupied orbitals, whereas in the NLS-CC approaches one calculates contributions to the reference energy for each subsystem, which leads to issues such as the proper separability of the NLS-CC wave functions [66] that do not enter the CIM-CC considerations. One should also point out the differences between the CIM-CC approaches advocated in this work and the DC-CC formalism of Ref. [91]. The key difference is in the fact that the CIM-CC formalism decomposes the CC correlation energy into contributions from the individual occupied LMOs, where each occupied orbital contribution to the total correlation energy is written in terms of the occupied and unoccupied orbitals that interact with a given occupied LMO, whereas the DC-CC approach divides the total CC correlation energy into contributions from the larger occupied and unoccupied orbital blocks defining the so-called central regions. In other words, the CIM-CC methods use a more detailed decomposition of the correlation energy when compared with the DC-CC technique. There also are substantial differences in the design of the orbital subsystems in the CIM-CC and DC-CC methods. Furthermore, the DC-CC approach makes use of the subsystem orbitals obtained in the DC variant of the HF method with sufficiently large buffer regions to enforce the converged results, whereas the CIM-CC methods pursued by us rely on the localized form of the HF orbitals obtained for the entire system. It is not clear yet how well the DC-CC formalism of Ref. [91] will perform in the context of the non-iterative CC calculations of the CCSD(T) or CR-CC(2,3) type. So far, the authors of Ref. [91] have limited themselves to the DC-CCSD approximation. Moreover, there may be formal issues related to the proper handling of the non-iterative triples corrections of the CCSD(T) or CR-CC(2,3) type within the DC-CC theoretical framework. In fact, the same remark applies to the NLS-CC approach of Ref. [66] which is easy to extend to the higher-order iterative

CC methods but may require nontrivial modifications resulting in a potential loss of accuracy or computational efficiency when the non-iterative CC approaches are considered. As shown in chapter, we can easily extend the CIM-CC considerations to the non-iterative CC methods of the CCSD(T) and CR-CC(2,3) type, reproducing the canonical CCSD(T) and CR-CC(2,3) correlation energies almost perfectly.

There also are differences between the CIM-CC approaches and the local correlation CC methods of Refs. [56–60], including the aforementioned splitting of the CC calculation for a large system into separate CC calculations for relatively small orbital subsystems characterizing the CIM-CC methodology that leads to the NLS of the CIM-CC schemes and the use of the orthonormal occupied and orthonormal or virtually orthonormal unoccupied orbitals in the CIM-CC case rather than the orthonormal occupied LMOs and the non-orthogonal unoccupied PAOs exploited in the local correlation CC methods of Refs. [56–60]. In analogy to the NLS-CC formalism of Ref. [66], the CIM-CC methods make use of the conventional CC codes developed for the orthonormal MO bases, whereas the local correlation CC approaches of Refs. [56–60] require a major reformulation of the CC equations to enable the use of non-orthogonal unoccupied orbitals. In particular, the local correlation CC methods of Refs. [56–60] replace the non-iterative corrections of CCSD(T) by the complicated and relatively expensive iterative steps that are needed to construct the approximate triply excited cluster amplitudes that formally enter the (T) energy correction of CCSD(T), since by switching from the canonical HF to the noncanonical, non-HF LMO bases one must incorporate additional terms involving the off-diagonal elements of the Fock matrix in the CCSD(T) expressions [57, 58]. The CIM-CCSD(T) and CIM-CR-CC(2,3) methods that we have developed and discuss in this work enable us to formulate the local triples corrections to the CCSD energies in a purely non-iterative fashion, i.e., without any need for iterative steps in the triples correction part exploited in the local CCSD(T) approach of Refs. [57, 58]. In developing the CIM-CCSD(T) and CIM-CR-CC(2,3) methods with the purely non-iterative local triples corrections we have exploited the intrinsic flexibility of the CIM ansatz that allows one to use the so-called quasi-canonical MOs (QCMOs) within each CIM subsystem instead of noncanonical LMOs once the CIM orbital subsystems are constructed.

The initial CIM-CC papers [38–40] showed some promise, demonstrating that the CIM-CC methods are capable of recovering 99% or so of the correlation energy calculated at the CCD level, but the molecular systems calculated at the CCD level had to be relatively small in size due to the explicit nested loop structure used in the underlying pilot CCD codes which slows down the required CC calculations for subsystems by a large factor. No other CC methods were examined in the earlier CIM-CC work either. This chapter reviews our recent effort toward the development and efficient implementation of the CIM-CCSD, CIM-CCSD(T), and CIM-CR-CC(2,3) approaches that addresses these important issues. Thus, unlike the initial implementations of the CIM-CCD approach [38–40], which utilized pilot CCD codes with explicit loops over excitations and orbital indices that enter the molecular integrals and cluster amplitudes in the CC equations, the CIM-CC methods discussed and examined in this work are characterized by high computational efficiency in both the CIM and the CC parts, enabling calculations for much larger molecular systems and basis sets and at the higher levels of CC theory, such as CCSD(T) or CR-CC(2,3), than previously possible. This is achieved by combining the NLS and trivial parallelism of the CIM ansatz with the fully vectorized CCSD, CCSD(T), and CR-CC(2,3) codes of Refs. [24, 26, 93], incorporated in the GAMESS package [94] and used in the CIM subsystem calculations. These codes owe their high efficiency to the use of diagram factorization, recursively generated intermediates, and fast matrix multiplication routines (cf., e.g., Refs. [95, 96]). By comparing the results of the canonical and CIM CCSD and CR-CC(2,3) calculations for a series of normal alkanes C_nH_{2n+2}, we demonstrate that the CIM-CC approaches recover the corresponding canonical CC correlation energies to within 0.1% or so, while offering the nearly linear scaling of the total CPU time with the system size and savings in the computer effort by a large factor. By examining the dissociation of dodecane into $C_{11}H_{23}$ and CH_3 and several lowest-energy structures of the $(H_2O)_n$ clusters, we show that the CIM-CCSD, CIM-CCSD(T), and CIM-CR-CC(2,3) methods accurately reproduce the relative energetics of the corresponding canonical CC calculations, often to within fractions of kcal/mol. The difficulties with achieving similarly high accuracies in the local CC calculations that employ basis sets with diffuse functions are addressed by suggesting a new, simplified, and at the same time improved variant of the CIM-CC theory whose performance is illustrated by the CIM-CCSD calculations for water clusters, as described by the 6-31++G(d, p) [97–99] and 6-311++G(d, p) [99, 100] basis sets. Although we mainly focus on benchmarking the CIM-CCSD, CIM-CCSD(T), and CIM-CR-CC(2,3) approaches by comparing them with the corresponding canonical CC results, which means that most of our molecular systems are such that we can still afford the required canonical CC calculations, the presented numerical tests explore much larger molecular sizes and larger basis sets when compared with the earlier CIM-CCD work, including water clusters as large as $(H_2O)_{20}$, normal alkanes as large as $C_{32}H_{66}$, and basis sets as large as 6-311++G(d, p).

2 Local Correlation CC Methods Employing the "Cluster-In-Molecule" Ansatz and Their Efficient Computer Implementation

In the following description, we review the single-reference CC methods used in our work and discuss the key elements of the CIM ansatz that lead to the CIM-CCSD, CIM-CCSD(T), and CIM-CR-CC(2,3) methods and their computer implementation. We begin with the overview of the canonical CCSD, CCSD(T), and CR-CC(2,3) approaches, which is followed by the description of the CIM-CCSD, CIM-CCSD(T), and CIM-CR-CC(2,3) methods and computer programs.

2.1 The Single-Reference CC Formalism and the Canonical CCSD, CR-CC(2,3), and CCSD(T) Approaches

The single-reference CC theory uses the following ansatz for the ground-state wave function:

$$|\Psi\rangle = e^T |\Phi\rangle, \tag{1}$$

where T is the cluster operator (a particle-hole excitation operator generating the connected components of $|\Psi\rangle$ through excitations from $|\Phi\rangle$) and $|\Phi\rangle$ is the reference determinant that defines the Fermi vacuum. In this chapter, we focus on the singlet ground states, and the reference determinant $|\Phi\rangle$ used in the canonical CC calculations is always the restricted HF (RHF) determinant.

The conventional CC approximations, such as CCSD, are based on truncating the many-body expansion of T at some excitation level. In the specific case of CCSD, T is approximated by

$$T^{\text{(CCSD)}} = T_1 + T_2, \tag{2}$$

where

$$T_1 = \sum_{i,a} t_a^i E_i^a \tag{3}$$

and

$$T_2 = \sum_{i<j,a<b} t_{ab}^{ij} E_{ij}^{ab}, \tag{4}$$

with $E_i^a = a^a a_i$ and $E_{ij}^{ab} = E_i^a E_j^b = a^a a^b a_j a_i$ representing the single and double excitation operators, defined in terms of the usual creation and annihilation operators, a^p and a_p, respectively, in a usual manner and t_a^i and t_{ab}^{ij} representing the corresponding singly and doubly excited cluster amplitudes. Here and elsewhere in this chapter, indices $i, j, k, \ldots$ designate the spin-orbitals occupied in $|\Phi\rangle$ and $a, b, c, \ldots$ designate the unoccupied spin-orbitals. Generic (occupied and unoccupied) spin-orbitals are designated by indices $p, q, r, \ldots$. If we have to discuss the one-electron spatial wave functions (orbitals) corresponding to the occupied, unoccupied, and generic spin-orbitals i, a, and p, we use the notation ϕ_i or $\phi_i(r)$, ϕ_a or $\phi_a(r)$, and ϕ_p or $\phi_p(r)$, respectively. Although in the actual computer implementation of the CIM-CC methods discussed in this work, we exploit the highly efficient spin-free formulation of the CC theory based on the orbitals that are functions of the spatial coordinates of an electron and the use of Goldstone diagrams, rather than the spin-orbitals that depend on the spatial and spin coordinates, with spin coordinates integrated in the final expressions, in the following presentation we employ the formally simpler spin-orbital form of the relevant CC equations. Throughout this chapter, we make the usual assumption that each MO ϕ_p is represented as a linear combination of the appropriate AOs, designated as χ_μ. In the canonical CC calculations, each MO is a linear combination of all AOs in the basis set. The design of the MO basis for the CIM-CC calculations limits the sets of AOs used to represent the relevant MOs to smaller AO subsets, as described in the next subsections.

The explicit equations for the cluster amplitudes t_a^i and t_{ab}^{ij} defining the CCSD approach are determined by projecting the electronic Schrödinger equation on the singly and doubly excited determinants, $|\Phi_i^a\rangle = E_i^a|\Phi\rangle$ and $|\Phi_{ij}^{ab}\rangle = E_{ij}^{ab}|\Phi\rangle$, respectively, to obtain

$$\langle\Phi_i^a|\bar{H}^{(\mathrm{CCSD})}|\Phi\rangle = 0, \tag{5}$$

$$\langle\Phi_{ij}^{ab}|\bar{H}^{(\mathrm{CCSD})}|\Phi\rangle = 0, \tag{6}$$

where

$$\bar{H}^{(\mathrm{CCSD})} = \left(He^{T_1+T_2}\right)_c = e^{-T_1-T_2}He^{T_1+T_2} \tag{7}$$

is the similarity-transformed Hamiltonian of CCSD, which is equivalent to the connected product of the Hamiltonian and the exponential wave operator defining the CCSD ansatz (designated by the subscript C). The compact, computationally efficient form of the above amplitude equations for the spin-free formulation of CCSD in terms of the spin-free analogs of the spin-orbital cluster amplitudes t_a^i and t_{ab}^{ij}, one- and two-electron integrals, $f_p^q = \langle p|f|q\rangle$ and $v_{pq}^{rs} = \langle pq|v|rs\rangle - \langle pq|v|sr\rangle$, where f is the Fock operator and v is the electron–electron interaction term, and the suitably defined set of recursively generated intermediates, which leads to a fully vectorized computer code that can be applied to any set of orthonormal orbitals, including canonical MOs, occupied LMOs and orthogonalized PAOs, and QCMOs used in this work, can be found in Ref. [93].

Once the system of CCSD equations, Eqs. (5) and (6), is solved for the cluster amplitudes t_a^i and t_{ab}^{ij}, the CCSD energy is calculated using the expression

$$E^{(\mathrm{CCSD})} = E^{(\mathrm{ref})} + \Delta E^{(\mathrm{CCSD})}, \tag{8}$$

where $E^{(\mathrm{ref})} = \langle\Phi|H|\Phi\rangle$ is the reference energy (in the canonical CC calculations for closed-shell systems, the RHF energy) and

$$\Delta E^{(\mathrm{CCSD})} = \sum_{i,a} f_i^a t_a^i + \sum_{i<j,a<b} v_{ij}^{ab}\tau_{ab}^{ij} = \sum_i \delta E_i^{(\mathrm{CCSD})} \tag{9}$$

is the CCSD correlation energy, where

$$\tau_{ab}^{ij} = t_{ab}^{ij} + t_a^i t_b^j - t_a^j t_b^i \tag{10}$$

and

$$\delta E_i^{(\text{CCSD})} = \sum_a f_i^a t_a^i + \frac{1}{2} \sum_{j,a<b} v_{ij}^{ab} \tau_{ab}^{ij}. \tag{11}$$

As explained in Section 2.2, the second form of the CCSD correlation energy in Eq. (9), which splits the total correlation energy into a sum of contribution $\delta E_i^{(\text{CCSD})}$, Eq. (11), associated with the individual occupied spin-orbitals employed in the correlated calculations, is particularly important for the CIM-CC considerations. In practice, when we use the spin-free CC formalism and spatial orbitals instead of spin-orbitals, the summations over spin-orbitals i, j, a, and b in Eqs. (9) and (11) are replaced by the shorter summations over the corresponding orbitals (the use of the spin-free formulation of CC theory must also be accompanied by the appropriate changes in the explicit mathematical expressions defining the CCSD energy in terms of cluster amplitudes and molecular integrals, which are not discussed here). In particular, the summation over i in the definition of the CCSD correlation energy that relates it to the individual contributions $\delta E_i^{(\text{CCSD})}$ is replaced by the summation over occupied spatial orbitals when the spin-free formulation of the CCSD approach is employed.

In this work, in addition to the CCSD approximation, we examine two different ways of correcting the CCSD energy for the effects of the connected triply excited clusters, namely, the CCSD(T) method and its completely renormalized CR-CC(2,3) extension. Since the CCSD(T) approach can be obtained as a natural approximation to CR-CC(2,3) [24, 25], we begin our brief description of both methods with the key equations of CR-CC(2,3).

In the CR-CC(2,3) method, one calculates the total electronic energy as

$$E^{(\text{CR-CC}(2,3))} = E^{(\text{ref})} + \Delta E^{(\text{CR-CC}(2,3))}$$

$$= E^{(\text{ref})} + \Delta E^{(\text{CCSD})} + \delta E^{(2,3)} = E^{(\text{CCSD})} + \delta E^{(2,3)}, \tag{12}$$

where the triples correction $\delta E^{(2,3)}$ to the CCSD total energy $E^{(\text{CCSD})}$ or correlation energy $\Delta E^{(\text{CCSD})}$ is defined by the following generic expression:

$$\delta E^{(2,3)} = \sum_{i<j<k,a<b<c} \ell_{ijk}^{abc} \mathfrak{M}_{abc}^{ijk} = \sum_i \delta E_i^{(2,3)}, \tag{13}$$

where

$$\delta E_i^{(2,3)} = \frac{1}{3} \sum_{j<k,a<b<c} \ell_{ijk}^{abc} \mathfrak{M}_{abc}^{ijk} \tag{14}$$

is the contribution to $\delta E^{(2,3)}$ associated with the individual spin-orbital i. The $\mathfrak{M}_{abc}^{ijk}$ coefficients entering Eqs. (13) and (14) are the triply excited moments of the CCSD equations which correspond to the projections of the CCSD equations on the triply excited determinants $|\Phi_{ijk}^{abc}\rangle$ [11, 12, 24–26],

$$\mathfrak{M}_{abc}^{ijk} = \langle \Phi_{ijk}^{abc} | \bar{H}^{(\text{CCSD})} | \Phi \rangle. \tag{15}$$

The ℓ_{ijk}^{abc} coefficients entering Eqs. (13) and (14) are the hole-particle de-excitation amplitudes that in the exact expansion of the correlation energy in terms of the generalized moments of the CCSD equations, which produces the full CI energy, define the three-body component of the left ground eigenstate $\langle \Phi | \mathcal{L}$ of the similarity-transformed Hamiltonian $\bar{H}^{(\text{CCSD})}$ obtained by diagonalizing $\bar{H}^{(\text{CCSD})}$ in the entire many-electron Hilbert space corresponding to a problem of interest. In practical applications, one has to find an approximate and computationally feasible way of determining the ℓ_{ijk}^{abc} amplitudes that avoids the diagonalization of $\bar{H}^{(\text{CCSD})}$ in the entire Hilbert space. Thus, in the CR-CC(2,3) method of Refs. [24–26] one makes the additional assumption that the triples-triples block of the matrix representing the similarity-transformed Hamiltonian $\bar{H}^{(\text{CCSD})}$ is diagonally dominant, as is the case when the canonical RHF MOs are used in the calculations, and that the one- and two-body components of the left $\langle \Phi | \mathcal{L}$ state are accurately represented by the left ground eigenstate of $\bar{H}^{(\text{CCSD})}$ resulting from the diagonalization of $\bar{H}^{(\text{CCSD})}$ in the subspace spanned by singly and doubly excited determinants, i.e., $\langle \Phi | L^{(\text{CCSD})}$, where

$$L^{(\text{CCSD})} = 1 + \Lambda_1 + \Lambda_2, \tag{16}$$

with

$$\Lambda_1 = \sum_{i,a} \lambda_i^a E_a^i \tag{17}$$

and

$$\Lambda_2 = \sum_{i,a} \lambda_{ij}^{ab} E_{ab}^{ij} \tag{18}$$

representing the one- and two-body components of the so-called 'lambda' operator of the analytic gradient CCSD theory [101] obtained by solving the linear system of equations

$$\langle \Phi | (1 + \Lambda_1 + \Lambda_2) \bar{H}^{(\text{CCSD})} | \Phi_i^a \rangle = E^{(\text{CCSD})} \lambda_i^a, \tag{19}$$

$$\langle \Phi | (1 + \Lambda_1 + \Lambda_2) \bar{H}^{(\text{CCSD})} | \Phi_{ij}^{ab} \rangle = E^{(\text{CCSD})} \lambda_{ij}^{ab}. \tag{20}$$

As long as there are no degeneracies among orbitals, this leads to the following formula for the ℓ_{ijk}^{abc} coefficients used in the CR-CC(2,3) calculations:

$$\begin{aligned}
\ell_{ijk}^{abc} &= \langle \Phi | (1 + \Lambda_1 + \Lambda_2) \bar{H}^{(\text{CCSD})} | \Phi_{ijk}^{abc} \rangle / D_{abc}^{ijk} \\
&= \left\langle \Phi \left| \left[\left(\Lambda_1 \bar{H}_2^{(\text{CCSD})} \right)_{DC} + \left(\Lambda_2 \bar{H}_1^{(\text{CCSD})} \right)_{DC} + \left(\Lambda_2 \bar{H}_2^{(\text{CCSD})} \right)_C \right] \right| \Phi_{ijk}^{abc} \right\rangle / D_{abc}^{ijk}.
\end{aligned} \tag{21}$$

Here, $\bar{H}_n^{(\text{CCSD})}$ designates the n-body component of $\bar{H}^{(\text{CCSD})}$, DC stands for the disconnected part of the corresponding operator product, and

$$D_{abc}^{ijk} = E^{(\text{CCSD})} - \langle \Phi_{ijk}^{abc} | \bar{H}^{(\text{CCSD})} | \Phi_{ijk}^{abc} \rangle = -\sum_{n=1}^{3} \bar{H}_n^{(\text{CCSD})} \tag{22}$$

are the Epstein–Nesbet-type perturbative denominators related to the diagonal matrix elements of the triples-triples block of $\bar{H}^{(\text{CCSD})}$ (if there are degeneracies among orbitals, one has to modify Eq. (21), as described, for example, in Ref. [26]).

As noted in Refs. [24–26] and as already alluded to above, Eq. (13) represents the generic form of the non-iterative triples corrections to the CCSD energy and encompasses other non-iterative triples CC approaches, not just CR-CC(2,3). In fact, one can obtain the entire variety of triples corrections to the CCSD energy formulated to date out of the single expression represented by Eq. (13) by simply using different formulas for the ℓ_{ijk}^{abc} amplitudes. For example, if we replace the $\langle \Phi_{ijk}^{abc} | \bar{H}^{(\text{CCSD})} | \Phi_{ijk}^{abc} \rangle$ diagonal in Eq. (22), which is expressed in terms of the one-, two- and three-body components of $\bar{H}^{(\text{CCSD})}$, by the approximate form of it using the one-body components of $\bar{H}^{(\text{CCSD})}$ only, we obtain the triples correction of the CCSD(2) theory of Gwaltney and Head-Gordon [20, 21]. If we approximate the $\langle \Phi_{ijk}^{abc} | \bar{H}^{(\text{CCSD})} | \Phi_{ijk}^{abc} \rangle$ diagonal even further and replace the Epstein–Nesbet-type denominator D_{abc}^{ijk}, Eq. (22), by the Møller–Plesset-type denominator defined by the spin-orbital energy difference for triples, $(\varepsilon_a + \varepsilon_b + \varepsilon_c - \varepsilon_i - \varepsilon_j - \varepsilon_k)$, we obtain the triples correction of the CCSD(2) approach of Hirata et al. [22, 23]. The widely used CCSD(T) approach, which interests us in this work, is obtained by replacing the Epstein–Nesbet-type denominator D_{abc}^{ijk}, Eq. (22), by the Møller–Plesset-type denominator $(\varepsilon_a + \varepsilon_b + \varepsilon_c - \varepsilon_i - \varepsilon_j - \varepsilon_k)$, neglecting the $(\Lambda_2 \bar{H}_1^{(\text{CCSD})})_{DC}$ term in the resulting equation, which is at least a fourth-order term in many-body perturbation theory (MBPT) when the HF reference is employed, while replacing the remaining $(\Lambda_1 \bar{H}_2^{(\text{CCSD})})_{DC}$ and $(\Lambda_2 \bar{H}_2^{(\text{CCSD})})_{C}$ terms, which appear in the third and second orders of MBPT, respectively, by $(T_1^\dagger V_N)_{DC}$ and $(T_2^\dagger V_N)_{C}$, where V_N is the two-body part of the Hamiltonian in the normal-ordered form, and approximating the triply excited moment $\mathfrak{M}_{abc}^{ijk}$, Eq. (15), by the lead contribution to $\mathfrak{M}_{abc}^{ijk}$ given by $\langle \Phi_{ijk}^{abc} | (V_N T_2)_C | \Phi \rangle$.

From the point of view of the CIM-CC considerations, the key observation is that by developing and implementing the CR-CC(2,3) approach using the generic form of the triples correction given by Eq. (13), we gain access to other triples corrections, including CCSD(T), without altering the overall structure of the underlying mathematical expressions, as explained above. Thus, by developing and implementing the CIM version of Eq. (13), we obtain the CIM codes for a wide variety of triples corrections, not just CR-CC(2,3), without much extra programming effort. The second important observation relevant to the CIM-CC work pursued in this chapter is that all non-iterative triples corrections to the CCSD energy defined by Eq. (13) naturally split into sums of contributions of the $\delta E_i^{(2,3)}$ type, Eq. (14), associated with the individual occupied spin-orbitals used in the post-SCF calculations.

Thus, the overall structure of the CIM-CC theory, in which one always relies on splitting the correlation energy into contributions associated with the individual occupied spin-orbitals, as previously discussed for CCSD, is preserved when using Eqs. (13) and (14) to define the triples corrections. As in the CCSD case, in practice, when we utilize the spin-free description and spatial orbitals instead of spin-orbitals, the summations over i, j, k, a, b, and c in Eqs. (13) and (14) are replaced by the shorter summations over orbitals and one has to make additional suitable changes in the actual mathematical expressions defining the triples corrections in terms of spin-free cluster amplitudes and molecular integrals. In the case of the triples corrections of CR-CC(2,3) and CCSD(T) discussed in this chapter, we rely on the fully vectorized expressions and codes that are based on the diagram factorization techniques and recursively generated intermediates described in Ref. [24] and the appropriate references cited therein, and in Ref. [93].

2.2 The "Cluster-In-Molecule" Approach to Local CC Calculations: Formal Concepts Behind the CIM-CCSD, CIM-CR-CC(2,3), and CIM-CCSD(T) Methods

As pointed out in the *Introduction*, the CIM ansatz of Refs. [38–40], which we have further developed and extended to the CCSD, CR-CC(2,3), and CCSD(T) methods in this and related work [102], originates from the observation that the total correlation energy of a large system can be determined by summing up the contributions from the occupied orthonormal LMOs and their associated occupied and unoccupied orbital domains as long as each of those domains provides an accurate description of the contributions to the correlation energy associated with each occupied LMO. Indeed, in the previous section we have already demonstrated that the CCSD correlation energy $\Delta E^{(\mathrm{CCSD})}$, Eq. (9), and the triples corrections $\delta E^{(2,3)}$ to the CCSD energy, given by Eq. (13) and defining the CR-CC(2,3), CCSD(T), and other non-iterative triples CC approximations, can be rewritten as sums of contributions associated with the individual occupied spin-orbitals. The CCSD correlation energy $\Delta E^{(\mathrm{CCSD})}$ is a sum of the $\delta E_i^{(\mathrm{CCSD})}$ contributions defined by Eq. (11), whereas the generic form of the triples correction to the CCSD energy, $\delta E^{(2,3)}$, is a sum of the $\delta E_i^{(2,3)}$ contributions given by Eq. (14). Clearly, up to this point, we have not done anything other than giving the CCSD correlation energy $\Delta E^{(\mathrm{CCSD})}$ and the triples corrections $\delta E^{(2,3)}$ a slightly different mathematical form that does not offer any computational advantages if we use delocalized orbitals and retain all contributions. However, we can now exploit the intrinsic flexibility of the CC theory in which the CC (e.g., CCSD) equations and the above energy expressions for $\Delta E^{(\mathrm{CCSD})}$ and $\delta E^{(2,3)}$, Eqs. (9) and (13), respectively, remain valid in both the delocalized MO basis of canonical RHF MOs and the LMO basis as long as both bases are orthonormal. As a matter of fact, the CCSD correlation energy $\Delta E^{(\mathrm{CCSD})}$, Eq. (9), and the generic form of the triples correction $\delta E^{(2,3)}$, Eq. (13), in which we do not yet approximate the ℓ_{ijk}^{abc} de-excitation amplitudes in any particular fashion (e.g., we use the exact ℓ_{ijk}^{abc} amplitudes, as discussed in the previous section), are invariant with respect to the separate orbital rotations within the occupied and unoccupied orbital blocks. Let us, then, assume that from now on we will use the occupied LMOs, obtained with one of the traditional MO localization schemes [88–90], to represent spin-orbitals $i, j, k, \ldots$ and the suitably constructed localized unoccupied orbitals to generate spin-orbitals $a, b, c, \ldots$ in the above expressions for $\Delta E^{(\mathrm{CCSD})}$ and $\delta E^{(2,3)}$ in terms of $\delta E_i^{(\mathrm{CCSD})}$ and $\delta E_i^{(2,3)}$. Throughout this chapter and related work [102], we rely on the Boys localization [88], available in the GAMESS package [94], with which our CIM-CCSD, CIM-CCSD(T), and CIM-CR-CC(2,3) computer programs are interfaced, although there is nothing fundamental in the CIM theory that would formally prevent us from using other localization schemes. The only exception is the Pipek–Mezey localization method [90] which relies on the separation of the σ and π orbitals, resulting in the zero values of the Fock matrix elements between the σ and the π LMOs independent of the distance between them. As in the earlier CIM work [38–40] and as described below, we use the values of the off-diagonal Fock matrix elements between the occupied LMOs to define the orbital domains that enter the CIM considerations. As further elaborated on below, the unoccupied LMOs are constructed by orthogonalizing and further localizing the suitable set of PAOs.

Following the above introductory remarks, let $i', j', k', \ldots$ and $a', b', c', \ldots$ be the occupied and unoccupied localized orthonormal spin-orbitals corresponding to the suitably constructed set of spatial LMOs. Since the CCSD correlation energy $\Delta E^{(\mathrm{CCSD})}$ and the generic triples correction $\delta E^{(2,3)}$, in which we do not approximate the ℓ_{ijk}^{abc} amplitudes in any specific manner, are invariant with respect to rotations of occupied and unoccupied orbitals, we can immediately rewrite the formulas for $\Delta E^{(\mathrm{CCSD})}$ and $\delta E^{(2,3)}$ in terms of these localized orthonormal spin-orbitals by replacing the indices $i, j, k, a, b,$ and c in Eqs. (9), (11), (13), and (14) by $i', j', k', a', b',$ and c', respectively. If we further realize that for a given occupied localized spin-orbital i' the only localized spin-orbitals $j', a',$ and b' that can substantially contribute to the CCSD correlation energy contribution $\delta E_{i'}^{(\mathrm{CCSD})}$, as in Eq. (11), and the only localized spin-orbitals $j', k', a', b',$ and c' that can substantially contribute to the triples correction contribution $\delta E_{i'}^{(2,3)}$, as in Eq. (14), are those that belong to the (spin)orbital domain that interacts with the spin-orbital i' in a significant manner, we can replace the resulting exact expressions for $\Delta E^{(\mathrm{CCSD})}$ and $\delta E^{(2,3)}$ by

$$\Delta \tilde{E}^{(\mathrm{CCSD})} = \sum_{i'} \delta \tilde{E}_{i'}^{(\mathrm{CCSD})} \tag{23}$$

and

$$\delta \tilde{E}^{(2,3)} = \sum_{i'} \delta \tilde{E}_{i'}^{(2,3)}, \tag{24}$$

respectively, where

$$\delta \tilde{E}_{i'}^{(\text{CCSD})} = \frac{1}{M_{i'}} \sum_{\{P_{i'}\}} \delta \tilde{E}_{i'}^{(\text{CCSD})}(\{P_{i'}\}), \tag{25}$$

with

$$\delta \tilde{E}_{i'}^{(\text{CCSD})}(\{P_{i'}\}) = \sum_{a' \in \{P_{i'}\}} f_{i'}^{a'} t_{a'}^{i'} + \frac{1}{2} \sum_{j', a' < b' \in \{P_{i'}\}} v_{i'j'}^{a'b'} \tau_{a'b'}^{i'j'}, \tag{26}$$

and

$$\delta \tilde{E}_{i'}^{(2,3)} = \frac{1}{M_{i'}} \sum_{\{P_{i'}\}} \delta \tilde{E}_{i'}^{(2,3)}(\{P_{i'}\}), \tag{27}$$

with

$$\delta \tilde{E}_{i'}^{(2,3)}(\{P_{i'}\}) = \frac{1}{3} \sum_{j' < k', a' < b' < c' \in \{P_{i'}\}} \ell_{i'j'k'}^{a'b'c'} \mathfrak{M}_{a'b'c'}^{i'j'k'}. \tag{28}$$

The symbol $\{P_{i'}\}$ in Eqs. (25), (26), (27), and (28) designates any (spin)orbital domain $\{P\}$, which we also refer to as the *subsystem* $\{P\}$, that contains the specific occupied localized spin-orbital $\phi_{i'}$ as a *central orbital*. The symbol $M_{i'}$ represents the number of subsystems $\{P\}$ that contain the specific spin-orbital i' or the corresponding orbital $\phi_{i'}$ as a central spin-orbital or orbital. By design, it may happen (and it does happen) that some correlated occupied localized spin-orbitals i' and the corresponding spatial LMOs $\phi_{i'}$ are central in multiple domains $\{P\}$; if we encounter this situation, we evaluate the correlation energy contributions $\delta \tilde{E}_{i'}^{(\text{CCSD})}$ and $\delta \tilde{E}_{i'}^{(2,3)}$ associated with a given spin-orbital i' by averaging the individual contributions $\delta \tilde{E}_{i'}^{(\text{CCSD})}(\{P_{i'}\})$, Eq. (26), and $\delta \tilde{E}_{i'}^{(2,3)}(\{P_{i'}\})$, Eq. (28), corresponding to the (spin)orbital domains $\{P_{i'}\}$ where i' or $\phi_{i'}$ is central, as in Eqs. (25) and (27), respectively. The precise definition of the term "central orbital" for a given subsystem $\{P\}$ or $\{P_{i'}\}$ is introduced in the next subsection. The only facts that matter at this point, which justify the above equations, are that each correlated occupied LMO and the corresponding α and β spin-orbitals are central for at least one subsystem, which, as explained in Section 2.3, is always the case, and that the (spin)orbital subsystems $\{P\}$ are chosen such that the $\delta \tilde{E}_{i'}^{(\text{CCSD})}$ and $\delta \tilde{E}_{i'}^{(2,3)}$ contributions to the correlation energy calculated using Eqs. (25), (26), (27), and (28), where j' in Eq. (26) and j' and k' in Eq. (28) run over the central and *environment* occupied spin-orbitals from the domain $\{P_{i'}\}$ and where a', b', and c' in Eqs. (26) and (28) are the suitably designed unoccupied localized spin-orbitals that are associated with the occupied spin-orbitals of $\{P_{i'}\}$, are accurate approximations to the exact contributions $\delta E_i^{(\text{CCSD})}$, Eq. (11), and $\delta E_i^{(2,3)}$, Eq. (14), respectively. As shown in this chapter, the careful design of (spin)orbital subsystems $\{P\}$ defining the CIM local correlation ansatz guarantees this.

We should also stress that at this formal stage the $\ell_{i'j'k'}^{a'b'c'}$ amplitudes in Eq. (28) are still unspecified. They could, for example, represent the exact amplitudes defining the three-body component of the left ground eigenstate $\langle\Phi|\mathcal{L}$ of the similarity-transformed Hamiltonian $\bar{H}^{(\text{CCSD})}$ obtained by diagonalizing $\bar{H}^{(\text{CCSD})}$ in the entire many-electron Hilbert space and written in the localized spin-orbital basis $\{p'\}$, but they cannot be automatically replaced by their approximate perturbative form given by Eq. (21) defining the CR-CC(2,3) approach in which one makes the additional assumption that the triples-triples block of the similarity-transformed Hamiltonian $\bar{H}^{(\text{CCSD})}$ is diagonally dominant, which is generally not the case when the localized spin-orbitals are employed. As a matter of fact, a naive attempt to use Eq. (21) to determine the $\ell_{i'j'k'}^{a'b'c'}$ amplitudes in the LMO bases that define the CIM subsystems may lead to substantial differences between the CIM-CR-CC(2,3) and the canonical CR-CC(2,3) energies. In order to make safe use of the perturbative expression for the ℓ_{ijk}^{abc} amplitudes, Eq. (21), in the CIM-CR-CC(2,3) and (after the suitable simplifications in the definitions of ℓ_{ijk}^{abc} and $\mathfrak{M}_{abc}^{ijk}$ discussed in the previous section) CIM-CCSD(T) calculations, and in order to retain the purely non-iterative and relatively inexpensive character of the CIM-CR-CC(2,3) and CIM-CCSD(T) local triples corrections, we must perform additional manipulations on the CIM formula for the contribution to the triples correction associated with a given localized spin-orbital i', Eq. (28), that invoke the concept of QCMOs. We discuss this very important aspect of the CIM-CC methodology below.

Suppose we have already determined the suitable (spin)orbital subsystems $\{P\}$, each consisting of a number of occupied and unoccupied orthonormal LMOs, designated as $\phi_{i'}$ and $\phi_{a'}$, respectively, and the corresponding spin-orbitals i' and a'. Let us consider the unitary transformation $\mathbf{R}$ of the LMOs $\phi_{i'}$ and $\phi_{a'}$ within each subsystem $\{P\}$ that does not mix the occupied and unoccupied orbitals defining $\{P\}$, i.e.,

$$\phi_{i'} = \sum_{i \in \{P\}} R_{ii'} \phi_i \tag{29}$$

and

$$\phi_{a'} = \sum_{a \in \{P\}} R_{aa'} \phi_a, \tag{30}$$

where ϕ_i and ϕ_a are the new occupied and unoccupied MOs belonging to subsystem $\{P\}$ that span exactly the same orbital subspace as the original LMOs $\phi_{i'}$ and $\phi_{a'}$ that have been used to construct $\{P\}$, and $R_{ii'}$ and $R_{aa'}$ are the corresponding transformation coefficients (we assume that $R_{ia'} = R_{ai'} = 0$). As in the rest of this chapter, we continue using the notation in which the spin-orbitals associated with the new MOs ϕ_i and ϕ_a are labeled as i and a, respectively. It is easy to show [102] that one can rewrite the expressions for the correlation energy contributions $\delta\tilde{E}_{i'}^{(\text{CCSD})}(\{P_{i'}\})$ and $\delta\tilde{E}_{i'}^{(2,3)}(\{P_{i'}\})$, Eqs. (26) and (28), respectively, associated with a given localized occupied spin-orbital i' and the specific (spin)orbital domain $\{P_{i'}\}$, which contains the spin-orbital i' (and the corresponding occupied LMO $\phi_{i'}$) as a central spin-orbital (orbital), in the following manner:

$$\delta \tilde{E}_{i'}^{(\mathrm{CCSD})}(\{P_{i'}\}) = \sum_{a \in \{P_{i'}\}} f_{i'}^{a} t_{a}^{i'} + \frac{1}{2} \sum_{j,a<b \in \{P_{i'}\}} v_{i'j}^{ab} \tau_{ab}^{i'j}, \tag{31}$$

$$\delta \tilde{E}_{i'}^{(2,3)}(\{P_{i'}\}) = \frac{1}{3} \sum_{j<k,a<b<c \in \{P_{i'}\}} \ell_{i'jk}^{abc} \mathfrak{M}_{abc}^{i'jk}. \tag{32}$$

Notice the presence of matrix elements $f_{i'}^{a}$ and $v_{i'j}^{ab}$, amplitudes $t_{a}^{i'}$, $\tau_{ab}^{i'j}$, and $\ell_{i'jk}^{abc}$, and moments $\mathfrak{M}_{abc}^{i'jk}$ of the mixed-index type in Eqs. (31) and (32) which carry index i' of the original localized spin-orbital used to design subsystem $\{P_{i'}\}$ and one or more unprimed indices labeling the new spin-orbitals resulting from the unitary transformation described by Eqs. (29) and (30). These mixed-index quantities are defined as follows:

$$f_{i'}^{a} = \sum_{i \in \{P_{i'}\}} R_{ii'}^{*} f_{i}^{a}, \tag{33}$$

$$v_{i'j}^{ab} = \sum_{i \in \{P_{i'}\}} R_{ii'}^{*} v_{ij}^{ab}, \tag{34}$$

$$t_{a}^{i'} = \sum_{i \in \{P_{i'}\}} R_{ii'} t_{a}^{i}, \tag{35}$$

$$\tau_{ab}^{i'j} = \sum_{i \in \{P_{i'}\}} R_{ii'} \tau_{ab}^{ij}, \tag{36}$$

$$\ell_{i'jk}^{abc} = \sum_{i \in \{P_{i'}\}} R_{ii'}^{*} \ell_{ijk}^{abc}, \tag{37}$$

and

$$\mathfrak{M}_{abc}^{i'jk} = \sum_{i \in \{P_{i'}\}} R_{ii'} \mathfrak{M}_{abc}^{ijk}, \tag{38}$$

respectively, where matrix elements of the Fock matrix f_{i}^{a}, two-electron integrals v_{ij}^{ab}, amplitudes t_{a}^{i}, τ_{ab}^{ij}, and ℓ_{ijk}^{abc}, and moments $\mathfrak{M}_{abc}^{ijk}$ are calculated in the new MO basis defining subsystem $\{P_{i'}\}$ spanned by orbitals ϕ_{i} and ϕ_{a}, Eqs. (29) and (30).

Equations (31) and (32), with $f_{i'}^{a}$, $v_{i'j}^{ab}$, $t_{a}^{i'}$, $\tau_{ab}^{i'j}$, $\ell_{i'jk}^{abc}$, and $\mathfrak{M}_{abc}^{i'jk}$ defined by Eqs. (33), (33), (34), (35), (36), (37), and (38), are fully equivalent to Eqs. (26) and (28) for $\delta \tilde{E}_{i'}^{(\mathrm{CCSD})}(\{P_{i'}\})$ and $\delta \tilde{E}_{i'}^{(2,3)}(\{P_{i'}\})$, and as such do not seem to help anything. However, since we have the freedom of choosing the unitary transformation $\mathbf{R}$ for each subsystem $\{P\}$, we can choose it such that the new MOs defined by Eqs. (29) and (30) diagonalize the occupied-occupied and unoccupied-unoccupied blocks of the Fock matrix in the orbital space of $\{P\}$. The resulting MOs, which are separately constructed for each CIM subsystem $\{P\}$, are called QCMOs (quasi-canonical MOs). Unlike the original subsystem orbitals $\phi_{p'}$, QCMOs are not localized any more, but this does not affect anything since once the CIM subsystems are determined, the QCMOs for a given subsystem $\{P\}$ span exactly the same orbital space as the original LMOs defining the same subsystem $\{P\}$, so that the cost of the CC calculations for a given subsystem $\{P\}$ in the QCMO basis (in the iterative CC methods, the cost per iteration) is exactly the same as the cost of the analogous CC calculations in the corresponding LMO basis. There are, however, substantial benefits resulting from the use of QCMOs rather than the original LMOs that have been used in the subsystem design in the CIM-CC calculations. The key benefit is the fact that similar to the canonical, RHF-based formulation of CR-CC(2,3), the triples-triples block of the similarity-transformed Hamiltonian of CCSD, $\bar{H}^{(\mathrm{CCSD})}$, becomes diagonally dominant and the CCSD cluster amplitudes t_a^i and t_{ab}^{ij}, and their left λ_i^a and λ_{ij}^{ab} counterparts closely resemble the analogous canonical amplitudes within the subsystem orbital subspace when the QCMO basis is employed, justifying the use of the perturbative formula for the ℓ_{ijk}^{abc} amplitudes, Eq. (21), in the CIM-CR-CC(2,3) calculations. The analogous remarks apply to CIM-CCSD(T), which is, as explained in the previous subsection, an approximation to CIM-CR-CC(2,3). All we need to do to perform the CIM-CR-CC(2,3) calculations is carry out first the CCSD and left CCSD calculations for each CIM subsystem in a QCMO basis, which we obtain by transforming the corresponding occupied and unoccupied LMOs defining the same subsystem through diagonalizations of the relevant occupied-occupied and unoccupied-unoccupied blocks of the Fock matrix, extract the resulting t_a^i, t_{ab}^{ij}, λ_i^a, and λ_{ij}^{ab} amplitudes, employ these amplitudes to calculate the triply excited moments $\mathfrak{M}_{abc}^{ijk}$, and the de-excitation amplitudes ℓ_{ijk}^{abc} in the QCMO basis using Eqs. (15) and (21), respectively, transform the resulting ℓ_{ijk}^{abc} amplitudes and moments $\mathfrak{M}_{abc}^{ijk}$ from a purely QCMO form to a mixed-index form involving the occupied localized spin-orbital i' and QCMO spin-orbitals j, k, a, b, and c with the help of Eqs. (37) and (38), construct the contributions $\delta\tilde{E}_{i'}^{(2,3)}(\{P_{i'}\})$ that are associated with the individual spin-orbitals i' and domains $\{P_{i'}\}$ where i' is central using Eq. (32), and calculate the final energy correction due to triples, $\delta\tilde{E}^{(2,3)}$, by summing up the $\delta\tilde{E}_{i'}^{(2,3)}(\{P_{i'}\})$ contributions, as in Eqs. (24) and (27). The CIM-CCSD(T) calculations are performed in the same manner. The only differences between the CIM-CCSD(T) and the CIM-CR-CC(2,3) calculations are the explicit expressions for ℓ_{ijk}^{abc} and $\mathfrak{M}_{abc}^{ijk}$ used in the CCSD(T) calculations in the QCMO subsystem bases, which are obtained by approximating Eqs. (15) and (21) and which do not require performing the left CCSD calculations for λ_i^a and λ_{ij}^{ab}, as explained in the previous subsection.

In addition to allowing us to formulate the well-defined local correlation CR-CC(2,3) and CCSD(T) schemes, the use of QCMOs in the CIM-CR-CC(2,3) and CIM-CCSD(T) calculations, as described above, offers a major advantage over other approaches to local triples corrections of the CCSD(T) type, namely, the triples corrections of CIM-CR-CC(2,3) and CIM-CCSD(T) are determined in a purely non-iterative fashion. This should be contrasted with the local CCSD(T) approach of Refs. [57, 58], in which the non-iterative steps of CCSD(T) are replaced by the considerably more complicated and generally more expensive iterative steps needed to construct the relevant approximate form of the triply excited cluster amplitudes that formally enter the (T) energy correction of CCSD(T). This is a consequence of the fact that by switching to the noncanonical, non-HF LMO basis and by doing nothing else, as in Refs. [57, 58], one has to incorporate additional terms involving the off-diagonal elements of the Fock matrix in the CCSD(T) method. Furthermore, unlike in Refs. [57, 58], our formulation of the CIM-CCSD(T) and CIM-CR-CC(2,3) approaches relies on the orthogonal (or virtually orthogonal) LMOs rather than the mixed set of orthogonal occupied LMOs and non-orthogonal unoccupied LMOs (i.e., PAOs). This allows us to rely on the highly efficient canonical CCSD(T) and CR-CC(2,3) codes described in Refs. [24, 93] rather than the specialized and much less transparent CC codes employing non-orthogonal (or biorthogonal) bases, exploited in Refs. [57, 58]. Because of the intrinsic loop structure of the triples correction parts of the highly efficient and vectorized CCSD(T) and CR-CC(2,3) codes described, for example, in Refs. [93, 103], the use of QCMOs in the CIM-CR-CC(2,3) and CIM-CCSD(T) calculations adds a prefactor of 3 to the CPU time of CR-CC(2,3) and CCSD(T) calculations for each CIM subsystem, but this is certainly a small price to pay considering the fact that the QCMO-based CIM-CR-CC(2,3) and CIM-CCSD(T) approaches retain the non-iterative character and high accuracies of the canonical CR-CC(2,3) and CCSD(T) methods, and the fact that the CIM subsystems consist of small numbers of orbitals independent of the system size, keeping the costs of CIM-CR-CC(2,3) and CIM-CCSD(T) calculations at a low level.

Interestingly, the iterative CIM-CCSD calculations benefit from the use of QCMOs as well. Since the CCSD energy is invariant with respect to separate rotations of occupied and unoccupied orbitals, we can, of course, solve the CCSD equations for each CIM subsystem $\{P\}$ directly in the LMO basis of orthonormal orbitals $\phi_{i'}$ and $\phi_{a'}$ used to construct these subsystems, and calculate the CIM-CCSD correlation energy using Eqs. (23), (25), and (26). This was the strategy used in the earlier CIM-CCD work [38–40]. However, we can also solve the CCSD equations for each CIM subsystem $\{P\}$ in the corresponding QCMO basis, as defined above, transform the resulting t_a^i and t_{ab}^{ij} amplitudes and the corresponding one- and two-electron integrals f_i^a and v_{ij}^{ab} from a purely QCMO form to a mixed-index form involving the occupied localized spin-orbital i' and QCMO spin-orbitals j, a, and b using Eqs. (33), (34), (35), and (36), calculate the correlation energy contributions $\delta \tilde{E}_{i'}^{(CCSD)}(\{P_{i'}\})$ that are associated with the individual spin-orbitals i' and domains $\{P_{i'}\}$ where i' is central using Eq. (31), and determine the final CIM-CCSD correlation energy $\Delta \tilde{E}^{(CCSD)}$ by summing up the $\delta \tilde{E}_{i'}^{(CCSD)}(\{P_{i'}\})$ contributions, as in Eqs. (23) and (25). Clearly, both methods lead to identical results if the subsystems used in the above two types of CIM-CCSD calculations are identical (we verified this numerically). However, the convergence rate of the CCSD iterations in the noncanonical LMO bases that are used to construct the CIM subsystems $\{P\}$ is significantly slower than the convergence rate of the CCSD iterations in the QCMO bases, which are obtained by diagonalizing the corresponding occupied-occupied and unoccupied-unoccupied blocks of the Fock matrix. For example, in the CIM-CCSD calculation for $(H_2O)_{10}$, as described by the 6-311++G(d, p) basis set, discussed in Section 4.2, one needs about 30 iterations on average per CIM subsystem to converge the CCSD subsystem calculations to within 10^{-5} hartree when the LMO basis is employed in the CCSD subsystem calculations. The use of the QCMO basis reduces the average number of the CCSD subsystem iterations in the same calculation to about 15, while allowing us to use the tighter convergence criterion of 10^{-6} hartree. This improved convergence and the resulting additional speedup of the CIM-CCSD computations, resulting from the use of QCMOs rather than LMOs in the CCSD subsystem calculations, can be easily understood if we look at the actual algorithm used to solve the CCSD equations. The highly efficient CCSD computer code utilized in this work and described in Ref. [93], which applies to the canonical as well as noncanonical MO bases, is based on writing the CCSD amplitude equations, Eqs. (5) and (6), in the following symbolic form:

$$^{[n+1]}t_a^i = (f_i^i - f_a^a)^{-1}\, F_a^i(\ldots, {}^{[n]}t_c^k, \ldots, {}^{[n]}t_{de}^{lm}, \ldots), \tag{39}$$

$$^{[n+1]}t_{ab}^{ij} = (f_i^i - f_a^a + f_j^j - f_b^b)^{-1}\, F_{ab}^{ij}(\ldots, {}^{[n]}t_c^k, \ldots, {}^{[n]}t_{de}^{lm}, \ldots), \tag{40}$$

where $F_a^i(\ldots, t_c^k, \ldots, t_{de}^{lm}, \ldots)$ and $F_{ab}^{ij}(\ldots, t_c^k, \ldots, t_{de}^{lm}, \ldots)$ are the quasi-linearized polynomial functions in cluster amplitudes that contain, in particular, terms that are linear in cluster amplitudes in which the off-diagonal Fock matrix elements of the f_i^j and f_a^b types multiply these cluster amplitudes. In writing Eqs. (39) and (40), we use the additional superscripts $[n+1]$ and $[n]$ to emphasize the new and the previous iterations. As in the HF-based canonical CCSD calculations, these linear terms in $F_a^i(\ldots, t_c^k, \ldots, t_{de}^{lm}, \ldots)$ and $F_{ab}^{ij}(\ldots, t_c^k, \ldots, t_{de}^{lm}, \ldots)$ that contain the off-diagonal elements of the Fock matrix are zero when the QCMO basis is employed, since f_i^j and f_a^b vanish for $i \neq j$ and $a \neq b$ in this case, speeding up the convergence, and are nonzero (often far from zero) when the noncanonical LMO basis is exploited, increasing the number of CCSD iterations based on Eqs. (39) and (40).

In order to complete our description of the theoretical and computational details behind the CIM-CCSD, CIM-CR-CC(2,3), and CIM-CCSD(T) approaches, we must provide information about the actual design of the occupied and unoccupied LMOs defining the CIM subsystems $\{P\}$. This is done in the next subsection. Obviously, if we do not introduce any approximations and there is only one subsystem or orbital domain that corresponds to all orbitals of the system, Eqs. (23)–(38) become equivalent to the exact expressions, Eqs. (9) and (11) for the CCSD correlation energy and Eqs. (13) and (14) for the triples correction to CCSD. In this case, the CIM and canonical CC calculations yield identical results, i.e., $\Delta\tilde{E}^{(CCSD)} = \Delta E^{(CCSD)}$ and $\delta\tilde{E}^{(2,3)} = \delta E^{(2,3)}$. The key idea of the CIM-CC formalism is to construct the orbital domains or subsystems $\{P\}$ such that one can determine the correlation energy contributions $\delta\tilde{E}_{i'}^{(CCSD)}$ and $\delta\tilde{E}_{i'}^{(2,3)}$ by solving the relevant CC equations in small orbital bases defining subsystems $\{P\}$ rather than in the large basis set of the entire system. Since the numbers of occupied and unoccupied orbitals, n_o and n_u, respectively, characterizing individual subsystems are in practice significantly smaller than the analogous orbital numbers characterizing the entire large system, and since the numbers of orbitals in individual subsystems are virtually independent of the system size, as they reflect on the immediate molecular environment within the large system that changes relatively little when the entire system is grown, the total CPU time of the CIM-CC calculations does not exceed M times the CPU time of the CC calculation for the largest CIM subsystem, where M is the total number of CIM subsystems. In particular, the CPU time required by the CIM-CCSD calculations does not exceed $M \times k_{\mathrm{iter}} n_{\mathrm{o}}^2 n_{\mathrm{u}}^4$, where n_{o} and n_{u} are the numbers of occupied and unoccupied orbitals in the largest CIM subsystem and k_{iter} is the maximum number of CCSD iterations. Similarly, the CPU time needed to obtain the triples corrections to the CCSD energy in the CIM-CR-CC(2,3) and CIM-CCSD(T) calculations does not exceed $M \times n_{\mathrm{o}}^3 n_{\mathrm{u}}^4$ times a small numerical prefactor related to the type of triples correction [CR-CC(2,3), CCSD(T), etc.] and the use of QCMOs which are needed to retain the non-iterative character of the local triples corrections [as already mentioned, the use of QCMOs results in a prefactor of 3; the use of CR-CC(2,3) introduces an additional prefactor of 2]. Once the molecular system under consideration becomes sufficiently large, these CPU time estimates characterizing the CIM-CC calculations become orders of magnitude smaller than those corresponding to the analogous canonical CC calculations, since the values of n_{o} and n_{u} characterizing the CIM subsystems are considerably smaller than the n_{o} and n_{u} values defining the basis set of the entire large system. Since the values of n_{o} and n_{u} characterizing different subsystems in a given CIM-CC calculation are often similar, the CPU times of the CIM-CC calculations scale nearly linearly with the number of subsystems and system size. Furthermore, since all CC calculations for individual subsystems, beginning with the required integral transformations from the AO to LMO bases defining subsystems and ending with the actual CCSD, CR-CC(2,3), and CCSD(T) calculations, are independent of one another, the resulting CIM-CC algorithms are trivially parallel. For example, if run on M processors, where M matches the number of CIM subsystems, the CPU time of any CIM-CC calculation does not exceed the time needed to complete the corresponding single CC calculation for the largest subsystem. The analogous remarks apply to memory and disk requirements. For example, in our implementation of the CIM-CC methods based on the CCSD, CCSD(T), and CR-CC(2,3) codes described in Refs. [24, 93], the memory requirements of a subsystem CCSD, CR-CC(2,3), or CCSD(T) calculation scale as $n_{\mathrm{o}} n_{\mathrm{u}}^3$ times a small prefactor that depends on the particular CC method [1 for CCSD and CCSD(T), 2 for CR-CC(2,3)], where n_{o} and n_{u} are the numbers of occupied and unoccupied orbitals defining a given subsystem. Thus, if one executes all of the subsystem CC computations in a sequential, subsystem-after-subsystem manner, using one processor only, the total memory requirements of the entire CIM-CC calculation will not exceed a small numerical prefactor, related to the CC method employed, times $n_{\mathrm{o}} n_{\mathrm{u}}^3$, where n_{o} and n_{u} are the numbers of occupied and unoccupied orbitals in the largest CIM subsystem. This means that the memory requirements of the CIM-CC calculations executed on a single processor are practically independent of the size of the system, because, as already mentioned, the individual subsystems are almost independent of the system size, reflecting on the nature of the immediate molecular environment within the large system that does not change much when the entire system is grown. If the CIM-CC calculations are executed in a parallel rather than sequential manner, where each subsystem calculation is executed on a different processor, the memory requirements per processor do not exceed a small prefactor associated with the CC method employed times $n_o n_u^3$, where n_o and n_u are the numbers of occupied and unoccupied orbitals in the largest CIM subsystem. All of these features make the CIM-CC calculations orders of magnitude less demanding than the corresponding canonical CC calculations, in terms of both the CPU time and the required memory. The analogous comments apply to disk requirements.

2.3 Determination of Local Orbital Domains for the CIM-CCSD, CIM-CR-CC(2,3), and CIM-CCSD(T) Calculations

From the above description, it immediately follows that the accuracy and computational efficiency of the CIM-CC calculations defined through Eqs. (23)–(27) and (31)–(38) largely depend on a suitable design of the orbital domains $\{P\}$ defining the individual CIM subsystems. Moreover, in order to keep the computer costs at a reasonably low level and in order to make sure that all post-RHF steps of the CIM-CC calculation, including the transformation of one- and two-electron integrals from the AO to the relevant LMO bases that precedes the CC steps, are split into independent, inexpensive calculations for orbital subsystems, so that the only calculation that has to be performed for the entire system is the initial RHF calculation, the final LMOs defining the individual subsystems $\{P\}$ used in the CIM-CC calculations must be represented as linear combinations of the AOs defining subsystems rather than linear combinations of all AOs in the basis set. This means that in addition to the suitable design of the orbital domains that define individual subsystems, we must project the occupied and unoccupied LMOs corresponding to each subsystem $\{P\}$ on the subspace spanned by AOs corresponding to this $\{P\}$. Last, but not least, in defining the occupied and unoccupied LMOs for each subsystem, we have to make sure that all orbitals used in the CIM calculations are orthogonal or, at the very least, orthogonal to a very good approximation, so that one can use Eqs. (23)–(27) and (31)–(38), and other conventional CC expressions derived for orthogonal MO bases in the CIM-CC considerations.

The CIM framework, originally introduced in Refs. [38–40], and further developed in Ref. [102] and this work, meets the above requirements. The key algorithmic steps that lead to the design of the orbital domains defining the individual CIM subsystems, relevant to the development and efficient computer implementation of the CIM-CCSD, CIM-CR-CC(2,3), and CIM-CCSD(T) methods discussed in this work, are as follows:

Step 1. First, we solve the RHF equations for the entire system and localize the core and correlated occupied orbitals obtained in the RHF calculations. The resulting occupied LMOs are designated as $\tilde{\phi}_{i'}$. We use tilde to emphasize that the LMOs obtained through the raw localization of the canonical RHF orbitals, which are used in the initial steps of the CIM subsystem design and which are linear combinations of all AOs defining the entire large system, are not yet the final and more compact LMOs $\phi_{i'}$ used in the actual CIM-CC calculations, which are spanned, thanks to suitable projections, by the relatively small subsets of AOs assigned to individual CIM subsystems via the corresponding AO domains defined in Step 7. As mentioned above, in the practical implementation of the CIM-CCSD, CIM-CR-CC(2,3), and CIM-CCSD(T) methods discussed in this chapter and related work [102], we use the Boys localization procedure to construct occupied LMOs (at least in principle, we could use any localization technique except for the Pipek–Mezey method, which relies on the separation of the σ and π orbitals that would lead to zero values of the Fock matrix elements between the σ and the π LMOs, independent of the distance between them, posing problems in our definition of orbital domains, as explained below).

Step 2. Next, we assign the AO domain $\Omega(i')$ to each occupied LMO $\tilde{\phi}_{i'}$ (including core MOs) obtained in Step 1. The AO domain $\Omega(i')$ corresponds to the atom or atoms that contribute most to the Mulliken charge of $\tilde{\phi}_{i'}$. We follow the recipe of Hampel and Werner [55] in which we sort all atoms in the system according to decreasing Mulliken orbital charges contributing to each $\tilde{\phi}_{i'}$. We define the AO domain $\Omega(i')$ by selecting the atoms with the largest atomic charges contributing to $\tilde{\phi}_{i'}$ until the sum of the atomic charges of selected atoms exceeds 1.98. Usually, when the orbitals are reasonably well localized, each AO domain $\Omega(i')$ contains one or two atoms (one atom if $\tilde{\phi}_{i'}$ is a core or a lone-pair LMO). The AO domains $\Omega(i')$ are used in Steps 7–12 to define the final sets of occupied and unoccupied orbitals employed in the CIM-CC calculations.

Step 3. After localizing the occupied MOs and assigning AO domains to each one of them, we construct *central LMO domains* $[i']$ which are associated with the individual correlated LMOs $\tilde{\phi}_{i'}$. The central domain $[i']$ consists of orbital $\tilde{\phi}_{i'}$, regarded as a central orbital, and its immediate environment, referred to as the *primary environment* of $\tilde{\phi}_{i'}$, which is obtained by analyzing the Fock matrix elements $\langle \tilde{\phi}_{i'} | f | \tilde{\phi}_{j'} \rangle$ for all values of j' that correspond to the correlated occupied LMOs of the entire system. The primary environment of $\tilde{\phi}_{i'}$ consists of those correlated LMOs $\tilde{\phi}_{j'}$ which satisfy the condition $|\langle \tilde{\phi}_{i'} | f | \tilde{\phi}_{j'} \rangle| > \zeta_1$, where ζ_1 is a suitably chosen numerical threshold. As advocated in Refs. [38–40], the primary environment of $\tilde{\phi}_{i'}$ defined in this manner consists of the occupied LMOs which are spatially close to $\tilde{\phi}_{i'}$ and which contribute to the correlation energy components of the $\delta \tilde{E}_{i'}^{(CCSD)}$ and $\delta \tilde{E}_{i'}^{(2,3)}$ type, defined by Eqs. (25) and (26) or (31) and Eqs. (27) and (32), respectively, in a substantial manner through the most significant cluster amplitudes that are sizable only when the occupied LMOs involved in their definition are spatially close (for example, the $t_{a'b'}^{i'j'}$ amplitudes are significant only when the $\tilde{\phi}_{i'}$ and $\tilde{\phi}_{j'}$ LMOs, or their final, more compact CIM counterparts $\phi_{i'}$ and $\phi_{j'}$ obtained in the subsequent steps discussed below, are spatially close). The off-diagonal matrix elements $\langle \tilde{\phi}_{i'} | f | \tilde{\phi}_{j'} \rangle$ provide a more robust measure of the spatial separation of $\tilde{\phi}_{i'}$ and $\tilde{\phi}_{j'}$ than the actual distance criterion (used, for example, in Ref. [55]). The central domain associated with the LMO $\tilde{\phi}_{i'}$ can be symbolically designated as $[i'] = (i'; j_1', \ldots, j_\kappa')$, where index i' represents the central orbital $\tilde{\phi}_{i'}$ and $j_1', \ldots, j_\kappa'$ correspond to the correlated

LMOs $\tilde{\phi}_{j_1'}, \ldots, \tilde{\phi}_{j_\kappa'}$ defining the primary environment of $\tilde{\phi}_{i'}$. Clearly, the number of central domains equals the number of all occupied LMOs used in the post-SCF calculations.

Step 4. It may happen (and often does happen) that some central domains are totally embedded in larger central domains. If this happens, we eliminate the smaller central domains and make the central orbitals of these smaller domains central orbitals of the larger domain they belong to. By repeating this process, we eventually arrive at the smallest possible number of larger central domains, containing one or more central orbitals, which are referred to as the *irreducible central domains*. For example, if Step 3 generates the following four central domains [17] = (17;20,25,27,38,50,57,64), [20] = (20;17,25,38,64), [38] = (38;17,20,27,64), and [64] = (64;17,20,38,57), it is immediately obvious that the smaller central domains [20], [38], and [64] are totally embedded in the central domain [17]. Thus, we eliminate the smaller domains [20], [38], and [64], and make orbitals 20, 38, and 64 central orbitals of the larger domain [17], which includes domains [20], [38], and [64] as orbital subsets. The resulting irreducible central domain is designated as [17,20,38,64] = (17,20,38,64;25,27,50,57). In general, the irreducible central domain $[i_1', \ldots, i_\alpha']$ is defined as $[i_1', \ldots, i_\alpha'] = (i_1', \ldots, i_\alpha'; j_1', \ldots, j_\beta')$, where $\tilde{\phi}_{i_1'}, \ldots, \tilde{\phi}_{i_\alpha'}$ are the corresponding central orbitals and $\tilde{\phi}_{j_1'}, \ldots, \tilde{\phi}_{j_\beta'}$ are the remaining orbitals from the primary environments of orbitals $\tilde{\phi}_{i_1'}, \ldots, \tilde{\phi}_{i_\alpha'}$. The number of irreducible central domains is often substantially smaller than the number of original central domains. Moreover, each correlated occupied LMO is central in exactly one irreducible central domain.

Step 5. Once the irreducible central domains are constructed, we add the *secondary environment* to each one of them. We do this by examining the off-diagonal elements of the Fock matrix $\langle \tilde{\phi}_{i'} | f | \tilde{\phi}_{j'} \rangle$, where $\tilde{\phi}_{i'}$ runs over all (central and primary environment) orbitals from a given irreducible central domain $[i_1', \ldots, i_\alpha']$ and $\tilde{\phi}_{j'}$ runs over all correlated occupied LMOs of the entire system that are not included in $[i_1', \ldots, i_\alpha']$. The secondary environment of the irreducible central domain $[i_1', \ldots, i_\alpha']$ consists of those correlated LMOs $\tilde{\phi}_{j'}$ that satisfy the condition $|\langle \tilde{\phi}_{i'} | f | \tilde{\phi}_{j'} \rangle| > \zeta_2$, where ζ_2 is an appropriate numerical threshold which is always at least as large as the threshold parameter ζ_1 defining the primary environments. By adding the secondary environment to the irreducible central domain, we obtain the *reducible full LMO domain*. If $\tilde{\phi}_{j_{\beta+1}'}, \ldots, \tilde{\phi}_{j_\gamma'}$ are the LMOs defining the secondary environment of the irreducible central domain $[i_1', \ldots, i_\alpha']$, the resulting reducible full LMO domain can be written as $\langle i_1', \ldots, i_\alpha' \rangle = (i_1', \ldots, i_\alpha'; j_1', \ldots, j_\beta'; j_{\beta+1}', \ldots, j_\gamma')$. The reducible full LMO domain consists of the central orbitals $\tilde{\phi}_{i_1'}, \ldots, \tilde{\phi}_{i_\alpha'}$, the primary environment orbitals $\tilde{\phi}_{j_1'}, \ldots, \tilde{\phi}_{j_\beta'}$ selected with the help of threshold ζ_1, and the secondary environment orbitals $\tilde{\phi}_{j_{\beta+1}'}, \ldots, \tilde{\phi}_{j_\gamma'}$, selected with the help of threshold $\zeta_2 (\geq \zeta_1)$, as described above. Examples discussed in the next sections illustrate the dependence of the CIM-CC results on the values of ζ_1 and ζ_2, which we typically set at 0.002–0.01 for ζ_1 and around 0.01 for ζ_2 (at least for the systems with light atoms that we have examined so far), although other choices are always possible and we continue to examine the issue of the optimum values of ζ_1 and ζ_2. Clearly, the larger ζ_1 and ζ_2, the smaller the computer cost of the CIM-

CC calculations. In the actual applications of the CIM-CCSD, CIM-CCSD(T), and CIM-CR-CC(2,3) methods, our preference is to choose the ζ_1 and ζ_2 values such that they enable us to recover practically all (to within 0.1% or so) of the corresponding correlation energies and such that they allow us to reproduce the relative energetics of different molecular structures on a potential energy surface to within 1 kcal/mol or a fraction thereof at a small fraction of the cost of the equivalent canonical CC calculations.

Step 6. In analogy to the central LMO domains, it often happens that some smaller reducible full LMO domains obtained in Step 5 are embedded in larger reducible full domains, sometimes in more than one. If we encounter such a situation, we automatically drop the smaller reducible full LMO domains and make the central orbitals of these smaller reducible full domains central in all of the larger domains that encompass the smaller ones. By repeating this process and inspecting the retained domains, we eventually end up with the minimum set of *irreducible full LMO domains*, which in general consist of the central, primary environment, and secondary environment orbitals that originate from the reducible full LMO domains obtained in Step 5. The composition of each irreducible full LMO domain can be symbolically represented as $(i_1{}', \ldots, i_\delta{}'; j_1{}', \ldots, j_\varepsilon{}')$, where $\tilde{\phi}_{i_1{}'}, \ldots, \tilde{\phi}_{i_\delta{}'}$ are the corresponding central orbitals and $\tilde{\phi}_{j_1{}'}, \ldots, \tilde{\phi}_{j_\varepsilon{}'}$ are the LMOs defining the environment (a combination of the primary and secondary environments). The shorthand notation for such an irreducible full LMO domain is $(i_1{}', \ldots, i_\delta{}')$. *The central and environment orbitals of a given irreducible full LMO domain define the occupied LMOs of the corresponding CIM subsystem* $\{P\}$. As implied by the steps described above, *each correlated occupied LMO is central in at least one subsystem* $\{P\}$, which is key for the validity of Eqs. (23)–(38) for the CIM-CCSD, CIM-CR-CC(2,3), and CIM-CCSD(T) correlation energies (and the analogous equations one could write for other single-reference CC methods). The construction of the unoccupied localized orbitals that are associated with a given subsystem $\{P\}$ as well as the final, more compact form of the occupied LMOs that define it, which are used in the CIM-CC calculations, requires a few additional steps which are summarized below as Steps 7–12.

Step 7. According to Eqs. (26), (31), and (32), in order to determine the specific correlation energy contribution $\delta\tilde{E}_{i'}^{(\mathrm{CCSD})}(\{P_{i'}\})$ or $\delta\tilde{E}_{i'}^{(2,3)}(\{P_{i'}\})$ associated with a given occupied spin-orbital i', which is one of the central spin-orbitals defining the subsystem $\{P_{i'}\}$ (in the closed-shell case discussed in this chapter, every central orbital $\phi_{i'}$ produces two central spin-orbitals corresponding to α and β spins), we have to construct the suitable set of unoccupied LMOs that define the most significant cluster amplitudes $t_{a'}^{i'}$ and $t_{a'b'}^{i'j'}$, and the resulting coefficients $\ell_{i'j'k'}^{a'b'c'}$ and $\mathfrak{M}_{a'b'c'}^{i'j'k'}$ (which are in reality, as discussed in the previous section, further transformed to a QCMO, mixed-index-type form) when $\phi_{j'}$ in $t_{a'b'}^{i'j'}$ and $\phi_{j'}$ and $\phi_{k'}$ in $\ell_{i'j'k'}^{a'b'c'}$ and $\mathfrak{M}_{a'b'c'}^{i'j'k'}$ are the occupied (central and environment) LMOs associated with $\{P_{i'}\}$. This is accomplished by determining first the projected occupied LMOs and the PAOs that correspond to the *extended subsystem* $\{\bar{P}_{i'}\}$ $(\{\bar{P}_{i'}\} \supseteq \{P_{i'}\})$, following Steps 8–10, which are subsequently orthogonalized, cleaned of redundancies, and projected back on the AO domain of $\{P_{i'}\}$ (Steps 11 and 12). Throughout this chapter, we use a notation in which the extended subsystem corresponding to subsystem $\{P\}$ is designated by writing a bar over the letter P. In general, the extended subsystem $\{\bar{P}\}$ is defined through the AOs of the corresponding system $\{P\}$ augmented by the AOs of the *buffer atoms* that are added to the set of atoms that correspond to $\{P\}$. The AO set that corresponds to subsystem $\{P\}$, which forms the *AO domain of subsystem* $\{P\}$ designated as $\Omega(P)$, is trivially obtained by forming a union of the AO domains $\Omega(i')$ determined in Step 2 and corresponding to all occupied LMOs $\tilde{\phi}_{i'}$ that belong to subsystem $\{P\}$. The buffer atoms associated with subsystem $\{P\}$ are defined as all non-hydrogen atoms within a distance of $4.0\,\text{Å}$ from any atom of subsystem $\{P\}$ plus all hydrogen atoms bonded to the non-hydrogen atoms of subsystem $\{P\}$ and its buffer, where the hydrogen atom is recognized as bound to the non-hydrogen atom X if the distance between X and this hydrogen atom is less than $r(\text{H}) + r(X) + 0.168\,\text{Å}$, with $r(\text{H})$ and $r(X)$ representing the van der Waals radii of H and X tabulated in the Supporting Information to Ref. [104]. Clearly, we could choose other values of the interatomic distances used to define the buffer atoms, but all of our numerical tests to date indicate that the above choices of the interatomic distances are sufficient for attaining high accuracies in the CIM-CC calculations. The AOs that correspond to the atoms of subsystem $\{P\}$ and to all of its buffer atoms form the *AO domain of the extended subsystem* $\{\bar{P}\}$, designated as $\Omega(\bar{P})$. In the following description, the number of all AOs in $\Omega(\bar{P})$ will be designated as $n(\bar{P})$. By design, $\Omega(\bar{P})$ contains the AO domain of the corresponding subsystem $\{P\}$, i.e., $\Omega(P)$, as a subset. In order to define the PAOs that are used to construct the unoccupied LMOs of the extended subsystem $\{\bar{P}\}$ and, eventually, the unoccupied LMOs of the corresponding subsystem $\{P\}$, we must first determine the occupied LMOs associated with the extended subsystem $\{\bar{P}\}$. This is done in Step 8.

Step 8. The occupied LMOs associated with the extended subsystem $\{\bar{P}\}$ contain the following orbitals: (i) all core LMOs that are associated with the atoms of $\{\bar{P}\}$, (ii) all correlated LMOs of the corresponding subsystem $\{P\}$, and (iii) the additional occupied LMOs other than those included in categories (i) and (ii) that can be assigned to the extended subsystem $\{\bar{P}\}$ based on the information about the AO domains generated in Step 2. Specifically, we assign the additional occupied LMO $\tilde{\phi}_{i'}$ to the extended subsystem $\{\bar{P}\}$ if the corresponding AO domain $\Omega(i')$, constructed in Step 2, is entirely included in the AO domain of $\{\bar{P}\}$. For example, if the AO domain of a given occupied LMO $\tilde{\phi}_{i'}$ consists of the AOs of two atoms, where one of the two atoms is in the extended subsystem $\{\bar{P}\}$ and another outside $\{\bar{P}\}$, this orbital $\tilde{\phi}_{i'}$ is not assigned to $\{\bar{P}\}$. In this particular case, both atoms defining the AO domain of LMO $\tilde{\phi}_{i'}$ would have to belong to the AO domain of the extended subsystem $\{\bar{P}\}$ for this LMO to be assigned to $\{\bar{P}\}$. In the following, the number of occupied LMOs associated with the extended subsystem $\{\bar{P}\}$ is designated as $n_o(\bar{P})$.

Step 9. Once the occupied LMOs $\tilde{\phi}_{i'}$ and all AOs χ_μ that are associated with each extended subsystem $\{\bar{P}\}$ are determined, as described in the previous two steps, we proceed with the construction of the unoccupied LMOs corresponding to each $\{\bar{P}\}$. This requires a few operations that are summarized in this step and in the following Steps 10 and 11. In the first operation, following the recipe suggested by Boughton and Pulay [105], we project each occupied LMO $\tilde{\phi}_{i'}$ corresponding to the extended subsystem $\{\bar{P}\}$, which up to this point has been represented as a linear combination of all AOs of the entire large system, on the extended AO domain $\Omega(\bar{P})$. In this way, all occupied LMOs assigned to the extended subsystem $\{\bar{P}\}$ are replaced by their $\Omega(\bar{P})$-projected counterparts which are linear combinations of the $n(\bar{P})$ AOs associated with this particular $\{\bar{P}\}$ only rather than all AOs in the basis describing the entire large system. The resulting orbitals $\bar{\phi}_{i'}$ are defined as follows:

$$\bar{\phi}_{i'} = \sum_{\mu \in \Omega(\bar{P})}^{n(\bar{P})} A_{\mu i'} \chi_\mu, \tag{41}$$

where the linear expansion coefficients $A_{\mu i'}$ are determined by minimizing the square of the norm of $(\bar{\phi}_{i'} - \tilde{\phi}_{i'})$,

$$||\bar{\phi}_{i'} - \tilde{\phi}_{i'}||^2 = \int [\bar{\phi}_{i'}(r) - \tilde{\phi}_{i'}(r)]^2 \, dr, \tag{42}$$

to make each new orbital $\bar{\phi}_{i'} \in \{\bar{P}\}$ as similar as possible to the original occupied LMO $\tilde{\phi}_{i'}$. This leads to a system of $n(\bar{P})$ linear equations for the $n(\bar{P})$ coefficients $A_{\mu i'}$, $\mu \in \Omega(\bar{P})$, for every i' (cf. Ref. [105]). Since the new occupied LMOs $\bar{\phi}_{i'}$ are no longer strictly orthonormal, and we need orthonormal LMOs in our algorithm, we orthogonalize and normalize them using the symmetric orthogonalization. This may lead to a slight modification of the values of the linear expansion coefficients $A_{\mu i'}$ in Eq. (41) compared with the values of these coefficients resulting from the minimization of $||\bar{\phi}_{i'} - \tilde{\phi}_{i'}||^2$. Interestingly, all of our experiences to date demonstrate that the LMOs $\bar{\phi}_{i'}$ obtained by projecting the original LMOs $\tilde{\phi}_{i'}$ assigned to the extended subsystem $\{\bar{P}\}$ on the AO domain $\Omega(\bar{P})$, although not strictly orthogonal, are to a very good approximation orthogonal, so that one can safely use them (after re-normalization) in practical CIM-CC calculations as if they were strictly orthogonal. Indeed, if we use the raw projected LMOs $\bar{\phi}_{i'}$ defined through Eqs. (41) and (42) in the CIM-CC calculations, without the re-orthogonalization after the projection of each $\tilde{\phi}_{i'}$ on $\Omega(\bar{P})$, instead of the symmetrically orthogonalized orbitals $\bar{\phi}_{i'}$, the calculated CIM-CCSD and CIM-CR-CC(2,3)/CIM-CCSD(T) correlation energies $\Delta \tilde{E}^{(\text{CCSD})}$ and $\delta \tilde{E}^{(2,3)}$, Eqs. (23) and (24), respectively, alter by only small fractions of a millihartree.

Step 10. Next, following the ideas of Pulay and Saebø [83–87], once the projected orthonormal (or, if we choose not to orthogonalize them, virtually orthogonal and normalized) occupied LMOs $\bar{\phi}_{i'}$ are constructed, we proceed toward the determination of the unoccupied orbital space associated with the extended subsystem $\{\bar{P}\}$ by calculating the corresponding PAOs $\bar{\chi}_\rho$,

$$|\bar{\chi}_\rho\rangle = \left(1 - \sum_{i'=1}^{n_o(\bar{P})} |\bar{\phi}_{i'}\rangle\langle\bar{\phi}_{i'}|\right) |\chi_\rho\rangle = \sum_{\mu \in \Omega(\bar{P})}^{n(\bar{P})} B_{\mu\rho} |\chi_\mu\rangle, \tag{43}$$

where, as mentioned earlier, $n_o(\bar{P})$ is the number of occupied LMOs in $\{\bar{P}\}$, including the corresponding core orbitals, and $n(\bar{P})$ is the total number of all AOs in the AO domain $\Omega(\bar{P})$ of the extended subsystem $\{\bar{P}\}$. We use the PAOs $\bar{\chi}_\rho$, Eq. (43), to design the unoccupied orbital space associated with the extended subsystem $\{\bar{P}\}$, since, as explained in Refs. [83–87], the projected atomic functions $\bar{\chi}_\rho(\boldsymbol{r})$ are localized on essentially the same atomic centers as the corresponding original AO functions $\chi_\rho(\boldsymbol{r})$. For comparison, the direct localization of unoccupied MOs by the same conventional algorithms as those used to define the occupied LMOs [88–90] is generally impossible due to the serious convergence problems encountered when larger molecules and larger AO sets are examined. One could consider alternative ways of determining unoccupied LMOs using, for example, the ideas described in Ref. [106] – and we may explore these ideas in our future studies – but in this and related work [102] we focus on employing the well-established methodology of designing the unoccupied orbitals associated with the extended subsystem $\{\bar{P}\}$ via the PAOs defined by Eq. (43), which are intrinsically local and rather easy to deal with. By substituting Eq. (41) into Eq. (43), one can demonstrate that the $n(\bar{P}) \times n(\bar{P})$ matrix $\mathbf{B}$ of the linear expansion coefficients $B_{\mu\rho}$ which define the PAOs $\bar{\chi}_\rho$ can be obtained in the following manner:

$$\mathbf{B} = \mathbf{1} - \mathbf{A} \bullet \mathbf{A}^\dagger \bullet \mathbf{S}, \tag{44}$$

where $\mathbf{1}$ is the $n(\bar{P}) \times n(\bar{P})$ unit matrix, $\mathbf{A}$ is the $n(\bar{P}) \times n_o(\bar{P})$ matrix of the linear expansion coefficients $A_{\mu i'}$ defining the projected occupied LMOs $\bar{\phi}_{i'}$, Eq. (41), and $\mathbf{S}$ is the $n(\bar{P}) \times n(\bar{P})$ overlap matrix,

$$S_{\mu\nu} = \langle \chi_\mu | \chi_\nu \rangle, \tag{45}$$

involving the AO orbitals χ_μ from the extended AO domain $\Omega(\bar{P})$.

Step 11. The PAOs $\bar{\chi}_\rho(\boldsymbol{r})$ defined by Eq. (43) are localized on essentially the same atomic centers as the original AOs $\chi_\rho(\boldsymbol{r})$ and, as such, can serve as a basis for constructing the unoccupied LMOs associated with the extended subsystem $\{\bar{P}\}$, but we must perform a few additional operations on these PAOs to make them usable in the CIM-CC calculations, since they are linearly dependent and non-orthogonal. The corresponding overlap matrix $\bar{\mathbf{S}}$,

$$\bar{S}_{\rho\sigma} = \langle \bar{\chi}_\rho | \bar{\chi}_\sigma \rangle, \tag{46}$$

which one can determine using the expression

$$\bar{\mathbf{S}} = \mathbf{B}^\dagger \bullet \mathbf{S} \bullet \mathbf{B}, \tag{47}$$

where $\mathbf{B}$ and $\mathbf{S}$ are defined by Eqs. (44) and (45), respectively, is not a diagonal matrix and has $n_o(\bar{P})$ zero eigenvalues. Thus, we transform the set of linearly dependent and non-orthogonal PAOs $\bar{\chi}_\rho$, given by Eq. (43), into an orthonormal and nonredundant set of $n_u(\bar{P}) \equiv n(\bar{P}) - n_o(\bar{P})$ unoccupied orbitals

$$\bar{\phi}_{a'} = \sum_{\rho \in \Omega(\bar{P})}^{n(\bar{P})} C_{\rho a'} \bar{\chi}_\rho = \sum_{\mu \in \Omega(\bar{P})}^{n(\bar{P})} (\mathbf{B} \bullet \mathbf{C})_{\mu a'} \chi_\mu, \tag{48}$$

where the unknown transformation coefficients $C_{\rho a'}$ are determined by performing the canonical orthonormalization of the PAOs $\bar{\chi}_\rho$ during which we eliminate the $n_o(\bar{P})$ redundant orbitals. In order to do this, we bring the overlap matrix $\bar{\mathbf{S}}$, Eq. (46), to the diagonal form represented by matrix Λ,

$$\Lambda = \mathbf{U}^\dagger \bullet \bar{\mathbf{S}} \bullet \mathbf{U}, \tag{49}$$

remove the $n_o(\bar{P})$ columns corresponding to the zero eigenvalues of $\bar{\mathbf{S}}$ in the transformation matrix $\mathbf{U}$ and the corresponding zero diagonal elements in Λ, which results, respectively, in the $n(\bar{P}) \times n_u(\bar{P})$ transformation matrix $\mathbf{U}'$ and the $n_u(\bar{P}) \times n_u(\bar{P})$ diagonal matrix Λ' that no longer has zero eigenvalues, and calculate the final $n(\bar{P}) \times n_u(\bar{P})$ matrix $\mathbf{C}$ of the transformation coefficients $C_{\rho a'}$ that define the orthonormal set of $n_u(\bar{P})$ orbitals $\bar{\phi}_{a'}$ according to Eq. (48) using the expression

$$\mathbf{C} = \mathbf{U}' \bullet (\Lambda')^{-1/2}. \tag{50}$$

The resulting orbitals $\bar{\phi}_{a'}$ are orthonormal, since the corresponding overlap matrix $\left\| \langle \bar{\phi}_{a'} \mid \bar{\phi}_{b'} \rangle \right\|_{a',b'=1}^{n_u(\bar{P})}$, obtained by forming the product of matrices $\mathbf{C}^\dagger$, $\bar{\mathbf{S}}$, and $\mathbf{C}$, satisfies

$$\mathbf{C}^\dagger \bullet \bar{\mathbf{S}} \bullet \mathbf{C} = (\Lambda')^{-1/2} \bullet (\mathbf{U}')^\dagger \bullet \bar{\mathbf{S}} \bullet \mathbf{U}' \bullet (\Lambda')^{-1/2} = (\Lambda')^{-1/2} \bullet \Lambda' \bullet (\Lambda')^{-1/2} = \mathbf{1}', \tag{51}$$

where $\mathbf{1}'$ is the corresponding $n_u(\bar{P}) \times n_u(\bar{P})$ unit matrix and where we used the fact that $(\mathbf{U}')^\dagger \bullet \bar{\mathbf{S}} \bullet \mathbf{U}' = \Lambda'$ [cf. Eq. (49)]. We complete this step by localizing orbitals $\bar{\phi}_{a'}$ using the same Boys localization that we used in localizing occupied LMOs.

Step 12. This is the final step in the design of the suitable sets of occupied and unoccupied LMOs, $\phi_{i'}$ and $\phi_{a'}$, respectively, associated with the individual subsystems $\{P\}$, which can be utilized in the CIM-CC calculations of the correlation energy, using Eqs. (23), (25), and (26) or (31) in the CCSD case and (24), (27), and (32) in the case of the triples corrections of CR-CC(2,3) and CCSD(T). Up to this point, we have two sets of orbitals for each extended domain $\{\bar{P}\}$, namely, (i) the set of $n_o(\bar{P})$ orthonormal, occupied LMOs $\bar{\phi}_{i'}$, Eq. (41) [or their normalized, nearly orthogonal analogs if we choose not to symmetrically orthogonalize $\bar{\phi}_{i'}$'s, as discussed in Step 9] and (ii) the non-redundant set of $n_u(\bar{P})$ orthonormal and localized PAOs $\bar{\phi}_{a'}$, Eq. (48), where $n_u(\bar{P})$ is defined as a difference between the number of all AOs in the extended AO domain $\Omega(\bar{P})$ and the number of occupied LMOs $\bar{\phi}_{i'}$. Both sets of orbitals are represented as linear combinations of the $n(\bar{P})$ AOs associated with the corresponding extended domain $\{\bar{P}\}$, which contains the AOs of the corresponding subsystem $\{P\}$ that form the AO domain $\Omega(P)$ as a subset. In order to complete the process of designing the occupied and unoccupied orbital spaces associated with the individual subsystems $\{P\}$, which can be used to determine the CIM-CCSD correlation energies $\Delta \tilde{E}^{(\text{CCSD})}$ and the CIM-CR-CC(2,3) and CIM-CCSD(T) triples corrections $\delta \tilde{E}^{(2,3)}$ using Eqs. (23), (24), (25), (26) or (31), (27), and (32), and in order to make sure that all post-RHF steps of the CIM-CC calculations, including the integral transformation from the AO to the relevant LMO bases that precedes the CC steps, are split into independent inexpensive calculations for subsystems $\{P\}$, we project orbitals $\bar{\phi}_{i'}$ and $\bar{\phi}_{a'}$ defining each extended domain $\{\bar{P}\}$ onto the corresponding AO domain $\Omega(P)$, while removing the core and the additional occupied LMOs that were added in Step 8 during the construction of the extended subsystems $\{\bar{P}\}$ from the original subsystems $\{P\}$. In this way, the resulting occupied and unoccupied LMOs, $\phi_{i'}$ and $\phi_{a'}$, respectively, which define a given subsystem $\{P\}$, become linear combinations of the relatively small set of AOs that are assigned to this subsystem via the AO domain $\Omega(P)$ introduced in Step 7. As in the case of orbitals $\bar{\phi}_{i'}$, we obtain the final orbitals $\phi_{i'}$ and $\phi_{a'}$ for the CIM-CC correlation energy calculations according to Eqs. (23), (25), and (26) or (31), in the CCSD case, and Eqs. (24), (27), and (32), in the case of triples corrections, by exploiting the idea of minimizing the squares of the norms of $(\phi_{i'} - \bar{\phi}_{i'})$ and $(\phi_{a'} - \bar{\phi}_{a'})$, respectively, to make the final LMOs $\phi_{p'}$, where $p' = i'$ or a', which span the occupied and unoccupied orbital spaces of each subsystem $\{P\}$, as close as possible to the corresponding LMOs $\bar{\phi}_{p'}$ that belong to the corresponding extended subsystem $\{\bar{P}\}$. Thus, we obtain

$$\phi_{p'} = \sum_{\mu \in \Omega(P)}^{n(P)} D_{\mu p'} \chi_\mu, \tag{52}$$

where $p' = i'$ and a' for the occupied and unoccupied orbitals, respectively, $n(P)$ is the number of AOs in the AO domain of subsystem $\{P\}$, $\Omega(P)$, and the linear expansion coefficients $D_{\mu p'}$, $\mu \in \Omega(P)$, are determined by minimizing

$$||\phi_{p'} - \bar{\phi}_{p'}||^2 = \int [\phi_{p'}(\boldsymbol{r}) - \bar{\phi}_{p'}(\boldsymbol{r})]^2 \, d\boldsymbol{r}, \tag{53}$$

or by solving the equivalent system of $n(P)$ linear equations for the $n(P)$ coefficients $D_{\mu p'}$, $\mu \in \Omega(P)$, for every p' that Eq. (53) leads to (see Ref. [105]). As mentioned earlier, in determining the final set of occupied LMOs $\phi_{i'}$ for CIM-CC calculations, we only focus on the occupied correlated orbitals associated with a given subsystem $\{P\}$, i.e., the core and the additional occupied LMOs that were added in Step 8 to construct the extended subsystems are no longer considered. We use the symbol $n_{\text{o}}(P)$ to designate the number of occupied correlated orbitals $\phi_{i'}$ associated with subsystem $\{P\}$. Clearly, $n_{\text{o}}(P) < n_{\text{o}}(\bar{P}) \ll n_{\text{o}}$, where $n_{\text{o}}(\bar{P})$ is the number of occupied orbitals in the extended subsystem $\{\bar{P}\}$ and n_{o} is the total number of occupied orbitals in the entire large system (in some cases, $n_{\text{o}}(P) \ll n_{\text{o}}(\bar{P})$ as well). Furthermore, in analogy to orbitals $\bar{\phi}_{i'}$, the occupied LMOs $\phi_{i'}$ determined via Eqs. (52) and (53) are, to a very good approximation, orthogonal and, as such, after normalization, can safely be used in the CIM-CC calculations as if they were rigorously orthonormal. We have, however, an option in the CIM-CCSD, CIM-CR-CC(2,3), and CIM-CCSD(T) computer codes developed in this work and related studies [102], which enables us to re-orthogonalize and then renormalize orbitals $\phi_{i'}$, obtained through the projection of orbitals $\bar{\phi}_{i'}$ on the AO domain $\Omega(P)$, in which we re-orthogonalize orbitals $\phi_{i'}$ defined by Eqs. (52) and (53) using the symmetric orthogonalization procedure. This may lead to tiny changes in the values of the linear expansion coefficients $D_{\mu i'}$ in Eq. (52), but all of our experiences to date indicate microhartree-like changes in the CIM-CC energies resulting from the additional symmetric orthogonalization of the occupied orbitals $\phi_{i'}$. The situation is somewhat different in the case of the unoccupied orbitals $\phi_{a'}$. Some of the corresponding unoccupied orbitals $\bar{\phi}_{a'}$ from the extended subsystem $\{\bar{P}\}$ substantially change when projected on the AO domain of the corresponding subsystem $\{P\}$, $\Omega(P)$, indicating that they do not adequately describe the unoccupied space of subsystem $\{P\}$. These unoccupied orbitals are removed from our considerations. We only retain those $\Omega(P)$-projected orbitals $\phi_{a'}$ which satisfy the condition $||\phi_{a'} - \bar{\phi}_{a'}|| < \eta$, where η is a suitably chosen numerical threshold, i.e., the orbitals that are well within the AO domain $\Omega(P)$. Typically, we use the η values around 0.2 (see Sections 3 and 4). If $n_u(P)$ designates the number of final unoccupied orbitals $\phi_{a'}$ associated with subsystem $\{P\}$ and retained in the CIM-CC calculations, we can write $n_u(P) < n_u(\bar{P}) \ll n_u$, where $n_u(\bar{P})$ is the number of unoccupied orbitals in the extended subsystem $\{\bar{P}\}$ and n_u is the total number of unoccupied orbitals in the entire large system (in many cases, $n_u(P) \ll n_u(\bar{P})$ as well). As in the case of the occupied orbitals $\phi_{i'}$, the unoccupied orbitals $\phi_{a'}$ obtained by projecting the orthonormal orbitals $\bar{\phi}_{a'}$ on the AO domain $\Omega(P)$ are, to a very good approximation, orthogonal to one another and to the occupied orbitals $\phi_{i'}$ and, as such, can safely be used in the CIM-CC calculations employing the CC codes developed for orthonormal bases, but we have an option in our CIM-CCSD, CIM-CR-CC(2,3), and CIM-CCSD(T) computer codes of enforcing the strict orthogonality of the unoccupied orbitals $\phi_{a'}$ to one another and to the occupied orbitals $\phi_{i'}$. In that case, after eliminating the orbitals $\phi_{a'}$ which do not satisfy the aforementioned condition $||\phi_{a'} - \bar{\phi}_{a'}|| < \eta$, we project out the tiny contamination due to the occupied orbitals $\phi_{i'}$ by redefining the unoccupied orbitals $\phi_{a'}$ retained in the CIM-CC considerations as $\left(1 - \sum_{i' \in \{P\}}^{n_o(P)} |\phi_{i'}\rangle \langle\phi_{i'}|\right) |\phi_{a'}\rangle$, and by orthogonalizing and normalizing the resulting orbitals $\phi_{a'}$ using the symmetric orthogonalization procedure. These operations lead to small changes in the values of the $D_{\mu a'}$ coefficients in Eq. (52) compared with the raw values of these coefficients resulting from the minimization of $||\phi_{a'} - \bar{\phi}_{a'}||^2$. Our experience to date indicates that when the enforcement of the strict orthogonality of the unoccupied orbitals $\phi_{a'}$ to one another and to the occupied orbitals $\phi_{i'}$, as described above, is invoked, it is preferable to use a somewhat less tight threshold η for selecting the unoccupied LMOs $\phi_{a'}$, such as $\eta = 0.5$, to retain the same levels of accuracy as those observed in the CIM-CC calculations reported in this work. Again, however, we have not yet encountered situations where the enforcement of the strict orthogonality of orbitals $\phi_{a'}$ is a necessity. The use of the nearly orthogonal unoccupied orbitals $\phi_{a'}$, obtained by minimizing $||\phi_{a'} - \bar{\phi}_{a'}||^2$, eliminating the orbitals that do not satisfy the $||\phi_{a'} - \bar{\phi}_{a'}|| < \eta$ condition, and normalizing the remaining orbitals $\phi_{a'}$ without an additional re-orthogonalization, has been sufficient in all of our benchmark calculations and tests performed to date, including those discussed in Sections 3 and 4.

After determining the orbital subsystems $\{P\}$, consisting of the central and environment-occupied LMOs $\phi_{i'}$ and the associated unoccupied LMOs $\phi_{a'}$ which are represented by linear combinations of the relatively small numbers of AOs defining the corresponding AO domains $\Omega(P)$, as in Eq. (52), and which form the orthonormal (or, if we do not enforce strict orthogonality after the respective projections, normalized and nearly orthogonal) LMO sets, as described in the above Steps 1–12, we can proceed to the CIM-CC calculations. It is important to emphasize that for each subsystem $\{P\}$ the corresponding orbitals $\phi_{i'}$ provide the best possible representation, within the space of AOs that are assigned to subsystem $\{P\}$, of the original occupied LMOs $\tilde{\phi}_{i'}$ obtained in Step 1. This is guaranteed by the appropriate projections described in Steps 9 and 12, in which we force the final occupied LMOs $\phi_{i'}$, written as linear combinations of the AOs from the AO domain $\Omega(P)$, to be as similar as possible to the original occupied LMOs $\tilde{\phi}_{i'}$ which are exploited in the initial Steps 1–7 and which are linear combinations of all AOs in the basis set. The analogous remarks apply to the final set of unoccupied LMOs $\phi_{a'}$ which provide the best representation within the AO domain $\Omega(P)$ of those unoccupied LMOs $\bar{\phi}_{a'}$ assigned to the extended subsystem $\{\bar{P}\}$ that do not significantly change upon the projection on $\Omega(P)$, as discussed in Steps 10–12. With the appropriate choice of the parameters ζ_1 and ζ_2, which define the occupied orbital spaces of subsystems $\{P\}$, as described in Steps 3 and 5, the above design of the occupied and unoccupied LMOs $\phi_{i'}$ and $\phi_{a'}$ that define each subsystem $\{P\}$ provides us with an accurate representation of the exact contributions $\delta E_i^{(\mathrm{CCSD})}$, Eq. (11), and $\delta E_i^{(2,3)}$, Eq. (14), to the CCSD correlation energy $\Delta E^{(\mathrm{CCSD})}$, Eq. (9), and, after transforming LMOs $\phi_{i'}$ and $\phi_{a'}$ to a QCMO form, the CR-CC(2,3)/CCSD(T) triples correction $\delta E^{(2,3)}$, Eq. (13). We simply replace the exact Eqs. (9), (10), and (11) for the CCSD correlation energy $\Delta E^{(\mathrm{CCSD})}$ and Eqs. (13), (14), (15) and (21) for the triples correction $\delta E^{(2,3)}$ by the CIM-CC expressions defining the approximate CCSD correlation energy $\Delta \tilde{E}^{(\mathrm{CCSD})}$, Eqs. (23), (25), and (26) or (31), and the approximate triples correction $\delta \tilde{E}^{(2,3)}$, Eqs. (24), (27), and (32), in which we use orbitals $\phi_{i'}$ and $\phi_{a'}$ designed in the above Steps 1–12 and, if required, their QCMO analogs (obtained from LMOs $\phi_{i'}$ and $\phi_{a'}$ by the rediagonalizations of the occupied-occupied and unoccupied-unoccupied blocks of the Fock matrix in subsystems, as described in the previous section), to determine the relevant molecular integrals $f_{i'}^a$ or $f_{i'}^{a'}$ and $v_{i'j}^{ab}$ or $v_{i'j'}^{a'b'}$, the CCSD cluster amplitudes $t_a^{i'}$ or $t_{a'}^{i'}$ and $\tau_{ab}^{i'j}$ or $\tau_{a'b'}^{i'j'}$, and their left CCSD "lambda" analogs, the de-excitation amplitudes $\ell_{i'jk}^{abc}$, and the triply excited moments $\mathfrak{M}_{abc}^{i'jk}$ for each subsystem that enter the CIM-CCSD and CIM-CR-CC(2,3)/CIM-CCSD(T) expressions. For each occupied spin-orbital i' associated with the $\Omega(P_{i'})$-projected representation of the LMO $\phi_{i'}$, which is used to define the individual correlation energy components $\delta \tilde{E}_{i'}^{(\mathrm{CCSD})}(\{P_{i'}\})$ according to Eq. (26), where $\{P_{i'}\}$ is the subsystem containing orbital $\phi_{i'}$ as a central orbital (by design, each correlated occupied LMO is central in at least one subsystem), the summation over j' in Eq. (26) runs over all occupied spin-orbitals that belong to $\{P_{i'}\}$ which correspond to the $\Omega(P_{i'})$-projected central and environment LMOs $\phi_{j'}$ and the summations over a' and b' run over the unoccupied spin-orbitals that correspond to the unoccupied orbitals $\phi_{a'}$ and $\phi_{b'}$ which are assigned to the same subsystem $\{P_{i'}\}$ and, in analogy to the occupied orbitals $\phi_{i'}$ and $\phi_{j'}$, are linear combinations of the AOs from $\Omega(P_{i'})$. If we choose the QCMO-based CIM-CCSD formulation, which is our preferred choice due to faster convergence, where we use Eq. (31) to define $\delta\tilde{E}_{i'}^{(\mathrm{CCSD})}(\{P_{i'}\})$ instead of Eq. (26), the summation over j and the summations over a and b in Eq. (31) run over the occupied and unoccupied spin-orbitals that belong to $\{P_{i'}\}$ and that correspond to the occupied and unoccupied QCMOs which are represented as linear combinations of the AOs from $\Omega(P_{i'})$ as well, since QCMOs ϕ_j, ϕ_a, and ϕ_b are obtained through the linear transformations of the occupied and unoccupied LMOs defining $\{P_{i'}\}$, as in Eqs. (29) and (30). Similarly, for each occupied spin-orbital i' associated with a given LMO $\phi_{i'}$, which defines the triples correction components $\delta\tilde{E}_{i'}^{(2,3)}(\{P_{i'}\})$ via Eq. (32), where again $\{P_{i'}\}$ is the subsystem containing $\phi_{i'}$ as a central orbital, the summations over j and k and the summations over a, b, and c in Eq. (32) run over the occupied and unoccupied spin-orbitals that belong to $\{P_{i'}\}$ and that correspond to the occupied and unoccupied QCMOs which are linear transformations of the small sets of occupied and unoccupied LMOs defining $\{P_{i'}\}$ spanned by the AOs from $\Omega(P_{i'})$. In this case, the use of QCMOs is necessary to benefit from the perturbative expressions defining the canonical CR-CC(2,3) and CCSD(T) methods.

2.4 The Remaining Information About the CIM-CCSD, CIM-CR-CC(2,3), and CIM-CCSD(T) Computer Codes

In this chapter, we focus on the highly efficient, vectorized, and parallel implementation of the CIM-CCSD, CIM-CCSD(T), and CIM-CR-CC(2,3) approaches that utilize the spin-free formulation of the underlying CCSD, CCSD(T), and CR-CC(2,3) methods and computer codes developed in Refs. [24, 93]. These codes are integrated in the GAMESS package [94] and we use them, after suitable modifications, in the CC subsystem calculations. Thus, for each CIM subsystem $\{P\}$ determined using Steps 1–12 discussed in the previous section, we first solve the spin-free CCSD equations corresponding to Eqs. (5) and (6) in the relatively small basis of the $n_o(P)$-occupied and $n_u(P)$-unoccupied LMOs or QCMOs that define the specific $\{P\}$, use the resulting cluster amplitudes to determine the correlation energy contributions $\delta\tilde{E}_{i'}^{(\mathrm{CCSD})}(\{P\})$ for each central LMO $\phi_{i'} \in \{P\}$, and repeat the same procedure for every subsystem $\{P\}$. If we need the non-iterative triples correction of CIM-CR-CC(2,3), we continue the subsystem CC calculations by solving the left CCSD equations, Eqs. (19) and (20), followed by the determination of the amplitudes ℓ_{ijk}^{abc} and moments $\mathfrak{M}_{abc}^{ijk}$ in the basis of $n_o(P)$-occupied and $n_u(P)$-unoccupied QCMOs of each $\{P\}$ which are subsequently used to calculate the mix-index amplitudes and moments $\ell_{i'jk}^{abc}$ and $\mathfrak{M}_{abc}^{i'jk}$ that enter the triples correction contributions $\delta\tilde{E}_{i'}^{(2,3)}(\{P\})$ for each central LMO $\phi_{i'} \in \{P\}$, again repeating these steps for every subsystem $\{P\}$. If we have an interest in the CIM-CCSD(T) energy, we skip the iterative steps of the left CCSD subsystem calculations and, after solving the standard CCSD equations for each subsystem $\{P\}$ in the corresponding QCMO basis, proceed directly to the determination of the appropriate form of the approximate amplitudes ℓ_{ijk}^{abc} and leading contributions to moments $\mathfrak{M}_{abc}^{ijk}$ that define the CCSD(T) scheme in the QCMO basis defining $\{P\}$, followed by the determination of the corresponding mix-index amplitudes and moments $\ell_{i'jk}^{abc}$ and $\mathfrak{M}_{abc}^{i'jk}$ that enter the triples correction contributions $\delta\tilde{E}_{i'}^{(2,3)}(\{P\})$ of CIM-CCSD(T) associated with the central LMOs $\phi_{i'} \in \{P\}$, repeating these steps for every subsystem $\{P\}$. Once all subsystem CC calculations are completed, the resulting correlation energy contributions $\delta\tilde{E}_{i'}^{(CCSD)}(\{P_{i'}\})$ and $\delta\tilde{E}_{i'}^{(2,3)}(\{P_{i'}\})$ that are associated with the central occupied LMOs of all subsystems are used to determine the final CIM-CCSD, CIM-CR-CC(2,3), and CIM-CCSD(T) energies, as explained in the earlier subsections. In all of the above considerations, we rely on the fact that the occupied and unoccupied LMOs $\phi_{i'}$ and $\phi_{a'}$, and the corresponding QCMOs ϕ_i and ϕ_a defining the CIM subsystems used in our algorithm, are either strictly orthonormal or, if we choose not to enforce strict orthogonality in the LMO case, normalized and orthogonal to a very good approximation, so that we can make safe use of the CCSD, CCSD(T), and CR-CC(2,3) computer codes developed for the orthonormal bases in Refs. [24, 93], which are interfaced with the integral routines available in the GAMESS package [94], to perform the relevant CC subsystem calculations.

Prior to setting up and carrying out the CC subsystem calculations, we must transform the one- and two-electron integrals from the AO to LMO or QCMO bases corresponding to each subsystem $\{P\}$. Since the occupied and unoccupied LMOs, $\phi_{i'}$ and $\phi_{a'}$, respectively, representing each subsystem $\{P\}$ are linear combinations of the AOs corresponding to the AO domain $\Omega(P)$ of the same subsystem, as in Eq. (52), the integral transformations that are needed to define the one- and two-electron integrals, $f_{p'}^{q'} = \langle p' | f | q' \rangle$ and $v_{p'q'}^{r's'} = \langle p'q' | v | r's' \rangle - \langle p'q' | v | s'r' \rangle$, respectively, in the LMO subsystem bases $\{\phi_{p'}, \ p' \in \{P\}\}$, and the analogous AO $\rightarrow$ MO transformations that are needed to determine the one- and two-electron integrals f_p^q and v_{pq}^{rs} in the QCMO subsystem bases $\{\phi_p, \ p \in \{P\}\}$, which enter the CC equations for the individual subsystems $\{P\}$, can be performed completely independent of one another, resulting in the near linear scaling and embarrassing parallelism of our CIM-CCSD, CIM-CR-CC(2,3), and CIM-CCSD(T) algorithms. For example, the transformed two-electron integrals $\langle \phi_{p'}\phi_{q'} | v | \phi_{r'}\phi_{s'} \rangle$, where $\phi_{p'}, \phi_{q'}, \phi_{r'}, \phi_{s'} \in \{P\}$ represent the LMOs needed in the CC calculations for a given CIM subsystem $\{P\}$, are determined using a four-index transformation

$$\langle \phi_{p'}\phi_{q'} | v | \phi_{r'}\phi_{s'} \rangle \equiv (\phi_{p'}\phi_{r'} | \phi_{q'} \phi_{s'})$$

$$= \sum_{\mu \in \Omega(P)}^{n(P)} D_{\mu p'} \left[\sum_{\rho \in \Omega(P)}^{n(P)} D_{\rho r'} \left[\sum_{\nu \in \Omega(P)}^{n(P)} D_{\nu q'} \left[\sum_{\sigma \in \Omega(P)}^{n(P)} D_{\sigma s'} \left(\chi_\mu \chi_\rho | \chi_\nu \chi_\sigma \right) \right] \right] \right], \tag{54}$$

where $\left(\chi_\mu \chi_\rho | \chi_\nu \chi_\sigma\right)$ are the two-electron atomic integrals, written in the Mulliken notation, defined using the AOs from the AO domain $\Omega(P)$ of the same subsystem $\{P\}$ and the transformation coefficients $D_{\mu p'}$ are taken from Eq. (52). The only calculations that have to be performed for the entire system are the initial RHF calculation and the localization of the resulting occupied orbitals. All post-RHF steps of the CIM-CCSD, CIM-CR-CC(2,3), and CIM-CCSD(T) calculations, starting with the AO $\to$ MO integral transformations and integral sorting that precede the CC work, are naturally split into independent calculations in small orbital bases that define the CIM subsystems. Since our CIM-CC codes are interfaced with the GAMESS package and use the suitably modified CCSD, CR-CC(2,3), and CCSD(T) routines incorporated in GAMESS, we use the GAMESS integral, RHF, and transformation subroutines to set up the CC subsystem calculations. The orbital localization routines, the routines that construct the CIM subsystems and the relevant LMO and QCMO bases, and the main CIM-CC driver that calls the appropriate GAMESS or modified GAMESS subroutines to carry out the initial RHF calculations and to perform the CC subsystem calculations are standalone codes outside the GAMESS system.

The high efficiency of our CIM-CCSD, CIM-CR-CC(2,3), and CIM-CCSD(T) computer programs is achieved in two ways. The first factor that greatly contributes to the high efficiency of the CIM-CC codes developed and benchmarked in this and related [102] work is the aforementioned vectorization of the spin-free CCSD, CR-CC(2,3), and CCSD(T) codes, which are used to perform the CC calculations for each subsystem and which were developed in Refs. [24, 93], resulting from the extensive use of diagram factorization, recursively generated intermediates, and fast matrix multiplication routines from the BLAS library. The second factor that significantly enhances the efficiency of our CIM-CC codes is the embarrassing parallelism of the CIM formalism which enables one to perform the CC subsystem calculations completely independent of one another. Thus, if we have access to more than one processor, we divide the CIM-CC calculation into smaller groups of subsystem CC calculations. In the present parallel implementation of the CIM-CCSD, CIM-CR-CC(2,3), and CIM-CCSD(T) methods developed in Ref. [102] and discussed in this work, we proceed as follows: If a given CIM-CC calculation splits the system of interest into M orbital subsystems $\{P\}$ and we have access to $K < M$ processors, we initiate the calculation by submitting the first K CC subsystem jobs to K processors (following the idea of one CC subsystem calculation per processor). When one of the subsystem CC calculations is completed and the corresponding processor is freed up, our parallel CIM-CC program submits the next CC run from the group of the remaining $(M - K)$ CC subsystem jobs awaiting execution to the available processor, and we repeat this procedure until all subsystem jobs are executed. In the optimum case of having access to $K = M$ processors, we simultaneously submit all M CC subsystem calculations to the M processors we have access to, completing the entire CIM-CC calculation in the time required by the CC calculation for the largest CIM subsystem. Since the numbers of orbitals defining the CIM subsystems $\{P\}$ are often quite similar, particularly when the entire system becomes very large, the parallel CIM-CC calculations are characterized by reasonable scalability. We have developed the parallel CIM-CCSD, CIM-CR-CC(2,3), and CIM-CCSD(T) codes using both the OpenMP and the MPI protocols for parallelization. Thus, we can run our CIM-CC calculations on both the multiprocessor shared-memory platforms and loosely coupled clusters. The efficiency of our parallel CIM-CC codes is not significantly affected by the computer architecture type, since all CC subsystem calculations are executed independent of one another using separate sets of molecular integrals that correspond to individual subsystems $\{P\}$, eliminating the need for communication among processors during CC subsystem calculations. Other factors that contribute to the efficiency of our CIM-CC codes, such as the use of QCMOs in defining the CC subsystem calculations which helps to reduce the number of CCSD iterations that are needed to obtain converged CCSD energies and cluster amplitudes, have been discussed in the previous subsections.

Finally, before discussing examples of the CIM-CCSD, CIM-CR-CC(2,3), and CIM-CCSD(T) calculations, we would like to add that our CIM-CC programs enable flexible definitions of the CIM subsystems in cases that may require special attention. One of the most important cases in this category is the calculation of a molecular potential energy surface involving bond breaking, where the individual subsystems $\{P\}$, obtained using Steps 1–12 described in the previous section, may change with the nuclear geometry. If the CIM subsystems $\{P\}$ are allowed to automatically adjust to the nuclear geometry, as is normally the case when we follow Steps 1–12, one may produce discontinuities on the resulting CIM-CC potential energy surfaces. Thus, our CIM-CC codes have an option that allows us to force a particular subsystem composition, corresponding to one of the nuclear geometries along the bond-breaking coordinate, for all nuclear geometries of interest, as in the fixed-domain local correlation approximations discussed, for example, in Ref. [107]. The fixed-domain CIM-CC calculations yield smooth molecular potential energy surfaces which are characterized, as shown in Section 3.2, by small nonparallelity errors (NPEs) relative to the potential energy surfaces obtained in the corresponding canonical CC calculations. The modular design of our CIM-CC computer programs is also flexible enough to enable us to modify Steps 1–12 used to determine the CIM subsystems without changing the overall structure of our CIM-CC codes. We benefit from this flexibility in the developments discussed in Section 4, where we examine an alternative and more robust version of the CIM-CC theory which will be fully described in the future work and which modifies the original Steps 1–12 described in Section 2.3 to achieve better performance in the CIM-CC calculations involving larger basis sets with diffuse functions.

3 Benchmark Calculations

In this section, we examine several examples of benchmark CIM-CCSD, CIM-CR-CC(2,3), and CIM-CCSD(T) calculations that illustrate the performance of our CIM-CC codes described in Section 2, which are interfaced with the GAMESS package. The corresponding canonical CCSD, CR-CC(2,3), and CCSD(T) calculations are performed using the CC routines developed in Refs. [24, 93] and integrated in GAMESS as standard options. All of the calculations discussed in this chapter were performed on the Altix 3700Bx2 system from SGI equipped with the 1.6-GHz Itanium2 processors, provided by the High Performance Computing Center at Michigan State University.

The examples included in this and the subsequent sections include the normal alkanes C_nH_{2n+2}, with $n = 12$, 16, 20, 24, 28, and 32, and the $(H_2O)_n$ clusters with $n = 10$, 12, 14, 16, and 20. All the CC calculations discussed in this chapter employ the RHF determinant as a reference and freeze the core orbitals correlating with the $1s$ orbitals of the C and O atoms. In each case, we use the six Cartesian components of the d orbitals present in the AO basis sets used in our calculations. In this section, we focus on the CIM-CC calculations performed with the 6-31G(d) and 6-31G(d, p) basis sets [97, 98]. The CIM-CCSD calculations for $(H_2O)_n$ clusters with the larger 6-31++G(d, p) [97–99] and 6-311++G(d, p) [99, 100] basis sets, which contain diffuse functions and create an interesting as well as challenging situation for the CIM methodology, are discussed in the next section.

In each case, the canonical and subsystem CCSD correlation energies were converged to 10^{-6} hartree, and QCMOs were employed in the CC subsystem calculations. As explained in the previous section, the use of QCMOs is necessary to determine the CIM-CR-CC(2,3) and CIM-CCSD(T) triples corrections, while leading to faster convergence of the CCSD subsystem calculations. In each case, we considered a few different levels of the CIM-CC theory which differ in the numerical values of the parameters ζ_1 and ζ_2 that define the primary and secondary orbital environments of the CIM subsystems (see Steps 3 and 5 of the CIM subsystem design discussed in the previous section). In all the CIM-CC calculations reported in this section, we set the threshold parameter η for selecting the unoccupied LMOs of each subsystem at 0.2. This choice of η seems optimum for the CIM-CC methodology using Steps 1–12 described in the previous section. The modified variant of the CIM-CC theory, which is particularly useful for basis sets containing diffuse functions and which is discussed in Section 4, uses a different design of the CIM subsystems, but $\eta = 0.2$ is a reasonable choice of η for it as well.

3.1 Correlation Energies of Normal Alkanes

We begin our discussion by examining the results obtained in the CIM and canonical CCSD and CR-CC(2,3) calculations for a series of normal alkanes C_nH_{2n+2} with $n = 12$, 16, 20, 24, 28, and 32, as described by the 6-31G(d) basis set. We do not discuss the analogous canonical and CIM CCSD(T) calculations since the CCSD(T) and CR-CC(2,3) results are virtually identical. The equilibrium geometries of all the C_nH_{2n+2} systems were obtained in the RHF/6-31G(d) optimizations (using the analytic RHF derivatives available in GAMESS). The resulting canonical and CIM CCSD and CR-CC(2,3) correlation energies and the information about the numbers of correlated occupied and unoccupied orbitals, n_o and n_u, respectively, defining the C_nH_{2n+2} systems of interest are included in Table 1. The corresponding CPU timings and the information about the numbers of CIM subsystems used in the CIM-CC calculations are summarized in Table 2. In reporting the timings of a particular CIM-CC calculation, we provide two values, namely, the total CPU time required by the CCSD iterations and calculations of the CR-CC(2,3) triples correction for all CIM subsystems, and the CPU times associated with the corresponding CC calculations for the largest CIM subsystem. A similar way of reporting the CPU timings of CIM-CC calculations is employed throughout the rest of this chapter. The CPU timings of the canonical calculations correspond to the CCSD iterations and calculations of triples energy corrections. Notice that because of the rapidly growing costs of the canonical CC calculations with the system size we were unable to perform the canonical CCSD calculations for the C_nH_{2n+2} alkanes with $n \geq 24$. For the same reason, we were unable to determine the canonical CR-CC(2,3) triples corrections for the C_nH_{2n+2} systems with $n \geq 20$. We could, however, estimate the CPU timings of the canonical CCSD and CR-CC(2,3) calculations for the larger alkanes using the theoretical scalings of the time-determining computational steps defining these methods in terms of n_o and n_u, which apply to our CC GAMESS codes, and the available timing data for the smaller C_nH_{2n+2} systems.

Two levels of the CIM-CC theory were adopted in the calculations for the C_nH_{2n+2} systems. In level 1 of the CIM-CCSD and CIM-CR-CC(2,3) calculations, we set the parameters ζ_1 and ζ_2 that define the primary and secondary orbital environments of the CIM subsystems at 0.01 and 0.05, respectively. In level 2 of CIM-CCSD and CIM-CR-CC(2,3), we used $\zeta_1 = 0.01$ and $\zeta_2 = 0.02$, i.e., we made the parameter ζ_2 defining the secondary environment somewhat tighter.

The results in Tables 1 and 2 show that the CIM-CCSD and CIM-CR-CC(2,3) calculations for the C_nH_{2n+2} systems recover the corresponding canonical CC correlation energies to within 0.14% at the CCSD level and 0.04% at the CR-CC(2,3) level, while offering the nearly linear scaling of the total CPU time with

Table 1 Canonical and CIM CCSD and CR-CC(2,3) correlation energies for normal alkanes C_nH_{2n+2} as described by the 6-31G(d) basis set. The geometries were obtained in the RHF/6-31G(d) optimizations. In level 1 of the CIM theory, $\zeta_1 = 0.01$ and $\zeta_2 = 0.05$; in level 2, $\zeta_1 = 0.01$ and $\zeta_2 = 0.02$. All energies are in hartree

			CCSD correlation energy			CR-CC(2,3) correlation energy		
n	n_o	n_u	Canonical	CIM (level 1)[a]	CIM (level 2)[a]	Canonical	CIM (level 1)[a]	CIM (level 2)[a]
12	37	183	−1.72201	−1.72032 (99.90%)	−1.72062 (99.92%)	−1.77148	−1.77160 (100.01%)	−1.77180 (100.02%)
16	49	243	−2.29168	−2.28888 (99.88%)	−2.28932 (99.90%)	−2.35724	−2.35779 (100.02%)	−2.35811 (100.04%)
20	61	303	−2.86170	−2.85778 (99.86%)	−2.85883 (99.90%)	–	−2.94426	−2.94518
24	73	363	–	−3.42639	−3.42772	–	−3.53049	−3.53167
28	85	423	–	−3.99501	−3.99662	–	−4.11673	−4.11817
32	97	483	–	−4.56362	−4.56551	–	−4.70297	−4.70466

[a] The numbers in parentheses represent the fractions of the canonical correlation energies recovered by the CIM approach.

Table 2 CPU timings of the canonical and CIM CCSD and CR-CC(2,3) calculations for normal alkanes C_nH_{2n+2}, as described by the 6-31G(d) basis set. The geometries were obtained in the RHF/6-31G(d) optimizations. In level 1 of the CIM theory, $\zeta_1 = 0.01$ and $\zeta_2 = 0.05$; in level 2, $\zeta_1 = 0.01$ and $\zeta_2 = 0.02$. All times are in CPU hours

| | CCSD iterations | | | CR-CC(2,3) triples correction | | | |
| | | CIM | CIM | | CIM | CIM | |
n	Canonical[a]	(level 1)[b]	(level 2)[b]	Canonical[a]	(level 1)[b]	(level 2)[b]	No. of subsystems[c]
12	8.8	3.3 (0.7)	3.6 (0.7)	49.0	29.3 (6.5)	46.0 (8.0)	7/7
16	41.3	3.1 (0.4)	8.0 (1.1)	330.9	43.0 (4.8)	92.9 (15.7)	11/11
20	143.2	5.8 (0.5)	24.3 (1.5)	[1530]	57.0 (5.2)	244.3 (13.8)	15/43
24	[418]	7.2 (0.4)	28.6 (1.5)	[5383]	69.9 (4.0)	289.3 (10.0)	19/55
28	[1038]	8.8 (0.6)	38.5 (1.0)	[15683]	111.1 (8.2)	421.5 (12.0)	23/67
32	[2300]	9.9 (0.6)	49.8 (1.2)	[39700]	101.1 (5.1)	499.4 (12.0)	27/79

[a] The numbers in square brackets represent the estimated CPU times that would be required by the canonical CCSD or CR-CC(2,3) calculations. In the case of CCSD calculations, which converge in 15 iterations for $n = 12$, 16, and 20, the least square linear regression was used to generate the estimated $n = 24$, 28, and 32 values, using the equation $y = 4.489x + 2.685$, where x represents the value of $10^{-12}n_o^2n_u^4$ and y is the CPU time in hours. In the case of the CR-CC(2,3) triples correction $\delta E^{(2,3)}$, the linear regression was used to generate the estimated $n = 20$, 24, 28, and 32 values, using the equation $y = 0.798x + 3.693$, where x represents the value of $10^{-12}n_o^2n_u^4$ and y is the CPU time in hours.
[b] The numbers in parentheses represent the CPU times required by the calculations for the largest CIM subsystems.
[c] The numbers before and after the slash represent the numbers of subsystems for the CIM levels 1 and 2, respectively.

the system size and savings in the computer effort by orders of magnitude when the normal alkanes under consideration become larger. Interestingly, the CIM-CC methodology offers significant savings in the CPU time, when compared with the corresponding canonical calculations, even when the system under investigation is relatively small. For example, the level-1 CIM-CC calculations are 2–3 times less time consuming than the corresponding canonical calculations when we examine a system as small as $C_{12}H_{26}$. The tighter level 2 does not offer significant savings in the computer effort when the CIM-CR-CC(2,3) method is applied to $C_{12}H_{26}$, but this situation changes as soon as we move to the next alkane listed in Tables 1 and 2, namely, $C_{16}H_{34}$, where the level-2 CIM-CCSD and CIM-CR-CC(2,3) calculations are 4–5 times faster than the corresponding canonical CC calculations. When the $C_{16}H_{34}$ system is examined, the level-1 CIM-CC calculations become faster than the corresponding canonical calculations by a factor of 13 in the CCSD case and 8 in the case of the triples correction of CR-CC(2,3). When the larger alkanes, such as $C_{24}H_{50}$, $C_{28}H_{58}$, and $C_{32}H_{66}$, are examined, the CIM-CC calculations become faster than the corresponding canonical calculations by orders of magnitude. This can be illustrated by the CPU time speedups on the order of 200–400 observed in the level-1 CIM-CCSD and CIM-CR-CC(2,3) calculations and 50–80 in the corresponding level-2 calculations. All of these are substantial savings, particularly considering the fact that it becomes extremely difficult or impossible to perform the CCSD and

CR-CC(2,3) calculations for the larger alkanes, not only due to CPU time, but also due to the enormous memory and disk requirements.

The results in Table 2 demonstrate that the time spent on the CC calculations for the largest CIM subsystem does not significantly change with the system size, confirming our earlier remarks that the numbers of orbitals defining individual subsystems are almost independent of the system size, clearly reflecting on the intuitive observation that the immediate environment of a given orbital within the large molecular system does not substantially change when the entire system is grown. What changes is the number of CIM subsystems, which grows more or less linearly with the size of the entire system, as shown in Table 2. If we can run the CIM-CC calculations in parallel using as many processors as the number of CIM subsystems, which is an easy thing to do since the CIM-CC methodology is trivially parallel, requiring virtually no communication among processors, we can complete the calculations for each alkane listed in Tables 1 and 2 in more or less the same time, which is 0.4–0.7 hours for the level-1 CIM-CCSD calculations, about 1 hour for the level-2 CIM-CCSD calculations, 4.0–8.2 hours for the level-1 calculations of the CIM-CR-CC(2,3) triples correction, and 8.0–15.7 hours for the level-2 calculations of the CIM-CR-CC(2,3) correction due to triples. Clearly, it is quite encouraging to observe that we are able to perform the calculations of the CR-CC(2,3) triples correction for the $C_{32}H_{66}$ system, which would normally require about 40,000 CPU hours or 4.5 CPU years when the canonical formulation of CC theory was employed, reproducing practically 100% of the canonical CR-CC(2,3) correlation energy, in as little as about 5 hours on 27 processors or 101 CPU hours total when level 1 of the CIM-CC theory is used and about half a day on 79 processors or 499 hours total when level 2 of the CIM-CC theory is made use of.

The results in Table 1 also show that level 1 of the CIM-CC theory is essentially as accurate as level 2, while offering additional nontrivial savings in the computer effort due to the use of the larger value of ζ_2. Although there are applications where the use of the tighter thresholds ζ_1 and ζ_2 may be necessary (for example, weakly bound molecular clusters, such as the water clusters examined in Sections 3.3 and 4.2, where the binding energies involved are much smaller), it is very encouraging to observe that we can select relatively loose thresholds ζ_1 and ζ_2, which offer huge savings in the computer effort, and recover the CC correlation energies in covalently bound systems to within about 0.1%. The fact that we recover the CCSD and CR-CC(2,3) correlation energies for the alkanes for which we could perform the corresponding canonical CC calculations to within 2–4 millihartree at the CCSD level and fractions of a millihartree at the CR-CC(2,3) level, where correlation energies are already on the order of 2–3 hartree, indicates that we can achieve high accuracies in the CIM-CC calculations for covalently bound molecules.

Undoubtedly, the CIM-CCSD and CIM-CR-CC(2,3) results for the normal alkanes shown in Tables 1 and 2 are very promising, but we must keep in mind that the large fraction of the correlation energy recovered in the CIM-CC calculations may be sometimes misleading, since even a small fraction of the correlation energy, such as 0.1%, can easily translate into differences between the CIM and the canonical CC energies on the order of a few kilocalories per mole when the molecular system of interest is large. In chemistry, one often deals with the relative energies between different molecular structures that one encounters in the course of reactions involving bond breaking or when examining isomeric forms of a given molecular system or cluster. It is, therefore, essential for the *ab initio electronic structure methods*, particularly for the high-level approaches based on the CC wave function ansatz, that they offer an accurate description of the relative energetics, with errors on the order of 1 kcal/mol or a fraction of kcal/mol. The ability of the CIM-CC methods to reproduce the relative energetics provided by the canonical CC approaches in calculations involving single bond breaking in a covalent molecular system is discussed next.

3.2 Bond Breaking in Dodecane

In order to examine the performance of the CIM-CC methodology in a situation involving single bond breaking, we carried out the CIM-CR-CC(2,3) calculations for the dissociation of $C_{12}H_{26}$ into $C_{11}H_{23}$ and CH_3, in which the terminal C–C bond, labeled as C_1–C_2, is broken (see Fig. 1). It is established that the canonical CR-CC(2,3) approach using the RHF determinant as a reference provides an accurate description of single bond-breaking situations [24–26, 108–116]. It is, therefore, interesting to examine how accurate the CIM-CR-CC(2,3) results are when compared with the corresponding canonical CR-CC(2,3) calculations when the $C_{12}H_{26} \rightarrow C_{11}H_{23} + CH_3$ reaction is examined. It is certainly useful to know what is the maximum unsigned error (MUE) in the calculated CIM-CR-CC(2,3) energies along the C_1–C_2 bond-breaking coordinate relative to the corresponding energies obtained in canonical CR-CC(2,3) calculations. Even more importantly, we would like to determine the NPEs defining the dissociation curve resulting from the CIM-CR-CC(2,3) calculations relative to the corresponding canonical CR-CC(2,3) curve [NPE is defined as the difference between the maximum and the minimum signed errors in the CIM-CR-CC(2,3) energies relative to the corresponding canonical CR-CC(2,3) energies along the bond-breaking coordinate].

In order to compare the results of the CIM and canonical CR-CC(2,3) calculations for the dissociation of $C_{12}H_{26}$ into $C_{11}H_{23}$ and CH_3, we considered several stretches of the C_1–C_2 bond displayed in Fig. 1, while keeping the remaining geometrical parameters at their equilibrium values determined in the RHF/6-31G(d) optimization. If the equilibrium C_1–C_2 distance in dodecane is designated as $R_e(C_1 - C_2)$, we calculated the CIM and canonical CR-CC(2,3) energies at the

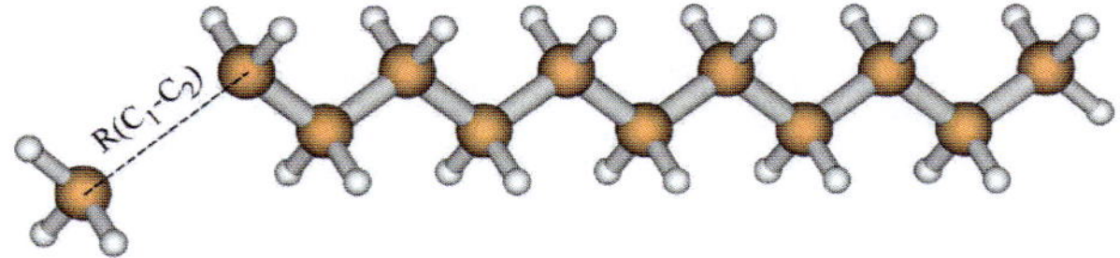

Fig. 1 The structures of dodecane with bond breaking along the C_1–C_2 bond

C_1–C_2 distances $R(C_1 - C_2)$ defined as the following multiples of $R_e(C_1 - C_2)$: $x R_e(C_1 - C_2)$, where $x = 0.7, 0.9, 1.0, 1.2, 1.5, 2.0, 2.5$, and 3.0. As in Section 3.1, the CIM and canonical CR-CC(2,3) calculations were performed with the 6-31G(d) basis set.

Four different levels of the CIM-CR-CC(2,3) theory were considered. In addition to levels 1 and 2 of CIM-CR-CC(2,3) introduced in the previous section, where the CIM parameters ζ_1 and ζ_2 are set at $(\zeta_1, \zeta_2) = (0.01, 0.05)$ and $(0.01, 0.02)$, respectively, we considered levels $1'$ and $2'$ which belong to the broader category of the fixed-domain CIM-CC methods mentioned at the end of the last section. In level $1'$ of CIM-CR-CC(2,3), the CIM subsystems at each geometry are precisely the same as at the equilibrium, $R(C_1 - C_2) = R_e(C_1 - C_2)$, and the subsystems at $R(C_1 - C_2) = R_e(C_1 - C_2)$ are defined by using $\zeta_1 = 0.01$ and $\zeta_2 = 0.05$. Thus, level $1'$ is a fixed-domain analog of the normal, variable-domain level 1. In level $2'$ of CIM-CR-CC(2,3), which is a fixed-domain counterpart of level 2, the CIM subsystems at each geometry are again exactly the same as at $R(C_1 - C_2) = R_e(C_1 - C_2)$, but the values of ζ_1 and ζ_2 defining the subsystems at $R(C_1 - C_2) = R_e(C_1 - C_2)$ are 0.01 and 0.02, respectively.

The results of the CIM-CR-CC(2,3) calculations using the above four levels of the CIM-CC theory are shown in Fig. 2. It is obvious from Fig. 2(a) that the CIM-CR-CC(2,3) energies are so close to the corresponding canonical energies that it is virtually impossible to see any difference between the potential energy curves resulting from the CIM-CR-CC(2,3) and canonical CR-CC(2,3) calculations when we plot them together. We have to examine the differences between the energies obtained in the CIM and canonical CR-CC(2,3) calculations as functions of $R(C_1 - C_2)$ [shown in Figs. 2(b) and (c)]. Depending on the C_1–C_2 distance, level 1 reproduces between 100.01 and 100.10% of the correlation energy obtained

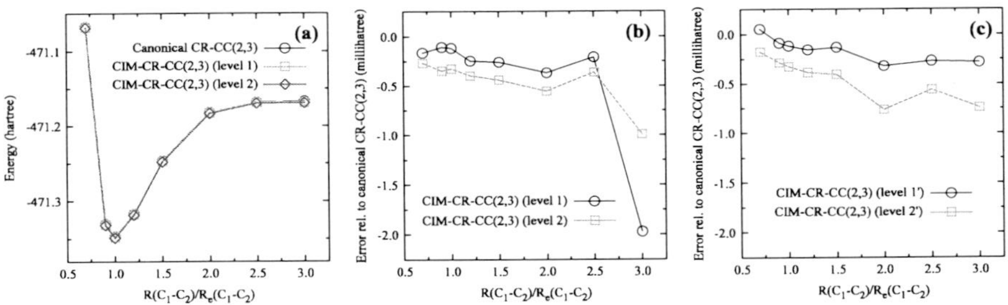

Fig. 2 (**a**) The canonical and CIM CR-CC(2,3) potential energy curves of dodecane, as described by the 6-31G(d) basis set, along the C_1–C_2 bond-breaking coordinate $R(C_1$–$C_2)$. (**b**) The errors in the CIM-CR-CC(2,3) energies relative to the corresponding canonical CR-CC(2,3) energies along the C_1–C_2 bond-breaking coordinate when the parameters ζ_1 and ζ_2 define the CIM subsystems at each nuclear geometry separately. (**c**) The errors in the CIM-CR-CC(2,3) energies relative to the corresponding canonical CR-CC(2,3) energies along $R(C_1$–$C_2)$ when the parameters ζ_1 and ζ_2 define the CIM subsystems only at the equilibrium geometry $R_e(C_1$–$C_2)$ and when the resulting subsystems at $R_e(C_1$–$C_2)$ are used for the remaining geometries. In levels 1 and $1'$, $\zeta_1 = 0.01$ and $\zeta_2 = 0.05$. In levels 2 and $2'$, $\zeta_1 = 0.01$ and $\zeta_2 = 0.02$

in the canonical CR-CC(2,3) calculations. Level 2 of CIM-CR-CC(2,3), which uses a somewhat tighter ζ_2 value, recovers 100.01–100.05% of the canonical CR-CC(2,3) correlation energies. From the point of view of the fractions of the canonical CR-CC(2,3) correlation energies recovered in the CIM-CR-CC(2,3) calculations, both levels seem very accurate. This is reflected in the small MUE values characterizing the level-1 and level-2 CIM-CR-CC(2,3) energies along the C_1–C_2 bond-breaking coordinate, which are summarized in Table 3.

We encounter, however, a potentially serious problem with the CIM-CR-CC(2,3) calculations performed with levels 1 and 2, in which the CIM subsystems are defined at each nuclear geometry separately, namely, the potential energy curves resulting from the level-1 and level-2 CIM-CR-CC(2,3) calculations display discontinuities around $R(C_1-C_2) = 2.5\,R_e(C_1-C_2)$. The local CIM domains remain unchanged up to $R(C_1 - C_2) = 2.5\,R_e(C_1 - C_2)$, but then they rearrange when the CH_3 fragment is at a larger distance from the remaining part of dodecane such that the orbitals localized on the C and H atoms of CH_3 form their own subsystem. The resulting discontinuities on the level-1 and level-2 CIM-CR-CC(2,3) potential energy curves are shown in Fig. 2(b). We can clearly see the sharp increase in the errors characterizing the CIM-CR-CC(2,3) energies obtained with levels 1 and 2 relative to the corresponding canonical CR-CC(2,3) energies in the $R(C_1 - C_2) > 2.5\,R_e(C_1 - C_2)$ region. The fixed-domain levels $1'$ and $2'$ eliminate the problem and produce smooth potential energy curves that are characterized by small variations in the errors describing the CIM-CR-CC(2,3) energies relative to the corresponding canonical CR-CC(2,3) calculations, as illustrated in Fig. 2(c). As shown in Table 3, the CIM levels $1'$ and $2'$ help to reduce the MUE values, particularly (as one might expect) in the level-1 CIM-CR-CC(2,3) calculations which use the less-tight ζ_2 value. In this case, the MUE value characterizing the level-1 CIM-CR-CC(2,3) calculation of 1.972 millihartree reduces to 0.320 millihartree. Perhaps more importantly, we also observe the reduction in the NPE values when going from levels 1 to $1'$ and 2 to $2'$. Again, the NPE reduction is particularly impressive when we go from level 1 to $1'$, where the NPE value of 1.866 millihartree characterizing the level-1 CIM-CR-CC(2,3) calculations decreases to 0.371 millihartree. We should recall that the level-1 (or level-$1'$) CIM-CR-CC(2,3) calculations for $C_{12}H_{26}$ are already 2–3 times less time consuming than the corresponding canonical CR-CC(2,3) calculations. We will continue examining the effect of switching from the variable-domain to fixed-domain CIM-CR-CC(2,3) calculations in bond-breaking situations involving larger alkanes and other molecules in a future work.

Table 3 The maximum unsigned errors (MUE) and the nonparallelity errors (NPE) relative to the canonical CR-CC(2,3) calculations characterizing the CIM-CR-CC(2,3) results (in millihartree) for the potential energy curves of dodecane along the C_1–C_2 bond-breaking coordinate $R(C_1-C_2)$

	Level 1	Level 2	Level $1'$	Level $2'$
MUE	1.972	0.988	0.320	0.761
NPE	1.866	0.725	0.371	0.588

It is certainly promising to observe that the CIM-CC methodology may enable us to accurately reproduce the potential energy curves along bond-breaking coordinates resulting from the canonical CC calculations. When covalent bonds are broken, as in the above example of the $C_{12}H_{26} \rightarrow C_{11}H_{23} + CH_3$ reaction, changes in the electronic energies along bond-breaking coordinates are quite large, so that errors in relative energies on the order of 1–2 millihartree are in most cases acceptable. There is, however, another category of applications where the accurate reproduction of relative energetics is important and where 1–2 millihartree errors in relative energies may no longer be regarded as small, namely, weakly bound molecular clusters, where one can encounter many different structures that differ in energies by only a few kilocalories per mole. Examples of such situations are discussed in the next subsection.

3.3 Relative Energies of Water Clusters

In order to examine the efficiency of the CIM-CC methods in calculations for molecular clusters involving non-covalent interactions, we performed a series of CIM and canonical CCSD, CR-CC(2,3), and CCSD(T) calculations for clusters containing 10, 12, 14, 16, and 20 water molecules. The main goal of these calculations was to learn how reliable the CIM-CC approaches might be in reproducing the relative energies of the lowest-energy structures of the $(H_2O)_n$ clusters with $n = 10 - 20$ where in each case there are several structures that lie within a few kilocalories per mole from the lowest-energy one.

The results of our calculations are summarized in Figs. 3, 4, 5, 6, 7, and 8. In each case, we are reporting the CIM-CC and, whenever available, the corresponding canonical CC energies of the ten lowest-energy structures resulting from the use of

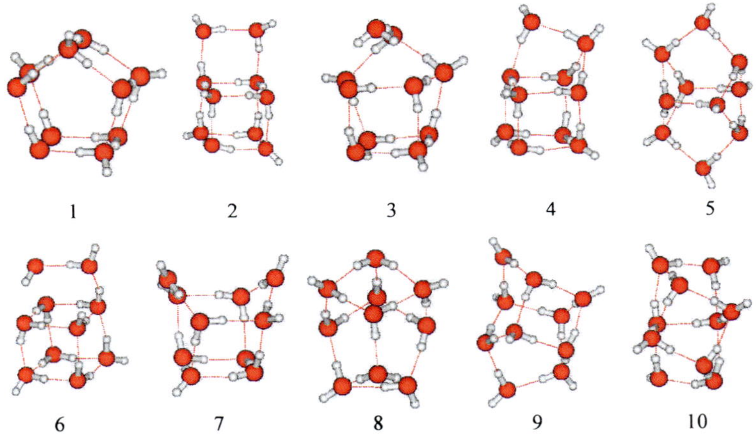

Fig. 3 Ten lowest-energy structures of $(H_2O)_{10}$, as obtained with the TIP4P force field

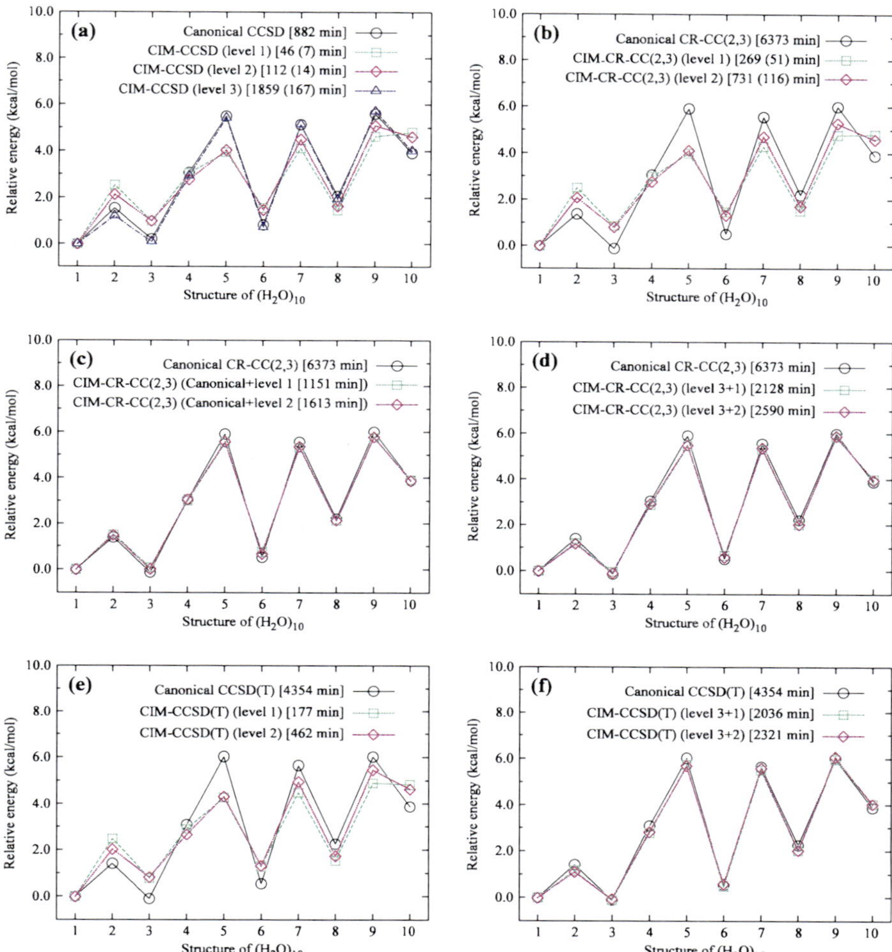

Fig. 4 A comparison of the canonical and CIM CCSD, CR-CC(2,3), and CCSD(T) relative energies (in kcal/mol) of the ten lowest-energy structures of $(H_2O)_{10}$, resulting from the use of the TIP4P force field, along with the corresponding CPU timings. (**a**) Canonical and CIM CCSD relative energies. (**b**) Canonical and CIM CR-CC(2,3) relative energies. (**c**) Canonical and mixed CIM CR-CC(2,3) relative energies, where the CCSD energies are obtained in the canonical calculations and the triples corrections of CR-CC(2,3) are calculated with CIM. (**d**) Canonical and mixed CIM CR-CC(2,3) relative energies, where the CCSD energies are obtained in level-3 CIM calculations and the triples corrections of CR-CC(2,3) are determined using levels 1 and 2. (**e**) Canonical and CIM CCSD(T) relative energies. (**f**) Canonical and mixed CIM CCSD(T) relative energies, where the CCSD energies are obtained in level-3 CIM calculations and the triples corrections of CCSD(T) are determined using levels 1 and 2. The 6-31G(d, p) basis set was used in all calculations. All energies are shifted such that the energies corresponding to the lowest-energy structure, labeled as structure 1, are all identical and set at zero. In level 1 of the CIM calculations, $\zeta_1 = 0.01$ and $\zeta_1 = 0.02$. In level 2, $\zeta_1 = 0.01$ and $\zeta_2 = 0.01$. In level 3, $\zeta_1 = 0.002$ and $\zeta_2 = 0.01$

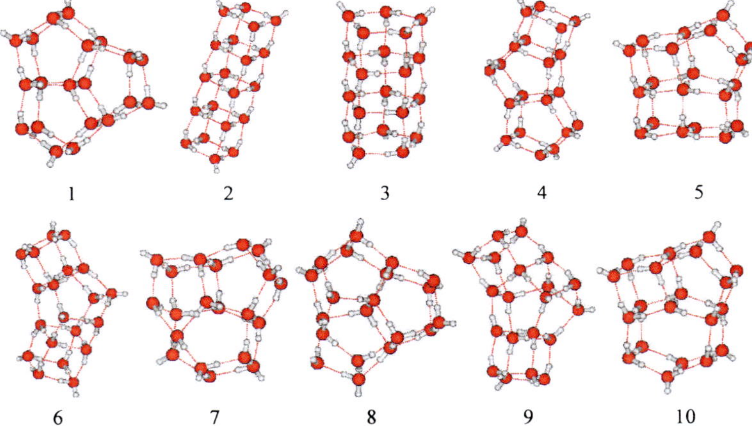

Fig. 5 Ten lowest-energy structures of $(H_2O)_{20}$, as obtained with the TIP4P force field

the TIP4P force field of Ref. [117]. For each water cluster, the energies are shifted such that the CIM and canonical CC energies corresponding to the lowest-energy structure, always designated as structure 1, are all identical and set at zero. The individual structures of each $(H_2O)_n$ system are labeled according to the increasing energy, as predicted by the TIP4P force field. Along with the relative energies, we provide information about the relevant CPU timings, including the total time required by all CC subsystem calculations and, if appropriate, the CPU time required by the CC calculations for the largest CIM subsystem. All of the reported timings

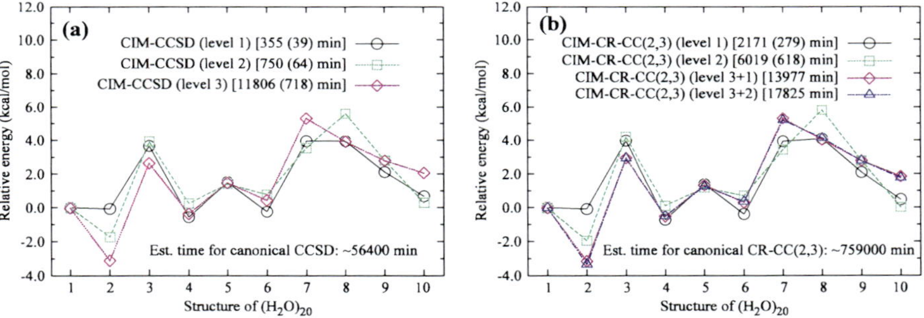

Fig. 6 A comparison of the canonical and CIM CCSD and CR-CC(2,3) relative energies (in kcal/mol) of the ten lowest-energy structures of $(H_2O)_{20}$ determined by the TIP4P force field along with the corresponding CPU timings. (a) CIM-CCSD relative energies. (b) CIM-CR-CC(2,3) and mixed CIM-CR-CC(2,3) relative energies, where the CCSD energies are obtained in level-3 CIM calculations and the triples corrections of CR-CC(2,3) are determined using levels 1 and 2. The 6-31G(d, p) basis set was used in all calculations. All energies are shifted such that the energies corresponding to the lowest-energy structure are all identical and set at zero. In level 1 of CIM calculations, $\zeta_1 = 0.01$ and $\zeta_1 = 0.02$. In level 2, $\zeta_1 = 0.01$ and $\zeta_2 = 0.01$. In level 3, $\zeta_1 = 0.002$ and $\zeta_2 = 0.01$

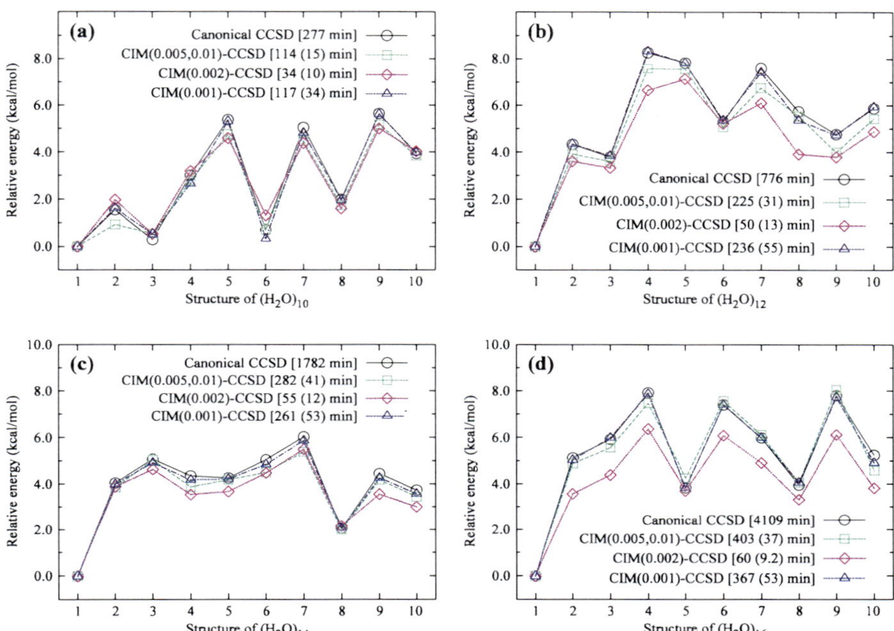

Fig. 7 A comparison of the canonical and CIM CCSD relative energies (in kcal/mol) of the ten lowest-energy structures of (**a**) $(H_2O)_{10}$, (**b**) $(H_2O)_{12}$, (**c**) $(H_2O)_{14}$, and (**d**) $(H_2O)_{16}$, determined by the TIP4P force field, along with the corresponding CPU timings. The 6-31G(d) basis set was used in all calculations. All energies are shifted such that the energies corresponding to the lowest-energy structure of a given water cluster, labeled as structure 1, are all identical and set at zero. The values in the square brackets represent the average total CPU times required by the canonical and CIM CCSD iterations. The values in parentheses represent the average CPU times required by the CCSD calculations for the largest CIM subsystems. CIM(0.005,0.01) stands for the earlier CIM approach based on Steps 1–12 of Section 2.3 with $\zeta_1 = 0.005$ and $\zeta_2 = 0.01$. CIM(0.002) indicates the modified, single-environment CIM scheme of Section 4.1 with $\zeta = 0.002$. CIM(0.001) indicates the modified CIM approach of the same type with $\zeta = 0.001$

refer to the CC steps, i.e., steps after the integral transformation from the AO to the appropriate MO bases. In order to facilitate this presentation, we only report the average timings per structure. The CPU timings corresponding to a given method vary somewhat for different structures of a given water cluster, but the variation is not significant and the average CPU time per structure is sufficient to provide the most important information. In this section, we primarily focus on the results summarized in Figs. 4 and 6, which deal with the CIM-CCSD, CIM-CR-CC(2,3), and CIM-CCSD(T) results for $(H_2O)_{10}$ and $(H_2O)_{20}$, and ways to improve or stabilize them through the concept of the so-called mixed CIM-CC approaches. In the next section, we examine the CIM-CCSD results for a series of $(H_2O)_n$ clusters with $n = 10$, 12, 14, and 16, and ways to improve these results when diffuse functions are present in the basis set via the alternative CIM-CC theory suggested in that section.

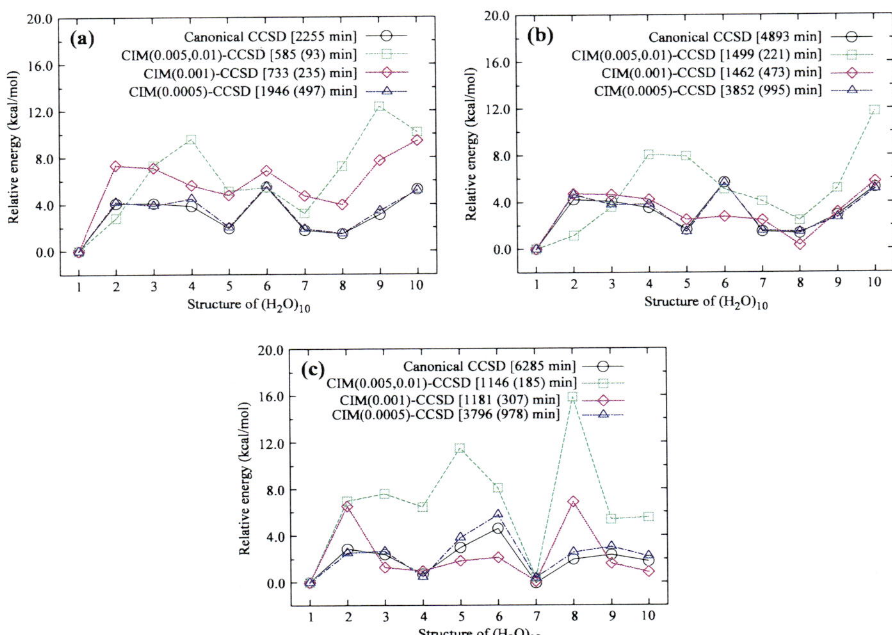

Fig. 8 A comparison of the canonical and CIM CCSD relative energies (in kcal/mol) of the ten lowest-energy structures of $(H_2O)_{10}$ [(**a**) and (**b**)] and $(H_2O)_{12}$ [(**c**)] determined by the TIP4P force field along with the corresponding CPU timings. The basis sets used in the calculations were 6-31++G(d, p) [(**a**) and (**c**)] and 6-311++G(d, p) [(**b**)]. All energies are shifted such that the energies corresponding to the lowest-energy structure of a given water cluster, labeled as structure 1, are all identical and set at zero. The values in the square brackets represent the average total CPU times required by the canonical and CIM CCSD iterations. The values in parentheses represent the average CPU times required by the CCSD calculations for the largest CIM subsystems. CIM(0.005,0.01) stands for the earlier CIM approach based on Steps 1–12 of Section 2.3 with $\zeta_1 = 0.005$ and $\zeta_2 = 0.01$. CIM(0.001) indicates the modified, improved CIM scheme of Section 4.1 with $\zeta = 0.001$. CIM(0.0005) indicates the modified CIM scheme of the same type with $\zeta = 0.0005$

We begin by discussing the CIM-CCSD, CIM-CR-CC(2,3), and CIM-CCSD(T) results for the ten lowest-energy structures of $(H_2O)_{10}$, shown in Fig. 3, which are summarized in Fig. 4. All of the calculations shown in Fig. 4 were performed with the 6-31G(d, p) basis set [so that we could still afford the canonical CCSD, CR-CC(2,3), and CCSD(T) calculations for comparison purposes]. We considered three levels of the CIM-CC theory, namely, level 1 in which $\zeta_1 = 0.01$ and $\zeta_2 = 0.02$ and which uses (depending on the structure) 23–27 CIM subsystems, a somewhat tighter level 2 in which $\zeta_1 = 0.01$ and $\zeta_2 = 0.01$, using 24–27 subsystems, and level 3 which uses the considerably tighter thresholds, particularly for defining the primary environment, of $\zeta_1 = 0.002$ and $\zeta_2 = 0.01$, characterized by 22–29 CIM subsystems.

As one can see from Fig. 4, levels 1 and 2 (particularly level 1) offer huge reductions of the CPU times compared with the corresponding canonical CC calculations despite the fact that $(H_2O)_{10}$ is a relatively small system. Indeed, one needs 882 and 6,373 CPU minutes per structure to perform the canonical CCSD and CR-CC(2,3) calculations, respectively, when the 6-31G(d, p) basis set is employed. The corresponding level-1 CIM-CCSD and CIM-CR-CC(2,3) calculations require only 46 and 269 minutes, respectively. This is a reduction of the CPU time needed to perform the canonical CCSD and CR-CC(2,3) calculations by a factor of about 20. If we execute these calculations on (depending on the molecular structure) 23–27 processors, i.e., each subsystem calculation is run on a separate processor, we can complete the level-1 CIM-CCSD and CIM-CR-CC(2,3) calculations in just 7 and 51 minutes, respectively. As one might expect, level 2 is more expensive, but it offers significant savings in the CPU time as well, requiring 112 and 731 minutes total, or 14 and 116 minutes, respectively, if each subsystem calculation is performed on a different processor, reducing the CPU times of the canonical CCSD and CR-CC(2,3) calculations by a factor of 8–9.

These savings in the computer effort offered by levels 1 and 2 of the CIM-CC theory, as defined above, are even more impressive if we analyze the fractions of the CCSD and CR-CC(2,3) correlation energies recovered by the CIM-CC calculations. The level-1 CIM-CC calculations recover, depending on the structure, 99.33–99.53% of the correlation energy in the CCSD case and 96.16–98.96% of the triples energy correction of CR-CC(2,3), which itself accounts for only about 1.5% of the CR-CC(2,3) correlation energy. The level-2 CIM-CC calculations improve these results further, recovering 99.46–99.63% of the correlation energy in the CCSD case and 97.05–99.29% of the CR-CC(2,3) triples correction.

It is apparent from Figs. 4(a) and (b) that the level-1 and level-2 CIM-CCSD and CIM-CR-CC(2,3) approaches provide a reasonably accurate description of the nontrivial relative energy patterns observed in the corresponding canonical calculations. Similar remarks apply to the CIM-CCSD(T) results obtained with levels 1 and 2 and summarized in Fig. 4(e). Thus, the results shown in Figs. 4(a), (b), and (e) create a positive impression. One has to be careful, however, in making final judgments, since our primary interest here is in the relative energies, not just in the correlation energies. Undoubtedly, the CIM-CCSD, CIM-CR-CC(2,3), and CIM-CCSD(T) calculations performed with levels 1 and 2 provide a qualitatively correct representation of the relative canonical energy patterns involving different structures of $(H_2O)_{10}$ at a small fraction of the cost of the corresponding canonical calculations, but we can see from Figs. 4(a), (b), and (e) that the agreement between the CIM and the canonical CC energies is not always perfect and errors in relative energies on the order of 1 kcal/mol remain [cf., e.g., structures 2, 3, 5, 6, 7, 9, and 10 in Figs. 4(a), (b), and (e)]. For example, if we evaluate the NPE values characterizing the CIM-CR-CC(2,3) versus canonical CR-CC(2,3) results for the ten lowest-energy structures of $(H_2O)_{10}$ shown in Fig. 4(b), we obtain 4.851 millihartree with level 1 and 4.321 millihartree with level 2. The questions are what causes these errors and how to improve the situation within the CIM-CC framework discussed in this chapter.

In order to understand the reasons for the observed NPE values of about 4–5 millihartree characterizing the level-1 and level-2 CIM-CR-CC(2,3) calculations, we conducted a numerical experiment in which we mixed the canonical CCSD results with the CIM-CR-CC(2,3) calculations. The results of this experiment are shown in Fig. 4(c). The idea here is that we add the triples correction $\delta \tilde{E}^{(2,3)}$ resulting from the level-1 or level-2 CIM-CR-CC(2,3) calculations to the canonical CCSD energy $E^{(CCSD)}$. As one can see from Fig. 4(c), there is hardly any difference between the relative energies obtained in the mixed (canonical + level 1 or canonical + level 2) CIM-CR-CC(2,3) calculations and the corresponding canonical energies. Indeed, the 4.851 and 4.321 millihartree NPE values resulting from pure levels 1 and 2 of CIM-CR-CC(2,3) reduce to 0.967 and 0.772 millihartree, respectively, i.e., to the NPE values characterizing the triples corrections of CIM-CR-CC(2,3), which obviously are much smaller than the corresponding CCSD correlation energies. Since most of the cost of the CR-CC(2,3) calculations is in the triples part, the total time of 6,373 minutes per structure reduces to 1,151 minutes when one adds the triples correction obtained with level-1 of CIM-CR-CC(2,3) to the canonical CCSD energy and 1,613 minutes when the triples correction obtained with level 2 of CIM-CR-CC(2,3) is added to the canonical CCSD energy. We can clearly see that the primary reason for the 4–5 millihartree NPE values characterizing pure levels 1 and 2 of CIM-CR-CC(2,3) is the somewhat inadequate treatment of the CCSD correlation energy by these CIM levels. Since the CCSD approach describes the bulk of the total correlation energy, the inadequacies in representing the CCSD correlation energies propagate into the CIM-CR-CC(2,3) calculations.

The idea of mixing the canonical CCSD and CIM-CR-CC(2,3) [or CIM-CCSD(T)] methods, as described above, can certainly be useful in applications of the CIM-CC theory, provided that we can afford the canonical CCSD calculations, but one will eventually run into the problem of prohibitively large computer costs of the canonical CCSD calculations for systems larger than $(H_2O)_{10}$ and for basis sets larger than 6-31G(d, p). For example, the average CPU time required by the canonical CCSD calculations for the ten lowest-energy structures of $(H_2O)_{20}$, which are shown in Fig. 5, as described by the 6-31G(d, p) basis set, is estimated at 56,400 minutes or 39 days per structure [cf. Fig. 6(a)]. This is much less than the 527 or so days needed to complete the corresponding canonical CR-CC(2,3) calculations [cf. Fig. 6(b)], but the issue remains. The question then is if one can reduce the costs of the mixed canonical + CIM CR-CC(2,3) and CCSD(T) calculations when running the underlying canonical CCSD calculations becomes prohibitively expensive. Clearly, we are interested in a solution which would enable us to reduce the computer costs of the mixed canonical + CIM CR-CC(2,3) and CCSD(T) calculations, while retaining their high accuracies.

The answer to this question is given in Figs. 4(d) and (f). In those two figures, we compare the results of the mixed CIM-CR-CC(2,3) and CIM-CCSD(T) calculations for the ten lowest-energy structures of $(H_2O)_{10}$, in which we add the triples corrections of CIM-CR-CC(2,3) and CIM-CCSD(T) resulting from the use of the relatively inexpensive CIM levels 1 and 2 to the energies obtained with the level-3 CIM-CCSD calculations. Let us recall that level 3 of CIM-CC uses a tight $\zeta_1 = 0.002$ value to define the primary environments of the CIM subsystems instead of $\zeta_1 = 0.01$ used in levels 1 and 2. In other words, in the resulting mixed CIM-CR-CC(2,3) and CIM-CCSD(T) approaches, which we abbreviate as levels

3+1 and 3+2, the CCSD calculation is treated at a higher level of the CIM theory than the triples parts of the CR-CC(2,3) and CCSD(T) calculations. For example, the fraction of the CCSD correlation energy reproduced by level 3 is more or less independent of the molecular structure of $(H_2O)_{10}$ and varies between 99.90 and 99.94%, as compared with 99.33–99.53 and 99.46–99.63% obtained with levels 1 and 2, respectively. In consequence, as demonstrated in Figs. 4(d) and (f), one can hardly notice the difference between the relative energies obtained in the CIM-CR-CC(2,3) and CIM-CCSD(T) calculations using the mixed levels 3+1 and 3+2 and the corresponding canonical energies. This is reflected in the great improvements to the NPE values resulting from the CIM-CR-CC(2,3) and CIM-CCSD(T) calculations with pure levels 1 and 2, relative to the corresponding canonical calculations, offered by the mixed levels 3+1 and 3+2 of CIM-CR-CC(2,3) and CIM-CCSD(T). Indeed, the NPE values of 4.851 and 4.321 millihartree that characterize pure levels 1 and 2 of CIM-CR-CC(2,3) reduce to 0.954 and 0.759 millihartree, respectively. This clearly shows that we can tremendously benefit from treating the bulk of the correlation energy provided by CCSD at a higher level of the CIM theory, while using lower levels of CIM level to determine triples corrections. Once the CCSD part of the correlation energy is stabilized within the CIM framework, it is sufficient to use less expensive CIM levels with looser thresholds ζ_1 and ζ_2 to calculate the triples corrections of CR-CC(2,3) and CCSD(T).

The only problem with the above CIM-CR-CC(2,3) and CIM-CCSD(T) calculations using mixed CIM levels 3+1 and 3+2 is that $(H_2O)_{10}$ is too small to present us with significant savings in the computer effort, particularly when a single processor is used. As demonstrated in Figs. 4(d) and (f), the mixed levels 3+1 and 3+2 of CIM-CR-CC(2,3) and CIM-CCSD(T) reduce the CPU timings of the canonical CR-CC(2,3) and CCSD(T) calculations for $(H_2O)_{10}$, as described by the 6-31G(d, p) basis set, by a factor of 2-3 only. We could, of course, use more processors and execute the individual CCSD, CR-CC(2,3), and CCSD(T) subsystem calculations on different processors, which would lead to substantially more significant speedups, but one could then argue in favor of carrying out the canonical CR-CC(2,3) and CCSD(T) parallel calculations on a larger number of processors to speedup the calculations. We have to consider larger systems, for which the canonical CC calculations become practically impossible, to experience more significant savings in the computer effort. In order to illustrate this, we performed the CIM-CR-CC(2,3) calculations using the mixed 3+1 and 3+2 levels for the ten lowest-energy clusters of $(H_2O)_{20}$, shown in Fig. 5 and described once again by the 6-31G(d, p) basis set. The resulting relative energies and CPU timings are summarized in Fig. 6(b). We show the results of the corresponding CIM-CCSD calculations with level 3 in Fig. 6(a).

The $(H_2O)_{20}$ cluster is too large for performing the canonical CCSD and CR-CC(2,3) calculations on our computers. Indeed, the estimated CPU times that would be required by the canonical CCSD and CR-CC(2,3) calculations are 56,400 and 759,000 minutes on average per structure. Obviously, the latter time is enormous. Furthermore, the canonical CCSD and CR-CC(2,3) calculations for $(H_2O)_{20}$ using the 6-31G(d, p) basis set are characterized by prohibitively large memory and disk requirements, which are estimated at about 70 GB RAM in the CCSD case and 160 GB RAM in the case of CR-CC(2,3). As one can see from Fig. 6(b), it takes 13,977 and 17,825 minutes per structure on average on a single processor to obtain the CIM-CR-CC(2,3) energies using the mixed levels 3+1 and 3+2, respectively, which, based on the benchmark calculations for $(H_2O)_{10}$ discussed above, are expected to offer a highly accurate representation of the relative energies of canonical CR-CC(2,3). Most of the time (11,806 minutes on average per structure on a single processor) is spent on the level-3 CCSD calculations [cf. Fig. 6(a)]. If we executed these calculations in parallel, using as many processors as the relevant numbers of the CIM subsystems, which are, depending on the structure of $(H_2O)_{20}$, 50–59 for level 1, 51–60 for level 2, and 52–73 for level 3 [as desired, more or less twice as many as in the $(H_2O)_{10}$ case], we would be able to complete the mixed CIM-CR-CC(2,3) calculations using levels 3+1 and 3+2 in about 1,000 and 1,340 minutes, respectively [these CPU timings are obtained by adding the time required by the CCSD calculation for the largest subsystem in level 3 of CIM and the CR-CC(2,3) calculation for the largest subsystems in levels 1 and 2 of CIM, respectively, which are provided in Fig. 6]. These are very short times compared with the unmanageable canonical CR-CC(2,3) calculations. We would have to use about 800 processors with a perfectly scalable, parallel, canonical CR-CC(2,3) code (which does not exist) and find a computer with a large total memory and disk to perform the canonical CR-CC(2,3) calculations for the $(H_2O)_{20}$ system described by the 6-31G(d, p) basis set. None of these extra measures are necessary when we make use of the mixed CIM-CR-CC(2,3) methodology discussed here. In addition to the considerable savings in the CPU time, we have absolutely no problem with fitting the mixed CIM-CR-CC(2,3) calculations for $(H_2O)_{20}$ in memory. Indeed, the level-3 CCSD calculations for the largest subsystem require 8.1 GB RAM, whereas the level-1 and level-2 calculations of the triples corrections of CR-CC(2,3) require 1.3 and 2.0 GB RAM, respectively. This is a consequence of the fact that the largest CIM subsystems in the level-3 CIM-CCSD calculations and level-1 and level-2 CIM-CR-CC(2,3) calculations are characterized by relatively small values of n_o and n_u compared with the entire $(H_2O)_{20}$ system, namely, 40 and 252 for level-3 of CIM-CCSD, 20 and 112 for level-1 of CIM-CR-CC(2,3), and 22 and 126 for level-2 of CIM-CR-CC(2,3). These orbital numbers should be compared to the values of n_o and n_u characterizing the entire $(H_2O)_{20}$ system, as described by the 6-31G(d, p) basis set employed here, which are $n_o = 80$ and $n_u = 400$, respectively. In other words, when executed on at most 73 processors (each subsystem calculation on a different processor), we can complete the CIM-CR-CC(2,3) calculations for the $(H_2O)_{20}$ system, as described by the 6-31G(d, p) basis set, using the mixed 3+1 and 3+2 levels discussed here that should provide an accurate representation of the relative energetics of the corresponding canonical CR-CC(2,3) calculations, dedicating no more than about 8 GB RAM in the largest subsystem calculation, in about 1,000–1,340 CPU minutes per structure. As shown in Fig. 6(b), one can hardly tell the difference between the relative energies obtained in the mixed CIM-CR-CC(2,3) calculations using levels 3+1 and 3+2. This confirms once again that it is less important what level of CIM theory is used in the non-iterative triples part of the CIM-CR-CC(2,3) [or CIM-CCSD(T)] calculation once the CCSD correlation energy is reasonably well stabilized.

We have performed a number of CIM-CCSD calculations for other water clusters and other basis sets, besides the 6-31G(d, p) basis set discussed so far, with the intention of learning if we can handle more realistic basis sets with diffuse functions, such as 6-311G++G(d, p), in the CIM-CC calculations for large weakly bound molecular clusters. This has led to the discovery of challenges facing the CIM-CC theory and the development of a modified variant of CIM-CC, which has shown considerable promise in calculations with larger basis sets involving diffuse functions. This modified variant of the CIM-CC theory is discussed in the next section.

4 Improved CIM-CC Theory for Calculations of Weakly Bound Molecular Clusters Involving Larger Basis Sets with Diffuse Functions

As shown in the previous section, the CIM-CC methods developed in this work can be quite successful in reproducing the correlation and relative energies resulting from the corresponding canonical CC calculations, but there are several issues that require further investigation. One of them is the use of two parameters ζ_1 and ζ_2 that define the primary and secondary environments in the CIM subsystem design, along with the parameter η which is used to select the final set of unoccupied LMOs. Although we have developed intuition about the values of ζ_1 and ζ_2 appropriate for a given application and although it is usually sufficient to keep η fixed at 0.2, it is useful to examine if one can reduce the number of variable CIM parameters without losing accuracy and without increasing the computer cost of the CIM-CC calculations. Another issue that needs to be examined is the issue of diffuse basis set functions which, as shown in this section, may create difficulties in the CIM-CC calculations for weakly bound molecular clusters. We believe that both of these issues can be addressed by modifying the design of the CIM subsystems, as discussed in the next section. The description of this improved variant of the CIM-CC theory will be followed by preliminary numerical examples that illustrate its performance.

4.1 The Proposed Changes in the Design of the CIM Subsystems

The analysis of the CIM-CC algorithm discussed in the earlier sections, particularly of Steps 1–12 used to design the CIM subsystems, combined with several test calculations, indicate that one may be able to improve the performance of the CIM-CC theory, particularly in calculations involving diffuse basis set functions, by modifying the initial steps of the CIM subsystem design in the following manner (W. Li and P. Piecuch, unpublished manuscript). First, after localizing the core and correlated occupied orbitals obtained in the RHF calculations (Step 1 in Section 2.3), we use the Mulliken charges of each occupied LMO to assign a *group of central occupied LMOs* $\tilde{\phi}_{i'}$ to each non-hydrogen atom and its adjacent hydrogen atoms. A given LMO $\tilde{\phi}_{i'}$ is assigned to the specific non-hydrogen atom or one of its adjacent hydrogen atoms, which define the corresponding group of central LMOs, if the Mulliken charge of this $\tilde{\phi}_{i'}$ associated with one of these atoms exceeds 0.3. As in Step 6 in Section 2.3, the hydrogen atom is recognized as bound to the non-hydrogen atom X if the distance between X and this hydrogen atom is less than $r(H) + r(X) + 0.168$ Å, where $r(H)$ and $r(X)$ are the van der Waals radii of H and X tabulated in the Supporting Information to Ref. [104]. Once a group of central LMOs $\tilde{\phi}_{i'}$ is assigned to each non-hydrogen atom and its adjacent hydrogen atoms, we add the additional *environment orbitals* to this group with the help of threshold ζ. The assignment of environment LMOs $\tilde{\phi}_{j'}$ to a group of central LMOs $\tilde{\phi}_{i'}$ associated with a given non-hydrogen atom and its adjacent hydrogen atoms is done in the same way as in the earlier Steps 3 and 5, except that now we have only one type of environment defined with one parameter ζ, not the primary and secondary environments used in the earlier CIM scheme which were defined using two parameters ζ_1 and ζ_2. Thus, the orbital environment of a given central LMO $\tilde{\phi}_{i'}$ from the group of central orbitals associated with a given non-hydrogen atom and its adjacent hydrogen atoms consists of those correlated LMOs $\tilde{\phi}_{j'}$ for which $|\langle \tilde{\phi}_{i'} | f | \tilde{\phi}_{j'} \rangle| > \zeta$, where ζ is a suitably chosen numerical threshold. As a result of these initial steps, we end up with the LMO subsets of central and environment LMOs, where each subset is assigned to a particular non-hydrogen atom and its adjacent hydrogen atoms and where each subset has the general form $[i_1', \ldots, i_\alpha'] = (i_1', \ldots, i_\alpha'; j_1', \ldots, j_\beta')$, where $\tilde{\phi}_{i_1'}, \ldots, \tilde{\phi}_{i_\alpha'}$ are the relevant central LMOs and $\tilde{\phi}_{j_1'}, \ldots, \tilde{\phi}_{j_\beta'}$ are the corresponding environment orbitals. It may sometimes happen that a particular subset $[i_1', \ldots, i_\alpha']$ is totally embedded in another subset that corresponds to another non-hydrogen atom and its adjacent hydrogen atoms. If this happens, we eliminate the smaller subset $[i_1', \ldots, i_\alpha']$ and make the central LMOs $\tilde{\phi}_{i_1'}, \ldots, \tilde{\phi}_{i_\alpha'}$ of these smaller subset central orbitals of the larger subset to which they belong. By repeating this process and analyzing the retained LMO subsets, we eventually end up with the minimum number of LMO subsets that are analogous to the irreducible full LMO domains of the earlier CIM scheme (cf. Step 6 in Section 2.3). Once we are at this stage, we define the central and environment orbitals of a given irreducible full LMO domain as the occupied LMOs of the corresponding subsystem $\{P\}$ and proceed toward the construction of the unoccupied LMOs that are associated with each subsystem $\{P\}$ as well as the final, compact form of the occupied LMOs that define it, which are subsequently used in the CIM-CC calculations, using Steps 7–12 of Section 2.3. As in the CIM scheme discussed in the earlier sections and as implied by the orbital manipulations described above, each correlated occupied LMO is central in at least one subsystem $\{P\}$, which is key for the validity of Eqs. (23)–(38) for the CIM-CCSD, CIM-CR-CC(2,3), and CIM-CCSD(T) correlation energies (and the analogous equations one could write for other CC approaches). The remaining details of the CIM-CC calculations, such as the construction of the subsystem QCMOs from the corresponding occupied and unoccupied LMOs, remain unchanged.

The modified CIM scheme, as described above (see W. Li and P. Piecuch, unpublished manuscript for further details), has a few advantages over the earlier scheme.

It uses only one parameter ζ to define the environment LMOs rather than the two parameters ζ_1 and ζ_2 employed in the previous CIM scheme. Since the initial groups of central LMOs are assigned to individual non-hydrogen atoms and adjacent hydrogen atoms, the design of the orbital subsystems in the modified CIM method described in this section is more intuitive and the resulting final subsystems $\{P\}$ do not unnecessarily vary with the nuclear geometry, creating a smoother and more balanced description of molecular potential energy surfaces. For example, the number of subsystems describing $(H_2O)_{10}$ in the calculations with level 1 of the earlier, dual-environment CIM scheme discussed in the previous sections varied, depending on the particular structure of $(H_2O)_{10}$, between 23 and 27. With level 3 discussed in the previous section, we had to deal with 22–29 subsystems, i.e., a rather significant variation in the number of subsystems and 2–3 times more subsystems than the number of water molecules in the cluster, when the earlier CIM scheme was employed. The modified, single-environment CIM scheme discussed in this section produces 9–10 subsystems for all ten lowest-energy structures of $(H_2O)_{10}$ examined in Section 3.3 in the analogous calculations, in agreement with the intuitive description of this water cluster. In general, for a cluster of n water molecules, the modified CIM method discussed here produces n or, sometimes, $n-1$ or $n-2$ subsystems, but we never end up with a situation where $(H_2O)_n$ is described by more than n subsystems, since the original groups of central orbitals that lead to new subsystems are assigned to the oxygen atoms and adjacent hydrogen atoms and there only are n oxygen atoms in $(H_2O)_n$. As shown below, all of this results in a more balanced description of different structures of the $(H_2O)_n$ clusters, both in terms of the fraction of the correlation energy recovered in the CIM-CC calculations and in terms of the resulting NPE values, even in the CIM-CC calculations with larger basis sets containing diffuse basis functions where the CIM-CC methodology discussed in the earlier sections encounters problems. The performance of the modified, single-environment CIM-CC theory introduced here is illustrated in the next subsection by a few examples of preliminary CIM-CCSD calculations for the $(H_2O)_n$ clusters.

4.2 Numerical Tests of the Modified, Single-Environment CIM-CC Approach: Relative Energies of Water Clusters

Before discussing the CIM-CC calculations for the $(H_2O)_n$ clusters with basis sets containing diffuse functions, where the earlier, dual-environment CIM scheme encounters problems with accurately reproducing the relative energies of the canonical CC calculations, we examine the performance of the modified, single-environment CIM scheme introduced in Section 4.1 (cf., also, W. Li and P. Piecuch, unpublished manuscript) in the calculations for the ten lowest-energy structures of the $(H_2O)_n$ clusters with $n = 10$, 12, 14, and 16, as described by the 6-31G(d) basis set. We chose the 6-31G(d) basis set, so that we could perform the canonical CCSD calculations for systems as big as $(H_2O)_{16}$. Due to the prohibitively large computer costs, we were unable to perform the canonical CR-CC(2,3) or CCSD(T) calculations for $(H_2O)_n$ with $n = 14$ and 16 that could provide the relevant reference energies for the corresponding CIM-CR-CC(2,3) and CIM-CCSD(T) calculations, so we limit ourselves to the canonical and CIM CCSD calculations. In all of the CIM-CCSD calculations employing both the previous (Steps 1–12 of Section 2.3) and the modified (this section) designs of the CIM subsystems, we used $\eta = 0.2$.

The results of the CIM-CCSD calculations employing the earlier and modified CIM schemes for the ten lowest-energy structures of the $(H_2O)_n$ clusters with $n = 10, 12, 14$, and 16, as described by the 6-31G(d) basis set, are shown in Fig. 7. As in Section 3.3, we used the TIP4P force field to determine the nuclear geometries of the ten lowest-energy structures of each $(H_2O)_n$ cluster under consideration (labeled, as before, as structures 1–10), but, to save space, we are not showing the actual molecular structures in a graphical form. Two levels of the modified, single-environment CIM-CCSD theory are considered, namely, CIM(0.002), in which $\zeta = 0.002$, and CIM(0.001), in which $\zeta = 0.001$. We compare these two types of the CIM-CCSD calculations based on the modified subsystem design described in the previous section with the results of the calculations employing the earlier, dual-environment CIM-CCSD scheme, in which we set ζ_1 at 0.005 and ζ_2 at 0.01 [designated as CIM(0.005,0.01)-CCSD], and the canonical CCSD results.

There are no diffuse functions in the 6-31G(d) basis set, so the earlier and modified CIM-CCSD schemes behave in a similar manner, although, as shown in Fig. 7, the modified design of the CIM subsystems discussed in this section offers additional improvements in the quality of the CIM-CCSD results without increasing the CPU time. For example, the CIM(0.001)-CCSD calculations that use the modified CIM subsystem design require the same amount of time as the CIM(0.005,0.01)-CCSD calculations based on Steps 1–12 of Section 2.3, while reducing the errors relative to the energies obtained with the canonical CCSD approach in a nontrivial manner. Indeed, the CIM(0.001)-CCSD calculations reproduce, depending on the molecular structure of the given water cluster, 99.83–99.88% of the canonical CCSD correlation energies for $(H_2O)_{10}$, 99.83–99.86% of the canonical CCSD correlation energies for $(H_2O)_{12}$, 99.81–99.82% of the canonical CCSD correlation energies for $(H_2O)_{14}$, and 99.80–99.82% of the canonical CCSD correlation energies for $(H_2O)_{16}$. The corresponding fractions of the canonical CCSD correlation energies obtained with the CIM(0.005,0.01)-CCSD scheme of the earlier type, which is characterized by virtually identical CPU timings, are somewhat lower, namely, 99.73–99.80, 99.72–99.77, 99.71–99.74, and 99.69–99.74%, respectively.

Improvements in the NPE values relative to the canonical CCSD energies of the ten lowest-energy structures of the water clusters under consideration offered by the CIM(0.001)-CCSD calculations using the modified subsystem design are even more impressive. Indeed, the CIM(0.001)-CCSD calculations produce the following NPE values relative to the corresponding canonical CCSD calculations: 1.000 millihartree for $(H_2O)_{10}$, 0.725 millihartree for $(H_2O)_{12}$, 0.338 millihartree for $(H_2O)_{14}$, and 0.693 millihartree for $(H_2O)_{16}$. The corresponding NPE values characterizing the CIM(0.005,0.01)-CCSD calculations, which have similar computer costs to those of CIM(0.001)-CCSD, of 1.401, 1.338, 1.023, and 1.729 millihartree, respectively, although quite reasonable, are not as good as those obtained with

CIM(0.001)-CCSD. Both the CIM(0.001)-CCSD and the CIM(0.005,0.01)-CCSD approaches offer significant savings in the computer effort, reducing the CPU time of the canonical CCSD calculations for the $(H_2O)_n$ clusters with $n = 10, 12, 14,$ and 16, described by the 6-31G(d) basis set, by factors of 2.4, 3.3–3.4, 6.3–6.8, and 10.2–11.2, respectively, but the CIM(0.001)-CCSD calculations based on the modified subsystem design described in Section 4.1 offer an improved description of the relative energies compared with the earlier design of the CIM subsystems defined by Steps 1–12 of Section 2.3. Interestingly enough, even the rougher CIM(0.002)-CCSD calculations that use the modified CIM subsystem design with $\zeta = 0.002$, which offer additional savings in the CPU time relative to the CIM(0.001)-CCSD calculations, by a factor of 6.1 in the $(H_2O)_{16}$ case, correctly reproduce the overall patterns of the relative energies observed in the canonical CCSD calculations, although the corresponding NPE values relative to canonical CCSD of 2.235, 2.888, 1.576, and 2.650 millihartree characterizing the CIM(0.002)-CCSD calculations for the $(H_2O)_n$ clusters with $n = 10, 12, 14,$ and 16, respectively, are not as impressive as those resulting from the CIM(0.001)-CCSD calculations.

The modified design of the CIM subsystems does not change the fundamental characteristics of all CIM-CC theories, such as the almost linear scaling of the CPU time with the system size, as reflected by the CPU timings of the CCSD calculations for the largest subsystems that do not significantly change when we go from $(H_2O)_{10}$ to $(H_2O)_{16}$ in the CIM(0.001)-CCSD and CIM(0.002)-CCSD calculations. What mostly changes as we increase the system size is the number of CIM subsystems which is, as already alluded to above, 10 for $(H_2O)_{10}$, 10–12 for $(H_2O)_{12}$, 14 for $(H_2O)_{14}$, and 16 for $(H_2O)_{16}$. We should reemphasize the fact that in each of the above calculations we end up with the number of subsystems that matches the number of water molecules in the given cluster in a natural manner, i.e., through the appropriate design of orbital subsystems $\{P\}$ characterizing the modified CIM scheme of Section 4.1, not through some ad hoc, a priori partitioning of water clusters under consideration into H_2O monomers.

Let us finally examine the performance of the modified (Section 4.1) versus earlier (Steps 1–12 of Section 2.3) CIM subsystem designs in the context of the CIM-CCSD calculations with two basis sets containing diffuse functions, namely, 6-31++G(d, p) and 6-311++G(d, p). We performed both types of the CIM-CCSD calculations for the ten lowest-energy structures of $(H_2O)_{10}$, as described by the 6-31++G(d, p) and 6-311++G(d, p) basis sets [Figs. 8(a) and (b), respectively], and $(H_2O)_{12}$ described by the 6-31++G(d, p) basis set [Fig. 8(c)]. Due to the prohibitively large computer costs, we were unable to perform the canonical CCSD calculations using the 6-311++G(d, p) basis set for $(H_2O)_{12}$, so we will use the 6-31G++(d, p) basis set instead in this case. We were not able to perform the canonical CR-CC(2,3) or CCSD(T) calculations for any of the above systems, again due to enormous costs of such calculations when the 6-31++G(d, p) and 6-311++G(d, p) basis sets are employed, so we limit ourselves to the discussion of the CIM-CCSD results. We compare the results of the CIM(0.005,0.01)-CCSD calculations, which are based on the earlier design of the CIM subsystems defined by Steps 1–12 of Section 2.3, with the results obtained with the CIM(0.001)-CCSD and

CIM(0.0005)-CCSD approaches using the modified subsystem design of Section 4.1, and with the canonical CCSD results. As in the case of the 6-31G(d) basis set, a comparison of the CIM(0.005,0.01)-CCSD and CIM(0.001)-CCSD results is particularly instructive, since both types of calculations are characterized by very similar CPU timings (see Fig. 8).

It is apparent from Fig. 8 that the use of diffuse functions in the CIM-CC calculations for the water clusters using the original subsystem design leads to severe difficulties in the description of the relative energies of different structures of $(H_2O)_{10}$ and $(H_2O)_{12}$. Although the CIM(0.005,0.01)-CCSD calculations employing Steps 1–12 of Section 2.3 recover more than 99% of the canonical CCSD correlation energy, the variation in the fractions of the recovered correlation energy, when we go from one structure of a given cluster to another, is too large to yield accurate relative energetics. Indeed, the fractions of the CCSD correlation energies recovered in the CIM(0.005,0.01)-CCSD calculations for different low-energy structures of the $(H_2O)_{10}$ and $(H_2O)_{12}$ clusters examined here vary between 99.13 and 99.91% for the $(H_2O)_{10}$/6-31++G(d, p) system, 99.21 and 99.85% for the $(H_2O)_{10}$/6-311++G(d, p) system, and 99.09 and 99.95% for the $(H_2O)_{12}$/6-31++G(d, p) system. As a result, the NPE values relative to the canonical CCSD energies of the ten lowest-energy structures of the water clusters under consideration characterizing the CIM(0.005,0.01)-CCSD calculations, which are 16.736 millihartree for the $(H_2O)_{10}$/6-31++G(d, p) system, 14.902 millihartree for the $(H_2O)_{10}$/6-311++G(d, p) system, and 22.173 millihartree for the $(H_2O)_{12}$/6-31++G(d, p) system, are very problematic. As shown in Fig. 8, the CIM(0.001)-CCSD calculations based on the modified subsystem design described in Section 4.1, which are characterized by the CPU timings that are almost the same as those characterizing the CIM(0.005,0.01)-CCSD calculations, offer significant improvements. Although the results of the CIM(0.001)-CCSD calculations are not perfect, the CIM(0.001)-CCSD approach provides a much better description of the relative energy patterns for the $(H_2O)_{10}$/6-31++G(d, p), $(H_2O)_{10}$/6-311++G(d, p), and $(H_2O)_{12}$/6-31++G(d, p) systems resulting from the canonical CCSD calculations. This is reflected in the considerably smaller variations in the fractions of the canonical CCSD correlation energies when going from one structure of a given water cluster to another, and the improved NPE values. Indeed, the fractions of the CCSD correlation energies recovered in the CIM(0.001)-CCSD calculations for different low-energy structures of the $(H_2O)_{10}$ and $(H_2O)_{12}$ clusters examined here vary between 99.80 and 100.15% for the $(H_2O)_{10}$/6-31++G(d, p) system, 99.71 and 99.98% for the $(H_2O)_{10}$/6-311++G(d, p) system, and 99.52 and 99.98% for the $(H_2O)_{12}$/6-31++G(d, p) system. The NPE values relative to canonical CCSD characterizing the CIM(0.001)-CCSD calculations for the water clusters under consideration are 7.431 millihartree for the $(H_2O)_{10}$/6-31++G(d, p) system, 6.265 millihartree for the $(H_2O)_{10}$/6-311++G(d, p) system, and 11.804 millihartree for the $(H_2O)_{12}$/6-31++G(d, p) system, which is a reduction by a factor of 2 when compared with the CIM(0.005,0.01)-CCSD calculations that have similar computer costs.

The question emerges if one can improve the performance of the CIM-CCSD approach relying on the modified design of the CIM subsystems described in

Section 4.1 even further and obtain a perfect agreement with the canonical CCSD results for the low-energy structures of the $(H_2O)_{10}/$ 6-31++G(d, p), $(H_2O)_{10}/$ 6-311++G(d, p), and $(H_2O)_{12}/$6-31++G(d, p) systems. Since the modified CIM scheme of Section 4.1 uses only a single parameter ζ in the subsystem design, we can improve the corresponding CIM-CCSD results by simply lowering ζ. As shown in Fig. 8, when we lower it to $\zeta = 0.0005$, the resulting CIM(0.0005)-CCSD energies are in an amazing agreement with the corresponding canonical CCSD energies. The fractions of the CCSD correlation energies recovered in the CIM(0.0005)-CCSD calculations for the ten lowest-energy structures of the $(H_2O)_{10}$ and $(H_2O)_{12}$ clusters examined here are not only extremely high, but also barely vary when we go from one structure to another, between 99.94 and 100.00% for the $(H_2O)_{10}/$ 6-31++G(d, p) system, 99.94 and 99.99% for the $(H_2O)_{10}/$6-311++G(d, p) system, and 99.94 and 100.03% for the $(H_2O)_{12}/$6-31++G(d, p) system. The NPE values relative to canonical CCSD characterizing the CIM(0.0005)-CCSD calculations for the water clusters under consideration are excellent as well, namely, 1.287 millihartree for the $(H_2O)_{10}/$6-31++G(d, p) system, 1.166 millihartree for the $(H_2O)_{10}/$ 6-311++G(d, p) system, and 2.406 millihartree for the $(H_2O)_{12}/$6-31++G(d, p) system. It is true that these improvements are accomplished at the expense of increasing the computer costs of the CIM-CCSD calculations and savings in the CPU time compared with the canonical CCSD calculations for the $(H_2O)_{10}/$ 6-31++G(d, p) and $(H_2O)_{10}/$6-311++G(d, p) systems are minimal. We must remember, however, that unlike the canonical CC calculations, which are characterized by the steep increase of the computer costs with the system size (N^6 in the CCSD case), the CIM-CC approaches are characterized by the almost linear scaling of the total CPU time with the system size. It is sufficient to go from $(H_2O)_{10}/$ 6-31++G(d, p) to $(H_2O)_{12}/$6-31++G(d, p), i.e., increase the system size by a small factor of 1.2 to begin to see the noticeable savings in the computer effort offered by the CIM-CC approaches based on the modified subsystem design. Indeed, the average total CPU time required to perform the CIM(0.0005)-CCSD calculations for $(H_2O)_{12}$, as described by the 6-31++G(d, p) basis set, which rely on the modified CIM scheme examined in this section, is approximately half that of the corresponding canonical CCSD calculations. If we increased the number of water molecules in the cluster, we would observe significantly larger CPU time savings. For example, based on the actual timings of the canonical CCSD calculations for the $(H_2O)_{10}/$ 6-31++G(d, p) and $(H_2O)_{12}/$6-31++G(d, p) systems, shown in Figs. 8(a) and (c), respectively, and the $n_o^2 n_u^4$ scaling of the CCSD CPU time with n_o and n_u, the estimated timing of the canonical CCSD calculation for $(H_2O)_{20}$ using the 6-31++G(d, p) basis set is 130,000 minutes (about 90.3 days). The corresponding CPU timing of the actual CIM(0.0005)-CCSD calculation for the lowest-energy structure of $(H_2O)_{20}$, as described by the 6-31++G(d, p) basis set, is 10,365 minutes (7.2 days) total or 1,832 minutes (30.5 hours) for the largest subsystem if the calculation for each of the 20 subsystems is run on a different processor. Further reduction of the computer time could be obtained if we used multiprocessor nodes and run each CCSD subsystem calculation on each node in parallel on as many processors as the given node provides, although we have not developed the CIM-CC codes of this type yet.

In conclusion, the modified design of the CIM subsystems described in Section 4.1 (see W. Li and P. Piecuch, unpublished manuscript for further details) offers considerable promise in the context of CC calculations for large molecular systems. Thus, we will continue examining this approach and perform additional calculations at various levels of CC theory, including CCSD(T) and CR-CC(2,3), in the future work.

5 Summary

In this chapter, we have reviewed our recent effort toward the extension of the linear scaling local correlation approach of Li and coworkers [38–40], abbreviated as CIM, to the standard CCSD approach and two CC methods with a non-iterative treatment of connected triply excited clusters, including the conventional CCSD(T) method and its completely renormalized CR-CC(2,3) analog [102] (see, also, W. Li and P. Piecuch, unpublished work). The local correlation formulation of the latter method based on the CIM formalism is particularly useful, since it enables one to obtain an accurate description of single bond breaking and biradicals, where CCSD(T) fails, with an ease of a "black-box" calculation of the CCSD(T) type [24–26, 109–117]. At the same time, CR-CC(2,3) is as accurate as CCSD(T) in applications involving closed-shell molecules near their equilibrium geometries.

As in the pioneering CIM work of Li and coworkers that dealt with the second-order MBPT and CCD approaches, the key idea of the CIM-CCSD, CIM-CCSD(T), and CIM-CR-CC(2,3) methods discussed in this chapter and related original papers [102] (see, also, W. Li and P. Piecuch, unpublished work) is the realization of the rather elementary fact that the total correlation energy of a large system can be rigorously represented as a sum of contributions from the occupied orthonormal LMOs and their respective occupied and unoccupied orbital domains. However, unlike in the earlier CIM-CCD work [38–40], which relied on the CCD codes with the nested explicit loop structure for projections on excited determinants and non-linear terms, the CIM-CCSD, CIM-CCSD(T), and CIM-CR-CC(2,3) methods and computer codes described in this chapter are characterized by high computational efficiency in both the CIM and the CC parts, enabling calculations for much larger systems and basis sets than previously possible. This is achieved by combining the NLS and trivial parallelism of the CIM ansatz with the vectorization of the CC codes used in the CIM subsystem calculations that rely on the heavy use of diagram factorization, recursively generated intermediates, and fast matrix multiplication routines. The highly efficient serial and parallel CIM-CCSD, CIM-CCSD(T), and CIM-CR-CC(2,3) codes discussed in this work, which are also fully vectorized in parts that deal with the CC subsystem calculations and which enable calculations for singlet ground states employing the RHF reference determinant, are interfaced with the GAMESS package.

Unlike in the previously reported local correlation CC work of Refs. [56–60], the CIM-CC methods rely on the conventional CC codes developed for the orthonormal MO sets, which are used in subsystem calculations. This results in considerable simplifications in the underlying computer algorithms compared with some other local CC efforts pursued in the past, including Refs. [56–60] where the authors were forced to reformulate the CC equations in a substantial manner to allow for the use of non-orthogonal unoccupied orbitals. In particular, the local correlation CC methods of Refs. [56–60] replace the non-iterative corrections of CCSD(T) by the substantially more complicated iterative steps that are needed to construct the relevant approximate form of the triply excited cluster amplitudes that formally enter the (T) energy correction of CCSD(T), since by switching to the noncanonical, non-HF LMO basis one has to incorporate additional terms involving the off-diagonal elements of the Fock matrix in the CCSD(T) scheme. The CIM-CCSD(T) and CIM-CR-CC(2,3) methods that we have developed enable us to formulate the local triples corrections to the CCSD energies in a purely non-iterative fashion, i.e., without the need for iterative steps in the triples part used in the local CCSD(T) approach of Refs. [57, 58]. In developing the CIM-CCSD(T) and CIM-CR-CC(2,3) methods with the non-iterative treatment of local triples corrections we have exploited the intrinsic flexibility of the CIM formalism that allows one to use MOs of the quasi-canonical type (QCMOs) within each subsystem instead of noncanonical LMOs once the CIM orbital subsystems are constructed. As discussed in this chapter, the use of QCMOs in the CIM-CC calculations offers other advantages, such as the substantially improved convergence of the CCSD subsystem calculations compared with the CCSD calculations for the CIM subsystems based on the LMOs employed in the subsystem design.

By comparing the results of the canonical and CIM CC calculations for normal alkanes and water clusters with the basis sets of the 6-31G(d) and 6-31G(d, p) type, we have shown that the CIM-CCSD, CIM-CCSD(T), and CIM-CR-CC(2,3) approaches recover the corresponding canonical CC correlation energies to within 0.1% or so, while offering nearly linear scaling of the CPU time with the system size and savings in the overall computer effort by orders of magnitude. By examining the dissociation of dodecane into $C_{11}H_{23}$ and CH_3 and several lowest-energy structures of the $(H_2O)_n$ clusters, we have demonstrated that the CIM-CC methods accurately reproduce the relative energetics of the corresponding canonical CC calculations. We have also shown that one can substantially benefit from exploiting the idea of mixed CIM-CC approaches, in which the underlying CIM-CCSD calculations that precede the evaluation of the CR-CC(2,3) and CCSD(T) corrections due to triples are performed with tighter thresholds for defining the CIM subsystems than the subsequent CR-CC(2,3) and CCSD(T) calculations.

Finally, we have addressed difficulties with achieving high accuracies in representing the relative energies in the CIM-CC calculations for weakly bound molecular clusters, such as clusters of water molecules, employing basis sets containing diffuse functions by describing a modified variant of the underlying CIM theory characterized by a different design of the CIM subsystems compared with the original subsystem design proposed in Refs. [38–40]. In the modified variant of the CIM formalism tested in this and related (W. Li and P. Piecuch, unpublished manuscript) work, we replace the original design of the CIM subsystems of Refs. [38–40], which is based on the ideas of central orbitals and the associated primary and secondary environments, by assigning central localized orbitals and the corresponding environment LMOs to each non-hydrogen atom and the adjacent hydrogen atoms that are bound to it. The performance of this modified variant of the CIM-CC theory has been illustrated by the CIM-CCSD calculations for water clusters of varying size, as described by the 6-31G(d), 6-31++G(d, p), and 6-311++G(d, p) basis sets. These calculations show that the modified CIM theory using central LMOs assigned to atoms and only one type of orbital environment (as opposed to the primary and secondary environments of the CIM subsystem design proposed in Refs. [38–40]) offers significant improvements in the description of the relative energies of water clusters when the AO basis sets contain diffuse functions, while providing additional small improvements in the CIM-CC calculations without diffuse functions at no extra cost compared with the CIM-CC calculations based in the original subsystem design of Refs. [38–40].

In addition to various molecular applications, our future work will focus on further development of the modified CIM-CC theory, in which central LMOs are assigned to atoms and only one type of orbital environment is needed, benchmarked in this chapter and related work (W. Li and P. Piecuch, unpublished manuscript), and extension of the CIM-CCSD and CIM-CR-CC(2,3) methods to non-singlet electronic ground states. In the latter case, we plan to combine the CIM-CC code infrastructure developed so far with the highly efficient open-shell CCSD and CR-CC(2,3) routines described in Ref. [26], which are integrated with GAMESS.

Acknowledgement We thank Professor Shuhua Li for useful discussions. This work has been supported by the Chemical Sciences, Geosciences and Biosciences Division, Office of Basic Energy Sciences, Office of Science, US Department of Energy (Grant No. DE-FG02-01ER15228; P.P) and the National Science Foundation's Graduate Research Fellowship (J.R.G). All calculations and development work reported in this chapter were carried out on the computer systems provided by the High Performance Computing Center and Department of Chemistry at Michigan State University.

References

1. F. Coester, Nucl. Phys. **7**, 421 (1958)
2. F. Coester, H. Kümmel, Nucl. Phys. **17**, 477 (1960)
3. J. Čížek, J. Chem. Phys. **45**, 4256 (1966)
4. J. Čížek, Adv. Chem. Phys. **14**, 35 (1969)
5. J. Čížek, J. Paldus, Int. J. Quantum Chem. **5**, 359 (1971)
6. R. J. Bartlett, in *Advanced Series in Physical Chemistry*, Vol. 2, *Modern Electronic Structure Theory, Part I*, ed. by D. R. Yarkony (World Scientific, Singapore, 1995), pp. 1047–1131
7. J. Gauss, in *Encyclopedia of Computational Chemistry*, ed. by P. v. R. Schleyer, N. L. Allinger, T. Clark, J. Gasteiger, P. A. Kollman, H. F. Schaefer III, P. R. Schreiner (Wiley, Chichester, UK, 1998), Vol. 1, pp. 615–636

8. J. Paldus, X. Li, Adv. Chem. Phys. **110**, 1 (1999)

9. T. D. Crawford, H. F. Schaefer III, Rev. Comput. Chem. **14**, 33 (2000)

10. J. Paldus, in *Handbook of Molecular Physics and Quantum Chemistry*, ed. by S. Wilson (Wiley, Chichester, 2003), Vol. 2, pp. 272–313

11. P. Piecuch, K. Kowalski, I. S. O. Pimienta, M. J. McGuire, Int. Rev. Phys. Chem. **21**, 527 (2002)

12. P. Piecuch, K. Kowalski, I. S. O. Pimienta, P.-D. Fan, M. Lodriguito, M. J. McGuire, S. A. Kucharski, T. Kuś, M. Musiał, Theor. Chem. Acc. **112**, 349 (2004)

13. G. D. Purvis III, R. J. Bartlett, J. Chem. Phys. **76**, 1910 (1982)

14. J. M. Cullen, M. C. Zerner, J. Chem. Phys. **77**, 4088 (1982)

15. G. E. Scuseria, A. C. Scheiner, T. J. Lee, J. E. Rice, H. F. Schaefer III, J. Chem. Phys. **86**, 2881 (1987)

16. P. Piecuch, J. Paldus, Int. J. Quantum Chem. **36**, 429 (1989)

17. M. Urban, J. Noga, S. J. Cole, R. J. Bartlett, J. Chem. Phys. **83**, 4041 (1985)

18. K. Raghavachari, G. W. Trucks, J. A. Pople, M. Head-Gordon, Chem. Phys. Lett. **157**, 479 (1989)

19. P. Piecuch, J. Paldus, Theor. Chim. Acta **78**, 65 (1990)

20. S. R. Gwaltney, M. Head-Gordon, Chem. Phys. Lett. **323**, 21 (2000)

21. S. R. Gwaltney, M. Head-Gordon, J. Chem. Phys. **115**, 2014 (2001)

22. S. Hirata, M. Nooijen, I. Grabowski, R. J. Bartlett, J. Chem. Phys. **114**, 3919 (2001); **115**, 3967 (2001) (Erratum)

23. S. Hirata, P.-D. Fan, A. A. Auer, M. Nooijen, P. Piecuch, J. Chem. Phys. **121**, 12197 (2004)

24. P. Piecuch, M. Włoch, J. Chem. Phys. **123**, 224105 (2005)

25. P. Piecuch, M. Włoch, J. R. Gour, A. Kinal, Chem. Phys. Lett. **418**, 467 (2006)

26. M. Włoch, J. R. Gour, P. Piecuch, J. Phys. Chem. A **111**, 11359 (2007)

27. A. P. Rendell, T. J. Lee, A. Komornicki, Chem. Phys. Lett. **178**, 462 (1991)

28. A. P. Rendell, T. J. Lee, R. Lindh, Chem. Phys. Lett. **194**, 84 (1992)

29. A. P. Rendell, M. F. Guest, R. A. Kendall, J. Comput. Chem. **14**, 1429 (1993)

30. R. Kobayashi, A. P. Rendell, Chem. Phys. Lett. **265**, 1 (1997)

31. P. Piecuch, J. I. Landman, Parallel Comp. **26**, 913 (2000)

32. S. Hirata, J. Phys. Chem. A **107**, 9887 (2003)

33. G. Baumgartner, A. Auer, D. E. Bernholdt, A. Bibireata, V. Choppella, D. Cociorva, X. Gao, R. J. Harrison, S. Hirata, S. Krishnamoorthy, S. Krishnan, C. Lam, Q. Lu, M. Nooijen, R. M. Pitzer, J. Ramanujam, P. Sadayappan, A. Sibiryakov, Proceedings of the IEEE, **93**, 276 (2005)

34. P. Piecuch, S. Hirata, K. Kowalski, P.-D. Fan, T. L. Windus, Int. J. Quantum Chem. **106**, 79 (2006)

35. T. Janowski, A. R. Ford, P. Pulay, J. Chem. Theory Comput. **3**, 1368 (2007)

36. R. M. Olson, L. J. Bentz, R. A. Kendall, M. W. Schmidt, M. S. Gordon, J. Chem. Theory Comput. **3**, 1312 (2007)

37. M. E. Harding, T. Metzroth, J. Gauss, A. A. Auer, J. Chem. Theory Comput. **4**, 64 (2008)

38. S. Li, J. Ma, Y. Jiang, J. Comput. Chem. **23**, 237 (2002)

39. S. Li, W. Li, J. Ma, Chin. J. Chem. **21**, 1422 (2003)

40. S. Li, J. Shen, W. Li, Y. Jiang, J. Chem. Phys. **125**, 074109 (2006)

41. O. Sinanoğlu, Adv. Chem. Phys. **6**, 315 (1964)

42. R. K. Nesbet, Adv. Chem. Phys. **9**, 321 (1965)

43. J. A. Pople, R. Krishnan, H. B. Schlegel, J. S. Binkley, Int. J. Quantum Chem. **14**, 545 (1978)

44. R. J. Bartlett, G. D. Purvis, Int. J. Quantum Chem. **14**, 561 (1978)

45. W. D. Laidig, G. D. Purvis III, R. J. Bartlett, Int. J. Quantum Chem. Symp. **16**, 561 (1982)

46. W. D. Laidig, G. D. Purvis III, R. J. Bartlett, Chem. Phys. Lett. **97**, 209 (1983)

47. W. D. Laidig, G. D. Purvis III, R. J. Bartlett, J. Phys. Chem. **89**, 2161 (1985)

48. W. Förner, J. Ladik, P. Otto, J. Čížek, Chem. Phys. **97**, 251 (1985)

49. W. Förner, Chem. Phys. **114**, 21 (1987)

50. M. Takahashi, J. Paldus, Phys. Rev. B **31**, 5121 (1985)

51. Y.-J. Ye, W. Förner, J. Ladik, Chem. Phys. **178**, 1 (1993)
52. R. Knab, W. Förner, J. Čížek, J. Ladik, J. Mol. Struct.: THEOCHEM **366**, 11 (1996)
53. W. Förner, R. Knab, J. Čížek, J. Ladik, J. Chem. Phys. **106**, 10248 (1997)
54. G. E. Scuseria, P. Y. Ayala, J. Chem. Phys. **111**, 8330 (1999)
55. C. Hampel, H.-J. Werner, J. Chem. Phys. **104**, 6286 (1996)
56. M. Schütz, H.-J. Werner, J. Chem. Phys. **114**, 661 (2000)
57. M. Schütz, J. Chem. Phys. **113**, 9986 (2000)
58. M. Schütz, H.-J. Werner, Chem. Phys. Lett. **318**, 370 (2000)
59. M. Schütz, J. Chem. Phys. **116**, 8772 (2002)
60. M. Schütz, Phys. Chem. Chem. Phys. **4**, 3941 (2002)
61. P. E. Maslen, M. S. Lee, M. Head-Gordon, Chem. Phys. Lett. **319**, 205 (2000)
62. J. E. Subotnik, M. Head-Gordon, J. Chem. Phys. **123**, 064108 (2005)
63. J. E. Subotnik, A. Sodt, M. Head-Gordon, J. Chem. Phys. **125**, 074116 (2006)
64. J. E. Subotnik, A. Sodt, M. Head-Gordon, J. Chem. Phys. **128**, 034103 (2008)
65. N. J. Russ, T. D. Crawford, J. Chem. Phys. **121**, 691 (2004)
66. N. Flocke, R. J. Bartlett, J. Chem. Phys. **121**, 10935 (2004)
67. A. A. Auer, M. Nooijen, J. Chem. Phys. **125**, 024104 (2006)
68. O. Christiansen, P. Manninen, P. Jørgensen, J. Olsen, J. Chem. Phys. **124**, 084103 (2006)
69. Q. Li, Y. Yi, Z. Shuai, J. Comput. Chem. **29**, 1650 (2008)
70. T. Korona, H.-J. Werner, J. Chem. Phys. **118**, 3006 (2003)
71. T. Korona, K. Pflüger, H.-J. Werner, Phys. Chem. Chem. Phys. **6**, 2059 (2004)
72. D. Kats, T. Korona, M. Schütz, J. Chem. Phys. **125**, 104106 (2006)
73. D. Kats, T. Korona, M. Schütz, J. Chem. Phys. **127**, 064107 (2007)
74. T. D. Crawford, R. A. King, Chem. Phys. Lett. **366**, 611 (2002)
75. D. Walter, A. B. Szilva, K. Niedelfeldt, E. A. Carter, J. Chem. Phys. **117**, 1982 (2002)
76. D. Walter, E. A. Carter, Chem. Phys. Lett. **346**, 177 (2001)
77. D. Walter, A. Venkatnathan, E. A. Carter, J. Chem. Phys. **118**, 8127 (2003)
78. T. S. Chwee, A. B. Szilva, R. Lindh, E. A. Carter, J. Chem. Phys. **128**, 224106 (2008)
79. D. G. Fedorov, K. Kitaura, J. Chem. Phys. **123**, 134103 (2005)
80. W. Li, S. Li, J. Chem. Phys. **121**, 6649 (2004)
81. H. M. Netzloff, M. A. Collins, J. Chem. Phys. **127**, 134113 (2007)
82. P. Pulay, Chem. Phys. Lett. **100**, 151 (1983)
83. S. Saebø, P. Pulay, Chem. Phys. Lett. **113**, 13 (1985)
84. P. Pulay, S. Saebø, Theor. Chim. Acta **69**, 357 (1986)
85. S. Saebø, P. Pulay, J. Chem. Phys. **86**, 914 (1987)
86. S. Saebø, P. Pulay, J. Chem. Phys. **88**, 1884 (1988)
87. S. Saebø, P. Pulay, Annu. Rev. Phys. Chem. **44**, 213 (1993)
88. S. F. Boys, Rev. Mod. Phys. **32**, 296 (1960)
89. C. Edmiston, K. Ruedenberg, Rev. Mod. Phys. **35**, 457 (1963)
90. J. Pipek, P. G. Mezey, J. Chem. Phys. **90**, 4916 (1989)
91. M. Kobayashi, H. Nakai, J. Chem. Phys. **129**, 044103 (2008)
92. F. Weinhold, in *Encyclopedia of Computational Chemistry*, ed. by P. v. R. Schleyer, N. L. Allinger, T. Clark, J. Gasteiger, P. A. Kollman, H. F. Schaefer III, P. R. Schreiner (Wiley, Chichester, UK, 1998), Vol. 3, pp. 1792–1811
93. P. Piecuch, S. A. Kucharski, K. Kowalski, M. Musiał, Comp. Phys. Commun. **149**, 71 (2002)
94. M. W. Schmidt, K. K. Baldridge, J. A. Boatz, S. T. Elbert, M. S. Gordon, J. H. Jensen, S. Koseki, N. Matsunaga, K. A. Nguyen, S. J. Su, T. L. Windus, M. Dupuis, J. A. Montgomery, J. Comput. Chem. **14**, 1347 (1993)
95. S. A. Kucharski, R. J. Bartlett, Theor. Chim. Acta **80**, 387 (1991)
96. M. D. Lodriguito, P. Piecuch, in *Progress in Theoretical Chemistry and Physics*, Vol. 18, *Frontiers in Quantum Systems in Chemistry and Physics*, ed. by S. Wilson, P. J. Grout, J. Maruani, G. Delgado-Barrio, P. Piecuch (Springer, Berlin, 2008), pp. 67–174
97. W. J. Hehre, R. Ditchfield, J. A. Pople, J. Chem. Phys. **56**, 2257 (1972)
98. P. C. Hariharan, J. A. Pople, Theor. Chim. Acta **28**, 213 (1973)

99. T. Clark, J. Chandrasekhar, G. W. Spitznagel, P. v. R. Schleyer, J. Comput. Chem. **4**, 294 (1983)
100. R. Krishnan, J. S. Binkley, R. Seeger, J. A. Pople, J. Chem. Phys. **72**, 650 (1980)
101. E. A. Salter, G. W. Trucks, R. J. Bartlett, J. Chem. Phys. **90**, 1752 (1989)
102. W. Li, P. Piecuch, J. R. Gour, S. Li, J. Chem. Phys. **131**, xxxx (2009), in press
103. M. Włoch, J. R. Gour, K. Kowalski, P. Piecuch, J. Chem. Phys. **122**, 214107 (2005)
104. W. Li, S. Li, Y. Jiang, J. Phys. Chem. A **111**, 2193 (2007)
105. J. W. Boughton, P. Pulay, J. Comput. Chem. **14**, 736 (1993)
106. J. E. Subotnik, A. D. Dutoi, M. Head-Gordon, J. Chem. Phys. **123**, 114108 (2005)
107. M. Schütz, G. Rauhut, H.-J. Werner, J. Phys. Chem. A **102**, 5997 (1998)
108. P. Piecuch, M. Włoch, A. J. C. Varandas, in *Progress in Theoretical Chemistry and Physics*, Vol. 16, *Topics in the Theory of Chemical and Physical Systems*, ed. by S. Lahmar, J. Maruani, S. Wilson, G. Delgado-Barrio (Springer, Berlin, 2007), pp. 63–121
109. M. Włoch, M. D. Lodriguito, P. Piecuch, J. R. Gour, Mol. Phys. **104**, 2149 (2006)
110. A. Kinal, P. Piecuch, J. Phys. Chem. A **111**, 734 (2007)
111. Y. Ge, M. S. Gordon, P. Piecuch, J. Chem. Phys. **127**, 174106 (2007)
112. J. J. Lutz, P. Piecuch, in *Nuclei and Mesoscopic Physics, Workshop on Nuclei and Mesoscopic Physics, WNMP 2007*, AIP Conference Proceedings, Vol. 995, ed. by P. Danielewicz, P. Piecuch, V. Zelevinsky (AIP Press, 2008), pp. 62–71
113. P. Piecuch, M. Włoch, A. J. C. Varandas, Theor. Chem. Acc. **120**, 59 (2008)
114. J. J. Lutz, P. Piecuch, J. Chem. Phys. **128**, 154116 (2008)
115. Y. Z. Song, A. Kinal, P. J. S. B. Caridade, A. J. C. Varandas, P. Piecuch, J. Mol. Struct.: THEOCHEM **859**, 22 (2008)
116. Y. Ge, M. S. Gordon, P. Piecuch, M. Włoch, J. R. Gour, J. Phys. Chem. A **112**, 11873 (2008)
117. W. L. Jorgensen, J. Chandrasekhar, J. D. Madura, R. W. Impey, M. L. Klein, J. Chem. Phys. **79**, 926 (1983)

The Correlation Consistent Composite Approach (ccCA): Efficient and Pan-Periodic Kinetics and Thermodynamics

Nathan J. DeYonker, Thomas R. Cundari, and Angela K. Wilson

Abstract The correlation consistent Composite Approach (ccCA) methodology is the most accurate MP2-based model chemistry developed to date and contains no optimized empirically based parameters. As the method has continued to evolve and found new areas of application, we provide in this chapter a historical context of the developments and discoveries, an overview of variations of the ccCA model chemistry, and some discourse on future research thrusts. The success of ccCA for chemical applications is reviewed.

Keywords: Model chemistry · Composite method · Correlation consistent · Thermochemistry

1 Introduction

In the last 20 years, the coupled cluster ansatz of the electronic Schrödinger wave equation has become a standard method used by the computational chemistry community to accurately treat the electron correlation problem. The current "gold standard" for determining accurate and systematically improvable energetic properties (i.e., isomerization, ionization, and atomization energies, bond-dissociation energies, and thermodynamic quantities) is generally based on using a coupled cluster

A.K. Wilson (✉)
Center for Advanced Scientific Computing and Modeling (CASCaM), Department of Chemistry, University of North Texas, 1155 Union Circle #305070, Denton, Texas 76203-5070, USA, e-mail: akwilson@unt.edu

N.J. DeYonker
Center for Advanced Scientific Computing and Modeling (CASCaM), Department of Chemistry, University of North Texas, Denton, Texas 76203-5070, USA, e-mail: ndeyonk@unt.edu

T.R. Kundari
Center for Advanced Scientific Computing and Modeling (CASCaM), Department of Chemistry, University of North Texas, Denton, Texas 76203-5070, USA, e-mail: tomc@unt.edu

P. Piecuch et al. (eds.), *Advances in the Theory of Atomic and Molecular Systems*, Progress in Theoretical Chemistry and Physics 19, DOI 10.1007/978-90-481-2596-8_9, © Springer Science+Business Media B.V. 2009

wavefunction with single, double, and perturbative triple excitations [CCSD(T)] in conjunction with reasonably complete basis sets (CBSs) [1–4]. The closed shell CCSD(T) method [5] typically scales as $N_o^3 N_v^4$, where N_o is the number of occupied orbitals and N_v is the number of virtual (unoccupied) orbitals. The N^7 "(T)" step requires converged CCSD amplitudes, which themselves scale as $I^* N_o^2 N_v^4$, where I is the number of iterations required for convergence.

Notable examples of recent achievements that reach the technological limits of standard single-reference coupled cluster methods are (but by no means limited to) massively parallelized implementations for computation of energies [6] and gradients [7], automated computer implementations of energies/gradients beyond full triple substitutions [8], "spin-flip" [9–11], and "completely renormalized" [12–15] methods for highly accurate determination of open-shell energies and bond-breaking mechanisms for ground and excited electronic states, determination of fullerene frequency-dependent polarizabilities [16], and accurate thermochemical properties for n-octane [17].

Unfortunately, the growth of computing power has not increased as anything like an N^7 function [18]. Thus, the utilization of coupled cluster methods remains limited in terms of the size of accessible chemical problems, in particular due to onerous memory and disk storage requirements. Even with the advent of parallel processing, many theoretical and experimental chemists must look toward methods more efficient than CCSD(T), such as the N^5-scaling second-order Møller–Plesset perturbation theory (MP2) method or the optimally N^3-scaling density functional theory (DFT) to account for electron correlation in a given wave function. It is these approximate methods that must first be used to explore novel chemical systems when coupled cluster-based approaches are intractable. In stark contrast to DFT, the strength of MO-based methods such as MP2 is that they are systematically improvable and deficiencies are quantifiable. Ab initio methods are inherently reliable tools to forge a lucid path toward quantitative gas-phase accuracy.

The typical benchmark of quantitative ab initio methods is "chemical accuracy," often defined as an energetic property (such as ionization energies or enthalpies of formation) being on average within $\pm 1.0 \, \text{kcal mol}^{-1}$ of a well-established experimental value. Chemical accuracy is not guaranteed even when using the CCSD(T) method in conjunction with valence-only electron correlation and a modest-sized basis set. Besides including higher angular momentum polarization functions within a basis set, other physical effects strongly contribute to the electronic energy and thus affect observable properties. These effects include relativistic effects (including scalar effects and spin-orbit coupling), core–valence electron correlation, the presence of low-lying or degenerate electronic states, non-Born–Oppenheimer effects, and geometric rotational/vibrational energy effects.

It has long been recognized that many of the variables (such as equilibrium geometry, vibrational effects, relativistic effects) in an ab initio electronic structure computation are not strongly coupled. Since analytic and numerical gradients of ab initio methods can be extremely costly, a modest level of theory (perhaps DFT) is often used to compute an approximate stationary point geometry, next with a more accurate single-point energy that is computed using, for example,

MP2 or CCSD(T) with a large basis set. This "double slash" (single-point level of theory//geometry optimization level of theory) idea was first popularized by Pople and coworkers [19, 20], and the early computational chemistry literature is replete with terminology such as MP2/6-311++G(d, p)//RHF/6-31G(d). From their work, it was also recognized that a standardized set of calculations, improving toward the basis set and electron correlation limits could be used to solve a more diverse array of chemical problems and achieve a consistent level of accuracy with a reduced computational expense, versus a CCSD(T)/large basis set calculation. The initial studies employing these so-called Gaussian-n (or Gn) methods coined the phrases "composite methods" or "model chemistries," to categorize a collection of efficient ab initio computations performed with the goal of reproducing the electronic energy of an expensive and often computationally intractable electronic structure computation [20–23]. This goal was accomplished using additive approximations of basis set and electron correlation effects obtained from multiple computations of both higher efficiency and lower accuracy. Figure 1 gives an example of a single-point computation performed in our laboratory [24] on gas-phase benzene using a very sophisticated level of theory [CCSD(T)/aug-cc-pVQZ-DK]. This high-level computation requires $\sim$110 hours of wall time to compute. If electron correlation, basis set size, and relativistic effects are not strongly coupled, then a MP2/large basis set computation could be combined with a CCSD(T)/small basis set and MP2-DK/small relativistic basis set computation to achieve a similar result as the CCSD(T)/aug-cc-pVQZ-DK "effective level of theory." These three computations also need not be run sequentially, and in the end, a savings of almost two orders of magnitude in CPU time is achieved. The strategy is computationally inexpensive relative to the effective level of theory, requiring a maximum of 5.5 hours of wall time on the same computer system.

Composite methods are usually compared against the proposed effective level of theory [2, 23, 25, 26], or a well-established experimental database. For example, the current training set that Gn methods are benchmarked against includes experimental enthalpies of formation (ΔH_f), dissociation energies (D_0), ionization potentials (IPs), electron affinities (EAs), and proton affinities (PAs). This test set, called the G3/05 set, contains 454 energetic properties overall [27]. Methods such as Gaussian-n have provided an additional important conceptual advance in quantitative computational chemistry, in that they have led to large, generally agreed-upon databases of molecular information against which both current and future methods may be fairly compared. However, while such databases ideally allow for calibration against as diverse a set of bonding schemes as is feasible, such tasks are necessarily focused on those portions of the periodic table for which data are more abundant, are deemed to be inherently more reliable, or are simply scientifically popular. As such, most thermodynamic databases are focused on organic molecules of small to medium size. For example, the G3/05 set does not contain atomic or energetic properties for transition metal-containing species. The size of molecules within the test set is also quite limited. For instance, the largest neutral molecules for which information is contained within this database are octane (C_8H_{18}) and azulene ($C_{10}H_8$), while the largest ion is the toluene cation ($C_7H_8^+$). While such molecules

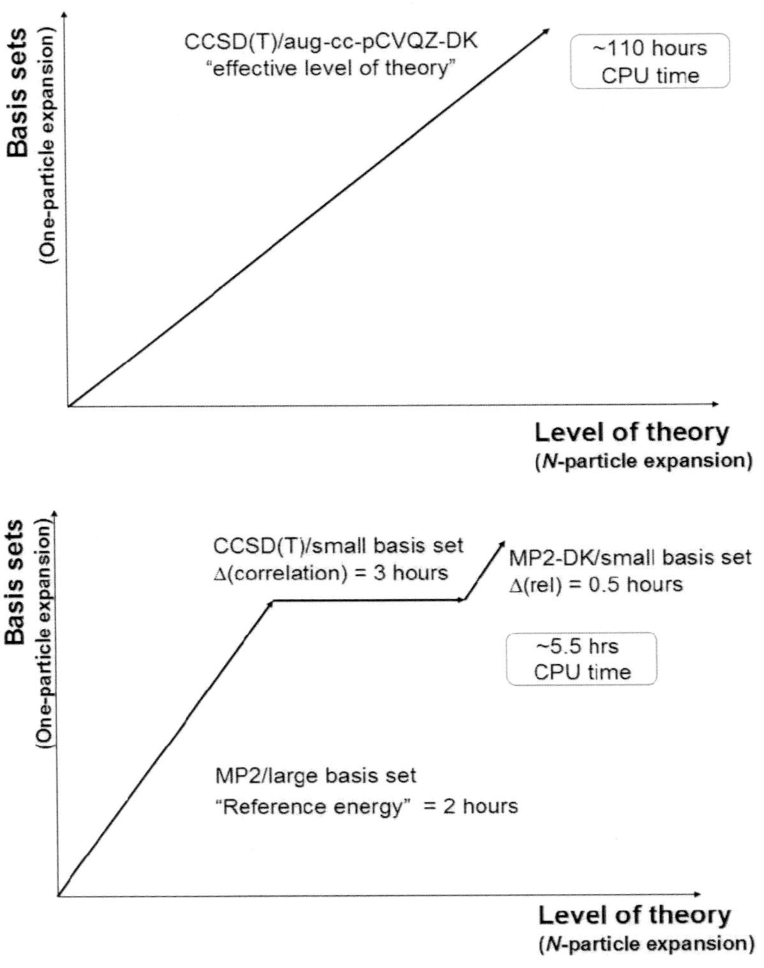

Fig. 1 An example of the time savings achieved when using additive corrections such as those employed in the ccCA methodology. When approximating a large basis set coupled cluster computation on the benzene molecule, a relative savings of almost two orders of magnitude in CPU time is achieved

may define the current extreme of high-level ab initio quantum theory, a perusal of any recent chemical journal makes it clear that chemists' elemental appetites are larger – both in variety and in size.

This review will briefly compare and contrast current popular model chemistries in order to describe the motivation for the correlation consistent Composite Approach (ccCA). Initial calibration studies using the Gn training sets and initial comparisons to Gn methods will highlight chemical systems for which ccCA is a reliable and preferable model chemistry. Accomplishments and current development of ccCA will be discussed.

2 Composite Methods

Until the advent of ccCA, molecular orbital (MO)-based composite methods generally could be divided into two categories. The first category of methods strives to compute energies at an accuracy of $\pm 1.0\,\text{kcal mol}^{-1}$ or better while retaining relatively high computational efficiency. Mostly these model chemistries build their additive corrections on MP2 (or sometimes MP4) energies. Due to the approximations that must be included to balance accuracy and efficiency, the model chemistries contained within this class of model chemistries generally include one or more semi-empirical parameters. Taking the Gn methods, for example, a high-level correction (HLC) is added to the atomic and molecular energies depending on the number of paired and unpaired valence electrons [22]. Initially, the HLC had a physical derivation; in the G1 method, the correction was obtained by taking the difference between the electronic energy of the H atom and the H_2 molecule and their near-exact analytical energies [20, 21]. In a molecular system, each unpaired electron and pair of electrons would require an equivalent energy difference. In a sense, the HLC was a sort of isogyric correction to the molecule. From the introduction of the G2 method [22], and up to the present time, the HLC is optimized in order to minimize the magnitude of the mean signed deviation (MSD) from experiment within the G3/05 training sets (and its predecessors for previous iterations of Gn). The HLC now accommodates for any systematic errors within the Gn methods. Besides the Gn methods, other model chemistries including a semi-empirical correction are the CBS-n methods of Petersson and coauthors [28–32] and versions of the multi-coefficient correlation method (MCCM) of Truhlar and coauthors [33, 34].

The second category of composite methods includes those that use the convergent hierarchy of coupled cluster methods (CCSDT, CCSDTQ, CCSDTQ5, etc.) to approximate the CBS, full configuration interaction limit. By employing sophisticated electron correlation corrections and avoiding parameterization, these methods generally strive for a target accuracy of better than $\pm 1.0\,\text{kJ mol}^{-1}$, and often thermochemical data obtained via these composite methods can be considered more reliable than rapidly aging experimental thermochemical studies. However, by eschewing parameterization, the high-accuracy coupled cluster-based approaches require daunting computational resources for molecules with more than a few nonhydrogen atoms. Examples of model chemistries in this category are the CCSD(T)/CBS-based method of Dixon, Feller, Peterson, and coworkers [2, 3, 17, 35–41], the focal point method of Allen, Császár, and coworkers [42–45], the Wn methods of Martin, Karton, and de Oliveira [46–49], and the HEAT method of Stanton, Gauss, and coworkers [50–52]. These coupled cluster-based methods have been employed in tandem with the self-consistent statistical analyses of the Active Thermochemical Tables (ATcT) developed by Ruscic and coauthors [53–57] and the Computational Results Database (CRDB) of Feller [58]. These efforts synthesize existing high-quality experimental and theoretical data, and it is likely that such compendia will become the standards for highly accurate ab initio benchmarks. A recent review by Helgaker and coauthors [4] lists the formulation of many popular model chemistries (including ccCA).

Gleaning wisdom from the first two decades of wide spread usage of model chemistries, our development of the ccCA model chemistry has focused on creating a third category of composite methods, combining the advantages of the two traditional approaches. At the onset of ccCA development, our research group had three goals in mind. The first was to create a cost-effective MP2-based model chemistry that achieved chemical accuracy with an overall competitive reliability to available composite methods, but *without any empirical parameters*. The second goal was to utilize methods that are common in every major ab initio computational chemistry software package. Moreover, the ccCA method should be included within open-source and free software packages, and have a transparent implementation. The third and final goal was to attempt to avoid the massive variation of current model chemistries, which can often lead to confusion or uncertainty about which variation is most appropriate for which specific chemical problem. Instead of fragmenting the model with variants whenever a deficiency arises, ccCA should smoothly evolve (but not horribly mutate) into a single recommended formula that is reliable for large chemical systems and which is not limited by the location of constituent elements in small subsets of the periodic table, i.e., it should be *pan-periodic*.

3 The Formulation of ccCA

The development of ccCA into its current recommended formulation will be discussed further in Section 4.1. Table 1 describes the algorithm and additive corrections involved, as well as the evolution of ccCA variants into its current form. We have come to recognize that while standard DFT methods are inadequate for highly accurate energetic properties, they are an invaluable compromise when used to determine high-quality molecular structures at a very low computational cost. Therefore, the recommended formulation for ccCA begins with obtaining an optimized geometry at the B3LYP/cc-pVTZ level of theory. This DFT methodology is also used to confirm that the stationary point obtained is a true minimum on the potential energy hypersurface (i.e., a positive definite Hessian). With the Hessian matrix, harmonic vibrational frequencies are computed and are scaled by a constant factor (0.989) to take into account for known deficiencies of the harmonic approximation. From the scaled harmonic vibrational frequencies, the zero-point energy $[\Delta E(\text{ZPE})]$ and the thermal enthalpy corrections are calculated. Once the equilibrium geometry and harmonic frequencies are determined, a series of single-point calculations are performed at more sophisticated levels of theory to obtain the ccCA energy. The reference energy of the system, to which the additive corrections are applied, is calculated by computing MP2 single-point energies at the aug-cc-pVDZ, aug-cc-pVTZ, and aug-cc-pVQZ basis set levels.

There are a number of approaches that can be used to determine CBS limits, and this choice can impact the accuracy of ccCA energies. Initially, the behavior of various CBS fits was examined via extrapolation of total energies (HF+MP2 correlation), and we found that reliable performance was obtained using either the mixed exponential/Gaussian functional devised by Peterson et al. [59], expressed as

Table 1 Additive corrections and description of the steps within the ccCA model chemistry. The evolution of the ccCA to its present algorithm is shown

	ccCA-TZ	ccCA-aTZ	ccCA CBS-1/ccCA CBS-2	ccCA-P/ccCA-S4	ccCA-PS3
Geometry optimization	B3LYP/6-31G(d)	B3LYP/6-31G(d)	B3LYP/6-31G(d)	B3LYP/6-31G($2df,p$)	B3LYP/cc-pVTZ
Harmonic vibrational frequencies	B3LYP/6-31G(d)	B3LYP/6-31G(d)	B3LYP/6-31G(d)	B3LYP/6-31G($2df,p$)	B3LYP/cc-pVTZ
Scale factor	0.9854	0.9854	0.9854	0.9854	0.9890
Reference energy E(ref)	MP4/cc-pVTZ	MP4/aug-cc-pVTZ	MP2/CBS aug-cc-pVnZ	MP2/CBS aug-cc-pVnZ	MP2/CBS aug-cc-pVnZ
Equation of MP2 CBS fit	–	–	Eq. (1)	Eq. (1) or Eq. (2)	(0.5) × Eq. (1) + Eq. (4)
$\Delta E(+)$	MP4/aug-cc-pVTZ – E(ref)	–	–	–	–
$\Delta E(2df, p)/\Delta E(\zeta)$	MP4/cc-pVQZ – MP4/cc-pVTZ	MP2/aug-cc-pVQZ – MP2/aug-cc-pVTZ	–	–	–
ΔE(QCI/CC)	QCISD(T)/ cc-pVTZ – E(ref)	QCISD(T)/ cc-pVTZ – MP4/ cc-pVTZ	QCISD(T)/ cc-pVTZ – MP2/ cc-pVTZ	CCSD(T)/ cc-pVTZ – MP2/ cc-pVTZ	CCSD(T)/ cc-pVTZ – MP2/ cc-pVTZ
ΔE(CV)	MP2(full)/ aug-cc-pCVTZ – MP2/ aug-cc-pVTZ	MP2(FC1)/ aug-cc-pCVTZ – MP2/ aug-cc-pVTZ	MP2(FC1)/ aug-cc-pCVTZ – MP2/ aug-cc-pVTZ	MP2(FC1)/ aug-cc-pCVTZ – MP2/ aug-cc-pVTZ	MP2(FC1)/ aug-cc-pCVTZ – MP2/ aug-cc-pVTZ
ΔE(DK)	–	–	–	MP2-DK/cc-pVTZ-DK – MP2/cc-pVTZ	MP2-DK/cc-pVTZ-DK – MP2/cc-pVTZ
Effective level of theory	QCISD(T,full) /aug-cc-pCVQZ	QCISD(T,FC1) /aug-cc-pCVQZ	QCISD(T,FC1) /aug-cc-pCV∞Z	CCSD-DK(T,FC1) /aug-cc-pCV∞Z-DK	CCSD-DK(T,FC1) /aug-cc-pCV∞Z-DK

$$E(x) = E_{CBS} + B \exp[-(x - 1)] + C \exp\left[-(x - 1)^2\right], \tag{1}$$

where x is the cardinal number of the basis set, or "zeta-level", i.e., $x = 2$ for aug-cc-pVDZ and $x = 3$ for aug-cc-pVTZ), and the CBS function using the inverse power of l_{max}^4 (where l_{max} is the highest angular momentum used by basis set functions) developed by Schwartz [60, 61], Kutzelnigg [62], Klopper, Wilson, and coworkers [63], and Martin [64], which is expressed as

$$E(l_{max}) = E_{CBS} + \frac{B}{(l_{max} + \frac{1}{2})^4}. \tag{2}$$

The ccCA energies obtained using Eq. (1) as reference are denoted as "ccCA-P," while ccCA energies using Eq. (2) are denoted as "ccCA-S4."

Currently, the recommended formula of ccCA utilizes separate HF and MP2 CBS extrapolations, though the necessity of separate extrapolations is still debated [2, 65, 66]. The CBS limit of the HF energy is calculated using the formula,

$$E(n) = E_{HF\text{-}CBS} + B \exp(-Cn). \tag{3}$$

The value of C is 1.63 and comes from a study by Halkier and coauthors, where an optimal value for two-point HF CBS extrapolations using triple-zeta and quadruple-zeta basis sets was based on comparison to numerical HF energies [67]. Using this value, HF energies for a test set of nine diatomics were within 1 mE$_h$ of the numerical HF limits. The $(1/l_{max})^3$ CBS fit of Schwartz is expressed as

$$E(l_{max}) = E_{CBS} + \frac{B}{(l_{max})^3}. \tag{4}$$

Using Eq. (3) for the HF CBS energy, and an average of Eqs. (1) and (4) for the MP2 CBS energy is denoted as "ccCA-PS3" and gave the lowest MAD for the G3/05 test set [68].

Once the HF CBS and MP2 CBS reference energies are computed, a correction $[\Delta E(CC)]$ for correlation effects beyond second-order perturbation theory is evaluated using CCSD(T) in combination with a smaller basis set, viz. CCSD(T)/cc-pVTZ,

$$\Delta E(CC) = E[CCSD(T)/cc - pVTZ] - E[MP2/cc - pVTZ].$$

Next, an adjustment $[\Delta E(CV)]$ for core–valence correlation effects is made through an MP2(FC1) aug-cc-pCVTZ computation

$$\Delta E(CV) = E[MP2(FC1)/aug - cc - pCVTZ] - E[MP2/aug - cc - pVTZ],$$

where "FC1" is used to indicate that all electrons of first-row atoms are correlated, all electrons of atoms Na–Ar are correlated except the $1s$ orbitals, and all electrons of atoms K–Kr are correlated except the $1s2s2p$ orbitals, i.e., the inner noble gas core MOs remain frozen. The next correction applied to ccCA is to account for scalar relativistic effects using the cc-pVTZ-DK basis sets and the spin-free, one-electron Douglas–Kroll–Hess (DKH) Hamiltonian [69–71]. The MP2 relativistic correction to the ccCA energy is formulated as

$$\Delta E(\text{SR-MP2}) = E[\text{MP2/cc-pVTZ-DK}] - E[\text{MP2/cc-pVTZ}].$$

Lastly, the reference energy $[E(\text{MP2/CBS}) + E(\text{HF/CBS})]$ and the four corrections from the previous steps are combined in an additive manner along with a spin-orbit correction $[\Delta E(\text{SO})]$ for atoms and molecules containing elements K–Kr with a Π ground electronic state to give the ccCA energy,

$$E_{\text{ccCA}} = E(\text{HF/CBS}) + E(\text{MP2/CBS}) + \Delta E(\text{CC}) + \Delta E(\text{CV})$$
$$+ \Delta E(\text{SR-MP2}) + \Delta E(\text{ZPE}) + \Delta E(\text{SO})$$

Note that all ccCA computations containing elements Na–Ar utilize the cc-pV(n+d)Z series of basis sets developed by Wilson and coworkers as the addition of so-called tight-d functions has been shown to be critical to the accurate modeling of these $3p$ elements [72–75].

Since the initial development of ccCA where we used the Gaussian03 ab initio software package [76] for all computations, we have computed ccCA energies on a wide variety of computer systems, using many different ab initio software packages, including NWChem 5.1 [77], Gaussian03, and Molpro 2002/Molpro 2006 [78]. Without resorting to a detailed discussion of relative CPU timings, we note that the most expensive step of the ccCA method is variably the MP2/aug-cc-pVQZ or CCSD(T)/cc-pVTZ computation, compared to MP4/6-31G($2df$, p) of the G3 method. Generally, computation of a ccCA energy will be approximately one order of magnitude more expensive than a G3 computation and approximately two orders of magnitude faster than a CCSD(T)/aug-cc-pVQZ computation (a typical reference energy of model chemistries such as HEAT or Wn).

4 Examples of Success Using ccCA

4.1 Initial Tests of ccCA Against G2-1 and G3/99 Training Sets

The first iteration of the ccCA methodology modeled the popular and successful Gn methods [79]. In fact, similar to the G3 method, the initial formulation of ccCA sought as its effective level of theory QCISD(T) with large basis sets and some account of core–valence correlation. However, ccCA utilized the larger and more flexible correlation consistent basis sets instead of the Pople-style sets adopted by the Gn authors, which we felt would enable the more systematic behavior of the correlation consistent basis sets to be incorporated into the model. We initially chose to test ccCA against the expanding suite of Gn/xx sets due to its diversity of energetic quantities [enthalpies of formation ($\Delta H_f^{298.15K}$), IPs, EAs, and PAs] and the size of the dataset, as well as the inclusion of experimental quantities that have an uncertainty of less than $1.0\,\mathrm{kcal\,mol^{-1}}$.

The first proposed algorithm of ccCA, called "ccCA-TZ" (Table 1), suffered from two major technical difficulties. The most severe was that the lack of diffuse functions in the reference energy basis set (cc-pVTZ) gave extremely poor results for EAs. The worst-case example is the EA of fluorine atom, where the zero-generation ccCA-TZ value deviated from experiment by $16.2\,\mathrm{kcal\,mol^{-1}}$! Secondly, the most resource-intensive step in the initial ccCA formulation involved an MP4/cc-pVQZ or MP4/aug-cc-pVTZ computation. With MP4 scaling as N^7, this became intractable for very large molecules. Additionally, efficiently parallelized closed- and open-shell variants of MP4 algorithms exist in very few ab initio software packages. Thus, MP4 computations were removed from the early ccCA formula. As well, diffuse functions were included throughout the reference energy computation, instead of a separate additive correction to account for increased basis set size (cf. the zero-generation ccCA-TZ and ccCA-aTZ implementations). Furthermore, the hierarchy of energies from double to quadruple-zeta quality could be computed affordably and CBS extrapolations of the MP2 energy were added to the ccCA methodology.

Interestingly, inclusion of a CBS extrapolation in the ccCA methodology gave the appearance of deteriorating the performance of ccCA. The first formulation "ccCA-aTZ" had a mean absolute deviation (MAD) of $0.99\,\mathrm{kcal\,mol^{-1}}$ for the 118 energetic quantities in the initial test set, while the formulation of ccCA including Eq. (1) for CBS extrapolations of total energies (ccCA CBS-1 in Table 1) had a MAD of $1.16\,\mathrm{kcal\,mol^{-1}}$. It was in fact a rare situation of getting the wrong answer for the right reason. While Gn methods fold scalar relativistic effects into the HLC, we found a fortuitous systematic bias in the initial ccCA method that was largely fixed by an additive correction accounting for scalar relativistic effects and inclusion of atomic spin-orbit coupling energies. When these two additive corrections were added to ccCA (at minimal added computational cost), the MAD for the initial test set fell to $0.81\,\mathrm{kcal\,mol^{-1}}$, a value safely within our definition of chemical accuracy. These improvements to the earliest ccCA formulations included no explicit parameterizations and were more accurate, overall, than the G3 and G3(MP2) methods [79].

The next improvement to ccCA [80] focused on several modifications: (1) using the coupled cluster method for extended valence electron correlation (rather than quadratic configuration interaction, known to perform unreliably for inorganic species and molecules with significant multireference character) [81–83], (2) utilizing the tight-d forms of the correlation consistent basis sets for $3p$ elements such as sulfur [74, 75, 84, 85] to improve CBS convergence, (3) improving the basis set chosen for geometry optimization, and (4) performing a closer examination of various CBS extrapolation functions. This next-generation ccCA method was then compared against the larger G3/99 test set [86], which included 223 enthalpies of formation, 88 adiabatic IPs, 58 adiabatic EAs, and 8 adiabatic PAs. In this study, improvements to ccCA resulted in an almost zero MSD for the G3/99 set (with a ccCA-S4 value of -0.08 kcal mol^{-1}) and a 0.96 kcal mol^{-1} MAD, which is equivalent to the accuracy of the most expensive and accurate MP4-based G3 variant (G3X) [87]. The near-zero MSD of ccCA is quite remarkable and a very important statistical metric in the absence of an HLC, considering the Gn HLC parameters are optimized via minimization of the MSD against their test sets (e.g., G2/97, G3/99). Additionally, significant statistical outliers within the G3/99 test set were examined, and clear evidence via sophisticated ab initio studies suggests that these quantities be reexamined by experiment (for example, the enthalpies of formation of C_2F_4 and C_2Cl_4) [17]. More than 20 questionable values remain in the Gn training sets, which have been incorporated into Gn models via HLC parameterization [88, 89]. Further analysis of these outliers – experimentally and computationally – would be of interest.

To provide a clear example of the substantial impact of parameterizations within model chemistries, Fig. 2 shows the magnitude of ccCA additive corrections versus G3B3 and G3B3(MP2) to the n-octane total atomization energy (TAE), used to compute thermochemical properties. As the ccCA reference energy, effectively at the MP2 aug-cc-pV∞Z level of theory, is quite sophisticated, the additive corrections enhance the quality of the ccCA energy without providing enormous compensation. For instance, the largest ccCA additive correction is the Δ(CC) correction, which contributes -17.2 kcal mol^{-1} to the TAE. The core valence correction counteracts the correlation correction to some degree, contributing $+6.3$ kcal mol^{-1} to the TAE. Examining the G3B3 and G3B3(MP2) results. G3 enthalpies of formation are extremely sensitive to the additive corrections, with the correction for extra polarization functions contributing $+137.0$ kcal mol^{-1} to the TAE. The ccCA Δ(CC) additive correction is notably smaller in magnitude versus the similar term in G3, as the G3 correction is the difference in QCISD(T) versus MP4 correlation energies.

As a testament to the success of Gn methods and the HLC, Fig. 2 shows the contribution of the G3B3 and G3B3(MP2) HLC toward the TAE of octane, $+48.8$ and $+66.8$ kcal mol^{-1}, respectively. In fact, without the HLC, the G3 MAD is 9.1 kcal mol^{-1} for the G3/05 test set [88]. Our intuition, based on Pople's original formulation of the HLC in Gaussian-1, was that the Gn HLC was necessary due to an incomplete treatment of basis set *and* electron correlation effects. After examining the performance of ccCA (with and without CBS extrapolations) in comparison to the G3 method with and without the HLC, we were led to the interesting conclusion that the HLC parameterization was primarily a one-electron basis set effect [79]. The results in Fig. 2 highlight the importance of using a high-quality basis set in ab initio investigations of energetic properties. While the basis sets used in ccCA are larger and more flexible than those used in Gn methods, it was our expectation, which has thus far been borne out, that the HLC-free theoretical foundation for the ccCA reference energy would result in a greater range of applicability for more complex chemical environments than those contained in the Gn/xx training sets, e.g., ground state and metal-containing molecules.

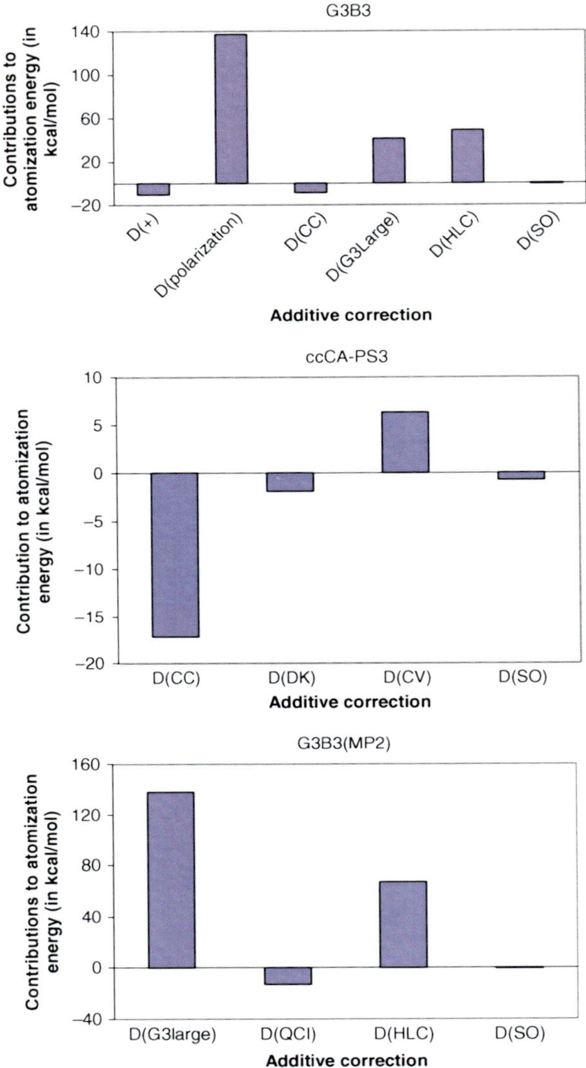

Fig. 2 Contributions of ccCA, G3, and G3(MP2) additive corrections to the atomization energy of octane

4.2 Calibration of ccCA Against Heavy p-block and Metal s-block Systems

At this point in the development of ccCA, correlation consistent basis sets for treatment of core–valence and core–core electron correlation effects did not exist for elements K–Kr. To further compare diverse chemical-bonding types, the Wilson laboratory at the University of North Texas, in collaboration with that of Professor Kirk Peterson at Washington State University, constructed the correlation consistent polarized *core* valence and *weighted core* valence basis sets (cc-pCVnZ and cc-pwCVnZ, respectively) for elements Ga–Kr [90]. Using these basis sets, along with the full suite of correlation consistent sets for elements K and Ca developed by Peterson (K.A. Peterson, personal communication) [91] the ccCA method was extended to study molecules containing elements K–Kr [92]. The accuracy of ccCA was initially assessed using the Ga–Kr containing molecules added to the G3/05 set [27]. This subset contained 51 atomic and molecular energetic properties.

The ccCA-P method achieved chemical accuracy with a MAD of 0.95 kcal mol^{-1}, and very little systematic bias with a MSD of -0.02 kcal mol^{-1}. Due to the large magnitude of total energies when computing systems with elements Ga–Kr using all-electron methods, a more significant difference was observed between ccCA-P and ccCA-S4 energies, as ccCA-S4 resulted in a MAD of 1.00 kcal mol^{-1} and a MSD of 0.07 kcal mol^{-1}. Blaudeau and Curtiss previously computed second-order spin-orbit coupling corrections for the $4p$ G3/05 test set, and showed changes in atomization energies of up to 0.8 kcal mol^{-1} [93, 94]. Clearly, a treatment of higher-order spin-orbit coupling is necessary for improvements beyond "chemical accuracy" of energetic properties for molecules containing elements heavier than gallium. When the second-order spin-orbit coupling values computed by Blaudeau and Curtiss were included in atomic and molecular ccCA energies, a substantial decrease in the MAD was observed. For example, the ccCA-P MAD improved from 0.95 kcal mol^{-1} to 0.88 kcal mol^{-1}. Indeed, it has been our observation that with the ccCA model, "better physics" generally gives "better numbers," thus highlighting the robust nature of the underlying physical model.

Alkali and alkaline earth metal oxides and hydroxides are widely used in a broad spectrum of chemical processes [95]. Moreover, the gas-phase thermochemical properties of these molecules have become an extraordinarily stringent test of new theories and composite methods. Due to unusual core–valence electron correlation and problems with basis set superposition error (BSSE), Gn and even coupled cluster-based W1 and W2 model chemistries have shown extremely poor performance. For example, the G3X ΔH_f value for Na_2O deviates from the recommended experimental result by nearly 14 kcal mol^{-1} and the W2 ΔH_f value for $Mg(OH)_2$ deviates from experiment by 59.1 kcal mol^{-1} [96, 97]! Computed enthalpies of formation within the experimental uncertainties were only obtained after G3X and W2 methods were modified to incorporate inclusion of core–valence correlation in *all* computations (substantially increasing the computational cost), along with significant changes to the quality of the geometry optimization. Nonstandard Gn variants have been proposed to obtain improved "s-block" energetics, such as the G2(thaw) method of Petrie [98–100], and the geometry counterpoise-corrected G3 method of Ma and coauthors [101–104].

The ccCA methodology was first used to compute the enthalpies of formation of oxides and hydroxides of Li, Be, Na, or Mg metal ions. The computations were performed using newly optimized correlation consistent basis sets for the metals

[104]. Due to the improved flexibility of the metal cc-pV_nZ basis sets, treatment of relativistic effects, and incorporation of extrapolations to the MP2 CBS limit (in turn minimizing the effect of BSSE), ccCA enthalpies of formation were within uncertainties in the recommended values for all eight compounds in this initial test set [105].

In spite of, or perhaps because of, the failure of standard Gn methods with the alkali and alkaline earth metal oxides and hydroxides, Gn methods have not been tested against a larger and more diverse set of s-block compounds. In fact, there are only ten s-block-containing enthalpies of formation contained within the G3/05 test set (of 270 total) and nine of those are diatomics. In our laboratories, we sought to understand the performance of ccCA on larger metal-containing species. Thus, a larger validation set was constructed that included the 42 s-block-containing systems within the 600 molecule training set created by Cioslowski and coauthors [106]. This set included species with more diverse bonding types than those within the G3/05 set, for example, dimers and trimers of ionic salts [$(NaF)_2$ and $(LiCl)_3$] and water-bound sodium ions [$Na(H_2O)_x, x = 1$–4]. Also included in this study were large organometallic complexes such as magnesocene [$Mg(C_5H_5)_2$] and beryllium diacetylacetonate [$Be(C_5H_7O_2)_2$] which would require intractable resources for coupled cluster-based model chemistries. From a chemical perspective, this s-block study is critically important as it is building into the model chemical-bonding types more prevalent in transition metal coordination chemistry.

For the 42 molecules contained in this set, ccCA-P and ccCA-S4 (MADs of 2.20 and 2.17 kcal mol^{-1}, respectively) performed better than G3 and G3B3 methods (MADs of 2.64 and 2.75 kcal mol^{-1}, respectively) [107]. Overall Gn performance for s-block molecules is not disturbing, given the circumstance that experimental uncertainties in the Cioslowski set are up to 5.0 kcal mol^{-1} (versus 1.0 kcal mol^{-1} in the G3/05 set). However, the excellent performance of ccCA demonstrates its robustness for computing thermochemical properties of metal-containing systems.

4.3 Calibration of ccCA Against the G3/05 Training Set

Continually increasing the scope of the ccCA methodology allowed us to assess even further improvements to the method. Modifications were considered in order to evolve ccCA toward the goal of a uniform implementation of the model chemistry for the entire periodic table. Thus, the following changes to ccCA were considered: (i) computing equilibrium geometries and vibrational frequencies with a correlation consistent basis set rather than a Pople-style basis set, (ii) extrapolating HF and MP2 energies separately to their respective CBS limits, (iii) utilizing updated theoretical atomic enthalpies of formation for C, B, Al, and Si that were recommended in recent studies by Tasi and coauthors [54], and Karton and Martin [53], (iv) using a UHF reference for the computation of scalar relativistic corrections, and (v) employing a theoretically based correction for anharmonic vibrational effects. None of these modifications significantly increased the computational time [68].

These new ccCA schemes were then tested on the entire G3/05 set of 454 energetic properties, and the ccCA-PS3 method had a MAD of $1.01\,kcal\,mol^{-1}$. While these modifications do not represent an overwhelming improvement over the previous ccCA methodology, this does not suggest that the changes were merely cosmetic. On the contrary, they are indicative of the substantial reliability of ccCA over a broad range of molecular species.

Due to statistical "noise" derived from experimental uncertainties in a large number of molecules within the G3/05 set, as well as the accuracy limits of an MP2-based composite method, $\pm 1.00\,kcal\,mol^{-1}$ may well be near the threshold of error for ccCA when applied toward the Gn test sets. In order to substantially improve the accuracy of ccCA for these test sets, changes to the basis set size or level of theory used in the additive corrections would likely be necessary, which will in turn greatly increase the computational requirements for ccCA energies. However, since ccCA does not include parameterizations, further improvements can easily be identified.

4.4 Successful Applications of ccCA

So far, our laboratories have primarily focused on the development, testing, and implementation of ccCA, but early success has been obtained in utilizing ccCA to solve chemical problems. For instance, a prototype version of ccCA was used by our group in collaboration with the experimental group of Professor T. Brent Gunnoe (then of North Carolina State University) to compute bond-dissociation enthalpies (BDEs) of ethylene, formaldehyde, methylene imine, carbodiimide, isocyanide, and their hydrogenated counterparts in order to probe useful correlations between the free BDEs of these model substrates and the proclivity of their insertion into the Ru(II)-phenyl bond of a catalyst-active species [108]. The BDEs of these small model systems provided a useful diagnostic to explain the thermochemical preferences of certain types of bond insertions.

Another early application of ccCA, using the implementation from Ref. [80], sought to resolve the large spread in tabulated enthalpies of formation of nitroaniline isomers. This intriguing problem was brought to our attention by Professor David H. Magers from Mississippi College. Due to the rather large size and lack of symmetry of these molecules, coupled cluster-based thermochemical studies are intractable. Also, experimental enthalpies of sublimation for the nitroanilines (upon which the gas-phase enthalpies are indirectly based) show large uncertainties. In fact, Politzer and coauthors have found literature values of the 3-nitroaniline $\Delta H_f(s)$ varying by a spread of $8.4\,kcal\,mol^{-1}$ [109]. The enthalpies of formation of 2-, 3-, and 4-nitroaniline were computed using a homodesmotic reaction scheme, with G3/G3B3, ccCA, and B3LYP/6-311+G(d, p) as the most sophisticated levels of theory [110]. The computed ccCA-P $\Delta H_f(g)$ for 3-nitroaniline was $16.9\,kcal\,mol^{-1}$, compared with the NIST literature value of $14.9 \pm 0.4\,kcal\,mol^{-1}$. Via cancellation of errors within the homodesmotic reaction scheme, it is unlikely that both the ccCA and the G3 methods have uncertainties of $1.0\,kcal\,mol^{-1}$ or greater. Even with worst-case accumulation of experimental uncertainties in the homodesmotic reaction scheme, the ccCA enthalpy of formation of 3-nitroaniline is computed to be 15.3 kcal mol^{-1}. In conclusion, the ccCA values certainly supported an experimental reassessment of both the enthalpy of formation and the enthalpy of sublimation for 3-nitroaniline. Moreover, the coupling of high-accuracy composite methods with traditional isodesmic analysis is an interesting concept, worthy of further consideration as per the recent work by Wheeler, Allen, and coworkers [111].

As sulfur-containing compounds participate in atmospheric, interstellar, and industrial chemical processes, a deeper understanding of their energetics and reaction mechanisms is vital. Wilson and coauthors have been actively involved in high-accuracy studies of sulfur-containing molecules [73, 74, 84]. The impetus for these studies came from recognition that standard correlation consistent basis sets for elements Na–Ar often showed poor agreement with experiment even after extrapolations to the CBS limit [112, 113]. Dunning, Peterson, and Wilson improved the CBS convergence of the correlation consistent basis sets by optimizing an extra tight-d function for elements Al–Ar [72]. In the first ccCA study, adoption of the cc-pV($n + d$)Z basis sets greatly improved the enthalpies of formation of sulfur-containing species such as SO_2 and SF_6 [79]. Inclusion of cc-pV($n + d$)Z and cc-pCV($n + d$)Z sets then became standard within the implementation of ccCA, and in part due to this modification, the ccCA methodology is substantially more accurate in the computation of inorganic enthalpies of formation than G3 and G4 methods [68, 80, 89]. Williams and Wilson tested ccCA against a set of 40 enthalpies of formation of sulfur-containing molecules, 9 BDEs, and 3 isomerization energies [114]. The performance of ccCA for the ΔH_f and BDE sets was an overwhelming improvement over Gn methods. For example, the ccCA MAD for the 40 enthalpies of formation was 0.86 kcal mol^{-1}, versus 2.29 kcal mol^{-1} using G3B3. As the MAD is a rather simplistic statistical metric preferred by many ab initio thermochemists, the difference in ccCA- versus G3B3-calculated 95% confidence intervals is even more staggering, 2.46 versus 6.57 kcal mol^{-1}. The ccCA MAD for the 9 BDEs was 0.62 kcal mol^{-1} compared with a G3B3 MAD of 1.03 kcal mol^{-1}. Using ccCA, isomerization energies were chemically accurate for the three isomeric pairs investigated, NSO/SNO, SSF$_2$/FSSF, and HSO/SOH.

Building on the success of using isodesmic reaction schemes to compute enthalpies of formation with ccCA [110], our laboratories have worked in tandem with laser photolysis experiments performed by Professor Paul Marshall of UNT to present a newly recommended value for the controversial enthalpy of formation of cyclohexadienyl radical (C_6H_7). Improved knowledge of the thermochemistry of the cyclohexadienyl radical is crucial to understanding the role of this species and related allyl radicals in mechanisms of metal catalyzed C–H bond activation, Bergman cyclizations, as well as oxidation of lipids and fatty acids. Overall, computed ccCA C_6H_7 enthalpies compared more favorably to both experimental results and values obtained using a coupled cluster-based model chemistry than did either DFT or Gn values.

Other tests of the ccCA methodology include a study of the thermochemistry of large hydrocarbons using a "keyword" implementation of ccCA in the Gaussian03 software package, an investigation of thermochemical properties of high-energy density materials containing nitro/amino [115], and a study of large molecules with organophosphorous [116] functional groups. While a recent study announced the first Gn study of a molecule with more than 30 atoms [117], our laboratories have recently doubled that target with successful computation of the enthalpy of formation of C_{60} [118]. This computation used a new, user-friendly keyword implementation of ccCA, which has been incorporated into NWChem. Additionally, the computation of the C_{60} ccCA energy did not overextend the resources of the Chinook supercomputer, which is a 600-node HP DL185 system equipped with dual dual-core AMD 2.6 GHz Opteron processors at Pacific Northwest National Laboratory. Clearly with ccCA, it will be possible to investigate quantitative energetics of molecular systems which possess relevance to biological and materials research thrusts, especially when coupled with current studies of ccCA performance improvement (vida infra).

4.5 Transition State Barrier Heights Using ccCA

The ccCA, like all past and current composite methods, has been successfully benchmarked against *ground state* energetic properties. A logical extension was to test the performance of the method for transition state energies, and thus reaction barrier heights. However, there are various problems which hinder theoretical attempts at reproducing experimental kinetic data, such as the exponential dependence of the chemical rate constant on the activation energy E_a indicated by the Arrhenius equation,

$$k = A \exp[-E_a/RT],$$

and that direct experimental measurement of a transition state barrier is not physically possible. In a simple treatment of gas-phase transition state theory, contributions from tunneling, barrier recrossing, coupling of ground and excited electronic states, the presence of multiple reaction channels, and potential van der Waals complexes in the reaction channels are also neglected. Still, two databases of transition state energies (the HTBH38/04 and the NHTBH38/04 sets) have been developed by Truhlar and coworkers [119]. The barrier heights contained in these test sets are either (a) adjusted against experimental rate constants combined with molecular dynamics simulations [120] or (b) from *ab* initio model chemistries (usually W1 or W2h) [121] that are generally "sub-chemically" accurate (i.e., energetic properties are reliably computed to within ± 1.0 kJ mol^{-1} of experimental values). We have used these sets to calibrate the performance of ccCA.

The ccCA-P and ccCA-S4 implementations from Ref. [80] were used to compare against 68 of the barrier heights contained in the HTBH38/04 and NHTBH38/04 databases. These barrier heights contain hydrogen transfer, heavy-atom transfer, bimolecular nucleophilic substitution, and unimolecular rearrangements and decompositions. Compared to G3B3 results, ccCA performs extremely well against Wn-computed barrier heights [122]. The ccCA-S4 implementation showed a

MSD of $-0.10\,\mathrm{kcal\,mol^{-1}}$ for 36 hydrogen transfer reactions, and a MSD of -0.19 $\mathrm{kcal\,mol^{-1}}$ for the remaining 32 barrier heights, versus $+1.25$ and $+1.41\,\mathrm{kcal\,mol^{-1}}$, respectively using G3B3. The ccCA-S4 MAD is $0.91\,\mathrm{kcal\,mol^{-1}}$ for hydrogen transfer barriers, and $0.98\,\mathrm{kcal\,mol^{-1}}$ for the remaining barriers. Overall, the G3B3 MAD was $1.98\,\mathrm{kcal\,mol^{-1}}$. Additionally, we found that hydrogen transfer reactions were quite susceptible to the quality of the geometry optimization, and if CCSD/6-31G(d) equilibrium geometries were used, the ccCA-S4 MAD for the 32 H-transfer reactions was reduced from 0.91 to $0.69\,\mathrm{kcal\,mol^{-1}}$! A similar improvement in accuracy was seen performing G3 single-point energies at the CCSD/6-31G(d) stationary points, where the MAD of the HTHB set was reduced from 1.71 to $1.49\,\mathrm{kcal\,mol^{-1}}$.

Also of note is that the G3/G3B3 HLC is determined via the number of paired and unpaired valence electrons. In a computation of barrier heights or reaction energies (unlike, for example, atomization energies), the HLC will thus mostly or completely cancel out, leading to a more appropriate comparison of Gn and ccCA methodologies. Thus, the very small magnitude of the MSD and the more comparable barrier heights to Wn methods show that ccCA is more able to approximate a similar effective level of theory to high-accuracy coupled cluster-based model chemistries than Gn methods.

Even more impressive is the comparison of ccCA versus DFT (including functionals specifically constructed for kinetic modeling) and other ab initio theory. In Fig. 3, the MADs of ccCA, G3B3, QCISD(T)/MG3, and four popular DFT functionals (using the MG3 basis set without diffuse functions on hydrogen) are shown

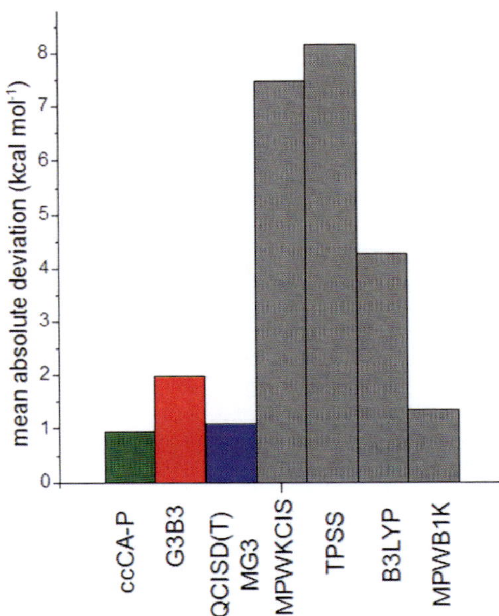

Fig. 3 Mean absolute deviations of ccCA, G3B3, and various functionals against the HTBH38/04 and NHTBH38/04 barrier height training sets

from the initial HTBH38/04 and NHTBH38/04 study of Truhlar [119]. The ccCA model chemistry outperforms QCISD(T)/MG3, and even the MPW1K functional, which was parameterized for superior performance when computing barrier heights [120, 123]! Other popular functionals, such as B3LYP, also result in unacceptable performance as compared with high-level ab initio computations.

4.6 Transition Metal Thermochemistry

Development of a complete suite of correlation consistent basis sets for the $3d$ transition metals (Sc–Zn) has been carried out by Balabanov and Peterson [124, 125]. Prior to these two studies, very limited work had been carried out on the construction of systematically improvable transition metal basis sets [126–128]. Despite the existence of large basis sets, such as atomic natural orbital basis sets [129, 130] and split-valence sets [131, 132] for transition metal elements, few systematic thermochemical studies on the scale of the Gn training sets have been carried out for transition metals. Notable examples include a study on 28 transition metal enthalpies of formation carried out in our laboratories using B3LYP and effective core potentials [133]. This study showed the deficiencies of using DFT methods designed for qualitative ab initio investigations to compute inorganic enthalpies of formation. Truhlar and coauthors have devised a set training set including 62 bond-dissociation energies of transition metal-containing molecules [134], which has been tested over a variety of functionals by Furche and Perdew [135]. Zhao and Truhlar have also updated the number of transition metal-containing species in their training sets during their examination and development of functionals [136]. Hyla-Kryspin and Grimme analyzed the performance of spin-component scaled MP2 and MP3 methods on a set of metal–carbonyl bond dissociations [137].

The largest systematic study of DFT performance on transition metals published thus far has been carried out by Riley and Merz [138]. Their investigation included a test set of 94 enthalpies of formation and 58 ionization energies. However, this study illustrates many of the challenges faced when modeling transition metal thermochemistry including obtaining the correct ground state symmetry and multiplicity, and utilizing data that may have been obtained at multiple temperatures. It concludes that TPSS1KCIS/TZVP gives the "best" thermochemical accuracy for this test set, with a MAD of $9.1\,\text{kcal}\,\text{mol}^{-1}$. Certain DFT functionals that Riley and Merz investigated showed remarkably poor performance for either early $3d$ metals (PBE1PBE, BB1K) and/or middle $3d$ metals (BLYP and BB95), with no clear physical or chemical trends identified to justify the widely varying performance of popular DFT functionals.

In general, highly accurate ab initio investigations of transition metal thermodynamics are plagued with severe difficulties, including the necessary description of dynamical (or short-range) electron correlation via inclusion of a sophisticated and expensive coupled cluster wave function, description of nondynamical (or long-range) correlation due to competing low-lying electronic states (or both) [139], increased scalar relativistic effects for the heavier transition metals, necessarily larger basis set requirements, increased importance of correlated valence and core electrons, etc. From a chemical perspective, the chemical diversity of transition metals, for example, their ability to stabilize multiple formal oxidation states, coordination numbers, and spin states makes the quantitative modeling of d-block metals particularly challenging. The competition between ground states in TM-containing molecules and increased "multireference character," especially in smaller inorganic radicals, suggests that wave functions involving perturbative energy corrections [MP2, CCSD(T)] may not be an appropriate starting point for a TM model chemistry. Also, additive corrections that are appropriate for accurate main group energies, as well as parameterizations (such as the HLC) are likely not globally transferable to obtain quantitative or "chemically accurate" TM energies. However, progress in examining thermochemical properties of transition metals with highly accurate methods is being made. Nielsen, Allendorf, and coauthors have succeeded in using isogyric reaction schemes with coupled cluster theory to examine ΔH_f values of a variety of chromium-containing compounds [140–142]. Dixon and coauthors have used a CBS-CCSD(T) model chemistry, such as the one described below, to examine binding and electronic excitation energies of transition metal oxides [143–146].

Our laboratories, along with Professor Kirk Peterson at Washington State University, sought to create a G3/05-like training set for transition metal species based on reliable and available experimental data. In our initial work, we selected 17 small inorganic species with enthalpies of formation contained in the JANAF thermochemical compendium [147]. However, relative to main group species, there is a severe dearth of reliable information on transition metal thermochemistry in the literature. Occasionally, tabulated inorganic enthalpies of formation in major thermochemical compendia can differ by $10–100\,\text{kcal mol}^{-1}$! Calibration and statistical analysis of ccCA, or any ab initio method, against these thermochemical values is extraordinarily difficult. Also, unlike the values contained in the main group G3/05 training set, where only experimental values with uncertainties less than $\pm 1.0\,\text{kcal mol}^{-1}$ are included, a similar training set for transition metals must be less stringent, given the error bars of $\pm 10\,\text{kcal mol}^{-1}$ for some species.

ccCA enthalpies of formation were computed for 17 molecules, ranging in size from diatomics to $Zn(CH_2CH_3)_2$ [41]. Note that in our initial study on transition metals, the only alteration in the ccCA methodology was the basis sets used for geometry optimizations and vibrational frequencies; the SBKJC(d) effective core potential [148] for TMs, and 6-31G(d) for main group elements. The harmonic frequencies were scaled by 0.9684 to account for anharmonicity in the ZPE and temperature corrections to enthalpy. Additionally, the CBS extrapolation behavior of the "Schwartz" CBS fits [Eqs. (2) and (4)] were closely examined. We showed that these extrapolation functions can give significantly different energies, as these extrapolations use l_{max} rather than the cardinal number ($n = 2$ for cc-pVDZ, $n = 3$ for cc-pVTZ, etc.) to model convergence behavior, and $l_{max} \neq n$ for the transition metal correlation consistent basis sets [124]. For the 17 molecules included in our initial test set, ccCA-S4, using l_{max} in the CBS energy fits [ccCA-S4(Q5)] provided the lowest errors, with a MSD of $-0.8\,\text{kcal mol}^{-1}$ and a MAD of $5.6\,\text{kcal mol}^{-1}$.

However, the two maximum deviations at this level of theory, for the enthalpies of formation of TiF_2 and TiF_3, were quite large, -26.5 and $-16.9\,kcal\,mol^{-1}$, respectively. These two molecules also possess the largest experimental error bars in the test set ($\pm 10\,kcal\,mol^{-1}$).

In order to assess the theoretical performance of ccCA, we computed energies for 10 of the molecules using a coupled cluster-based model chemistry employing the DKH Hamiltonian throughout. Geometries were optimized using the CCSD(T) with a DK-CCSD(T) aug-cc-pVTZ-DK level of theory (zero-point energies were taken from ccCA results). The CBS extrapolation was performed using DK-CCSD(T) with basis sets from aug-cc-pVTZ to aug-cc-pV5Z in size. Additionally, an additive correction for core–valence correlation effects was performed at the CCSD(T,FC1) aug-cc-pwCVTZ level of theory. For well-behaved main group systems, this DK-CCSD(T) CBS approach would generally be accurate to $\sim \pm 0.5\,kcal\,mol^{-1}$ [3, 17, 26, 40]. Interestingly, the DK-CCSD(T) CBS enthalpies of formation for TiF_2 and TiF_3 were significantly worse than ccCA, with deviations of -33.2 and -23.8 $kcal\,mol^{-1}$, respectively. As the experimental $TiF_2\Delta H_f$ in the JANAF tables is derived from reactions involving TiF_3, and theoretically these molecules do not possess significant multireference character (via examination of the magnitude of coupled cluster T_1 and T_2 amplitudes), we suggested an updated experimental investigation of these two species and removed them from the training set.

For the remaining eight molecules, the DK-CCSD(T) CBS method had a MAD of $3.1\,kcal\,mol^{-1}$, versus $3.4\,kcal\,mol^{-1}$ using ccCA-S4(Q5). At least for the systems benchmarked in this study, their accuracies are somewhat comparable. When TiF_2 and TiF_3 are removed from the larger test set of 17 molecules, the ccCA-S4(Q5) MSD is $2.0\,kcal\,mol^{-1}$ and the MAD is $3.4\,kcal\,mol^{-1}$. Compared to the average of the experimental errors bars for these 15 molecules ($\pm 2.2\,kcal\,mol^{-1}$), the ccCA methodology is successful for computing enthalpies of formation for small molecules. Given the typical tabulated uncertainties in TM enthalpies of formation as well as the initial performance of both the ccCA and the DK-CCSD(T) CBS methods, our group has set as an initial goal a MAD of $\pm 3.0\,kcal\,mol^{-1}$ (dubbed "transition metal thermochemical accuracy") for an expanded test set of transition metal-containing species.

During the course of this investigation, however, the first catastrophic failure of the ccCA methodology was identified. For larger carbonyl complexes, such as $Ni(CO)_4$ or $Mn(CO)_5Cl$, ccCA performed quite poorly, with errors of 30–100 $kcal\,mol^{-1}$. For example, the ΔH_f value of $Ni(CO)_4$ using ccCA-S4(Q5) had a deviation from the experimental value ($-144.0 \pm 2.5\,kcal\,mol^{-1}$) of $+45.6\,kcal\,mol^{-1}$, versus $-7.4\,kcal\,mol^{-1}$ using the DK-CCSD(T) CBS model chemistry. As ccCA does not rely on parameterization to achieve accurate results, systematic errors in the method can be examined and strategies can be developed for elimination or amelioration. In this situation, we found significant overcorrection in the $\Delta(CV)$ additive correction for the larger TM species, notably those that contained carbon monoxide as a ligand. Modifications to the core–valence addition correction have substantially improved the accuracy of ccCA for larger inorganic complexes. This modified ccCA is chemically accurate for enthalpies of formation of main group species and achieves "transition metal chemical accuracy" for a large set of smaller TM radicals, as well as for enthalpies of formation of carbonyl complexes and ferrocene; yet, the computational requirements are not significantly increased [149].

For metal-containing species, there is clearly a lack of highly accurate computations in the literature [125], though the capabilities certainly exist for the elucidation of the near-full CI CBS limit electronic structure of small TM-containing molecules, and Dixon's group has started to make important contributions in this regard [143, 144]. Also, despite the significant efforts of experimentalists like Hildenbrand [150, 151] and Ebbinghaus [152, 153], new and revised gas-phase thermochemical data on transition metal species are appearing in the literature at a slow pace. Though the relationship between basis set sophistication and treatment of electron correlation is more complex for transition metals than for main group systems, we are the first to observe that non-parameterized, single-reference ab initio composite methods *can* attain an acceptable level of reliability for $3d$ transition metal complexes. Since it is an efficient MP2-based method, ccCA can be used to compute quantitatively accurate energies for systems pertinent to inorganic, organometallic, and bioinorganic research thrusts.

5 Future Directions and Conclusions

We have coded and are now testing user-friendly implementations of ccCA within both the popular Gaussian03 software package and the massively parallelized open-source NWChem package. Of course, ccCA energies can also be obtained in any ab initio software package that has the capability of computing DFT, MP2, and CCSD(T) energies (such as Molpro and GAMESS, for example).

Recent projects have been completed to further reduce the computational requirements of ccCA without resorting to empirical parameters or substantially changing the methods employed in the additive corrections. The first has been to use the "resolution of the identity" correlated methods in ccCA [154]. By employing the resolution of the identity, or RI approximation, the four-index two-electron Coulomb integrals can be represented as linear combinations of three- or two-index integrals, via insertion of a projection operator and the usage of small "auxiliary" basis sets [155]. Computational scaling of matrix operations involving these new Coulomb integrals is reduced from N^4 to $N^2 m$, where m is the number of auxiliary basis functions. Obviously this reduction in scaling propagates into substantial savings of disk and time requirements for electron correlation methods such as MP2 [156, 157] and CCSD(T) [158], where the two-electron integrals are utilized.

As auxiliary RI correlation consistent basis sets have been optimized, an RI-ccCA implementation was created by Prascher and Wilson [154], replacing all MP2 steps with their equivalent RI-MP2 computations, and the CCSD(T)-additive correction with the local RI-CCSD(T) method developed by Schütz and Werner [159–161]. Correlation consistent auxiliary basis sets developed by Weigend [162] were utilized for the various ccCA steps. Benchmark computations on 102 closed shell molecules, the largest being benzene, showed that using RI-ccCA gave time savings of 83% on average, versus the standard ccCA implementation, and disk-space requirements were generally reduced by almost two orders of magnitude. The sacrifice in terms of energetic accuracy was also minimal for the molecules tested. For the 102 molecules examined in this study, the average total energy difference between RI-ccCA and ccCA (assuming the same CBS extrapolation scheme is used) is only $0.27\,\text{kcal mol}^{-1}$. Additionally, time and disk-requirement savings improved as the molecular size increased. To our knowledge, the RI-ccCA is the first composite method to successfully exploit the resolution of the identity approximation, and provides a promising avenue toward achievement of quantitative large-molecule energies.

To conclude, progress is rapid on many fronts in order to increase the accessibility and usefulness of the ccCA. While it is more expensive than Gn model chemistries (when the RI version of ccCA is not utilized), ccCA has been used to provide reliable results in situations where atypical chemical bonds are present and more popular composite methods fail. We have shown that although it may not be as accurate as high-level post-CCSD(T) model chemistries, its physical sophistication is just as strong. At the current rate of technological progress, the prohibitive scaling of single-reference, post-CCSD(T) computations ensures that it will be many years before such methods can be commonly used to explore the electronic structure of molecules with 25 or more heavy atoms. Approximate, but nonempirically parameterized, methods such as ccCA will be essential for moving beyond the confines of organic-based training sets and into the realm of new and more challenging chemical problems.

Research thrusts in our laboratories have provided a new and useful tool for spectroscopists and physical chemists to help validate, predict, and design new chemistry. The ccCA methodology is the first MP2-based model chemistry to approach quantitative accuracy of transition metal energies, and represents a significant milestone beyond qualitative DFT applications. Clearly, ccCA is the most "pan-periodic" and accurate MP2-based model chemistry, and we encourage ccCA to be adopted by the quantum chemistry community for computing reliable energies of large molecules without reliance on the limitations of empirical parameters.

References

1. L. A. Curtiss, P. C. Redfern, D. J. Frurip, in *Reviews of Computational Chemistry*, ed. by K. B. Lipkowitz, D. B. Boyd (Wiley-VCH, New York, 2000), Vol. 15, p. 147
2. D. Feller, K. A. Peterson, D. A. Dixon, J. Chem. Phys. **129**, 204105 (2008)
3. D. Feller, K. A. Peterson, J. Chem. Phys. **126**, 114105 (2007)
4. T. Helgaker, W. Klopper, D. P. Tew, Mol. Phys. **106**, 2107 (2008)
5. K. Raghavachari, G. W. Trucks, J. A. Pople, M. Head-Gordon, Chem. Phys. Lett. **157**, 479 (1989)
6. T. Janowski, P. Pulay, J. Chem. Theory Comput. **4**, 1585 (2008)
7. M. E. Harding, T. Metzroth, J. Gauss, A. A. Auer, J. Chem. Theory Comput. **4**, 64 (2008)
8. S. Hirata, P. D. Fan, A. A. Auer, M. Nooijen, P. Piecuch, J. Chem. Phys. **121**, 12197 (2004)

9. A. I. Krylov, Chem. Phys. Lett. **338**, 375 (2001)
10. A. I. Krylov, Acc. Chem. Res. **39**, 83 (2006)
11. A. I. Krylov, Annu. Rev. Phys. Chem. **59**, 433 (2008)
12. P. Piecuch, S. A. Kucharski, K. Kowalski, Chem. Phys. Lett. **344**, 176 (2001)
13. K. Kowalski, P. Piecuch, J. Chem. Phys. **113**, 5644 (2000)
14. S. Nangia, D. G. Truhlar, M. J. McGuire, P. Piecuch, J. Phys. Chem. A **109**, 11643 (2005)
15. J. J. Zheng, J. R. Gour, J. J. Lutz, M. Wloch, P. Piecuch, D. G. Truhlar, J. Chem. Phys. **128**, 044108 (2008)
16. K. Kowalski, J. R. Hammond, W. A. de Jong, A. J. Sadlej, J. Chem. Phys. **129**, 226101 (2008)
17. L. Pollack, T. L. Windus, W. A. de Jong, D. A. Dixon, J. Phys. Chem. A **109**, 6934 (2005)
18. C. L. Janssen, I. M. B. Nielsen *Parallel Computing in Quantum Chemistry* (CRC Press, Boca Raton, 2008)
19. J. A. Pople, B. T. Luke, M. J. Frisch, J. S. Binkley, J. Chem. Phys. **89**, 2198 (1985)
20. J. A. Pople, M. Head-Gordon, D. J. Fox, K. Raghavachari, L. A. Curtiss, J. Chem. Phys. **90**, 5622 (1989)
21. L. A. Curtiss, C. Jones, G. W. Trucks, K. Raghavachari, J. A. Pople, J. Chem. Phys. **93**, 2537 (1990)
22. L. A. Curtiss, K. Raghavachari, G. W. Trucks, J. A. Pople, J. Chem. Phys. **94**, 7221 (1991)
23. L. A. Curtiss, J. E. Carpenter, K. Raghavachari, J. A. Pople, J. Chem. Phys. **96**, 9030 (1992)
24. Run on one node containing a four quad-core 2.40 GHz Intel processor, with Gaussian03 and 128 GB RAM.
25. L. A. Curtiss, K. Raghavachari, P. C. Redfern, G. S. Kedziora, J. A. Pople, J. Phys. Chem. A **105**, 227 (2001)
26. D. Feller, K. A. Peterson, J. Chem. Phys. **108**, 154 (1998)
27. L. A. Curtiss, P. C. Redfern, K. Raghavachari, J. Chem. Phys. **123**, 124017 (2005)
28. G. P. F. Wood, L. Radom, G. A. Petersson, E. C. Barnes, M. J. Frisch, J. A. Montgomery, J. Chem. Phys. **125**, 094106 (2006)
29. G. A. Petersson, D. K. Malick, W. G. Wilson, J. W. Ochterski, J. J. A. Montgomery, M. J. Frisch, J. Chem. Phys. **109**, 10570 (1998)
30. G. A. Petersson, M. A. Al-Laham, J. Chem. Phys. **94**, 6081 (1991)
31. G. A. Petersson, A. Bennett, T. G. Tensfeldt, M. A. Al-Laham, W. A. Shirley, J. Mantzaris, J. Chem. Phys. **89**, 2193 (1988)
32. J. W. Ochterski, G. A. Petersson, J. J. A. Montgomery, J. Chem. Phys. **104**, 2598 (1996)
33. P. L. Fast, D. G. Truhlar, J. Phys. Chem. A **104**, 6111 (2000)
34. P. L. Fast, J. C. Corchado, M. L. Sanchez, D. G. Truhlar, J. Phys. Chem. A **103**, 5129 (1999)
35. D. Feller, D. A. Dixon, J. S. Francisco, J. Phys. Chem. A **107**, 1604 (2003)
36. D. A. Dixon, D. Feller, K. A. Peterson, J. Chem. Phys. **115**, 2576 (2001)
37. D. Feller, D. A. Dixon, J. Phys. Chem. A **104**, 3048 (2000)
38. D. Feller, J. Chem. Phys. **98**, 7059 (1993)
39. T. H. Dunning Jr. K. A. Peterson, J. Chem. Phys. **113**, 7799 (2000)
40. D. Feller, K. A. Peterson, T. D. Crawford, J. Chem. Phys. **124**, 054107 (2006)
41. N. J. DeYonker, K. A. Peterson, G. Steyl, A. K. Wilson, T. R. Cundari, J. Phys. Chem. A **111**, 11269 (2007)
42. W. D. Allen, A. L. L. East, A. G. Császár, in *Structures and Conformations of Non-Rigid Molecules*, ed. by J. Laane, M. Dakkouri, B. van der Vecken, H. Oberhammer, (Kluwer, Dordrecht, 1993), p. 343
43. A. G. Császár, W. D. Allen, H. F. Schaefer, J. Chem. Phys. **108**, 9751 (1998)
44. M. S. Schuurman, S. R. Muir, W. D. Allen, H. F. Schaefer, J. Chem. Phys. **120**, 11586 (2004)
45. S. E. Wheeler, K. A. Robertson, W. D. Allen, I. Schaefer, Y. J. Bomble, J. F. Stanton, J. Phys. Chem. A **111**, 3819 (2007)
46. A. Karton, P. R. Taylor, J. M. L. Martin, J. Chem. Phys. **127**, 064104 (2007)
47. A. D. Boese, M. Oren, O. Atasoylu, J. M. L. Martin, M. Kallay, J. Gauss, J. Chem. Phys. **120**, 4129 (2004)
48. S. Parthiban, J. M. L. Martin, J. Chem. Phys. **114**, 6014 (2001)

49. A. Karton, E. Rabinovich, J. M. L. Martin, B. Ruscic, J. Chem. Phys. **125**, 144108 (2006)

50. Y. J. Bomble, J. Vazquez, M. Kallay, C. Michauk, P. G. Szalay, A. G. Császár, J. Gauss, J. F. Stanton, J. Chem. Phys. **125**, 064108 (2006)

51. M. E. Harding, J. Vazquez, B. Ruscic, A. K. Wilson, J. Gauss, J. F. Stanton, J. Chem. Phys. **128** (2008)

52. A. Tajti, P. G. Szalay, A. G. Császár, M. Kallay, J. Gauss, E. F. Valeev, B. A. Flowers, J. Vazquez, J. F. Stanton, J. Chem. Phys. **121**, 11599 (2004)

53. A. Karton, J. M. L. Martin, J. Phys. Chem. A **111**, 5936 (2007)

54. G. Tasi, R. Izsak, G. Matisz, A. G. Császár, M. Kallay, B. Ruscic, J. F. Stanton, Chem. phys. chem. **7**, 1664 (2006)

55. B. Ruscic, J. E. Boggs, A. Burcat, A. G. Császár, J. Demaison, R. Janoschek, J. M. L. Martin, M. L. Morton, M. J. Rossi, J. F. Stanton, P. G. Szalay, P. R. Westmoreland, F. Zabel, T. Berces, J. Phys. Chem. Ref. Data **34**, 573 (2005)

56. B. Ruscic, R. E. Pinzon, M. L. Morton, N. K. Srinivasan, M. C. Su, J. W. Sutherland, J. V. Michael, J. Phys. Chem. A **110**, 6592 (2006)

57. B. Ruscic, J. V. Michael, P. C. Redfern, L. A. Curtiss, K. Raghavachari, J. Phys. Chem. A **102**, 10889 (1998)

58. D. Feller. J. Comp. Chem. **17**, 1571 (1996)

59. K. A. Peterson, D. E. Woon, T. H. Dunning Jr., J. Chem. Phys. **100**, 7410 (1994)

60. C. Schwartz, Phys. Rev. A **126**, 1015 (1962)

61. C. Schwartz, in *Methods in Computational Physics*, ed. by B. J. Alder, S. Fernbach, M. Rotenberg (Academic, New York, 1963), Vol. 2, p. 262

62. W. Kutzelnigg, J. D. Morgan, J. Chem. Phys. **96**, 4484 (1992)

63. A. Halkier, T. Helgaker, P. Jorgensen, W. Klopper, H. Koch, J. Olsen, A. K. Wilson, Chem. Phys. Lett. **286**, 243 (1998)

64. J. M. L. Martin, Chem. Phys. Lett. **259**, 669 (1996)

65. F. Jensen, Theor. Chem. Acc. **113**, 267 (2005)

66. A. Karton, J. M. L. Martin, Theor. Chem. Acc. **115**, 330 (2006)

67. A. Halkier, T. Helgaker, P. Jorgensen, W. Klopper, J. Olsen, Chem. Phys. Lett. **302**, 437 (1999)

68. N. J. DeYonker, B. R. Wilson, A. W. Pierpont, T. R. Cundari, A. K. Wilson, Mol. Phys., **107**, 1107 (2009)

69. M. Douglas, N. M. Kroll. Ann. Phys. **82**, 89 (1974)

70. B. A. Hess, Phys. Rev. A **33**, 3742 (1986)

71. B. A. Hess, Phys. Rev. A **32**, 756 (1985)

72. T. H. Dunning Jr. K. A. Peterson, A. K. Wilson, J. Chem. Phys. **114**, 9244 (2001)

73. N. X. Wang, A. K. Wilson, J. Phys. Chem. A **107**, 6720 (2003)

74. A. K. Wilson, T. H. Dunning Jr., J. Phys. Chem. A **108**, 3129 (2004)

75. S. Yockel, A. K. Wilson, Abstr. Pap. Am. Chem. S. **231** (2006)

76. M. J. Frisch, G. W. Trucks, H. B. Schlegel, G. E. Scuseria, M. A. Robb, J. R. Cheeseman, J. A. Montgomery Jr. T. Vreven, K. N. Kudin, J. C. Burant, J. M. Millam, S. S. Iyengar, J. Tomasi, V. Barone, B. Mennucci, M. Cossi, G. Scalmani, N. Rega, G. A. Petersson, H. Nakatsuji, M. Hada, M. Ehara, K. Toyota, R. Fukuda, J. Hasegawa, M. Ishida, T. Nakajima, Y. Honda, O. Kitao, H. Nakai, M. Klene, X. Li, J. E. Knox, H. P. Hratchian, J. B. Cross, V. Bakken, C. Adamo, J. Jaramillo, R. Gomperts, R. E. Stratmann, O. Yazyev, A. J. Austin, R. Cammi, C. Pomelli, J. W. Ochterski, P. Y. Ayala, K. Morokuma, G. A. Voth, P. Salvador, J. J. Dannenberg, V. G. Zakrzewski, S. Dapprich, A. D. Daniels, M. C. Strain, O. Farkas, D. K. Malick, A. D. Rabuck, K. Raghavachari, J. B. Foresman, J. V. Ortiz, Q. Cui, A. G. Baboul, S. Clifford, J. Cioslowski, B. B. Stefanov, G. Liu, A. Liashenko, P. Piskorz, I. Komaromi, R. L. Martin, D. J. Fox, T. Keith, M. A. Al-Laham, C. Y. Peng, A. Nanayakkara, M. Challacombe, P. M. W. Gill, B. Johnson, W. Chen, M. W. Wong, C. Gonzalez, J. A. Pople. Gaussian 03, Revision D.02, Gaussian, Inc.: Wallingford CT, 2004

77. T. P. Straatsma, E. Aprà, T. L. Windus, E. J. Bylaska, W. de Jong, S. Hirata, M. Valiev, M. Hackler, L. Pollack, R. Harrison, M. Dupuis, D. M. A. Smith, J. Nieplocha, V. Tipparaju, M. Krishnan, A. A. Auer, E. Brown, G. Cisneros, G. Fann, H. Früchtl, J. Garza, K. Hirao, R. Kendall, J. Nichols, K. Tsemekhman, K. Wolinski, J. Anchell, D. Bernholdt, P. Borowski, T. Clark, D. Clerc, H. Dachsel, M. Deegan, K. Dyall, D. Elwood, E. Glendening, M. Gutowski, A. Hess, J. Jaffe, B. Johnson, J. Ju, R. Kobayashi, R. Kutteh, Z. Lin, R. Littlefield, X. Long, B. Meng, T. Nakajima, S. Niu, M. Rosing, G. Sandrone, M. Stave, H. Taylor, G. Thomas, J. van Lenthe, A. Wong, Z. Zhang NWChem, A Computational Chemistry Package for Parallel Computers, Version 5.1 (2008), Pacific Northwest National Laboratory, Richland, Washington 99352–0999, USA
78. MOLPRO, version 2006.1, a package of ab initio programs, H.-J. Werner, P. J. Knowles, R. Lindh, F. R. Manby, M. Schütz, P. Celani, T. Korona, A. Mitrushenkov, G. Rauhut, T. B. Adler, R. D. Amos, A. Bernhardsson, A. Berning, D. L. Cooper, M. J. O. Deegan, A. J. Dobbyn, F. Eckert, E. Goll, C. Hampel, G. Hetzer, T. Hrenar, G. Knizia, C. Köppl, Y. Liu, A. W. Lloyd, R. A. Mata, A. J. May, S. J. McNicholas, W. Meyer, M. E. Mura, A. Nicklass, P. Palmieri, K. Pflüger, R. Pitzer, M. Reiher, U. Schumann, H. Stoll, A. J. Stone, R. Tarroni, T. Thorsteinsson, M. Wang, and A. Wolf, see http://www.molpro.net
79. N. J. DeYonker, T. R. Cundari, A. K. Wilson, J. Chem. Phys. **124**, 114104 (2006)
80. N. J. DeYonker, T. Grimes, S. Yockel, A. Dinescu, B. Mintz, T. R. Cundari, A. K. Wilson, J. Chem. Phys. **125**, 104111 (2006)
81. J. Hrusak, S. Tenno, S. Iwata, J. Chem. Phys. **106**, 7185 (1997)
82. J. D. Watts, M. Urban, R. J. Bartlett, Theor. Chim. Acta **90**, 341 (1995)
83. J.-P. Blaudeau, M. P. McGrath, L. A. Curtiss, L. Radom, J. Chem. Phys. **107**, 5016 (1997)
84. R. D. Bell, A. K. Wilson. Chem. Phys. Lett. **394**, 105 (2004)
85. A. K. Wilson, D. E. Woon, K. A. Peterson, T. H. Dunning Jr., J. Chem. Phys. **110**, 7667 (1999)
86. L. A. Curtiss, K. Raghavachari, P. C. Redfern, J. A. Pople, J. Chem. Phys. **112**, 7374 (2000)
87. L. A. Curtiss, P. C. Redfern, K. Raghavachari, J. A. Pople, J. Chem. Phys. **114**, 108 (2001)
88. L. A. Curtiss, P. C. Redfern, K. Raghavachari, J. Chem. Phys. **126**, 084108 (2007)
89. L. A. Curtiss, P. C. Redfern, K. Raghavachari, J. Chem. Phys. **127**, 124105 (2007)
90. N. J. DeYonker, K. A. Peterson, A. K. Wilson, J. Phys. Chem. A **111**, 11383 (2007)
91. J. Koput, K. A. Peterson, J. Phys. Chem. A **106**, 9595 (2002)
92. N. J. DeYonker, B. Mintz, T. R. Cundari, A. K. Wilson, J. Chem. Theory Comput. **4**, 328 (2008)
93. J. P. Blaudeau, M. P. McGrath, L. A. Curtiss, L. Radom, J. Chem. Phys. **107**, 5016 (1997)
94. J. P. Blaudeau, L. A. Curtiss, Int. J. Quantum Chem. **61**, 943 (1997)
95. N. N. Greenwood, A. Earnshaw, *Chemistry of the Elements*, 2nd ed. (Butterworth-Heinemann, Oxford, UK 1997)
96. A. Schulz, B. J. Smith, L. Radom, J. Phys. Chem. A **103**, 7522 (1999)
97. M. B. Sullivan, M. A. Iron, P. C. Redfern, J. M. L. Martin, L. A. Curtiss, L. Radom, J. Phys. Chem. A **107**, 5617 (2003)
98. S. Petrie, J. Phys. Chem. A **102**, 6138 (1998)
99. S. Petrie, Chem. Phys. Lett. **283**, 131 (1998)
100. S. Petrie, J. Phys. Chem. A **105**, 9931 (2001)
101. F. M. Siu, N. L. Ma, C. W. Tsang, J. Chem. Phys. **114**, 7045 (2001)
102. S. Abirami, N. L. Ma, N. K. Goh, Chem. Phys. Lett. **359**, 500 (2002)
103. F. M. Siu, N. L. Ma, C. W. Tsang, Chem-Eur. J. **10**, 1966 (2004)
104. B. P. Prascher, K. A. Peterson, D. E. Woon, T. H. Dunning, Jr., A. K. Wilson, unpublished
105. D. S. Ho, N. J. DeYonker, A. K. Wilson, T. R. Cundari, J. Phys. Chem. A **110**, 9767 (2006)
106. J. Cioslowski, M. Schimeczek, G. Liu, V. Stoyanov, J. Chem. Phys. **113**, 9377 (2000)
107. N. J. DeYonker, D. S. Ho, A. K. Wilson, T. R. Cundari, J. Phys. Chem. A **111**, 10776 (2007)
108. J. P. Lee, K. A. Pittard, N. J. DeYonker, T. R. Cundari, T. B. Gunnoe, J. L. Petersen, Organometallics **25**, 1500 (2006)
109. P. Politzer, P. Lane, M. C. Concha, Struct Chem. **15**, 469 (2004)

110. N. J. DeYonker, T. R. Cundari, A. K. Wilson, C. A. Sood, D. H. Magers, J. Mol. Struc.: THEOCHEM **775**, 77 (2006)
111. S. E. Wheeler, K. N. Houk, P. v. R. Schleyer, and W. D. Allen, J. Am. Chem. Soc. **131**, 2547 (2009)
112. C. W. Bauschlicher, H. Partridge, Chem. Phys. Lett. **240**, 533 (1995)
113. J. M. L. Martin, O. Uzan, Chem. Phys. Lett. **282**, 16 (1998)
114. T. G. Williams, A. K. Wilson. J. Sulf. Chem. **29**, 353 (2008)
115. K. Jorgenson, G. Oyedepo, A. K. Wilson, to be submitted
116. K. M. Holmes, N. J. DeYonker, E. C. Garrett, T. R. Cundari, A. K. Wilson, to be submitted
117. D. Bond. J. Phys. Chem. A **112**, 1656 (2008)
118. N. J. DeYonker, W. A. de Jong, T. Grimes, T. R. Cundari, A. K. Wilson, to be submitted
119. Y. Zhao, N. Gonzalez-Garcia, D. G. Truhlar, J. Phys. Chem. A **109**, 2012 (2005)
120. B. J. Lynch, P. L. Fast, M. Harris, D. G. Truhlar, J. Chem. Phys. A **104**, 4811 (2000)
121. J. M. L. Martin, G. de Oliveira. J. Chem. Phys. **111**, 1843 (1999)
122. T. V. Grimes, A. K. Wilson, N. J. DeYonker, T. R. Cundari, J. Chem. Phys. **127**, 154117 (2007)
123. A. D. Becke, J. Chem. Phys. **112**, 4020 (2000)
124. N. B. Balabanov, K. A. Peterson, J. Chem. Phys. **123**, 064107 (2005)
125. N. B. Balabanov, K. A. Peterson, J. Chem. Phys. **125**, 074110 (2006)
126. A. Ricca, C. W. Bauschlicher, Theor. Chem. Acc. **106**, 314 (2001)
127. C. W. Bauschlicher, Theor. Chem. Acc. **103**, 141 (1999)
128. K. G. Dyall, Theor. Chem. Acc. **112**, 403 (2004)
129. R. Pou-amérigo, M. Merchán, I. Nebotgil, P. O. Widmark, B. O. Roos, Theor. Chim. Acta **92**, 149 (1995)
130. C. W. Bauschlicher, Theor. Chem. Acc. **92**, 183 (1995)
131. F. Weigend, R. Ahlrichs, Phys. Chem. Chem. Phys. **7**, 3297 (2005)
132. F. Weigend, F. Furche, R. Ahlrichs, J. Chem. Phys. **119**, 12753 (2003)
133. T. R. Cundari, H. A. R. Leza, T. Grimes, G. Steyl, A. Waters, A. K. Wilson, Chem. Phys. Lett. **401**, 58 (2005)
134. N. E. Schultz, Y. Zhao, D. G. Truhlar, J. Phys. Chem. A **109**, 11127 (2005)
135. F. Furche, J. P. Perdew, J. Chem. Phys. **124**, 044103 (2006)
136. Y. Zhao, D. G. Truhlar, Theor. Chem. Acc. **120**, 215 (2008)
137. I. Hyla-Kryspin, S. Grimme, Organometallics **23**, 5581 (2004)
138. K. E. Riley, K. M. Merz, J. Phys. Chem. A **111**, 6044 (2007)
139. N. J. DeYonker, Y. Yamaguchi, W. D. Allen, C. Pak, H. F. Schaefer, K. A. Peterson, J. Chem. Phys. **120**, 4726 (2004)
140. I. M. B. Nielsen, M. A. Allendorf, J. Phys. Chem. A **109**, 928 (2005)
141. I. M. B. Nielsen, M. D. Allendorf, J. Phys. Chem. A **110**, 4093 (2006)
142. E. J. Opila, D. L. Myers, N. S. Jacobson, I. M. B. Nielsen, D. F. Johnson, J. K. Olminsky, M. D. Allendorf, J. Phys. Chem. A **111**, 1971 (2007)
143. S. G. Li, D. A. Dixon, J. Phys. Chem. A **112**, 6646 (2008)
144. H. J. Zhai, S. Li, D. A. Dixon, L. S. Wang, J. Am. Chem. Soc. **130**, 5167 (2008)
145. S. G. Li, D. A. Dixon, J. Phys. Chem. A **110**, 6231 (2006)
146. S. G. Li, D. A. Dixon, J. Phys. Chem. A **111**, 11908 (2007)
147. J. M. W. Chase, C. A. Davies, J. J. R. Downey, D. J. Frurip, R. A. McDonald, A. N. Syverud, NIST-JANAF Tables (4th ed.), J. Phys. Chem. Ref. Data, Mono. **9**, Suppl. 1 ed. 1998
148. W. J. Stevens, H. Basch, M. Krauss, J. Chem. Phys. **81**, 6026 (1984)
149. N. J. DeYonker, T. G. Williams, A. E. Imel, A. K. Wilson, T. R. Cundari, J. Chem. Phys. **131**, 024106 (2009)
150. D. L. Hildenbrand, High Temp Mat Sci. **35**, 151 (1996)
151. D. L. Hildenbrand, J. Chem. Phys. **103**, 2634 (1995)
152. B. B. Ebbinghaus, Combust. Flame **93**, 119 (1993)
153. B. B. Ebbinghaus, Combust. Flame **101**, 311 (1995)
154. B.P. Prascher, J. Lai, A.K. Wilson, J. Chem. Phys., submitted

155. H. A. Fruchtl, R. A. Kendall, R. J. Harrison, K. G. Dyall, Int. J. Quantum Chem. **64**, 63 (1997)
156. O. Vahtras, J. Almlöf, M. W. Feyereisen, Chem. Phys. Lett. **213**, 514 (1993)
157. M. Feyereisen, G. Fitzgerald, A. Komornicki, Chem. Phys. Lett. **208**, 359 (1993)
158. A. P. Rendell, T. J. Lee, J. Chem. Phys. **101**, 400 (1994)
159. M. Schütz, J. Chem. Phys. **113**, 9986 (2000)
160. M. Schütz, H. J. Werner, Chem. Phys. Lett. **318**, 370 (2000)
161. M. Schütz, H. J. Werner, J. Chem. Phys. **114**, 661 (2001)
162. F. Weigend, Phys. Chem. Chem. Phys. **8**, 1057 (2006)

On the Performance of a Size-Extensive Variant of Equation-of-Motion Coupled Cluster Theory for Optical Rotation in Chiral Molecules

T. Daniel Crawford and Hideo Sekino

Abstract The modified equation-of-motion coupled cluster approach of Sekino and Bartlett is extended to computations of the mixed electric-dipole/magnetic-dipole polarizability tensor associated with optical rotation in chiral systems. The approach – referred to here as a linearized equation-of-motion coupled cluster (EOM-CC$_L$) method – is a compromise between the standard EOM method and its linear response counterpart, which avoids the evaluation of computationally expensive terms that are quadratic in the field-perturbed wave functions, but still yields properties that are size-extensive/intensive. Benchmark computations on five representative chiral molecules, including (P)-hydrogen peroxide, (S)-methyloxirane, (S)-2-chloropropionitrile, (R)-epichlorohydrin, and $(1S,4S)$-norbornenone, demonstrate typically small deviations between the EOM-CC$_L$ results and those from coupled cluster linear response theory, and no variation in the signs of the predicted rotations. In addition, the EOM-CC$_L$ approach is found to reduce the overall computing time for multi-wavelength-specific rotation computations by up to 34%.

Keywords: Coupled cluster theory · linear response theory · equation-of-motion coupled cluster theory · optical rotation · chirality

1 Introduction

The ab initio computation of molecular properties – including those associated with time-dependent external electric and magnetic fields – has advanced significantly in the last several decades, yielding accurate models for linear, quadratic, and higher-order response functions. When electron correlation effects play a pivotal

T.D. Crawford (✉)
Department of Chemistry, Virginia Tech, Blacksburg, Virginia 24061, USA,
e-mail: crawdad@vt.edu

H. Sekino
Department of Knowledge-Based Information Engineering, Toyohashi University of Technology,
Toyohashi 441-8580, Japan, e-mail: sekino@tutkie.tut.ac.jp

P. Piecuch et al. (eds.), *Advances in the Theory of Atomic and Molecular Systems*,
Progress in Theoretical Chemistry and Physics 19, DOI 10.1007/978-90-481-2596-8_10,

role in the prediction of such properties, the quantum chemical method of choice is coupled cluster theory[1–3], as demonstrated by numerous chemical applications appearing in the literature[4–8]. One such application is optical rotation in chiral molecules[9–14], where high-accuracy theoretical predictions are essential for the goal of constructing new, robust models of optical activity and ultimately the reliable assignment of absolute stereochemical configurations of chiral compounds.

There are two perturbation-based formulations of coupled cluster theory for determining molecular properties: the equation-of-motion (EOM-CC)[15, 16, 5, 17] and linear-response (CCLR)[18–25] approaches. The key difference between the two arises in the choice of the functional form of the perturbed wave function expansion – a linear, CI-like parametrization in the EOM-CC case, and an exponential in CCLR (vide infra). This choice leads to a desirable simplicity in the EOM-CC approach in that properties may be computed from generalized expectation values, i.e., that the perturbed wave function appears only linearly in each order in the expression for the property in question. However, the resulting disadvantage of the EOM-CC approach is that second- and higher-order properties are not size-extensive. Properties derived from the CCLR approach, on the other hand, may be viewed as quasienergy derivatives, and thus retain the same size-extensivity as the unperturbed ground-state energy. However, the cost of the CCLR framework is the lack of a simple, generalized expectation value structure for the perturbed energies and the appearance of computationally expensive terms that are at least quadratic in the perturbed wave functions for second- and higher-order properties.

In 1999, Sekino and Bartlett[26] developed a modified form of the EOM-CC approach that retains the convenient, linear structure to the perturbed energies, but also eliminates the unlinked diagrams that destroy the size-extensivity of the perturbed energies. This approach, which they referred to as "Model III" involved the elimination of specific disconnected terms from the linear equations defining the coupled cluster left-hand ground-state wave function (the well known Λ equations). As a result, the method is computationally more efficient than CCLR, but is no longer formally exact in the full-CC limit, unlike both EOM-CC and CCLR. Benchmark computations using this new approach yielded dynamic polarizabilities and transition moments for LiH as well as spin–spin coupling constants and polarizabilities of ethane that were essentially unchanged from their CCLR counterparts.

The purpose of this work is to extend Sekino and Bartlett's approach – which we will refer to as a linearized EOM coupled cluster (EOM-CC$_L$) method – to computations of the frequency-dependent optical rotations of chiral molecules. The development of coupled cluster methods in this field has been dedicated to the implementation of streamlined models of chiroptical properties that are applicable to large molecules[27, 28], and this work represents a possible step toward that goal. We will compare the performance of the EOM-CC$_L$ approach to its linear-response counterpart – both in terms of theoretical predictions and computational efficiency – for the rigid chiral molecules (S)-methyloxirane, (S)-2-chloropropionitrile, and (1S,4S)-norbornenone, as well as the conformationally flexible species (R)-epichlorohydrin.

2 Theoretical Approach

In this section we outline the fundamental differences in molecular properties evaluated using the EOM-CC and CCLR approaches. Although our exposition will make use of a frequency-independent (static-field) formalism, the practical extension of the final equations to frequency-dependent perturbations is trivial, requiring only the insertion of appropriate field frequencies in the perturbed wave function equations.

2.1 Equation-of-Motion Coupled Cluster Theory

The CC energy functional is given by[2]

$$E_{\text{CC}} = \langle 0 | (1 + \Lambda) e^{-T} H_0 e^{T} | 0 \rangle = \langle 0 | (1 + \Lambda) \overline{H}_0 | 0 \rangle = E^{(0)}, \tag{1}$$

where $|0\rangle$ is a single-determinant reference wave function (typically a Hartree–Fock wave function), and H_0 is the electronic Hamiltonian in the absence of the external perturbation, V. The cluster excitation, T, and de-excitation, Λ, operators are determined, respectively, from the stationary conditions

$$\langle \phi | \overline{H}_0 | 0 \rangle = 0 \tag{2}$$

and

$$\langle 0 | (1 + \Lambda) \left(E^{(0)} - \overline{H}_0 \right) | \phi \rangle = 0, \tag{3}$$

where $|\phi\rangle$ represents the set of excited determinants generated by T from $|0\rangle$ [e.g., for the coupled cluster singles and doubles (CCSD) model, $T = T_1 + T_2$].

The EOM-CC method[15, 16] makes use of the similarity-transformed perturbed Hamiltonian,

$$\overline{H} = e^{-T} H e^{-T} = e^{-T} H_0 e^{T} + e^{-T} V e^{T} = \overline{H}_0 + \overline{V}, \tag{4}$$

as well as a linear (CI-like) perturbational expansion of the wave function:

$$|\psi_{\text{EOM–CC}}\rangle = \left(1 + T^{(1)} + T^{(2)} + \ldots\right) |0\rangle, \tag{5}$$

where the superscripts denote the order of perturbation and the excitation operators, $T^{(n)}$, generate the same set of determinants from $|0\rangle$ as $T = T^{(0)}$. Thus, the EOM-CC zeroth-order right- and left-hand wave functions are, respectively,

$$|\psi^{(0)}\rangle = |0\rangle \tag{6}$$

and

$$\langle \psi^{(0)} | = \langle 0 | (1 + \Lambda). \tag{7}$$

First-order properties may be determined via the corresponding first-order Rayleigh–Schrödinger equation,

$$\left(E^{(0)} - \overline{H}_0\right) |\psi^{(1)}\rangle = \left(\overline{V} - E^{(1)}\right) |\psi^{(0)}\rangle, \tag{8}$$

which, after substitution of the EOM-CC perturbed wave functions and projection onto the zeroth-order left-hand state becomes

$$E^{(1)} = \langle \psi^{(0)} | \overline{V} | \psi^{(0)} \rangle = \langle 0 | (1 + \Lambda) \overline{V} | 0 \rangle. \tag{9}$$

Equation (9) is a generalized expectation value and identical to the deriative formulation.

Second-order properties require the corresponding second-order Rayleigh–Schrödinger equation, viz.

$$E^{(2)} |\psi^{(0)}\rangle = \left(\overline{V} - E^{(1)}\right) |\psi^{(1)}\rangle - \left(E^{(0)} - \overline{H}_0\right) |\psi^{(2)}\rangle, \tag{10}$$

which, after substitution and projection, becomes

$$\begin{aligned}
E^{(2)} &= \langle \psi^{(0)} | \left(\overline{V} - E^{(1)}\right) |\psi^{(1)}\rangle - \langle \psi^{(0)} | \left(E^{(0)} - \overline{H}_0\right) |\psi^{(2)}\rangle \\
&= \langle 0 | (1 + \Lambda) \left(\overline{V} - E^{(1)}\right) T^{(1)} |0\rangle - \langle 0 | (1 + \Lambda) \left(E^{(0)} - \overline{H}_0\right) T^{(2)} |0\rangle \\
&= \langle 0 | (1 + \Lambda) \left(\overline{V} - E^{(1)}\right) T^{(1)} |0\rangle.
\end{aligned} \tag{11}$$

The second term on the right-hand side is zero because $T^{(2)}$ generates only determinants in the same space as T, leading to matrix elements satisfying Eq. (3). Thus, the evaluation of a second-order property in the EOM-CC formulation requires only the first-order perturbed wave function, which is obtained by projection of Eq. (8) onto the set of determinants generated by T:

$$\langle \phi | \left(E^{(0)} - \overline{H}_0\right) T^{(1)} |0\rangle = \langle \phi | \overline{V} |0\rangle. \tag{12}$$

Properties computed using Eq. (11) are not size-extensive, however, because of the appearance of unlinked diagrams arising from disconnected terms implicit in Eq. (3). Such terms naturally cancel if the T and $T^{(n)}$ operators are not truncated, implying that the EOM-CC method is formally exact in the full-CC limit.

2.2 Coupled Cluster Linear-Response Theory

In the CCLR formulation[18–20, 22], the untransformed Hamiltonian is partitioned into zeroth- and first-order components,

$$H = H_0 + V, \tag{13}$$

and the wave function is written in terms of an exponential parametrization,

$$|\psi_{\text{CCLR}}\rangle = e^{T^{(0)}+T^{(1)}+T^{(2)}+\cdots}|0\rangle. \tag{14}$$

In this case, the CCLR right- and left-hand zeroth-order wave functions are naturally chosen to be

$$|\psi^{(0)}\rangle = e^{T}|0\rangle \tag{15}$$

and

$$\langle\psi^{(0)}| = \langle 0|\,(1+\Lambda)\,e^{-T}, \tag{16}$$

respectively, and the right-hand first- and second-order wave functions may be written as

$$|\psi^{(1)}\rangle = e^{T}\,T^{(1)}|0\rangle \tag{17}$$

and

$$|\psi^{(2)}\rangle = e^{T}\left[\frac{1}{2}\left(T^{(1)}\right)^2 + T^{(2)}\right]|0\rangle. \tag{18}$$

Projection of the first-order Rayleigh–Schrödinger in the CCLR formulation onto the zero-order left-hand wave function gives

$$\langle\psi^{(0)}|\left(E^{(0)} - H_0\right)|\psi^{(1)}\rangle = \langle\psi^{(0)}|\left(V - E^{(1)}\right)|\psi^{(0)}\rangle$$
$$\langle 0|\,(1+\Lambda)\,e^{-T}\left(E^{(0)} - H_0\right)e^{T}\,T^{(1)}|0\rangle = \langle 0|\,(1+\Lambda)\,e^{-T}\left(V - E^{(1)}\right)e^{T}|0\rangle$$
$$\langle 0|\,(1+\Lambda)\left(E^{(0)} - \overline{H}_0\right)T^{(1)}|0\rangle = \langle 0|\,(1+\Lambda)\left(\overline{V} - E^{(1)}\right)|0\rangle. \tag{19}$$

The left-hand side above is zero because $T^{(1)}$ generates only determinants in the space for which Eq. (3) is satisfied, leading to exactly the same result as EOM-CC given in Eq. (9).

In the second order, however, the EOM-CC and CCLR formulations differ due to the appearance of quadratic terms in Eq. (18). Starting from the CCLR second-order Rayleigh–Schrödinger equation projected onto the zeroth-order left-hand wave function, we obtain

$$E^{(2)} = \langle\psi^{(0)}|\left(V - E^{(1)}\right)|\psi^{(1)}\rangle - \langle\psi^{(0)}|\left(E^{(0)} - H_0\right)|\psi^{(2)}\rangle$$
$$= \langle 0|\,(1+\Lambda)\,e^{-T}\left(V - E^{(1)}\right)e^{T}\,T^{(1)}|0\rangle$$
$$\quad - \langle 0|\,(1+\Lambda)\,e^{-T}\left(E^{(0)} - H_0\right)e^{T}\left[\frac{1}{2}\left(T^{(1)}\right)^2 + T^{(2)}\right]|0\rangle$$
$$= \langle 0|\,(1+\Lambda)\left(\overline{V} - E^{(1)}\right)T^{(1)}|0\rangle$$
$$\quad - \langle 0|\,(1+\Lambda)\left(E^{(0)} - \overline{H}_0\right)\left[\frac{1}{2}\left(T^{(1)}\right)^2\right]|0\rangle. \tag{20}$$

The first term on the right-hand side above is identical to its EOM-CC counterpart, but the second term is unique to CCLR. As before, the term containing $T^{(2)}$ in the above equation is zero because Eq. (3) is satisfied in the space of determinants generated by T. However, $\left(T^{(1)}\right)^2$ generates higher determinants, and thus may yield nonzero contributions. It may be shown[26] that the unlinked diagrams generated by the first term on the right-hand side of Eq. (20) are exactly cancelled by corresponding terms arising from the second, quadratic term. Thus, Eq. (20) may be written in its more common form as[21]

$$E^{(2)} = \langle 0 | (1 + \Lambda) \left(\left[\overline{V}, T^{(1)} \right] + \frac{1}{2} \left[\left[\overline{H}_0, T^{(1)} \right], T^{(1)} \right] \right) |0\rangle. \tag{21}$$

The appearance of commutators between $T^{(1)}$ and both $\overline{V}$ and $\overline{H}_0$ in this expression emphasizes the size-extensive/intensive nature of CCLR properties. However, the quadratic terms increase the computational expense of the CCLR approach relative to EOM-CC.

3 The Sekino–Bartlett Approach and Optical Rotation

In their 1999 paper[26], Sekino and Bartlett defined several models to ameliorate the size-extensivity error of the EOM-CC approach, while simultaneously avoiding the evaluation of the expensive quadratic term in the full second-order CCLR expression in Eq. (21). One of these, dubbed "Model III," eliminated the size-extensivity error of EOM-CC completely by (a) retaining only the term linear in $T^{(1)}$ in Eq. (21), (b) dropping the commutator,

$$E^{(2)} = \langle 0 | (1 + \Lambda) \overline{V} T^{(1)} |0\rangle; \tag{22}$$

and (c) using a modified form of Eq. (3), viz.

$$\langle 0 | \overline{H}_0 | \phi \rangle + \langle 0 | \left(\Lambda \overline{H}_0 \right)_c | \phi \rangle = 0, \tag{23}$$

where the subscript c denotes that only diagrams in which Λ and $\overline{H}_0$ are connected are included. While the removal of the commutator introduces terms that do not scale linearly with the size of the system, the removal of the disconnected components of Λ restores size-extensivity to the model, which we will refer to as a linearized EOM-CC$_L$ method [29]. The drawback to this approach is that it is no longer formally exact, even if the cluster operators are not truncated. Nevertheless, Sekino and Bartlett demonstrated close numerical agreement between the CCLR, the EOM-CC, and the EOM-CC$_L$ approaches by benchmark polarizability calculations on LiH and ethane at the CCSD level of theory.

The second-order property on which we focus in this work is optical rotation, i.e., the rotation of plane-polarized light induced by non-racemic sample of a chiral compound. As demonstrated first in 1928 by Rosenfeld[30], this rotation is related to the trace of the frequency-dependent electric-dipole/magnetic-dipole polarizability tensor[31, 32, 10, 12, 14],

$$\mathbf{G}'(\omega) = -\frac{2\omega}{\hbar} \sum_{j \neq 0} \frac{\mathrm{Im}[\langle \psi_0 | \boldsymbol{\mu} | \psi_j \rangle \langle \psi_j | \mathbf{m} | \psi_0 \rangle]}{\omega_{j0}^2 - \omega^2}, \tag{24}$$

where $\boldsymbol{\mu} = -\mathbf{r}$ and $\mathbf{m} = -\frac{1}{2}\mathbf{L} = -\frac{1}{2}\mathbf{r} \times \mathbf{p}$ are the (atomic-unit) electric- and magnetic-dipole vector operators, respectively, and ω denotes the frequency of the incident polarized radiation field. The summation includes all excited electronic states, $|\psi_j\rangle$, and ω_{j0} is the excitation energy of the j-th state, which is typically assumed to be far from resonance.

In order to evaluate $\mathbf{G}'(\omega)$ efficiently using the CCLR and EOM-CC$_\mathrm{L}$ approaches, we cast Eq. (24) above in a double-perturbation form directly comparable to Eq. (21) involving both the electric- and the magnetic-dipole operators:

$$\mathbf{G}'(\omega) = -\mathrm{Im}\Big\{ \hat{C}^{\pm\omega}\hat{P}(\boldsymbol{\mu}(-\omega), \mathbf{m}(\omega)) \times$$

$$\Big[\langle 0|(1 + \Lambda)[\overline{\mu}, T_m^\omega]|0\rangle + \frac{1}{2}\langle 0|(1 + \Lambda)[[\overline{H}_0, T_\mu^\omega], T_m^{-\omega}]|0\rangle \Big]\Big\}, \tag{25}$$

where the overbar denotes the similarity transformation of the given operator as defined in the previous section. The permutation operator $\hat{C}^{\pm\omega}$ simultaneously changes the signs on the chosen field frequency and takes the complex conjugate of the expression, and $\hat{P}$ symmetrizes the expression with respect to the perturbations $\boldsymbol{\mu}$ and $\mathbf{m}$. The first-order perturbed cluster operators are obtained from the frequency-dependent analog of Eq. (12),

$$\langle \phi|(\omega - \overline{H}_0)T_m^\omega|0\rangle = \langle \phi|\overline{m}|0\rangle, \tag{26}$$

where the subscripts on T denote the associated perturbation.

Following Sekino and Bartlett[26], the EOM-CC$_\mathrm{L}$ approach to optical rotation involves elimination of (a) the quadratic terms in Eq. (25); (b) the commutator between the perturbation operator and the perturbed cluster operators in the linear term; and (c) the disconnected contributions to the Λ equations consistent with Eq. (23), resulting in the simpler expression,

$$\mathbf{G}'(\omega) = -\mathrm{Im}\Big\{ \hat{C}^{\pm\omega}\hat{P}\,(\boldsymbol{\mu}(-\omega), \mathbf{m}(\omega))\big[\langle 0|(1 + \Lambda)\overline{\mu}\,T_m^\omega|0\rangle\big]\Big\}. \tag{27}$$

4 Computational Details

The CCLR and EOM-CC$_\mathrm{L}$ approaches to the $\mathbf{G}'(\omega)$ tensor were computed at the CCSD level of theory for hydrogen peroxide (H_2O_2) and four representative chiral molecules: (S)-methyloxirane[33–36], (S)-2-chloropropionitrile[37],

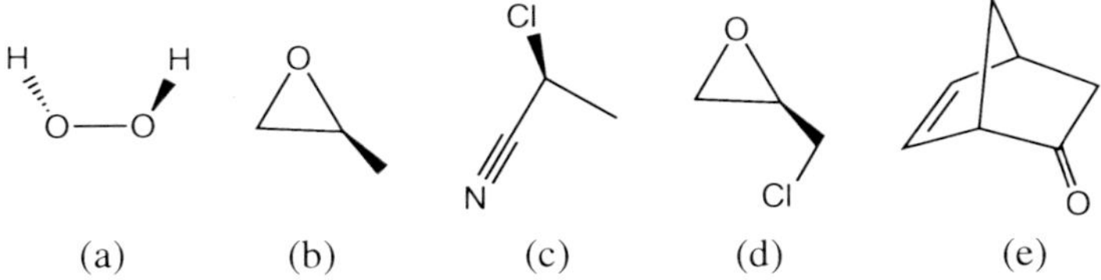

(a) (b) (c) (d) (e)

Fig. 1 Structures of the four molecular test cases considered in this work: (**a**) (*P*)-hydrogen peroxide; (**b**) (*S*)-methyloxirane; (**c**) (*S*)-2-chloropropionitrile; (**d**) (*R*)-epichlorohydrin; and (**e**) (*1S,4S*)-norbornenone

(*R*)-epichlorohydrin[38], and (*1S,4S*)-norbornenone[33, 14, 39], as shown in Fig. 1. All computations were carried out using the correlation-consistent basis sets of Dunning and coworkers[40], including diffuse functions[41], which have been shown to be important for chiroptical properties[42]. The B3LYP/cc-pVTZ optimized geometry was used for each test case[43–45], computed using the Gaussian03 program package[46]. Core orbitals were frozen in the coupled cluster calculations: $1s$ for C and O, $1s2s2p$ for Cl. All coupled cluster specific rotations were obtained using the PSI3 program package[47].

Two choices of representation of the electric-dipole operator were used in this work: the origin-dependent length-gauge ($\mu = -r$), with the molecular center-of-mass as the computational origin, and the origin-independent modified velocity-gauge ($\mu = -p$)[48], in which $\mathbf{G}'(\omega)$ is shifted by its static-limit ($\omega \to 0$) value. In addition, we have computed the origin-dependence vector of the length-gauge approach, which requires the mixed r/p polarizability (computed within the corresponding CCLR and EOM-CC$_L$ approaches), which serves as a diagnostic of the validity of the length-gauge rotations.

5 Results and Discussion

The size-extensivity/intensivity of the EOM-CC$_L$ approach for optical rotation is demonstrated by the CCSD results in Table 1 for the (*P*) enantiomer of the hydrogen peroxide monomer and dimer. Both the trace of the $\mathbf{G}'(\omega)$ tensor at 355 nm and the corresponding specific rotations [the total rotation of plane-polarized light normalized for path length (in dm) and concentration (in g/mL)] are reported. The trace of $\mathbf{G}'(\omega)$ is a size-extensive quantity (much like the dipole-polarizability), and when the monomers are placed 10^5 Å apart (and are thus non-interacting), both the CCLR and the EOM-CC$_L$ approaches yield precisely twice the trace obtained for a single H_2O_2 molecule, as expected. The specific rotation, on the other hand, is a size-intensive quantity, by definition, and thus both the monomer and the dimer exhibit identical rotations. The difference between the length and modified velocity representations of the electric-dipole operator is significant for this case, and results primarily from the arbitrary origin dependence of the former. In addition, the CCLR

Table 1 The trace of the 355 nm Rosenfeld tensor of Eq. (24) (in atomic units) and the corresponding specific rotation [in deg dm^{-1} (g/mL)$^{-1}$] for the (P) enantiomer of the hydrogen peroxide monomer and dimer in the CCSD/aug-cc-pVDZ linear reponse (LR) and EOM-CC$_L$ approximations[a]

| | Length gauge | | | |
| | Tr[$\mathbf{G}'(355)$] | | [α]$_{355}$ | |
	LR	EOM-CC$_L$	LR	EOM-CC$_L$
Monomer	−0.0544	−0.0531	−442	−432
Dimer	−0.1088	−0.1062	−442	−432
	Modified velocity gauge			
Monomer	−0.0105	−0.0112	−666	−710
Dimer	−0.0210	−0.0224	−666	−710

[a]Specific rotations computed at the B3LYP/cc-pVTZ-optimized geometry of the monomer. The dimer consists of two monomers placed 10^5 Å apart.

and EOM-CC$_L$ approaches give very similar rotations for both choices of gauge, with the largest difference (6%) occurring for the modified velocity representation.

Table 2 reports CCSD-specific rotations for the rigid molecules (S)-methyloxirane (also known as propylene oxide) and (S)-2-chloropropionitrile for several typical wavelengths of the polarized field. Both molecules have been studied extensively by coupled cluster theory[33–37] and density-functional methods[33, 49, 50] in part because of the availability of gas-phase experimental specific rotation data[51, 50, 52]. (S)-Methyloxirane, in particular, has been shown to be a particularly difficult case because of its sensitivity to vibrational[49, 35, 36] and solvent[53, 51, 52, 35, 54] effects. For (S)-2-chloropropionitrile, the CCLR and EOM-CC$_L$ approximations yield essentially identical results for all wavelengths, with the largest deviation (8%) between the two occurring for the aug-cc-pVTZ basis set using the modified velocity gauge at 355 nm. Comparison to the available gas-phase experimental data is possible for (S)-2-chloropropionitrile: at 633 and 355 nm, the cavity ring-down polarimetry (CRDP) measurements of Wilson et al.[52] yield −6.8 ± 2.3 and −37.9 ± 2.9° dm^{-1} (g/mL)$^{-1}$, respectively, in good agreement with the CCLR and EOM-CC$_L$ results in Table 2.

For (S)-methyloxirane, the two methods again give identical results for the longer wavelengths, but larger differences (up to 40%) appear for the modified velocity gauge at shorter wavelengths. Furthermore, for both molecules the norm of the origin-dependence vector for the dipole-length representation increases for all wavelengths in the EOM-CC$_L$ approach relative to CCLR, even though the specific rotations themselves tend to decrease. For example, for (S)-methyloxirane, the 355 nm origin-dependence vector increases in length from 15.1° dm^{-1} (g/mL)$^{-1}$ (CCLR) to 23.4° dm^{-1} (g/mL)$^{-1}$ (EOM-CC$_L$) for the aug-cc-pVDZ basis set. Comparison with gas-phase experiment is less satisfactory in the case of (S)-methyloxirane, with 633 and 355 nm CRDP values of −8.39 ± 0.20 and +7.49 ± 0.30° dm^{-1} (g/mL)$^{-1}$, respectively[52]. Recent studies by Ruud and Zanasi[49], by Kongsted et al.[36], and by Crawford et al.[55] indicate that vibrational and

Table 2 Specific rotations [in deg dm^{-1} (g/mL)$^{-1}$] of (S)-methyloxirane and (S)-2-chloropropionitrile, computed using the CCSD method within the linear response (LR) and EOM-CC$_L$ formalisms[a]

	(S)-Methyloxirane				aug-cc-pVTZ			
	aug-cc-pVDZ							
	Length gauge		Modified velocity gauge		Length gauge		Modified velocity gauge	
Wavelength (nm)	LR	EOM-CC$_L$	LR	EOM-CC$_L$	LR	EOM-CC$_L$	LR	EOM-CC$_L$
633	−24	−23	−26	−23	−16	−15	−16	−13
589	−27	−26	−29	−26	−18	−17	−18	−15
436	−43	−42	−47	−40	−27	−26	−28	−21
355	−50	−47	−55	−43	−27	−25	−27	−16
	(S)-2-Chloropropionitrile							
633	−8	−8	−6	−5	−10	−10	−8	−7
589	−9	−9	−7	−6	−11	−11	−9	−8
436	−19	−18	−14	−13	−23	−23	−19	−18
355	−32	−32	−25	−23	−40	−39	−34	−31

[a] Specific rotations computed at the B3LYP/cc-pVTZ-optimized geometry of each molecule.

temperature corrections must be included in the model to obtain even qualitative agreement with experiment in this case.

(R)-Epichlorohydrin serves as an example of a conformationally flexbile molecule exhibiting three low-lying conformers, distinguished from one another by the position of the chlorine atom on the methyl group relative to the oxirane ring. In 2003, Polavarapu and coworkers[56] demonstrated that the lowest-lying conformer, *gauche-II*, which places the Cl atom farthest from the ring, and the highest-energy conformer, *cis*, both have large positive rotations, while the next highest conformer, *gauche-I*, exhibits a comparably large negative rotation. In 2006, Tam and Crawford[38] reported CCSD-LR results for this molecule and found good agreement with experimental gas-phase-specific rotations – within 1% for the dipole-length representation and 7% for the modified velocity representation. Using the complete basis set (CBS)-extrapolated CCSD(T) populations of Ref. [38], the CCSD LR- and EOM-CC$_L$-specific rotations reported in Table 3 are nearly identical at all four wavelengths. However, the origin-dependence vector of the length-gauge approach again increases for the individual conformers: 37% for the *cis* conformer, 65% for *gauche-I*, and 75% for *gauche-II*, regardless of wavelength.

Table 4 summarizes the CCSD-specific rotations for ($1S,4S$)-norbornenone, a rigid molecule that has proved problematic for CC methods because of their apparent underestimation of the sodium D-line (589 nm)-specific rotation, as compared to the liquid-phase experimental result of $-1146°\,\mathrm{dm}^{-1}\,(\mathrm{g/mL})^{-1}$[33, 14, 39]. In this case, the difference between the LR and the EOM-CC$_L$ approaches is significantly larger for the modified velocity gauge (15%) than for the length gauge results (2%). In addition, unlike the previous test cases, the EOM-CC$_L$ approach produces a slightly larger specific rotation than the LR approach in the length-gauge representation. However, neither approach yields any improvement in the 589-nm specific rotation as compared to experiment, and thus the source of the disagreement between theory and experiment for this difficult case remains elusive.

Both the EOM-CC$_L$ and the CCLR approaches require the iterative computation of the perturbed wave functions given by Eq. (26) for each perturbation operator ($\boldsymbol{\mu}$, $\boldsymbol{p}$, and $\boldsymbol{m}$) and for both positive and negative field frequencies – a total of 12 perturbed wave functions for each wavelength. In addition, the modified velocity gauge

Table 3 Conformationally averaged specific rotations [in deg dm^{-1} (g/mL)$^{-1}$] of (R)-epichlorohydrin computed using the CCSD method within the linear response (LR) and EOM-CC$_L$ formalisms[a]

	Length gauge		Modified velocity gauge	
Wavelength/nm	LR	EOM-CC$_L$	LR	EOM-CC$_L$
633	54	57	46	46
589	64	67	54	55
436	134	139	115	116
355	235	244	205	206

[a] Specific rotations of each conformer computed using the aug-cc-pVDZ basis set at the B3LYP/cc-pVTZ optimized geometry. Gas-phase CBS CCSD(T) populations taken from Ref. [38].

Table 4 Specific rotations [in deg dm^{-1} (g/mL)$^{-1}$] of (*1S,4S*)-norbornenone computed using the CCSD method within the linear response (LR) and EOM-CC$_L$ formalisms[a]

Wavelength/nm	Length gauge		Modified velocity gauge	
	LR	EOM-CC$_L$	LR	EOM-CC$_L$
633	−596	−607	−453	−381
589	−722	−736	−549	−462
436	−1861	−1898	−1417	−1198
355	−4807	−4905	−3677	−3128

[a] Specific rotations computed using the aug-cc-pVDZ(C,O)/cc-pVDZ(H) basis set at the B3LYP/cc-pVTZ optimized geometry.

representation requires the static-field ($\omega \rightarrow 0$) limit of the velocity-gauge $\mathbf{G}'$ tensor, which entails an additional six perturbed wave functions. For each wavelength, the evaluation of the $\mathbf{G}'(\omega)$ tensor itself involves the linear and quadratic terms appearing in Eq. (25) for CCLR or only the linear term in Eq. (27) for EOM-CC$_L$. It should be noted that, although similar quadratic terms appear in Eq. (2) for the unperturbed T amplitudes, they are more computationally significant in the CCLR calculation for optical rotation because they must be evaluated at least 18 times for each $\mathbf{G}'$ tensor (once for each of the nine combinations of Cartesian components of the perturbations and for $+\omega$ and $-\omega$ field frequencies). In the dual-gauge, four-wavelength calculations presented here, the quadratic contributions must be evaluated a total of 225 times. Thus, the computational impact of such terms on the evaluation of frequency-dependent properties is significantly larger than for the ground-state energy alone.

Table 5 reports wall timings for each of these major components of the CCSD-specific rotation computations for several of the test cases above. As expected, the majority of the computing time is consumed by the computation of the 54 perturbed wave functions needed for all four wavelengths considered here. However, the time required for the quadratic contributions in Eq. (25) is still significant – ranging from 19 to 34% of the total wall time. In addition, the evaluation of the quadratic terms involves substantial disk storage (and thus more I/O) because of the need to store intermediate quantites of size ov^3 (where o is the number of occupied orbitals and v is the number of virtual orbitals). Thus, elimination of those terms as prescribed by the EOM-CC$_L$ approach results in significant computational savings, with very little apparent impact on the final specific rotations.

6 Conclusions

The efficient computation of molecular response properties such as optical rotation is of paramount importance, and schemes for reducing the computational effort required for high-accuracy methods such as coupled cluster theory become even more crucial for larger, chemically relevant molecules. However, algorithmic improvements must not come at the expense of the overall accuracy of the theory, and the EOM-CC$_L$ approach of Sekino and Bartlett provides a reasonable com-

Table 5 Timings (in hours) for four-wavelength CCSD specific rotation calculations[a]

Molecule	Basis set (# Functions)	Perturbed wave functions[b]	Linear terms[c]	Quadratic terms[d]	Total
(S)-Methyloxirane	aug-cc-pVDZ (146)	7.4	0.28	2.2	9.9
(S)-Methyloxirane	aug-cc-pVTZ (322)	81.6	2.6	19.4	103.2
(S)-2-Chloropropionitrile	aug-cc-pVDZ (155)	5.8	0.41	3.1	9.4
(S)-2-Chloropropionitrile	aug-cc-pVTZ (326)	103.2	3.7	26.2	134.4
(1S,4S)-Norbornenone	aug-cc-pVDZ(C,O) / cc-pVDZ(H) (224)	88.9	2.7	28.0	120.5

[a] Wall times recorded on isolated 64-bit 3.0 GHz Xeon processor with U320 SCSI disks.

[b] Time required to compute 54 perturbed wave functions: 12 for each wavelength plus 6 for the static velocity-gauge tensor.

[c] Time required to compute the linear terms in Eqs. (25) or (27).

[d] Time required to compute the quadratic terms in Eq. (25).

promise between the simplicity of EOM-CC and the size-extensivity/intensivity of CCLR. In this work we have extended the EOM-CC$_L$ formalism to specific rotations of chiral molecules. By elimination of terms in the property tensors that are quadratic in the perturbed wave functions, the EOM-CC$_L$ method reduces the overall computing time by up to 34%, with concomitant reduction in disk storage, yet with little to no loss of accuracy as compared to the CCLR approach, as illustrated by benchmark computations on a series of representative chiral compounds.

Acknowledgement The authors thank Prof. Rodney J. Bartlett for helpful discussions. T.D.C. was supported by a grant from the US National Science Foundation (CHE-0715185) and a subcontract from Oak Ridge National Laboratory by the Scientific Discovery through Advanced Computing (SciDAC) program of the US Department of Energy, the division of Basic Energy Science, Office of Science, under contract number DE-AC05-00OR22725.

References

1. R. J. Bartlett, in *Modern Electronic Structure Theory*, Vol. 2 of *Advanced Series in Physical Chemistry*, ed. by D. R. Yarkony (World Scientific, Singapore, 1995), Chap. 16, pp.1047–1131
2. R. J. Bartlett, M. Musial, Rev. Mod. Phys. **79**, 291 (2007)
3. T. D. Crawford, H. F. Schaefer, in *Reviews in Computational Chemistry*, ed. by K. B. Lipkowitz and D. B. Boyd (VCH Publishers, New York, 2000), Vol. 14, Chap. 2, pp. 33–136
4. J. E. Rice, G. E. Scuseria, T. J. Lee, P. R. Taylor, J. Almlöf, Chem. Phys. Lett. **191**, 23 (1992)
5. J. F. Stanton, R. J. Bartlett, J. Chem. Phys. **98**, 7029 (1993)
6. S. A. Perera, M. Nooijen, R. J. Bartlett, J. Chem. Phys. **104**, 3290 (1996)
7. S. A. Perera, R. J. Bartlett, J. Am. Chem. Soc. **118**, 7849 (1996)
8. C. Hättig, P. Jørgensen, J. Chem. Phys. **109**, 2762 (1998)
9. R. K. Kondru, P. Wipf, D. N. Beratan, J. Am. Chem. Soc. **120**, 2204 (1998)
10. M. Pecul, K. Ruud, Adv. Quantum Chem. **50**, 185 (2005)
11. P. J. Stephens, D. M. McCann, J. R. Cheeseman, M. J. Frisch, Chirality **17**, S52 (2005)
12. T. D. Crawford, Theor. Chem. Acc. **115**, 227 (2006)
13. P. L. Polavarapu, Chem. Rec. **7**, 125 (2007)
14. T. D. Crawford, M. C. Tam, M. L. Abrams, J. Phys. Chem. A **111**, 12057 (2007)
15. H. Sekino, R. J. Bartlett, Int. J. Quantum Chem. Symp. **18**, 255 (1984)
16. J. Geertsen, M. Rittby, R. J. Bartlett, Chem. Phys. Lett. **164**, 57 (1989)
17. J. F. Stanton, R. J. Bartlett, J. Chem. Phys. **99**, 5178 (1993)
18. H. J. Monkhorst, Int. J. Quantum Chem. Symp. **11**, 421 (1977)
19. D. Mukherjee, P. K. Mukherjee, Chem. Phys. **39**, 325 (1979)
20. E. Dalgaard, H. J. Monkhorst, Phys. Rev. A **28**, 1217 (1983)
21. R. Kobayashi, H. Koch, P. Jørgensen, Chem. Phys. Lett. **219**, 30 (1994)
22. A. E. Kondo, P. Piecuch, J. Paldus, J. Chem. Phys. **102**, 6511 (1995)
23. P. Piecuch, A. E. Kondo, V. Špirko, J. Paldus, J. Chem. Phys. **104**, 4699 (1996)
24. A. E. Kondo, P. Piecuch, J. Paldus, J. Chem. Phys. **104**, 8566 (1996)
25. O. Christiansen, P. Jørgensen, C. Hättig, Int. J. Quantum Chem. **68**, 1 (1998)
26. H. Sekino, R. J. Bartlett, Adv. Quantum Chem. **35**, 149 (1999)
27. N. J. Russ, T. D. Crawford, Chem. Phys. Lett. **400**, 104 (2004)
28. N. J. Russ, T. D. Crawford, Phys. Chem. Chem. Phys. **10**, 3345 (2008)
29. P. B. Rozyczko, S. A. Perera, M. Nooijen, R. J. Bartlett, J. Chem. Phys. **107**, 6736 (1997)
30. L. Rosenfeld, Z. Physik **52**, 161 (1928)
31. L. D. Barron, *Molecular Light Scattering and Optical Activity*, 2nd edition (Cambridge University Press, Camridge, UK, 2004)

32. D. J. Caldwell, H. Eyring, *The Theory of Optical Activity* (Wiley, New York, 1971)
33. K. Ruud, P. J. Stephens, F. J. Devlin, P. R. Taylor, J. R. Cheeseman, M. J. Frisch, Chem. Phys. Lett. **373**, 606 (2003)
34. M. C. Tam, N. J. Russ, T. D. Crawford, J. Chem. Phys. **121**, 3550 (2004)
35. J. Kongsted, T. B. Pedersen, M. Strange, A. Osted, A. E. Hansen, K. V. Mikkelsen, F. Pawlowski, P. Jørgensen, C. Hättig, Chem. Phys. Lett. **401**, 385 (2005)
36. J. Kongsted, T. B. Pedersen, L. Jensen, A. E. Hansen, K. V. Mikkelsen, J. Am. Chem. Soc. **128**, 976 (2006)
37. T. D. Kowalczyk, M. L. Abrams, T. D. Crawford, J. Phys. Chem. A **110**, 7649 (2006)
38. M. C. Tam, T. D. Crawford, J. Phys. Chem. A **110**, 2290 (2006)
39. T. D. Crawford, P. J. Stephens, J. Phys. Chem. A **112**, 1339 (2008)
40. T. H. Dunning, J. Chem. Phys. **90**, 1007 (1989)
41. R. A. Kendall, T. H. Dunning, R. J. Harrison, J. Chem. Phys. **96**, 6796 (1992)
42. J. R. Cheeseman, M. J. Frisch, F. J. Devlin, P. J. Stephens, J. Phys. Chem. A **104**, 1039 (2000)
43. A. D. Becke, J. Chem. Phys. **98**, 5648 (1993)
44. C. Lee, W. Yang, R. G. Parr, Phys. Rev. B **37**, 785 (1988)
45. P. J. Stephens, F. J. Devlin, C. F. Chabalowski, M. J. Frisch, J. Phys. Chem. **98**, 11623 (1994)
46. M. J. Frisch, G. W. Trucks, H. B. Schlegel, G. E. Scuseria, M. A. Robb, J. R. Cheeseman, J. A. Montgomery, Jr., T. Vreven, K. N. Kudin, J. C. Burant, J. M. Millam, S. S. Iyengar, J. Tomasi, V. Barone, B. Mennucci, M. Cossi, G. Scalmani, N. Rega, G. A. Petersson, H. Nakatsuji, M. Hada, M. Ehara, K. Toyota, R. Fukuda, J. Hasegawa, M. Ishida, T. Nakajima, Y. Honda, O. Kitao, H. Nakai, M. Klene, X. Li, J. E. Knox, H. P. Hratchian, J. B. Cross, C. Adamo, J. Jaramillo, R. Gomperts, R. E. Stratmann, O. Yazyev, A. J. Austin, R. Cammi, C. Pomelli, J. W. Ochterski, P. Y. Ayala, K. Morokuma, G. A. Voth, P. Salvador, J. J. Dannenberg, V. G. Zakrzewski, S. Dapprich, A. D. Daniels, M. C. Strain, O. Farkas, D. K. Malick, A. D. Rabuck, K. Raghavachari, J. B. Foresman, J. V. Ortiz, Q. Cui, A. G. Baboul, S. Clifford, J. Cioslowski, B. B. Stefanov, G. Liu, A. Liashenko, P. Piskorz, I. Komaromi, R. L. Martin, D. J. Fox, T. Keith, M. A. Al-Laham, C. Y. Peng, A. Nanayakkara, M. Challacombe, P. M. W. Gill, B. Johnson, W. Chen, M. W. Wong, C. Gonzalez, J. A. Pople, GAUSSIAN-03, Gaussian Inc., Pittsburg, PA USA, 2003
47. T. D. Crawford, C. D. Sherrill, E. F. Valeev, J. T. Fermann, R. A. King, M. L. Leininger, S. T. Brown, C. L. Janssen, E. T. Seidl, J. P. Kenny, W. D. Allen, J. Comp. Chem. **28**, 1610 (2007)
48. T. B. Pedersen, H. Koch, L. Boman, A. M. J. S. de Meras, Chem. Phys. Lett. **393**, 319 (2004)
49. K. Ruud, R. Zanasi, Angew. Chem. Int. Ed. Engl. **44**, 3594 (2005)
50. K. B. Wiberg, Y. G. Wang, S. M. Wilson, P. H. Vaccaro, J. R. Cheeseman, J. Phys. Chem. A **109**, 3448 (2005)
51. T. Müller, K. B. Wiberg, P. H. Vaccaro, J. Phys. Chem. A **104**, 5959 (2000)
52. S. M. Wilson, K. B. Wiberg, J. R. Cheeseman, M. J. Frisch, P. H. Vaccaro, J. Phys. Chem. A **109**, 11752 (2005)
53. Y. Kumata, J. Furukawa, T. Fueno, Bull. Chem. Soc. Japan **43**, 3920 (1970)
54. P. Mukhopadhyay, G. Zuber, M. R. Goldsmith, P. Wipf, D. N. Beratan, Chem. Phys. Chem. **7**, 2483 (2006)
55. T. D. Crawford, M. C. Tam, M. L. Abrams, Mol. Phys. **105**, 2607 (2007)
56. P. L. Polavarapu, A. Petrovic, F. Wang, Chirality **15**, S143 (2003)

Performance of Block Correlated Coupled Cluster Method with the CASSCF Reference Function for Carbon–Carbon Bond Breaking in Hydrocarbons

Jun Shen, Tao Fang, and Shuhua Li

Abstract The block correlated coupled cluster method, with the complete active-space self-consistent-field reference function (CAS-BCCC), has been applied to investigate the bond-breaking potential energy surfaces (PESs) for a C—C bond in two alkanes (ethane and 2,3-dimethyl-butane) and a C=C bond in two alkenes (ethylene and 2,3-dimethyl-2-butene). The results are compared with those from other multireference methods (CASPT2, MR-CISD, and MR-CISD+Q). It is demonstrated that the CAS-BCCC method can provide more accurate PESs for C—C bond-breaking PESs than CASPT2 and MR-CISD. The overall performance of CAS-BCCC is shown to be comparable to that of MR-CISD+Q for systems under study.

Keywords: Block correlated coupled cluster method · Bond breaking · Potential energy surface · Size-extensivity error

1 Introduction

The potential energy surfaces (PES) of carbon–hydrogen and C—C bond breaking in hydrocarbons have been investigated with a variety of theoretical methods [1–7]. In order to obtain highly accurate PESs, a number of multireference electronic structure methods based on the multi-configuration self-consistent-field (MCSCF) wave function (especially, the complete active space self-consistent-field (CASSCF) wave function) have been used. These methods include (1) multireference configuration interaction with single and double excitations (MR-CISD) [8];

J. Shen, T. Fang, and S. Li (✉)
School of Chemistry and Chemical Engineering, Key Laboratory of Mesoscopic Chemistry of Ministry of Education, Institute of Theoretical and Computational Chemistry, Nanjing University, Nanjing 210093, P. R. China
e-mail: shuhua@nju.edu.cn

P. Piecuch et al. (eds.), *Advances in the Theory of Atomic and Molecular Systems*, Progress in Theoretical Chemistry and Physics 19, DOI 10.1007/978-90-481-2596-8_11,

241

(2) multireference perturbation theory (MRPT), such as CASSCF-based second-order perturbation theory (CASPT2) [9, 10]; and (3) multireference coupled cluster (MRCC) methods [11–46]. Among these approaches, the formalisms of the first two types have been well established and documented, nevertheless, there is no unique CC ansatz based on a multiconfigurational reference function. The existing genuine MRCC approaches can be divided into three major categories: the Fock-space or valence-universal (VU) approach, the Hilbert-space or state-universal (SU) approach, and the state-specific (SS) approach. These three formalisms are all based on an effective Hamiltonian spanned by the reference determinants (the model space) but differ in the forms of the wave operators that generate a set of the target states for VU and SU approaches, or one state for SS approaches, when acting on a model space of states. Although significant progress has been made in the implementation of various MRCC methods [40–42], MRCC methods have not been established as practical tools for routine uses, like MR-CISD or CASPT2. On the other hand, MR-CISD has considerable size-extensivity errors that increase with the system size, so that its applications are limited to small molecules. To overcome this problem, the simplest and most commonly used approach is to add a size-extensivity correction with the Davidson formula [47–49] to MR-CISD (together denoted as MR-CISD+Q). This correction is quite accurate for small molecules such that MR-CISD+Q is nearly size-extensive for small molecules. However, the size-extensivity error of MR-CISD+Q still becomes significant for relatively large molecules, as shown previously [50–53].

It should be mentioned that within the single-reference coupled cluster (SRCC) framework, some effective approaches for treating quasi-degenerate electronic states have also been developed [54–95]. The common feature of these approaches is to include higher excitations (in addition to single and double excitations) in the SRCC wave function through some approximate or perturbative ways. For example, active-space CC or complete active space CC approaches [43–46, 86–95] are extensions of the single-reference CCSD with the inclusion of dominant triple and quadruple (or even higher) excitations that are related to the nature of the studied electronic state, which are selected by using the concept of active orbitals. Other approaches, such as the spin-flip method [61–66], the orbital-optimized CC approach [69–76], and the completely renormalize CC approaches [77–85] have the advantage of being "black-box" and cost effective, and they have been demonstrated to provide accurate descriptions on bond-breaking PESs and quasi-degenerate electronic states in many cases. However, when there is an avoided surface crossing leading to dramatic changes of the leading configurations, the accuracy of these approaches may decrease significantly.

In our previous work [96], we have developed an alternative multireference method, block correlated coupled cluster (BCCC) approach, for the electronic structure calculation of systems with noticeable multireference character. In this approach, the reference function is expressed as the tensor product of the most important many-electron states in each block, assuming that all orbitals in a system are divided into blocks, and the cluster operator is introduced to incorporate electron correlation between blocks. If the CASSCF wave function is taken as the reference function of the BCCC method, the scheme is called CAS-BCCC. In our previous work [97], the CAS-BCCC scheme with the cluster operator truncated up to the four-block correlation level (CAS-BCCC4 in short) has been developed and implemented. The CAS-BCCC4 method has been shown to be quite successful in calculating the PESs of bond-breaking processes in some small molecules, the singlet–triplet gaps in diradicals, activation barriers for some isomerization reactions, spectroscopic constants in diatomic molecules, and excitation energies for low-lying excited states in small molecules [97–103].

In this chapter we apply this CAS-BCCC4 approach to calculate the bond-dissociation PESs for a C—C bond in two alkanes (ethane and 2,3-dimethyl-butane) and a C=C bond in two alkenes (ethylene and 2,3-dimethyl-2-butene). The CAS-BCCC4 results are compared with those from the corresponding internally contracted CASPT2 [104, 105], MR-CISD [106–109], and MR-CISD+Q methods. As demonstrated before [97], the CAS-BCCC4 method is not exactly size-extensive, but its size-extensivity error does not increase with the total number of electrons (only depends on the size of the active space). So, it is very interesting to compare the performance of the CAS-BCCC4 method with those of the other three methods in calculating the C—C (or C=C) bond-breaking PESs of two structurally analogous molecules with different substituents (e.g., ethane and 2,3-dimethyl-butane).

2 Methodology

The exact ground-state wave function in the CAS-BCCC framework is formulated as

$$\Psi_{BCCC} = e^{T}\Phi_0, \tag{1}$$

where

$$\Phi_0 = A_0^{+}i^{+}j^{+}\cdots|0\rangle . \tag{2}$$

In this second-quantized form of the CASSCF wave function, A_0^{+} represents the creation operator for the reference state of block **A**, which contains all orbitals in the active space, and i^{+} stands for the creation operator in the ith occupied spin orbital. For convenience, the active space (or block **A**) is denoted as (N_0, M) (N_0 electrons in M spatial orbitals). The reference state of block **A** corresponds to the lowest energy state in the N_0-electron subspace. Other N_0-electron block states and block states with different numbers of electrons are considered as "excited" block states [97], which are also required for CAS-BCCC calculations. Except for block **A**, each of the other blocks is defined to contain only one spin orbital.

When the total cluster operator is truncated up to the four-block correlation level, i.e.,

$$T \approx T_1 + T_2 + T_3 + T_4, \tag{3}$$

the approximate CAS-BCCC method is abbreviated as CAS-BCCC4. In general, the n-block correlation operators can be divided into three types: the external type that describes the correlation among spin orbitals, the internal type that describes the relaxation effect of the reference state of block $\mathbf{A}$, and the semi-internal type describing the correlation between block $\mathbf{A}$ and all spin orbitals. The explicit expressions of all cluster operators up to the four-block correlation level have been given elsewhere [97–100]. For instance, the T_4 operator has the following form:

$$T_4 = T_{4A} + T_{4B} + T_{4C} + T_{4D} + T_{4E}, \tag{4}$$

with

$$T_{4A} = \frac{1}{2} \sum_{U}^{N_0-1} \sum_{a,b}^{vir} \sum_{i}^{occ} A_U^+ A_0^- i^- a^+ b^+ t_{4A}\left(U, i, a, b\right), \tag{4a}$$

$$T_{4B} = \frac{1}{2} \sum_{U}^{N_0+1} \sum_{i,j}^{occ} \sum_{a}^{vir} A_U^+ A_0^- a^+ i^- j^- t_{4B}\left(U, a, i, j\right), \tag{4b}$$

$$T_{4C} = \frac{1}{4} \sum_{i,j}^{occ} \sum_{a,b}^{vir} a^+ b^+ i^- j^- t_{4C}\left(i, j, a, b\right), \tag{4c}$$

$$T_{4D} = \frac{1}{6} \sum_{U}^{N_0-3} \sum_{a,b,c}^{vir} A_U^+ A_0^- a^+ b^+ c^+ t_{4D}\left(U, a, b, c\right), \tag{4d}$$

$$T_{4E} = \frac{1}{6} \sum_{U}^{N_0+3} \sum_{i,j,k}^{occ} A_U^+ A_0^- i^- j^- k^- t_{4E}\left(U, i, j, k\right). \tag{4e}$$

Here the capital letter U is used to represent the electronic states of block A, which may belong to subspaces with different numbers of electrons. For example, the summation over U in Eq. (4a) runs over only the subspace with $(N_0 - 1)$ electrons. Labels $i, j, k, \ldots$ are used to denote occupied spin orbitals and $a, b, c, \ldots$ for virtual spin orbitals. $A_U^+ A_0^-$ represents the replacement operator that replaces the reference state of block $\mathbf{A}$ with the Uth "excited" block state. In addition, $t_{4A}(U, i, a, b)$ (and so on) are the excitation amplitudes to be determined in CAS-BCCC4 calculations. As in the traditional CC methods, one can project the Schrödinger equation onto Φ_0 and all excited configuration functions to obtain a set of coupled equations for determining the excitation amplitudes, as described previously [96–99],

$$\langle \Phi_0 | H | \Psi_{BCCC} \rangle = E_{BCCC} \langle \Phi_0 | \Psi_{BCCC} \rangle = E_{BCCC}, \tag{5}$$

$$\langle \Phi^V | H | \Psi_{BCCC} \rangle = E_{BCCC} \langle \Phi^V | \Psi_{BCCC} \rangle, \tag{6}$$

$$\langle \Phi^{V,a} | H | \Psi_{BCCC} \rangle = E_{BCCC} \langle \Phi^{V,a} | \Psi_{BCCC} \rangle, \tag{7}$$

$$\cdots$$

In the equations above, $\Phi^V = A_V^+ A_0^- \Phi_0$, $\Phi^{V,a} = A_V^+ A_0^- a^+ \Phi_0$, etc., are excited configuration functions. Although the final working equations are very complicated, these nonlinear equations can still be efficiently solved in the iterative manner with the updated equation below proposed by Hirata and Bartlett [110] for the amplitudes,

$$t(u)^{k+1} = t(u)^k - \frac{\langle \Phi(u)|H|\Psi_{\mathrm{BCCC}}\rangle - E_{\mathrm{BCCC}}\langle \Phi(u) | \Psi_{\mathrm{BCCC}}\rangle}{\Delta E_{\mathrm{shift}} + \langle \Phi(u)|H|\Phi(u)\rangle - E_{\mathrm{BCCC}}}. \tag{8}$$

Here a compound index u is used for simplicity to represent a set of indices that define a particular excitation amplitude. As a result, $\Phi(u)$ is a shorthand notation for a given "excited" configuration function. A positive value, $\Delta E_{\mathrm{shift}}$, is added in the denominator to improve the convergence (in some cases).

To efficiently implement the CAS-BCCC4 method, we have transformed the spin orbital formulations into the spatial orbital formulations and introduced a number of intermediate arrays. In the present implementation, the most time-consuming step scales as $n_o^2 n_v^4$, with $n_o(n_v)$ being the number of occupied (virtual) spatial orbitals, if the active space is relatively small (say, with no more than six active orbitals). Thus, the CAS-BCCC4 method shares the same computational scaling with the traditional CCSD method, although the former has a large prefactor (a detailed discussion is given in Ref. [98]). Thus, with relatively small active spaces, CAS-BCCC4 calculations can be done routinely for medium-sized molecules with moderate basis sets. On the other hand, since the CAS-BCCC4 wave function is invariant to the separate unitary transformation among occupied, active, and virtual orbitals, respectively, any orbitals (canonical or localized orbitals) obtained from the CASSCF calculation can be employed for the subsequent CAS-BCCC4 calculation.

As addressed previously [97], the CAS-BCCC4 approach is not exactly size-extensive. This is because the CAS-BCCC4 approach is somewhat like the configuration interaction method in describing the correlation effects between block **A** and spin orbitals. However, due to the fact that the products between any external type excitation operators are allowed in the CAS-BCCC wavefunction, most dynamical correlation can be effectively recovered. As a result, the size-extensivity error of CAS-BCCC4 will not increase with increasing the total number of electrons. As demonstrated later, the size-extensivity error of CAS-BCCC4 will be smaller than that of MR-CISD+Q in relatively large molecules.

3 Results and Discussions

In this section, the CAS-BCCC4 approach is applied to study the PESs of the C–C bond breaking in ethane and 2,3-dimethyl-butane, and the PESs of the C=C bond breaking in ethylene and 2,3-dimethyl-2-butene. The structures of these four molecules are shown in Fig. 1. In the PES scan, the central C–C bond is varied from 1.2 to 5.0 Å for two alkane molecules and from 1.1 to 5.0 Å for two alkene molecules. The bonding and antibonding orbitals associated with the central C–C bond (i.e., σ and σ^* for the C–C bond, and σ, σ^*, π, and π^* for the C=C bond) are chosen as the active orbitals. So, the (2,2) active space is used for two alkane molecules, and the (4,4) active space is used for two alkene molecules. At each C–C distance, all of the other geometrical freedoms are optimized at the corresponding CASSCF level. Then, our program, linked to the GAMESS program [111], is used to obtain single-point CAS-BCCC4 energies for all points. As shown previously [97–103], the external single excitations (T_{2C}) and semi-internal triple excitations (T_{4D} and T_{4E}) usually have very minor effects on the relative energies, and thus will be neglected in this work to save the computational time. For comparison, CASPT2, MR-CISD, and MR-CISD+Q calculations with the same active space as in the corresponding CAS-BCCC4 calculation are carried out with the MOLPRO package [112]. In all the post-CASSCF calculations, the core orbitals of the carbon atoms are kept frozen. The 6-31G** basis set (with six Cartesian d-like functions) is employed for all calculations.

Fig. 1 Structures of the four molecules studied. (**a**) ethane; (**b**) 2,3-dimethyl-butane; (**c**) ethylene; (**d**) 2,3-dimethyl-2-butene

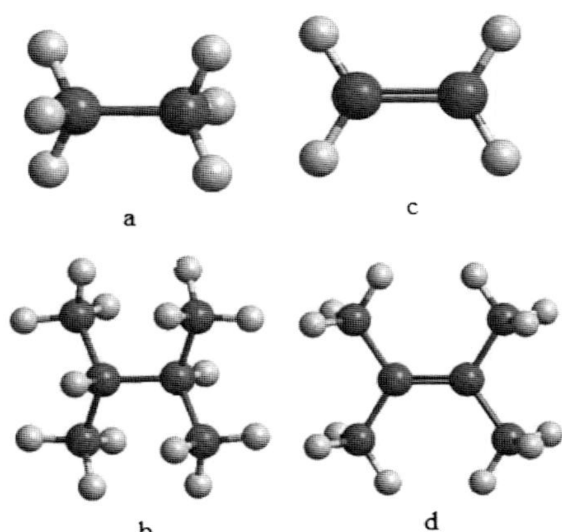

3.1 C–C Bond Breaking in Alkanes

3.1.1 Ethane

The PESs of CAS-BCCC4, CASPT2, MR-CISD, and MR-CISD+Q are shown in Fig. 2. For this molecule, all of these four PES curves are parallel to each other quite

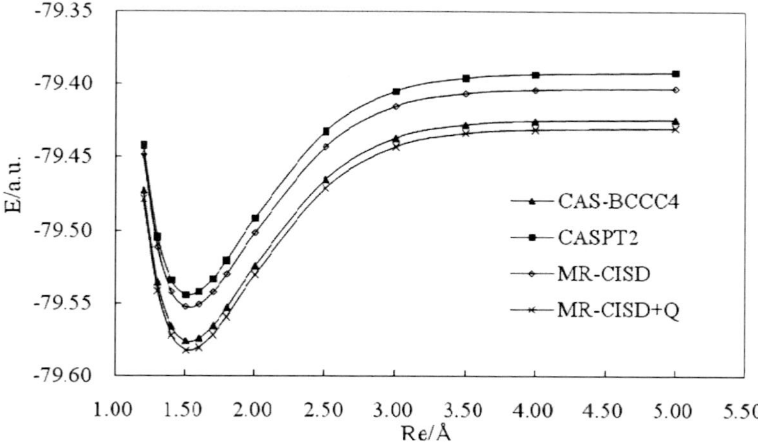

Fig. 2 Potential energy curves for the C—C bond breaking in ethane

well, and the CAS-BCCC4 curve is in excellent agreement with the MR-CISD+Q curve. Each of the four curves shows that the equilibrium C—C bond length (the lowest point on each curve) is about 1.5 Å, and the energy changes little after about 4.0 Å. At each point on the PES, the energies obtained from the four methods are in the order MR-CISD+Q < CAS-BCCC4 < MR-CISD < CASPT2. So CAS-BCCC4 can recover more dynamic correlation than CASPT2 and MR-CISD. For better comparison, the energies of MR-CISD+Q and the deviations of the other three methods against MR-CISD+Q values are collected in Table 1. The maximum and minimum energy deviations and nonparallelity errors (NPEs, defined as the absolute value of the difference between the maximum and the minimum energy deviations) are listed in Table 2. From these results, one can see that the maximum energy

Table 1 MR-CISD+Q energies (a.u.) and the deviations against MR-CISD+Q (kcal/mol) for CAS-BCCC4, CASPT2, and MR-CISD for ethane. The CASSCF(2,2) reference function and the 6-31G** basis set are used for all theoretical methods

R(Å)	MR-CISD+Q	CAS-BCCC4	CASPT2	MR-CISD
1.20	−79.478342	3.93	22.89	18.94
1.30	−79.541188	3.96	23.34	18.75
1.40	−79.571486	3.99	23.68	18.61
1.50	−79.581790	4.03	23.93	18.50
1.60	−79.580053	4.07	24.12	18.41
1.70	−79.571329	4.09	24.26	18.32
1.80	−79.558869	4.12	24.38	18.25
2.00	−79.530155	4.12	24.51	18.09
2.50	−79.470952	4.02	24.26	17.71
3.00	−79.442881	3.93	23.88	17.42
3.50	−79.433792	3.90	23.88	17.28
4.00	−79.431242	3.89	23.96	17.23
5.00	−79.430150	3.89	24.02	17.19

Table 2 Maximum and minimum energy deviations and NPE (kcal/mol) against MR-CISD+Q energies for ethane

Method	ΔE_{max}	ΔE_{min}	NPE
CAS-BCCC4	4.12	3.89	0.23
CASPT2	24.51	22.89	1.62
MR-CISD	18.94	17.19	1.74

deviation of CAS-BCCC4 is only 4.12 kcal/mol, which is much smaller than that of CASPT2 (24.51 kcal/mol) and MR-CISD (18.94 kcal/mol). The NPE of CAS-BCCC4 is only 0.23 kcal/mol, while the NPEs of CASPT2 and MR-CISD are 1.62 and 1.74 kcal/mol, respectively. Thus, for the C—C bond breaking in ethane, one can see that the performance of CAS-BCCC4 is comparable to that of MR-CISD+Q, better than that of MR-CISD or CASPT2.

3.1.2 2,3-Dimethyl-butane

For this molecule, MR-CISD+Q energies and relative energies from CAS-BCCC4, CASPT2, and MR-CISD are given in Table 3, and the corresponding PESs of all theoretical methods are shown in Fig. 3. Different from that in ethane, the CAS-BCCC4 energies are now lower than MR-CISD+Q energies, and CASPT2 energies are lower than MR-CISD energies. From the results shown in Table 4, one can see that the NPEs (against CAS-BCCC4) for MR-CISD+Q, CASPT2, and MR-CISD are 1.38, 2.59, and 6.08 kcal/mol, respectively. Compared with CAS-BCCC4, MR-CISD+Q and CASPT2 still work fairly well, but MR-CISD gives less satisfactory descriptions.

Table 3 MR-CISD+Q energies (a.u.) and the deviations against MR-CISD+Q (kcal/mol) for CAS-BCCC4, CASPT2, and MR-CISD for 2,3-dimethyl-butane. The CASSCF(2,2) reference function and the 6-31G** basis set are used for all theoretical methods

R(Å)	MR-CISD+Q	CAS-BCCC4	CASPT2	MR-CISD
1.20	−236.219372	−13.22	31.50	102.12
1.30	−236.283240	−13.04	32.60	101.60
1.40	−236.314890	−12.89	33.39	101.20
1.50	−236.326786	−12.77	33.94	100.88
1.60	−236.326785	−12.68	34.32	100.62
1.70	−236.319860	−12.60	34.58	100.39
1.80	−236.309168	−12.54	34.74	100.20
2.00	−236.283764	−12.46	34.86	99.86
2.50	−236.229765	−12.36	34.47	99.03
3.00	−236.202754	−12.16	34.24	98.22
3.50	−236.193286	−11.97	34.55	97.75
4.00	−236.190236	−11.89	34.82	97.53
5.00	−236.189217	−11.84	35.00	97.42

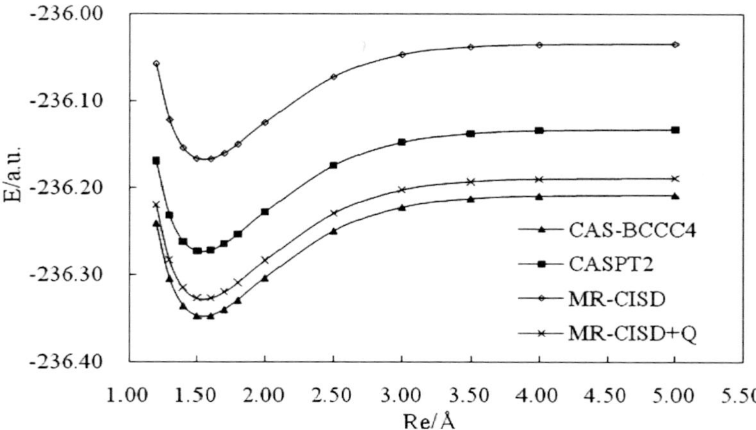

Fig. 3 Potential energy curves for the C—C bond breaking in 2,3-dimethyl-butane

Table 4 Maximum and minimum energy deviations and NPE (kcal/mol) against CAS-BCCC4 energies for 2,3-dimethyl-butane

Method	ΔE_{max}	ΔE_{min}	NPE
CASPT2	47.32	44.73	2.59
MR-CISD	115.34	109.26	6.08
MR-CISD+Q	13.22	11.84	1.38

It is instructive to compare the size-extensivity errors inherent in CAS-BCCC4, MR-CISD, and MR-CISD+Q for two alkanes under study. Here, the size-extensivity error for a given method is defined as the difference between the energy of the molecule in the dissociation limit (10 Å) and the sum of energies of two separate fragments calculated by the same method. It should be noted that, when in treating a fragment with an unpaired electron, a CAS-BCCC4 calculation with the active space (1,1) (all orbitals are from a ROHF calculation, and the active orbital is the open-shell orbital [97]) should be done, and ROHF-based CISD (or CISD+Q) should be used to replace MR-CISD (or MR-CISD+Q). For ethane, the size-extensivity errors for CAS-BCCC4, MR-CISD, and MR-CISD+Q are 1.80, 8.65, and 1.62 kcal/mol, respectively. For the larger molecule, 2,3-dimethyl-butane, the size-extensivity errors are 1.21 kcal/mol for CAS-BCCC4, 48.85 kcal/mol for MR-CISD, and 19.09 kcal/mol for MR-CISD+Q. Thus, the size-extensivity error of MR-CISD increases rapidly with increasing the total number of electrons, while that of MR-CISD+Q also shows a significant increase. However, CAS-BCCC4 behaves very differently from MR-CISD or MR-CISD+Q: its size-extensivity error does not increase from ethane to 2,3-dimethyl-butane (in fact, the size-extensivity error of CAS-BCCC4 is mainly dependent on the size of the active space). Thus, although CAS-BCCC4 and MR-CISD+Q have comparable size-extensivity errors in small molecules, CAS-BCCC4 will have much smaller size-extensivity errors than MR-CISD+Q in large molecules. This difference has a significant influence

on the energies calculated with CAS-BCCC4 and MR-CISD+Q. The fact that the
CAS-BCCC4 energy is lower than the MR-CISD+Q energy in 2,3-dimethyl-butane
is a reflection of this point. As a result, it is expected that for the C—C bond breaking
in molecules larger than 2,3-dimethyl-butane, CAS-BCCC4 would provide more
accurate descriptions than MR-CISD+Q.

3.2 C=C Bond Breaking in Alkenes

3.2.1 Ethylene

The PESs from all theoretical methods are shown in Fig. 4, and energy deviations
of three other methods against MR-CISD+Q values are given in Fig. 5. It is clear
from Fig. 4 that in ethylene the relative order of energies from different methods
in the entire range of the C=C distances is the same as that in ethane. As shown
in Table 5, against MR-CISD+Q values CAS-BCCC4 has the smallest NPE (only
0.18 kcal/mol), while CASPT2 and MR-CISD have significantly larger NPEs (2.78
and 1.35 kcal/mol, respectively). Thus, CAS-BCCC4 is very competitive with MR-
CISD+Q in describing the C=C bond breaking in ethylene. The performance of
CAS-BCCC4 in ethylene is consistent with that in ethane.

3.2.2 2,3-Dimethyl-2-butene

The calculated PESs from different methods are shown in Fig. 6, and the energy
deviation curves are shown in Fig. 7. Similar to the results obtained for
2,3-dimethyl-butane, we find that for 2,3-dimethyl-2-butene CAS-BCCC4 predicts lower energies than MR-CISD+Q, and CASPT2 predicts lower energies than
MR-CISD. The maximum energy deviation between CAS-BCCC4 and MR-CISD+Q

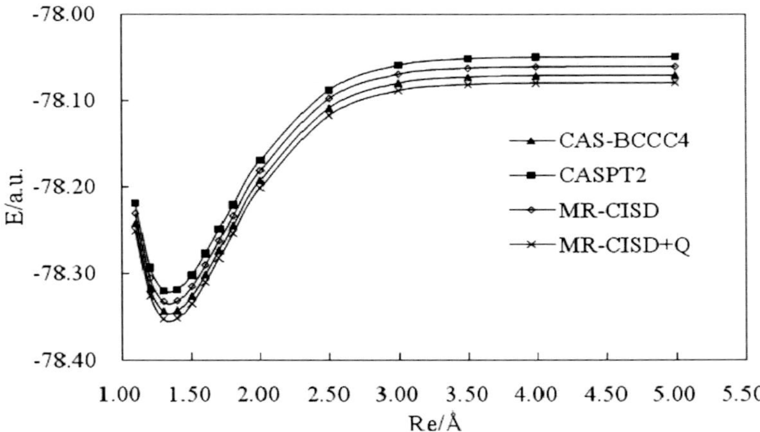

Fig. 4 Potential energy curves for the C=C bond breaking in ethylene

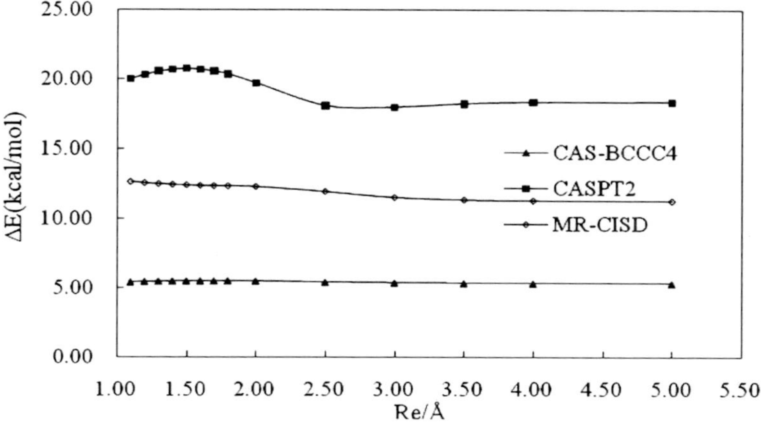

Fig. 5 Energy deviations of various methods against MR-CISD+Q values for ethylene

Table 5 Maximum and minimum energy deviations and NPE (kcal/mol) against MR-CISD+Q energies for ethylene

Method	$\Delta E_{\max}$	$\Delta E_{\min}$	NPE
CAS-BCCC4	5.54	5.36	0.18
CASPT2	20.78	18.00	2.78
MR-CISD	12.64	11.29	1.35

is 6.87 kcal/mol for this molecule, being somewhat less than that in 2,3-dimethyl-butane (13.22 kcal/mol). Relative to the CAS-BCCC4 energies, MR-CISD+Q, CASPT2, and MR-CISD have NPEs of 1.98, 5.59, 5.96 kcal/mol, respectively (Table 6). Thus, the overall performance of CAS-BCCC4 in ethylene and 2,3-dimethyl-2-butene is clearly better than CASPT2 and MR-CISD, being comparable to that of MR-CISD+Q.

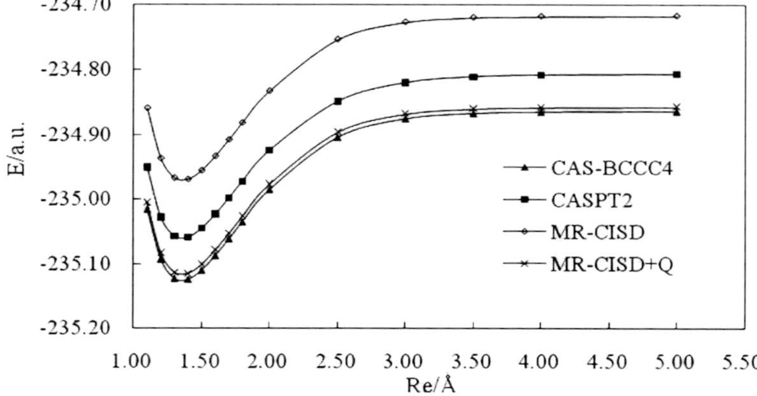

Fig. 6 Potential energy curves for the C=C bond breaking in 2,3-dimethyl-2-butene

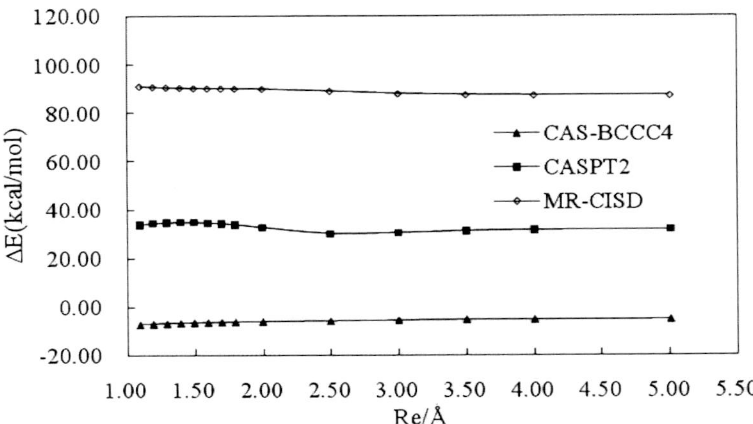

Fig. 7 Energy deviations of various methods against MR-CISD+Q values for 2,3-dimethyl-2-butene

Table 6 Maximum and minimum energy deviations and NPE (kcal/mol) against CAS-BCCC4 energies for 2,3-dimethyl-2-butene

Method	ΔE_{max}	ΔE_{min}	NPE
CASPT2	41.33	35.74	5.59
MR-CISD	98.23	92.27	5.96
MR-CISD+Q	6.87	4.89	1.98

4 Conclusions

The CAS-BCCC4 approach has been applied to investigate the bond-breaking PESs for a C—C bond in two alkane molecules (ethane and 2,3-dimethyl-butane) and a C=C bond in two alkene molecules (ethylene and 2,3-dimethyl-2-butene). Our calculations demonstrate that CAS-BCCC4 energies are usually lower than the corresponding MR-CISD or CASPT2 energies, indicating that more dynamic correlation is recovered by the CAS-BCCC4 approach. For small molecules like ethane and ethylene, CAS-BCCC4 energies are slightly higher than MR-CISD+Q energies in the entire range of the bond distances under study, but the CAS-BCCC4 PESs are very parallel to the MR-CISD+Q PESs. However, in two relatively larger molecules (2,3-dimethyl-butane and 2,3-dimethyl-2-butene), CAS-BCCC4 predicts slightly lower energies than MR-CISD+Q does, although the two PESs from CAS-BCCC4 and MR-CISD+Q are still quite close to each other. Thus, for systems under study, the overall performance of CAS-BCCC4 is comparable to that of MR-CISD+Q.

As demonstrated in this work, CAS-BCCC4 has a smaller size-extensivity error than MR-CISD+Q if the system size becomes larger. Hence, if large systems (significantly larger than 2,3-dimethyl-2-butene) are investigated, CAS-BCCC4 should possess a higher accuracy than MR-CISD+Q. On the other hand, we point out that the computational cost of our present CAS-BCCC4 code is still much higher than that of the MR-CISD+Q code implemented in the MOLPRO package. In principle, the computational cost of the CAS-BCCC4 approach should be comparable to that of the traditional CCSD method, if the active space is no larger than the (4,4) space (a detailed discussion on the computational cost of CAS-BCCC4 has been provided elsewhere [98]), while the cost of a MR-CISD (or MR-CISD+Q) calculation is roughly the cost of the corresponding CCSD calculation times the number of determinants in the active space. In reality, the implementation of the MR-CISD (or MR-CISD+Q) method in the MOLPRO package using the internal contraction approximation is very efficient, while our CAS-BCCC4 code still needs extensive optimization (due to the very complicated formulations). We hope that, in the near future, a fully optimized CAS-BCCC4 code will allow CAS-BCCC4 calculations affordable for many medium-size or relatively large systems. In summary, our calculations presented here show that CAS-BCCC4 provides accurate descriptions for the C—C bond-breaking PESs in systems under study.

Acknowledgement This work was supported by the National Natural Science Foundation of China (Grant Nos. 20625309 and 20833003) and the National Basic Research Program (Grant No. 2004CB719901).

References

1. A. Dutta, C. D. Sherrill, J. Chem. Phys. **118**, 1610 (2003)
2. M. L. Abrams, C. D. Sherrill, J. Phys. Chem. A **107**, 5611 (2003)
3. L. B. Harding, Y. Georgievskii, S. J. Klippenstein, J. Phys. Chem. A **109**, 4646 (2005)
4. A. A. Golubeva, A. V. Nemukhin, S. J. Klippenstein, L. B. Harding, A. I. Krylov, J. Phys. Chem. A **111**, 13264 (2007)
5. L. B. Harding, S. J. Klippenstein, L. B. Harding, Phys. Chem. Chem. Phys. **9**, 4055 (2007)
6. Y. Ge, M. S. Gordon, P. Piecuch, J. Chem. Phys. **127**, 174106 (2007)
7. Y. Ge, M. S. Gordon, P. Piecuch, M. Wloch, J. R. Gour, J. Phys. Chem. A **112**, 11873 (2008)
8. B. O. Roos, P. Bruna, S. D. Peyerimhoff, R. Shepard, D. L. Cooper, J. Gerratt, M. Raimondi, *Ab Initio Methods in Quantum Chemistry, II* (Wiley, New York, 1987)
9. K. Andersson, P.-Å. Malmqvist, B. O. Roos, A. J. Sadlej, and K. Wolinski, J. Phys. Chem. A **94**, 5483 (1990)
10. K. Andersson, P.-Å. Malmqvist, B. O. Roos, J. Chem. Phys. **96**, 1218 (1992)
11. D. Mukherjee, R. K. Moitra, A. Mukhopadhyay, Mol. Phys. **33**, 955 (1977)
12. A. Haque, D. Mukherjee, J. Chem. Phys. **80**, 5058 (1984)
13. I. Lindgren, Int. J. Quantum Chem. **S12**, 33 (1978)
14. W. Kutzelnigg, J. Chem. Phys. **77**, 3081 (1982)
15. A. Haque, U. Kaldor, Chem. Phys. Lett. **117**, 347 (1986)
16. S. R. Hughes, U. Kaldor, Chem. Phys. Lett. **194**, 99 (1992)
17. R. Offerman, W. Ey, H. Kummel, Nucl. Phys. A **273**, 349 (1976)
18. R. Offerman, Nucl. Phys. A **273**, 368 (1976)
19. W. Ey, Nucl. Phys. A **296**, 189 (1978)
20. B. Jeziorski, H. J. Monkhorst, Phys. Rev. A **24**, 1668 (1981)
21. L. Stolarczyk, H. J. Monkhorst, Phys. Rev. A **32**, 743 (1985)
22. B. Jeziorski, J. Paldus, J. Chem. Phys. **88**, 5673 (1988)
23. L. Meissner, K. Jankowski, J. Wasilewski, Int. J. Quantum Chem **34**, 535 (1988)

24. J. Paldus, J. Pylypow, B. Jeziorski, in *Many Body Methods in Quantum Chemistry, Lecture Notes in Chemistry*, vol. 52, ed. by U. Kaldor (Springer, Berlin, 1989)
25. P. Piecuch, J. Paldus, Theor. Chim. Acta **83**, 69 (1992)
26. A. Balkova, S. A. Kucharski, L. Meissner, R. J. Bartlett, Theor. Chim. Acta **80**, 335 (1991)
27. X. Li, J. Paldus, J. Chem. Phys. **119**, 5320 (2003)
28. X. Li, J. Paldus, J. Chem. Phys. **119**, 5346 (2003)
29. X. Li, J. Paldus, J. Chem. Phys. **124**, 034112 (2006)
30. M. Hanrath, J. Chem. Phys. **123**, 084102 (2005)
31. U. S. Mahapatra, B. Datta, B. Bandyopadhyay, D. Mukherjee, Adv. Quantum Chem. **30**, 163 (1998)
32. U. S. Mahapatra, B. Datta, D. Mukherjee, J. Phys. Chem. A **103**, 1822 (1999)
33. U. S. Mahapatra, B. Datta, D. Mukherjee, J. Chem. Phys. **110**, 6171 (1999)
34. S. Chattopadhyay, U. S. Mahapatra, D. Mukherjee, J. Chem. Phys. **112**, 7939 (2000)
35. S. Chattopadhyay, D. Pahari, D. Mukherjee, U. S. Mahapatra, J. Chem. Phys. **120**, 5968 (2004)
36. I. Hubac, P. Neogrady, Phys. Rev. **50**, 4558 (1994)
37. J. Pittner, J. Chem. Phys. **118**, 10876 (2003)
38. J. Pittner, O. Demel, J. Chem. Phys. **122**, 181101 (2005)
39. I. Hubac, J. Pittner, P. Carsky, J. Chem. Phys. **112**, 8779 (2000)
40. F. A. Evangelista, W. D. Allen, H. F. Schaefer III, J. Chem. Phys. **125**, 154113 (2006)
41. F. A. Evangelista, W. D. Allen, H. F. Schaefer III, J. Chem. Phys. **127**, 024102 (2007)
42. F. A. Evangelista, A. C. Simmonett, W. D. Allen, H. F. Schaefer III, J. Gauss, J. Chem. Phys. **128**, 124104 (2008)
43. N. Oliphant, L. Adamowicz, J. Chem. Phys. **94**, 1229 (1991)
44. P. Piecuch, N. Oliphant, L. Adamowicz, J. Chem. Phys. **99**, 1875 (1993)
45. P. Piecuch, S. A. Kucharski, R. J. Bartlett, J. Chem. Phys. **110**, 6103 (1999)
46. L. Adamowicz, J.-P. Malrieu, V. V. Ivanov, J. Chem. Phys. **112**, 10075 (2000)
47. S. R. Langhoff, E. R. Davidson, Int. J. Quantum Chem. **8**, 61 (1974)
48. E. R. Davidson, D. W. Silver, Chem. Phys. Lett. **52**, 403 (1977)
49. W. Duch, G. H. F. Diercksen, J. Chem. Phys. **101**, 3018 (1994)
50. K. A. Peterson, A. K. Wilson, D. W. Woon, J. T. H. Dunning, Theor. Chem. Acc. **97**, 251 (1997)
51. I. Shavitt, Mol. Phys. **94**, 3 (1998)
52. J. Cabrero, R. Caballol, J.-P. Malrieu, Mol. Phys. **100**, 919 (2002)
53. V. Vallet, P. Macak, U. Wahlgren, I. Grenthe, Theor. Chem. Acc. **115**, 145 (2006)
54. X. Li, J. Paldus, J. Chem. Phys. **107**, 6257 (1997)
55. X. Li, J. Paldus, J. Chem. Phys. **108**, 637 (1998)
56. X. Li, J. Paldus, J. Chem. Phys. **110**, 2844 (1999)
57. X. Li, J. Paldus, J. Chem. Phys. **113**, 9966 (2000)
58. X. Li, J. Paldus, Mol. Phys. **98**, 1185 (2000)
59. X. Li, J. Paldus, J. Chem. Phys. **124**, 174101 (2006)
60. X. Li, J. Paldus, J. Chem. Phys. **125**, 164107 (2006)
61. A. I. Krylov, Chem. Phys. Lett. **338**, 375 (2001)
62. A. I. Krylov, Chem. Phys. Lett. **350**, 522 (2001)
63. A. I. Krylov, C. D. Sherrill, J. Chem. Phys. **116**, 3194 (2002)
64. L. V. Slipchenko, A. I. Krylov, J. Chem. Phys. **117**, 4694 (2002)
65. J. S. Sears, C. D. Sherrill, A. I. Krylov, J. Chem. Phys. **118**, 9084 (2003)
66. Y. H. Shao, M. Head-Gordon, A. I. Krylov, J. Chem. Phys. **118**, 4807 (2003)
67. T. Van Voorhis, M. Head-Gordon, Chem. Phys. Lett. **317**, 575 (2000)
68. T. Van Voorhis, M. Head-Gordon, J. Chem. Phys. **112**, 5633 (2000)
69. A. I. Krylov, C. D. Sherrill, E. F. C. Byrd, M. Head-Gordon, J. Chem. Phys. **109**, 10669 (1998)
70. A. I. Krylov, C. D. Sherrill, M. Head-Gordon, J. Chem. Phys. **113**, 6509 (2000)
71. C. D. Sherrill, A. I. Krylov, E. F. C. Byrd, M. Head-Gordon, J. Chem. Phys. **109**, 4171 (1998)

72. R. C. Lochan, M. Head-Gordon, J. Chem. Phys. **126**, 164101 (2007)
73. S. R. Gwaltney, M. Head-Gordon, Chem. Phys. Lett. **323**, 21 (2000)
74. S. R. Gwaltney, C. D. Sherrill, M. Head-Gordon, A. I. Krylov, J. Chem. Phys. **113**, 3548 (2000)
75. S. R. Gwaltney, M. Head-Gordon, J. Chem. Phys. **115**, 2014 (2001)
76. S. R. Gwaltney, E. F. C. Byrd, T. Van Voorhis, M. Head-Gordon, Chem. Phys. Lett. **353**, 359 (2002)
77. P. Piecuch, K.Kowalski, I. S. O. Pimienta, M. J. McGuire, Int. Rev. Phys. Chem. **21**, 527 (2002)
78. K. Kowalski, P. Piecuch, J. Chem. Phys. **116**, 7411 (2002)
79. M. Włoch, J. R. Gour, K. Kowalski, P. Piecuch, J. Chem. Phys. **122**, 214107 (2005)
80. P. Piecuch, K. Kowalski, I. S. O. Pimienta, P. D. Fan, M. Lodriguito, M. J. McGuire, S. A. Kucharski, T. Kuś, M. Musiał, Theor. Chem. Acc. **112**, 349 (2004)
81. P. Piecuch, M. Włoch, J. Chem. Phys. **123**, 224105 (2005)
82. A. Kinal, P. Piecuch, J. Phys. Chem. A **111**, 734 (2007)
83. M. Włoch, J. R. Gour, P. Piecuch, J. Phys. Chem. A **111**, 11359 (2005)
84. P. Piecuch, M. Włoch, K. Kowalski, A. J. C. Varandas, Theor. Chem. Acc. **120**, 59 (2008)
85. J. Zheng, J. R. Gour, J. J. Lutz, M. Włoch, P. Piecuch, D. G. Truhlar, J. Chem. Phys. **128**, 044108 (2008)
86. P. Piecuch, L. Adamowicz, J. Chem. Phys. **100**, 5792 (1994)
87. P. Piecuch, L. Adamowicz, Chem. Phys. Lett. **221**, 121 (1994)
88. K. Kowalski, S. Hirata, M. Włoch, P. Piecuch, T. L. Windus, J. Chem. Phys. **123**, 074319 (2005)
89. K. Kowalski, P. Piecuch, J. Chem. Phys. **113**, 8490 (2000)
90. K. Kowalski, P. Piecuch, Chem. Phys. Lett. **344**, 165 (2001)
91. K. Kowalski, P. Piecuch, J. Chem. Phys. **115**, 643 (2001)
92. P. Piecuch, S. Hirata, K. Kowalski, P. D. Fan, T. L. Windus, Int. J. Quantum Chem. **106**, 79 (2006)
93. J. R. Gour, P. Piecuch, M. Włoch, J. Chem. Phys. **123**, 134113 (2005)
94. D. I. Lyakh, V. V. Ivanov, L. Adamowicz, J. Chem. Phys. **122**, 024108 (2005)
95. P. Piecuch, S. A. Kucharski, V. Špirko, J. Chem. Phys. **111**, 6679 (1999)
96. S. H. Li, J. Chem. Phys. **120**, 5017 (2004)
97. T. Fang, S. Li, J. Chem. Phys. **127**, 204108 (2007)
98. J. Shen, T. Fang, W. Hua, S. Li, J. Phys. Chem. A **112**, 4703 (2008)
99. T. Fang, J. Shen, S. Li, J. Chem. Phys. **128**, 224107 (2008)
100. J. Shen, T. Fang, S. Li, Y. Jiang, J. Phys. Chem. A **112**, 4703 (2008)
101. T. Fang, J. Shen, S. Li, J. Chem. Phys. **129**, 234106 (2008)
102. J. Shen, T. Fang, S. Li, Sci. China Ser. B-Chem. **51**, 1197 (2008)
103. J. Shen, T. Fang, S. Li, Y. Jiang, Chem. J. Chinese Uni. **29**, 2341 (2008)
104. H.-J. Werner, Mol. Phys. **89**, 645 (1996)
105. P. Celani, H.-J. Werner, J. Chem. Phys. **112**, 5546 (2000)
106. H.-J. Werner, E. A. Reinsch, J. Chem. Phys. **76**, 3144 (1982)
107. H.-J. Werner, P. J. Knowles, J. Chem. Phys. **89**, 5803 (1988)
108. P. J. Knowles, H.-J. Werner, Chem. Phys. Lett. **145**, 514 (1988)
109. P. J. Knowles, H.-J. Werner, Theor. Chim. Acta **84**, 95 (1992)
110. S. Hirata, R. J. Bartlett, Chem. Phys. Lett. **321**, 216 (2000)
111. M. W. Schmidt, K. K. Baldridge, J. A. Boatz, S. T. Elbert, M. S. Gordon, J. H. Jensen, S. Koseki, N. Matsunaga, K. A. Nguyen, S. J. Su, T. L. Windus, M. Dupuis, J. A. Montgomery, J. Comput. Chem. **14**, 1347 (1993)
112. H.-J. Werner, P. J. Knowles, R. Lindh et al., MOLPRO,Version2006.1, a package of *ab initio* programs

Fermi-Vacuum Invariance in Multiconfiguration Perturbation Theory

Ágnes Szabados and Péter R. Surján

Abstract We investigate the dependence of multiconfigurational perturbation theory framework on the choice of the Fermi-vacuum. A new formulation, based on a posteriori averaging is suggested. The averaged theory is invariant with respect to Fermi-vacuum choice but enhances the intruder effect. The performance of the averaged formulation is illustrated on the ethylene rotational potential curve.

Keywords: Multireference · Perturbation theory · Fermi-vacuum dependence · Ethylene torsional barrier

1 Introduction

Dissociation of covalent bonds in molecules, description of electronic excited states, or transition metal compounds belong to current problems of theoretical chemistry, usually addressed as "multiconfiguration cases." Basic quantum chemical paradigms, such as configuration interaction, coupled-cluster (CC), and perturbation theory (PT) have lived diverse multireference (MR) extensions of the single-reference approach, in order to treat the above systems. Among MR theories one may distinguish genuine MR methods [1–5] as well as essentially single-reference-type multiconfiguration (MC) approaches [6–8]. This categorization – which has been most used in CC methodology – is of relevance from the point of view of this PT study. Purely single-reference CC methods, called completely renormalized CC approaches, which do not use any MR or MC concepts, also have the potential to describe bond breaking and biradicals [9, 10]. A common

Á. Szabados (✉)
Laboratory of Theoretical Chemistry, Loránd Eötvös University, H-1518 Budapest,
POB 32, Hungary,
e-mail: szabados@chem.elte.hu

Péter R. Surján
Laboratory of Theoretical Chemistry, Loránd Eötvös University, H-1518 Budapest,
POB 32, Hungary,
e-mail: surjan@chem.elte.hu

P. Piecuch et al. (eds.), *Advances in the Theory of Atomic and Molecular Systems*,
Progress in Theoretical Chemistry and Physics 19, DOI 10.1007/978-90-481-2596-8_12,
© Springer Science+Business Media B.V. 2009

characteristic of purely single-reference- and essentially single-reference-type MC methods is the presence of a formal reference determinant in the theory, which facilitates the particle/hole orbital assignment and involves one well-defined excitation amplitude set. As opposed to this, genuine MRCC methods consider all determinants present in a complete or incomplete model space as Fermi-vacua, and consequently involve multiple amplitude sets to parametrize the exact wavefunction. We adopt the convention of referring to this latter strategy as Jeziorski–Monkhorst parametrization.

Multiple amplitude sets have been applied in PT methodology as well. Perturbative approximations of the Jeziorski–Monkhorst parametrized SS-MRCC [11] show outstanding properties: size-consistency, size-extensivity, and the potential to be intruder free at the same time. It is however not widespread to consider multiple amplitude sets within a PT framework for the electron correlation problem. A usual categorization of MRPT frameworks is the distinction of effective Hamiltonian strategies [12–14] and single-but-multi philosophies [15, 16]. The former can target many electronic states of a system while the latter focuses on just one state at a time. Both approaches have their own advantages and shortcomings. In rough terms effective Hamiltonian theories are good candidates for fulfilling the size-extensivity requirement; they however struggle with serious intruder state problem [17]. Single-but-multi approaches like CASPT [18, 19] or MRMP [20] usually violate size-consistency, but the sensitivity to intruders may be less severe. There are several exceptions to the above rule. The intermediate Hamiltonian theory developed by Malrieu et al. [21], e.g., is size extensive and avoids intruders via the application of multiple partitionings when building the effective Hamiltonian matrix.

Multiconfiguration PT (MCPT) [22] is another exception to the above categorization, since it can provide rigorously size-consistent energy at second order [23] although it applies a single-but-multi framework. In MCPT there appears a Fermi-vacuum which can be any of the determinants of nonzero weight in the multiconfiguration zero-order function. The necessity to pinpoint a Fermi-vacuum is a common feature of MCPT and single-reference-type MRCC approaches. Neither MCPT nor affected MRCC theories show invariance with respect to the Fermi-vacuum choice. Switching from one Fermi-vacuum to another when following a potential energy surface has been shown to cause discontinuities. Though the error due to this effect has been shown to be small in MRCC [24], the non-invariance can become a qualitative problem when causing symmetry breaking. This issue has been addressed recently in the context of XCASCCSD and cured by symmetry-adaptation at a suitable step of the algorithm [25].

In this study we consider the same problem in the framework of MCPT and propose a modification which restores the invariance to the choice of Fermi-vacuum. This involves calculating the perturbed quantities by all possible choices and constructing a weighted average. The number of parameters in the theory agrees with that of a Jeziorski–Monkhorst-type MRCC parametrization. The redundancy of a Jeziorski–Monkhorst parametrization however does not show up in the present approach due to the fact that perturbational amplitudes corresponding to different

Fermi-vacua are determined separately. The linear combination of thus produced functions is constructed a posteriori, the weights taken from the unrelaxed zero-order wavefunction.

In this account we first briefly summarize MCPT followed by an analysis on the extent of Fermi-vacuum non-invariance in different versions of the theory. We continue by presenting the approach to ensure Fermi-vacuum invariance and close with a numerical illustration.

2 Theory

2.1 Zero-Order Hamiltonians

The perturbational framework designated as MCPT has been developed in two variants: the original formulation, hereafter called projected-MCPT [22] (p-MCPT) and a version which can give size-consistent correction at second order, hereafter called unprojected-MCPT [23] (u-MCPT). Both variants start with a multiconfiguration zero-order function written as

$$|0\rangle = c_{\mathrm{HF}}|\mathrm{HF}\rangle + \sum_{K \in R \backslash \{\mathrm{HF}\}} c_K |K\rangle, \tag{1}$$

where $|\mathrm{HF}\rangle$ and $|K\rangle$ denote determinants. Determinant $|\mathrm{HF}\rangle$ is used as Fermi-vacuum. It can be any determinant which appears with nonzero weight in $|0\rangle$, practically it is the one associated with the largest coefficient squared. Determinants appearing in the zero-order reference function form a subspace in the configuration space, denoted by R. The sum for K in Eq. (1) runs over elements of subspace R, except the pinpointed Fermi-vacuum.

In MCPT theories a zero-order Hamiltonian is constructed in spectral form, taking $|0\rangle$ as ground-state zero-order eigenvector and determinants different from $|\mathrm{HF}\rangle$ as excited state zero-order eigenvectors. These functions represent a non-orthogonal basis in the FCI space which necessitates the treatment of overlap. It is at this stage where p-MCPT and u-MCPT versions deviate. In p-MCPT

1. excited vectors $|K\rangle$ are first Schmidt-orthogonalized to $|0\rangle$ to get $|K'\rangle$;
2. a reciprocal set is constructed to $|K'\rangle$, denoted by $\langle \widetilde{K'}|$.

In u-MCPT version step 1 is missing, reciprocal vectors to the set $|0\rangle$ and $|K\rangle$'s are directly constructed, giving $\langle \widetilde{0}|$ and $\langle \widetilde{K}|$. The spectral form of the zero-order Hamiltonian is nonsymmetric in both formulations, due to the use of bi-orthogonal vector sets:

$$\hat{H}^{(0)}_{\mathrm{p\text{-}MCPT}} = E_0|0\rangle\langle 0| + \sum_{K \in R \backslash \{\mathrm{HF}\}} E_K |K'\rangle\langle \widetilde{K'}| \tag{2}$$

Table 1 Left- and right-hand eigenvectors of zero-order Hamiltonians in p-MCPT and u-MCPT

	Ground state		Excited state	
	Right hand	Left hand	Right hand	Left hand
$\hat{H}^{(0)}_{\text{p-MCPT}}$	$|0\rangle$	$\langle 0|$	$|K'\rangle = |K\rangle - |0\rangle c_K$	$\langle \widetilde{K'}| = \langle K| - \frac{c_K}{c_{\text{HF}}}\langle \text{HF}|$
$\hat{H}^{(0)}_{\text{u-MCPT}}$	$|0\rangle$	$\langle \widetilde{0}| = \frac{1}{c_{\text{HF}}}\langle \text{HF}|$	$|K\rangle$	$\langle \widetilde{K}| = \langle K| - \frac{c_K}{c_{\text{HF}}}\langle \text{HF}|$

$$\hat{H}^{(0)}_{\text{u-MCPT}} = \eta_0 |0\rangle\langle\widetilde{0}| + \sum_{K \in R\setminus\{\text{HF}\}} \eta_K |K\rangle\langle\widetilde{K}| . \tag{3}$$

Comparing zero-order operators of Eqs. (2) and (3) one may observe that an advantage of Schmidt-orthogonalization is getting the zero-order Hamiltonian symmetric at least in the one-dimensional reference space spanned by $|0\rangle$. Left- and right-hand zero-order eigenvectors expressed in terms of determinants $|\text{HF}\rangle$, $|K\rangle$, and MR function $|0\rangle$ are listed in Table 1 for completeness. Detailed derivation of the reciprocal vectors has been shown in an earlier report [23].

When defining zero-order ground-state eigenvalues, projection of the Schrödinger equation is taken either with $\langle 0|$ or with $\langle\widetilde{0}|$ to get

$$E^{(0)}_{\text{p-MCPT}} = E_0 = \langle 0|\hat{H}|0\rangle \tag{4}$$

or

$$E^{(0)}_{\text{u-MCPT}} = \eta_0 = \langle\widetilde{0}|\hat{H}|0\rangle , \tag{5}$$

respectively. By this definition, the first-order energy correction is zero in both versions. It is to be mentioned at this point that the symmetric and nonsymmetric energy formulae (4) and (5) are equivalent if the zero-order function $|0\rangle$ is obtained from a variational solution of the Schrödinger equation. This is the case, e.g., if $|0\rangle$ is a complete active space (CAS) wavefunction or stems from a configuration-interaction procedure.

Zero-order excited state energies can be chosen in both variants, e.g., in the spirit of Davidson and Kapuy (DK) [26–29]

$$E_K = E_0 + \Delta\varepsilon_K$$

or

$$\eta_K = \eta_0 + \Delta\varepsilon_K,$$

where $\Delta\varepsilon_K$ is formed as sums and differences of one-particle energies defined as diagonals of a generalized Fockian. This partitioning is different from Møller–Plesset, since the generalized Fockian is not necessarily diagonal.

Alternatively, excited state energies may be chosen to obtain a generalized Epstein–Nesbet (EN) partitioning

$$E_K = \eta_K = \langle K|\hat{H}|K\rangle \; . \tag{6}$$

A nonsymmetric generalization of EN partitioning is also conceivable in the form $E_K = \langle \widetilde{K}'|\hat{H}|K'\rangle$ or $\eta_K = \langle \widetilde{K}|\hat{H}|K\rangle$. For our present purpose the definition of Eq. (6) is more appealing because it avoids any reference to the Fermi-vacuum. A detailed discussion on partitionings in MCPT can be found in Ref. [30].

2.2 Non-Invariance to the Choice of Fermi-Vacuum

The first-order wavefunction in p-MCPT as obtained by bi-orthogonal Rayleigh–Schrödinger theory looks

$$\Psi^{(1)}_{\text{p-MCPT}}(\text{HF}) = - \sum_{K \in R \backslash \{\text{HF}\}} |K'\rangle \frac{\langle \widetilde{K}'|\hat{H}|0\rangle}{E_K - E_0}$$

while the second-order energy is given by

$$\begin{aligned}
E^{(2)}_{\text{p-MCPT}}(\text{HF}) &= - \sum_{K \in R \backslash \{\text{HF}\}} \frac{\langle 0|\hat{H}|K'\rangle \langle \widetilde{K}'|\hat{H}|0\rangle}{E_K - E_0} \\
&= - \sum_{K \in R \backslash \{\text{HF}\}} \frac{\left(\langle 0|\hat{H}|K\rangle - c_K E_0\right)\left(\langle K|\hat{H}|0\rangle - c_K \eta_0\right)}{E_K - E_0} \; .
\end{aligned} \tag{7}$$

Notation HF in round braces on the left-hand side refer to the dependence of the quantities on the Fermi-vacuum choice. Examining the second-order energy, dependence on $|\text{HF}\rangle$ is least severe in EN partitioning if $E_0 = \eta_0$. In this case neither the numerator nor the denominator of the energy expression is affected by changing the choice for $|\text{HF}\rangle$. Dependence on the Fermi-vacuum only stems from the restriction on the sum $K \neq \text{HF}$. In contrast to this, denominators show explicit dependence on $|\text{HF}\rangle$ in DK partitioning, because particle/hole assignment is based on $|\text{HF}\rangle$. This means that changing $|\text{HF}\rangle$ may change the actual orbital indices which contribute to a $\Delta\varepsilon_K$. Whether the value ε_i itself is affected depends on the definition. One may use the diagonals of the generalized Fockian appearing in multiconfigurational self-consistent field (MCSCF) theories [31]

$$\varepsilon_i = h_i + \sum_{jk} P_{jk} \left(\langle ij|ik\rangle - \langle ij|ki\rangle\right), \tag{8}$$

where h_i denotes a one-electron integral incorporating kinetic energy and nuclear–electron attraction, and the two-electron integral $\langle ij|ik\rangle$ is written in the $\langle 12|12\rangle$

convention. Ordinarily the one-particle density matrix is defined as

$$P_{jk} = \langle 0|a_k^+ a_j|0\rangle \ . \tag{9}$$

Using Eq. (9) in the expression for ε_i, orbital energies are independent from the Fermi-vacuum choice. It may be however appealing to substitute the Hartree–Fock density matrix

$$P_{jk} = \langle \mathrm{HF}|a_k^+ a_j|\mathrm{HF}\rangle \tag{10}$$

into Eq. (8), because ε_is obtained this way do not show degeneracies which may stem from the spatial symmetry of the system and may lead to zero excitation energy denominators in the PT expressions. Substituting Eq. (10) in Eq. (8) has the consequence that ε_is become Fermi-vacuum dependent.

Turning our attention to u-MCPT formulation, the first-order wavefunction can be written as

$$\Psi^{(1)}_{\text{u-MCPT}}(\mathrm{HF}) = - \sum_{K \in R \setminus \{\mathrm{HF}\}} |K\rangle \frac{\langle \widetilde{K}|\hat{H}|0\rangle}{\eta_K - \eta_0}$$

while the second-order energy looks

$$\begin{aligned}
E^{(2)}_{\text{u-MCPT}}(\mathrm{HF}) &= - \sum_{K \in R \setminus \{\mathrm{HF}\}} \frac{\langle \widetilde{0}|\hat{H}|K\rangle\langle \widetilde{K}|\hat{H}|0\rangle}{\eta_K - \eta_0} \\
&= -\frac{1}{c_{\mathrm{HF}}} \sum_{K \in R \setminus \{\mathrm{HF}\}} \frac{\langle \mathrm{HF}|\hat{H}|K\rangle \left(\langle K|\hat{H}|0\rangle - c_K \eta_0\right)}{\eta_K - \eta_0} \ .
\end{aligned} \tag{11}$$

In contrast to the p-MCPT version, second-order u-MCPT energy shows explicit dependence on the Fermi-vacuum in the numerator. As illustrated in Section 3, dependence of second-order u-MCPT on the choice of $|\mathrm{HF}\rangle$ is more expressed in numerical terms than $E^{(2)}$ of p-MCPT.

2.3 A Fermi-Vacuum Invariant Treatment

A simple way to remove dependence of the PT expressions on the Fermi-vacuum is to deliberately make every possible choice and form an average of the quantities obtained. Theoretical formulation may start by a linearized Jeziorski–Monkhorst-type parametrization of the wavefunction

$$\Psi = |0\rangle + \sum_{K \in R} \hat{T}_K |K\rangle c_K \tag{12}$$

with

$$\hat{T}_K = \sum_{L \neq K} t_K^L |L'\rangle\langle K| , \qquad K \in R ,$$

if working with the vector set of p-MCPT. Substitution of the excitation operator $\hat{T}_K$ into the wavefunction Eq. (12) reveals that the number of parameters associated with an excited determinant L not present in $|0\rangle$, equals the number of determinants appearing in $|0\rangle$. This parametrization is redundant, since it is not possible to determine the amplitudes by the usual projection of the nth-order Schrödinger equation by excited functions.

Presently we do not attempt to take down sufficient conditions for determining all the amplitudes from a coupled set of equations. Instead, we apply an a posteriori procedure and determine amplitudes so that the nth order wavefunction becomes

$$\overline{\Psi}^{(n)} = \sum_{K \in R} c_K^2 \, \Psi^{(n)}(K) , \qquad n \geq 1 , \tag{13}$$

a weighted average of separately determined nth order corrections, obtained by a given choice for the Fermi-vacuum, K. To reach this goal, t_K^Ls are to be evaluated as

$$t_K^{L\,(1)} = -c_K \frac{\langle \widetilde{L}' | \hat{H} | 0 \rangle}{E_L - E_0} , \qquad K \in R, \tag{14}$$

where $\langle \widetilde{L}' | = \langle L | - c_L/c_K \langle K |$ is the reciprocal vector to $|L\rangle$, constructed while K plays the role of the principal determinant.

Terms of the energy may be obtained either by averaging according to

$$\overline{E}_{\text{p-MCPT}}^{(n)} = \sum_{K \in R} c_K^2 \, E_{\text{p-MCPT}}^{(n)}(K)$$

or by calculating the usual projection

$$\overline{E}_{\text{p-MCPT}}^{(n)} = \langle 0 | \hat{H} - \hat{H}_{\text{p-MCPT}}^{(0)} | \overline{\Psi}^{(n-1)} \rangle ,$$

with $n \geq 2$. By either formula one arrives at the second-order energy

$$\overline{E}_{\text{p-MCPT}}^{(2)} = - \sum_{K \in R} c_K^2 \sum_{L \neq K} \frac{\langle 0 | \hat{H} | L' \rangle \langle \widetilde{L}' | \hat{H} | 0 \rangle}{E_L - E_0} . \tag{15}$$

Note, that $\langle \widetilde{L}' |$ depends on K, and $E_L - E_0$ may also be K dependent (e.g-in DK partitioning). For this reason the outer sum on K cannot be evaluated irrespective of L. Equation (15) is one of our working formulae, which ensures Fermi-vacuum independence of the energy, within the p-MCPT framework.

It is worthwhile to examine a Fermi-vacuum invariant formulation with the use of basis vectors of u-MCPT also, e.g-to obtain a size-consistent second-order energy. This variant of the theory does not alter form (12) of the exact wavefunction, it only affects the expression of the excitation operator:

$$\hat{T}_K = \sum_{L \neq K} t_K^L |L\rangle\langle K| , \qquad K \in R . \tag{16}$$

Parameters t_K^L at first order, fulfilling criterion Eq. (13) in u-MCPT look

$$t_K^{L\,(1)} = -c_K \frac{\langle \widetilde{L}|H|0\rangle}{\eta_L - \eta_0} , \qquad K \in R ,$$

which equals Eq. (14) if the excitation energy denominators are the same.

At difference with p-MCPT, energy terms obtained by averaging or by projection with $\langle 0|$ are different, since the reciprocal function to $|0\rangle$ varies with the Fermi-vacuum choice. In this work we investigate the averaging procedure defined by

$$\overline{E}_{\text{u-MCPT}}^{(n)} = \sum_{K \in R} c_K^2 \; E_{\text{u-MCPT}}^{(n)}(K) ,$$

which now applies to $n = 0$ as well, to give

$$\overline{E}_{\text{u-MCPT}}^{(0)} = \sum_{K \in R} c_K^2 \; \langle \widetilde{K}|\hat{H}|0\rangle = \langle 0|\hat{H}|0\rangle = E_{\text{p-MCPT}}^{(0)} .$$

The second-order energy obtained by averaging looks

$$\overline{E}_{\text{u-MCPT}}^{(2)} = -\sum_{K \in R} c_K \sum_{L \neq K} \frac{\langle K|\hat{H}|L\rangle\langle \widetilde{L}|\hat{H}|0\rangle}{\eta_L - \eta_0} , \tag{17}$$

where c_K appears in the first power since the reciprocal counterpart of $|0\rangle$ is $c_K^{-1}\langle K|$ when K plays the role of the Fermi-vacuum. This latter expression is the second working formula of this study which is tested in Section 3.

Comparing averaged energy formulae one may observe that relative simplicity of u-MCPT expressions with respect to p-MCPT is diminished. Averaged zero order in u-MCPT becomes exactly equal to $E_{\text{p-MCPT}}^{(0)}$, which is unaffected by averaging. At second order, determinants interacting with any determinant in $|0\rangle$ contribute to Eq. (17). This is not the case in the original formulation, where only determinants interacting with the Fermi-vacuum may appear when summing for K in Eq. (11).

Before proceeding to applications, let us discuss another important aspect of averaged second-order energies Eqs. (15) and (17). Both formulae are of multi-partitioning nature: the zero-order operator varies with principal determinant $|K\rangle$. This affects not only the bi-orthogonal vector set, but also the zero-order excitation energies. The latter fact may become a disadvantage, since it may enhance sensitivity to intruders. If there is just one pair of determinants appearing in $|0\rangle$ which are close in energy at zero order, it is going to produce an almost zero denominator, when either of them is taken as $|HF\rangle$. An extreme example is a pair of open-shell determinants, e.g., $|K_1\rangle = |j_\beta i_\alpha \ldots\rangle$ and $|K_2\rangle = |j_\alpha i_\beta \ldots\rangle$ which are exactly degenerate in DK partitioning. This particular problem may be solved by turning to a spin-adapted formulation, but it does not provide a solution if there are two or more spin functions belonging to a given multiplicity.

To suppress intruders, one may- e.g.-turn to intermediate Hamitonian theory, where multi-partitioning has been applied successfully as a remedy [21]. Quasi-degeneracy may also be handled by partial averaging, i.e., omitting the problematic determinant pair from the summation for K in Eqs. (15) and (17). As long as the affected determinants do not become principal weight in the problem considered, partial averaging may be acceptable. However, to apply the theory in the general case, further considerations on intruder avoidance are necessary. We do not address this question in this work. Our aim is only to illustrate the usefulness of averaging, when intruder states do not influence the situation. For this end we restrict ourselves to two-determinantal multiconfiguration reference function, built exclusively of closed shell determinants.

3 Numerical Illustration

A test case showing spectacular failure of MCPT is provided by the torsional potential curve of the ethylene molecule. In equilibrium the system possesses D_{2h} symmetry. Upon rotating the CH_2 groups with respect to each other, the symmetry reduces to D_2. At the top of the barrier (at 90° dihedral angle) the point group becomes D_{2d}, non-Abelian.

The system is computed in Dunning's double-zeta polarized basis [32]. A multiconfigurational reference function is provided by a CAS function with 2-electrons on two orbitals. The Full-CI solution being exclusive due to the large system size, state-selective MRCCSDT[2+2] method [8] was computed. Since this method incorporates full triples, it is highly superior to the second-order PT methods we wish to evaluate, and serves as a good benchmark. Notation 2+2 refers to active indices (two-hole, two-particle) which define a reference space for MRCC.

Inspecting the coefficients squared of the two-determinantal CAS wavefunction as a function of the dihedral angle (Fig. 1), one can see that the two determinants exchange the principal role for dihedral angles smaller and larger than 90°. At 90° the weights of the two determinants become opposite value, and orbital degeneracies build up in the spectrum of the generalized Fockian [Eqs. (8) and (9)], in accordance with the non-Abelian symmetry. In Fig. 2 we present rotational barriers obtained by either p-MCPT or u-MCPT in both EN and DK partitioning. Due to the degeneracy problem, Eq. (10) was applied when computing the generalized Fockian to produce orbital energies for DK partitioning. In Fig. 2 the Fermi-vacuum choice follows

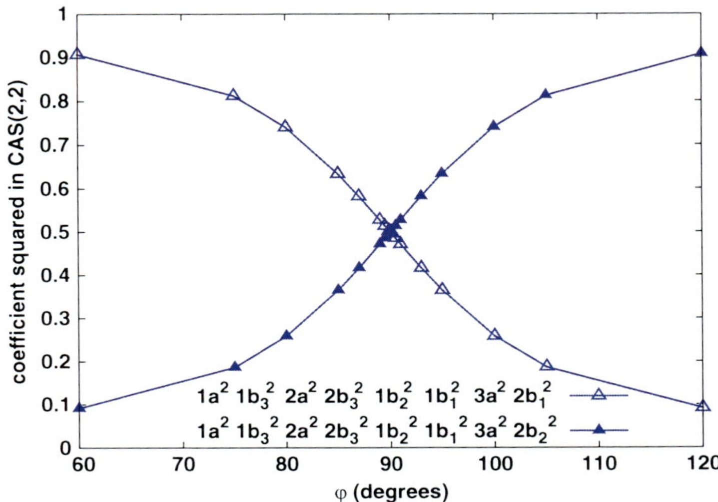

Fig. 1 Coefficient squared of the determinants constituting the CAS(2,2) function, for the C_2H_4 molecule, in double zeta-polarized basis set

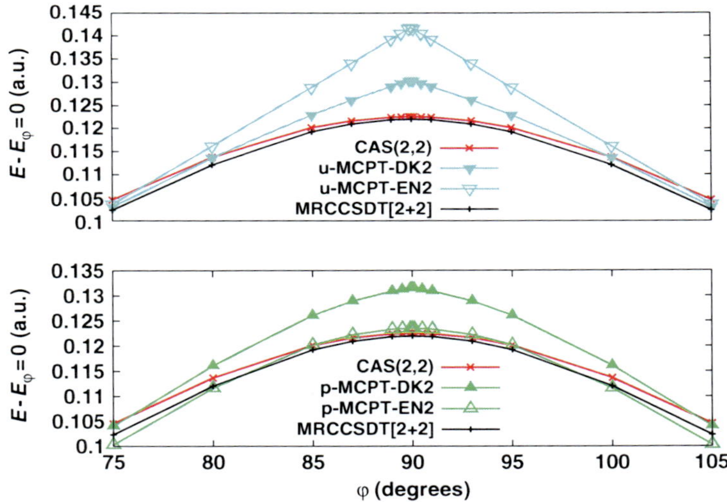

Fig. 2 Rotational barrier top as the CH_2 moieties are twisted in the ethylene molecule. Methods applied are CAS(2,2) and subsequent Fermi-vacuum-dependent perturbative corrections Eqs. (11) or (7), in two partitionings. The curve by MRCCSDT$[2+2]$ is shown for reference

the principal determinant, hence it is changed at 90°. Figure 2 does not indicate ill-behavior of p-MCPT neither in EN nor in DK partitioning. The same holds for MRCCSDT[2+2] which – in principle – also depends on the Fermi-vacuum choice. In contrast to the above, u-MCPT shows a completely erroneous, cusp-like barrier

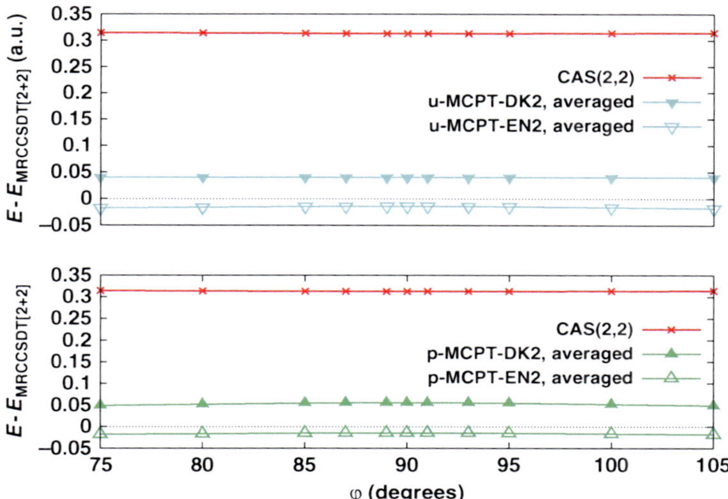

Fig. 3 Energy difference taken with MRCCSDT[2 + 2] around 90° dihedral angle of the ethylene molecule. Methods applied are CAS(2,2) and subsequent Fermi-vacuum-dependent perturbative corrections, averaged according to Eqs. (15) or (17). Abbreviations DK and EN refer to the partitioning (see text)

top. The source of this phenomenon is the crossing of two continuous energy curves obtained by one or the other determinant taken as Fermi-vacuum. The crossing occurs in p-MCPT as well, but in u-MCPT the crossing angle is considerably sharp as the derivatives with respect to the dihedral angle differ a lot from zero at 90°. In p-MCPT, the crossing at 90° remains unnoticed in Fig. 2.

To show the improvement in averaged theories, difference with MRCCSDT[2+2] energy is plotted in Fig. 3. The straight curves obtained for all averaged theories indicate the removal of the cusp-like crossing at 90°. The usual behavior of EN partitioning overshooting the exact energy is also probable in this case, based on Fig. 3.

Comparing the values for the rotational barrier collected in Table 2, one sees that neither of the averaged MCPT methods manage to improve the barrier of CAS(2,2), which fortuitously shows extraordinary accuracy. It is only u-MCPT in DK parti-

Table 2 Barrier height of the ethylene molecule in DZP basis, obtained by the CAS(2,2) method and subsequent Fermi-vacuum-dependent perturbative corrections, averaged according to Eqs. (15) or (17). Abbreviations DK and EN refer to the partitioning. Method MRCCSDT[2 + 2] serves as a basis of comparison

	Barrier (Hartree)
CAS(2,2)	0.1226
p-MCPT-DK2, averaged	0.1387
p-MCPT-EN2, averaged	0.1313
u-MCPT-DK2, averaged	0.1226
u-MCPT-EN2, averaged	0.1313
MRCCSDT[2+2]	0.1220

tioning which gives a barrier of same accuracy at second order. However, MCPT also improves cca. 250 milliHartee in total energy error, according to Fig. 3. This altogether means that averaged u-MCPT in DK partitioning at second order does represent an improvement over the CAS(2,2) wavefunction in this system.

Acknowledgement This work was supported by the Hungarian National Grant Agency OTKA, grant number NI67702 and by the Indo-Hungarian project IND 04/2006. The authors appreciate discussions with Professor D. Mukherjee (Kolkata, India).

References

1. J. Silverstone, O. Sinanoğlu, J. Chem. Phys. **44**, 3608 (1966)
2. B. Jeziorski, H.J. Monkhorst, Phys. Rev. A **24**, 1668 (1981)
3. J. Paldus, in *Methods in Computational Molecular Physics*, Vol. 293, ed. by S. Wilson, G. Dierksen (Plenum, New York, 1992), pp. 99–104
4. X. Li, J. Paldus, J. Chem. Phys. **120**, 5890 (2004)
5. N. Bera, S. Gosh, D. Mukherjee, J. Phys. Chem. A **109**, 11462 (2005)
6. N. Oliphant, L. Adamowicz, J. Chem. Phys. **94**, 1229 (1991)
7. P. Piecuch, N. Oliphant, L. Adamowicz, J. Chem. Phys. **99**, 1875 (1993)
8. M. Kállay, P. Szalay, P. Surján, J. Chem. Phys. **117**, 980 (2002)
9. P. Piecuch, M. Włoch, J.R. Gour, A. Kinal, Chem. Phys. Lett. **418**, 467 (2006)
10. P. Piecuch, M. Włoch, J. Chem. Phys. **123**, 224105 (2005)
11. P. Ghosh, S. Chattopadhyay, D. Jana, D. Mukherjee, Int. J. Mol. Sci. **3**, 733 (2002)
12. D. Hegarty, M.A. Robb, Mol. Phys. **37**, 1455 (1979)
13. G. Hose, U. Kaldor, J. Phys. B **12**, 3827 (1979)
14. Y. Khait, J. Song, M.R. Hoffmann, J. Chem. Phys. **117**, 4133 (2002)
15. K. Wolinski, H. Sellers, P. Pulay, Chem. Phys. Lett. **140**, 225 (1987)
16. K. Wolinski, P. Pulay, J. Chem. Phys. **90**, 3647 (1989)
17. P. Durand, J. Malrieu, in *Ab Initio Methods in Quantum Chemistry*, Vol. 2, ed. by K. Lawley (Wiley, New York, 1987), pp. 321–104
18. K. Andersson, P.Å. Malmqvist, B.O. Roos, A.J. Sadlej, K. Wolinski, J. Phys. Chem. **94**, 5483 (1990)
19. K. Andersson, P.Å. Malmqvist, B.O. Roos, J. Chem. Phys. **96**, 1218 (1992)
20. K. Hirao, Chem. Phys. Lett. **201**, 59 (1993)
21. A. Zaitevskii, J.P. Malrieu, Theor. Chim. Acta **96**, 269 (1997)
22. Z. Rolik, Á. Szabados, P.R. Surján, J. Chem. Phys. **119**, 1922 (2003)
23. Á. Szabados, Z. Rolik, G. Tóth, P.R. Surján, J. Chem. Phys. **122**, 114104 (2005)
24. M. Kállay, P.G. Szalay, P.R. Surján, J. Chem. Phys. **117**, 980 (2002)
25. D.I. Lyakh, V.V. Ivanov, L. Adamowicz, J. Chem. Phys. **128**, 074101 (2008)
26. E.R. Davidson, J. Chem. Phys. **57**, 1999 (1972)
27. E. Kapuy, F. Bartha, F. Bogár, C. Kozmutza, Theor. Chim. Acta **72**, 337 (1987)
28. E. Kapuy, F. Bartha, F. Bogár, Z. Csépes, C. Kozmutza, Int. J. Quantum Chem. **37**, 139 (1990)
29. J. Pipek, F. Bogár, Top. Curr. Chem. **203**, 43 (1999)
30. P. Surján, Z. Rolik, Á. Szabados, D. KHohalmi, Ann. Phys. (Leipzig) **13**, 223 (2004)
31. R. McWeeny, *Methods of Molecular Quantum Mechanics* (Academic, London, 1989)
32. T.H. Dunning Jr., J. Chem. Phys. **53**, 2829 (1970)

On the Wave Function of Coulson and Fischer:
A Third Way in Quantum Chemistry

Stephen Wilson

Abstract The wave function of Coulson and Fischer is examined within the context of recent developments in quantum chemistry. It is argued that the Coulson–Fischer ansatz establishes a 'third way' in quantum chemistry, which should not be confused with the traditional molecular orbital and valence bond formalisms. The Coulson–Fischer theory is compared with 'modern' valence bond approaches and also modern multireference correlation methods. Because of the non-orthogonality problem which arises when wave functions are constructed from arbitrary orbital products, the application of the Coulson–Fischer method to larger molecules necessitates the introduction of approximation schemes. It is shown that the use of hierarchical orthogonality restrictions has advantages, combining a picture of molecular electronic structure which is an accord with simple, but nevertheless empirical, ideas and concepts, with a level of computational complexity which renders practical applications to larger molecules tractable. An open collaborative virtual environment is proposed to foster further development.

Keywords: Coulson–Fischer wave function · Coulson–Fischer analysis · Coulson–Fischer theory · 'Modern' valence bond theory · Multireference correlation problem · Collaborative virtual environment

1 Preamble

In the proceedings of the *Quantum Systems in Chemistry & Physics XII* workshop [1], Wilson and Hubač [2] described "A Collaborative Virtual Environment for Molecular Electronic Structure Theory" involving eight scientists from six countries which was created in order to develop many-body methods based on Brillouin–Wigner theory under the auspices of the EU COST programme.

S. Wilson (✉)
Physical and Theoretical Chemistry Laboratory, University of Oxford, South Parks Road, Oxford OX1 3QZ, England
e-mail: quantumsystems@gmail.com

P. Piecuch et al. (eds.), *Advances in the Theory of Atomic and Molecular Systems*, Progress in Theoretical Chemistry and Physics 19, DOI 10.1007/978-90-481-2596-8_13, © Springer Science+Business Media B.V. 2009

In this chapter, we propose the creation of a new collaborative virtual environment for the development of the Coulson–Fischer method for molecular wave functions. It is proposed that this environment should be open. This chapter gives some background to the project.

2 Introduction

Fifty years ago, in 1959, Mulliken and Roothaan [3] began a review of some of the then new developments in molecular quantum mechanics by recalling that

> Dirac once stated that, in principle, the whole of chemistry is implicit in the laws of quantum mechanics [4]. In other words, quantum mechanics offers the possibility that all quantities of chemical interest – the sizes, shapes, and energies of molecules in their ground states and in activated states, and their electric, magnetic, and thermodynamic properties – may eventually be computed purely theoretically.

The prospects for practical computational quantum chemistry had changed radically in the mid-twentieth century with the advent of the electronic computer. The Mulliken–Roothaan review was concerned with "broken bottlenecks" and, in particular, the then recently developed techniques for the automated evaluation of molecular integrals. Although their review was mainly concerned with integrals over Slater-type orbitals,[1] Mulliken and Roothaan clearly recognized the potential of "machine calculations":

> It can now be predicted with confidence that machine calculations will lead gradually toward a really fundamental quantitative understanding of the rules of valence and the exceptions to these; toward a real understanding of the dimensions and detailed structures, force constants, dipole moments, ionization potentials, and other properties of stable molecules and equally unstable radicals, anions, and cations, and chemical reaction intermediates; toward a basic understanding of activated states in chemical reactions, and of triplet and other excited states which are important in combustion and explosion processes and in photochemistry and in radiation chemistry; and also of intermolecular forces; further, of the structure and stability of metals and other solids; of those parts of molecular wave functions which are important in nuclear magnetic resonance, nuclear quadrupole coupling, and other interaction involving electrons and nuclei; and of very many other aspects of the structure of matter which are now understood only qualitatively or semi-empirically.

However, the first electronic digital computers were too slow and had too small a memory to realize the dream of substituting the computer for the chemistry laboratory. It was only in the late 1970s and early 1980s when computers with the power to make significant progress began to appear.[2]

---

[1] The now ubiquitous Gaussian-type functions had been introduced into quantum chemistry in 1950 by Boys [5] and independently by McWeeny [6], but the advantages of these functions had not been widely recognized during the 1950s.

[2] The volume *Chemistry by Computer* [7] provides an overview of the applications of computers in chemistry in the mid-1980s.

The past 30 or 40 years have witnessed a relentless increase in the power of computing machines. It has been observed [8] that the processing power of computers seems to double every eighteen months. This has been dubbed "Moore's Law" after G.E. Moore, one of the co-founders of Intel. As the historian J.M. Roberts points out [9]

> No other technology has ever improved so rapidly for so long.

Moore's Law, coupled with advances in theory and computational algorithms, has led to computational quantum chemistry becoming increasingly competitive amongst the methods available for studying the structure and behaviour of matter on a molecular scale.

It should be emphasized that modern science has a battery of complementary probes of matter which are increasingly being applied in a 'problem-based,' as opposed to 'technique-based,' approach to a given system, each technique providing a different perspective. Different techniques give complementary information about the studied system; the sum of the information gained by exploiting different methodologies is often greater than that obtained by one technique alone. Furthermore, the complexity inherent in many new forms of matter suggests that single probe is unlikely to provide a complete understanding of a given problem. Complexity carries with it the need for complementary probes. Computational quantum chemistry provides a key probe of matter in the modern research environment as Mulliken and Roothaan predicted in their 1959 paper [3].

In contemporary quantum chemical research, two conceptually different approaches can be recognized in the application of ab initio methodology. In their well-known text *Ab initio Molecular Orbital Theory*, Pople and his co-authors [10] describe the first approach as that in which

> each problem is examined at the highest level of theory currently feasible for a system of its size

In the second approach identified by Pople and his co-authors

> a level of theory is first clearly defined, after which it is applied *uniformly* to molecular systems of *all* sizes up to the maximum determined by available computational resources.

They continue

> Such a theory, if prescribed uniquely for any configuration of the nuclei and any number of electrons, may be termed a *theoretical model* within which all structures, energies, and other physical properties can be explored once the mathematical procedure has been implemented through a computer program.

This second approach leads to what Pople and his co-workers term a "theoretical model chemistry." In practical applications, the complete basis set limit for full configuration interaction cannot be achieved with finite computing resources, except for the very smallest systems. Compromises have to be made in order to achieve a wide range of applicability. Geometry optimization may, for example, be carried out with some 'lower level' theory and/or basis set of 'moderate' size followed by 'more accurate' calculations using 'higher level' theory and/or an 'extended' basis set at the optimized geometry. Obviously, this approach introduces a degree of empiricism into an otherwise ab initio calculation, but such procedures can lead to an accuracy which could not otherwise be achieved. The G3 composite theoretical model chemistry, described by Pople et al. [11, 12], is a typical procedure of this type. Systematic comparison of the results, supported by a given model, with corresponding data derived from experiment may give a model a predictive capability in situations where experiment is difficult or impossible, or simply too expensive. It is this second approach, based on Pople's concept of theoretical model chemistries, that has become preeminent in applied computational quantum chemistry today and the 1998 Nobel Prize for Chemistry was shared by Sir John Pople FRS *for his development of computational methods in quantum chemistry* [13].

Today, there is a 'standard' ab initio model of molecular electronic structure. This consists of an independent particle model followed by a many-body description of correlation effects.[3] The Hartree–Fock molecular orbital model is almost invariably employed as the first stage. The most widely used correlation method [14] is second-order many-body perturbation theory, that is, the "MP2" method. Higher accuracy is often pursued by combining coupled cluster theory with a perturbative description of the "triple excitation" component of the correlation energy. This is the hybrid "CCSD(T)" method. These single-reference methods are robust and implemented in a wide range of quantum chemical computer packages, such as GAMESS [15, 16] and GAUSSIAN [17].[4]

For problems such as molecular dissociation requiring the use of a multireference formalism, the methods developed over the past 40 years are not yet widely accepted as robust and reliable. The importance of using a "many-body" formalism is recognized – that is, a formalism in which the energy scales linearly with the number of electrons in the system being studied. In practice, it is found that multireference many-body formalisms suffer from a number of problems, the most demanding of which is that associated with the so-called intruder states. These difficulties suggest that it is time to re-examine some of the basic models employed in quantum chemical studies.

The purpose of this essay is to examine the Coulson–Fischer wave function [18] and the approach of Coulson and Fischer from the perspective of contemporary

---

[3] For overviews of molecular electronic structure theory see, for example,

1. R. McWeeny, *Methods of Molecular Quantum Mechanics*, 2nd edition, Academic Press, London (1992)
2. T. Helgaker, P. Jørgensen and J. Olsen, *Molecular Electronic Structure Theory*, John Wiley, Chichester (2000)
3. S. Wilson, P.F. Bernath and R. McWeeny, *Handbook of Molecular Physics and Quantum Chemistry*, volume 2: *Molecular Electronic Structure*, John Wiley, Chichester (2003)
4. S. Wilson, *Electron Correlation in Molecules*, Dover, New York (2007).

[4] See the author's recent report to the Specialist Periodical Reports series *Chemical Modelling: Applications and Theory* [14] for an overview of many of the quantum chemical computer packages currently available.

quantum chemistry. Some background to the current work is given in Section 2. Section 3 recalls the fundamental ingredients of the Coulson–Fischer notes. In Sections 4 and 5, we briefly survey modern valence bond theory and the multireference correlation problem, respectively, before, in Section 6 suggesting that the Coulson–Fischer approach provides a third way in quantum chemistry which combines the advantages of molecular orbital and valence bond theories whilst avoiding many of their weaknesses. In the final section, Section 8, a collaborative virtual environment for the study of Coulson–Fischer theory is proposed. It is suggested that an *open* environment of this type could facilitate further development of Coulson–Fischer methodology as it has the potential to involve groups and individuals located in geographically distributed sites.

3 Background

Soon after the first application of quantum mechanics to the description of the structure of molecules by Heitler and London [19] in 1927 (see also ref. [20]), two rival theories emerged, which, although recognized by their proponents as equivalent when sufficiently refined, provided markedly different pictures of the electronic structure of molecular systems. The molecular orbital theory, which was proposed by Hund [21–23] and by Mulliken [24–27], built on the atomic model used in spectroscopy. The valence bond theory was proposed by Pauling [28–31] and by Slater [32–34], who constructed the molecular wave function from atomic components. Pauling regarded valence bond theory as the quantum mechanical realization of the theory of valency published by Lewis in 1916 [35]. Although Slater [34], in particular, and Van Vleck and Sherman [36] recognized the equivalence of the two theories when refined, rival schools quickly emerged. The rivalry continues, to some extent, to the present time. As recently as 2003, *Accounts of Chemical Research* published [37] a *Conversation on VB vs MO Theory* between the Nobel Laureate Roald Hoffmann and two advocates of 'modern' valence bond methods, Sason Shaik and Philippe Hiberty. These latter authors suggest *A Never-Ending Rivalry?* [38].

In his book, *A Terrible Beauty: The People and Ideas that Shaped the Modern Mind* [39], which was published in 2000, Peter Watson writes

> The greatest breakthrough in theoretical chemistry in the twentieth century was achieved by one man, Linus Pauling, whose idea about the nature of the chemical bond was as fundamental as the gene and the quantum because it showed how physics governed molecular structure and how that structure was related to the properties, and even the appearance, of the chemical elements. Pauling explained the logic of why some substances were yellow liquids, others white powders, still others red solids. The physicist Max Perutz's verdict was that Pauling's work transformed chemistry into 'something to be understood and not just memorised' [40]

This quotation demonstrates the influence that valence bond theory and its principal proponent Linus Pauling had in chemistry and the molecular sciences.

In 2003, Hoffmann writes [37]

> I think one reason chemists believed Pauling (and his apostle/expositor to the organic community, Wheland) is because he not only capitalized on what was already there – the idea of covalent and ionic bonds. But also, and this is largely forgotten, Pauling was not just a theoretician - he was America's premier structural chemist. I would guess that at the time that great Cornell book, 'The Nature of the Chemical Bond', was written, Pauling and his students had done half the crystal structures known. And they made electron diffraction a practical technique. Pauling spoke to chemists of the physical structure of the molecules they care about, and even though he was a theorist, he spoke with unparalleled experimental authority

However, Hoffmann continues [37]

> he [Pauling] ignored MO theory to a degree that was clearly perceived by the community as blind, if not unethical. His interests (and great, great creative powers) also shifted to biological problems.

Molecular orbital theory began to assume the dominant position that it enjoys today with the introduction of the electronic computer. The computational tractability of the matrix Hartree–Fock formalism developed by Roothaan [41] and by Hall [42] in the 1950s, followed by the formulation within the algebraic approximation of "many-body" methods for handling the electron correlation problem in the 1970s [43–45], underpins much of contemporary molecular electronic structure theory and practice. When suitably refined modern molecular orbital-based theory can achieve an accuracy which goes a long way towards realizing the potential of "machine calculations" predicted by Mulliken and Roothaan [3] in 1959.

Valence bond theory can be refined by admitting additional structures. This leads to the approach termed 'multistructure valence bond' theory. On the other hand, molecular orbital theory can be refined by including additional configurations. The 'traditional' approach to the electron correlation problem based on this methodology was called "configuration interaction." ("Many-body" approaches provide an analysis of the configuration interaction expansion which remains useful when a truncated expansion is employed in the description of an extended system [46].) Discussing these two approaches, Hoffmann explains that [37]

> They are equivalent for H_2, i.e. VB + ionic structures = MO + configuration interaction = different orbitals for different spins, as we teach our students. But go a tad beyond H_2 and they become at the zeroth level – that's the level practicing chemists find theory of use – nonequivalent.

Hoffmann concludes by stressing the need for a balanced approach [37]:

> Taken together, MO and VB theories constitute not an arsenal, but a tool kit, simple gifts from the mind to the hands of chemists. Insistent on a journey through the perfervid bounty of modern chemistry equipped with one set of tools and not the other puts one at a disadvantage. Discarding any one of the two theories undermines the intellectual heritage of chemistry.

In the following section, we turn our attention to a third perspective on the molecular electronic structure problem which combines many of the advantages of MO and VB theories whilst avoiding some of their weaknesses.

4 The Coulson–Fischer Notes

Sixty years ago, in 1949, Coulson and Fischer published a seminal paper [18] in the *Philosophical Magazine*, entitled *Notes on the Molecular Orbital Treatment of the Hydrogen Molecule*. In this note, they presented a wave function for the hydrogen molecule, which, whilst retaining a simple physical picture, combines the advantages of the two rival theories of molecular electronic structure, MO and VB theories.

Let us briefly summarize the discussion given by Coulson and Fischer.

They consider two forms of an approximate wave function for the ground state of the hydrogen molecule which depend on a single parameter. These wave functions correspond to the VB and MO wave function for a particular choice of the parameter. The first wave function considered by Coulson and Fischer is written as

$$\psi = N\{\psi_{cov} + k\psi_{ion}\}, \tag{1}$$

where ψ_{cov} is the covalent function

$$\psi_{cov} = \phi_a(1)\phi_b(2) + \phi_b(1)\phi_a(2) \tag{2}$$

and ψ_{ion} is the ionic function

$$\psi_{ion} = \phi_a(1)\phi_a(2) + \phi_b(1)\phi_b(2). \tag{3}$$

The factor N in Eq. (1) is a normalizing factor.

The second form of the approximate wave functions considered by Coulson and Fischer is

$$\psi = N'\left\{(\sigma_{1s})^2 - \mu\left(\sigma_{1s}^*\right)^2\right\}, \tag{4}$$

where σ_{1s}

$$\sigma_{1s} = \phi_a + \phi_b \tag{5}$$

and σ_{1s}^*

$$\sigma_{1s}^* = \phi_a - \phi_b. \tag{6}$$

Again the factor N' in Eq. (4) is a normalization factor.

In the above approximate wave functions ϕ_a and ϕ_b are normalized $1s$-atomic orbitals centred on the two nuclei a and b and have the form

$$\phi_a = \sqrt{\frac{\zeta}{\pi}} \exp\{-\zeta r_a\}, \tag{7}$$

and

$$\phi_b = \sqrt{\frac{\zeta}{\pi}} \exp\{-\zeta r_b\}. \tag{8}$$

ζ is the usual screening constant.

Wave function (1) is the standard covalent-ionic resonance of VB theory. The parameter k can take values from $k = 0$, which corresponds to the 'pure covalent' description of the hydrogen molecule ground state (this is the wave function first considered by Wang [47] when he introduced a variable screening constant into the Heitler-London wave function), through to $k = \infty$, which corresponds to a "pure ionic" description. The 'best' wave function of the form (1), determined by invoking the variation theorem to determine the optimal screening constant, was first reported by Weinbaum [48] in 1933.

Wave function (4) is the standard configuration interaction expansion in MO theory. The parameter μ can take values from -1 to $+1$. Putting $\mu = 0$ gives the 'pure molecular orbital' description first considered by Coulson [49] in 1937. Table 1 summarizes the behaviour of the approximate wave functions as a function of the parameters k and μ in the Coulson–Fischer analysis. (This table is taken from the work of Coulson and Luz [50].)

Table 1 Parameters in the Coulson–Fischer analysis[a]

k	0	0.26	1.0	∞
μ	1	0.59	0	-1
description	Pure covalent	'Best' function	Pure orbital	Molecular pure ionic
	Over-correlated		Under-correlated	Under-correlated
ζ	1.165	1.193	1.195	1.065

[a] Taken from the work of Coulson and Luz [50] "A Note on Electron Correlation in the Hydrogen Molecule."

Coulson and Fischer demonstrated that the two forms of the approximate wave function, (1) and (4), are equivalent if

$$k = \frac{1 - \mu}{1 + \mu}. \tag{9}$$

Substituting (5) and (6) into the right-hand side of (4) gives (neglecting the normalization factor)

$$(\phi_a(1) + \phi_b(1))(\phi_a(2) + \phi_b(2)) - \mu(\phi_a(1) - \phi_b(1))(\phi_a(2) - \phi_b(2)). \tag{10}$$

Rearranging this equation gives

$$(1 - \mu)(\phi_a(1)\phi_a(2) + \phi_b(1)\phi_b(2)) + (1 + \mu)(\phi_a(1)\phi_b(2) + \phi_b(1)\phi_a(2)).$$
(11)

Using the functions (2) and (3), the wave function (4) can then be written as

$$\psi = (1 + \mu)\psi_{cov} + (1 - \mu)\psi_{ion}.$$
(12)

The relation (9) is evident by comparing (1) and (12).

Coulson and Fischer also considered a third approximate wave function which has the form

$$\psi = N''(\varphi_a(1)\varphi_b(2) + \varphi_b(1)\varphi_a(2)),$$
(13)

in which the orbitals are written in the form

$$\varphi_a(1) = \phi_a(1) + \lambda\phi_b(1)$$
(14)

and

$$\varphi_b(2) = \phi_b(2) + \lambda\phi_a(2)$$
(15)

with the orbitals φ_a and φ_b related by a reflection in the plane perpendicular to the internuclear axis and passing through its mid-point, i.e.

$$\varphi_a = \sigma_h\varphi_b.$$
(16)

When $\lambda = 0$, $\varphi_a = \phi_a$ and $\varphi_b = \phi_b$ so that the wave function (13) immediately becomes equal to ψ_{cov}. The wave function (13) is equivalent to (1) with k set to zero. When $\lambda = 1$, $\varphi_a = \varphi_b$ and $\varphi_a = \sigma_{1s}$ so that (13) is equivalent to (4) when $\mu = 0$. Neglecting the normalization factor N'', the right-hand side of the approximate wave function (13) may be written as

$$(\phi_a(1) + \lambda\phi_b(1))(\phi_b(2) + \lambda\phi_a(2))(\phi_b(1) + \lambda\phi_a(1))(\phi_a(2) + \lambda\phi_b(2)),$$
(17)

which can be rearranged to give

$$\left(1 + \lambda^2\right)(\phi_a(1)\phi_b(2)\phi_b(1)\phi_a(2)) + 2\lambda(\phi_a(1)\phi_a(2) + \phi_b(1)\phi_b(2)).$$
(18)

Comparing the second line of (18) with (2) and (3), we see that (13) may be written as

$$\psi = N\left\{\psi_{cov} + \frac{2\lambda}{1 + \lambda^2}\psi_{ion}\right\}.$$
(19)

The approximate wave function (13) is equivalent to the form (1) if

$$k = \frac{2\lambda}{1 + \lambda^2} \tag{20}$$

and thus to the form (4) when

$$\frac{1 - \mu}{1 + \mu} = \frac{2\lambda}{1 + \lambda^2}. \tag{21}$$

The parameters k, μ and λ are related by (9), (20) and (21). Table 2 summarizes the relations between these three parameters.

Table 2 Relations between the parameters k, μ and λ

Parameter	k	μ	λ
k	–	$\frac{1-\mu}{1+\mu}$	$\frac{2\lambda}{1+\lambda^2}$
μ	$\frac{1-k}{1+k}$	–	$\left(\frac{\lambda-1}{\lambda+1}\right)^2$
λ	$\frac{1}{2}k \pm 2\sqrt{1 - k^2}$	$\frac{\mu+2\sqrt{\mu}+1}{1-\mu}$	–

Hence the approximation (1), which is a prototype of multistructure valence bond theory, is equivalent to approximation (4), a prototype of molecular orbital configuration interaction theory, and both (1) and (4) are equivalent to (13), the Coulson–Fischer wave function.

We submit that the Coulson–Fischer ansatz affords a third way in quantum chemistry that is distinct from the traditional valence bond and molecular orbital theories. In the next two sections, we very briefly survey the current state of the art in valence bond theory and in the multireference correlation problem based on the molecular orbital theory, before considering the Coulson–Fischer theory in more detail in Section 6.

5 'Modern' Valence Bond Theory

Equation (1) defines the valence bond theory for the ground state of the hydrogen molecule. McWeeny [51] describes how

Valence bond (VB) theory, in any of the modern ab initio forms now available, is capable of giving an excellent account of both localized and nonlocalized bonding, using wave functions which are compact, accurate, and easily visualized. By representing a molecular wave function as a weighted mixture of 'VB structures', each relating to a classical bonding scheme, it is possible to describe the electronic structure of a molecule in 'chemical' language and also the course of a chemical reaction in terms of the same structures and their changing weights as the reaction proceeds.

In a recent review entitled "Advances in valence bond theory", Karadakov [52] writes

While VB theory will certainly remain a source of highly visual qualitative interpretations of chemical bonding and reactivity which are arguably more versatile than their MO counterparts, from a quantitative viewpoint, with the proliferation of ab initio approaches to molecular electronic structure VB, has been gradually relegated to a somewhat backstage role. The main reasons for this are the much higher computational costs associated with VB calculations, and the fact that VB wave functions often have to be 'hand-tailored' to a particular problem which makes them difficult to use by non-specialists and may introduce an understandable strong dependence of the outcome of a calculation on the construction of the wave function.

In spite of this somewhat pessimistic assessment, Karadakov [52] continues

The rapid developments in computer technology within the last two decades have enabled a number of previously unfeasible VB calculations. In turn, this has stimulated a series of theoretical developments and brought about a resurgence of interest in VB theory, which some authors see as a VB 'renaissance'.

Evidence for this "renaissance" is seen in the number of monographs [53–55], edited volumes [56–58] and review articles [52, 59, 60, 62, 63, 61, 64–69] on valence bond theory published in recent years. These works display a rich variety of theoretical machinery inspired by the valence bond picture of molecular structure. Some of the methodologies – particularly in the so-called modern valence bond theories introduced by Gerratt and Lipscomb [70, 71] under the name "spin-coupled wave functions" and developed by Gerratt [72, 73] and his collaborators [68, 74–86] over the past 40 years – exploit the Coulson–Fischer ansatz. As we have seen in Section 3 and will consider further in Section 6, the Coulson–Fischer theory presents a third way of constructing approximate molecular wave functions which combine many of the advantages of both molecular orbital theory and valence bond theory.

In a paper published in 1953 as part of a series under the general title "The molecular orbital theory of chemical valency," Hurley, Lennard–Jones and Pople [87] presented "A theory of paired electrons in polyatomic molecules." The pair function model of Hurley et al. employed a Coulson–Fischer-type wave function to describe each pair of electrons in a polyatomic molecule. Orthogonality constraints were imposed between orbitals associated with different pairs of electrons in order to render the theory practical, i.e. computationally tractable. Hurley presented the corresponding orbital equations in a subsequent paper [88] which was published in 1956.

By the early 1970s, the Hurley–Lennard–Jones–Pople pair function model had been applied to a range of simple polyatomic molecules, such as water [78], methane [79] and diborane [76]. For example, in Fig. 1 we show two valence orbitals in the pair function description of the ground state of the methane molecule. One of these has the form of a 'distorted sp^3 hybrid' and the other can be described as a 'distorted hydrogen $1s$ function.' The remaining valence orbitals of this system are related to those shown in Fig. 1 by symmetry operations. Goddard and his co-workers, working at Caltech, presented an entirely equivalent theory under the name "generalized valence bond theory" [89]. Code for performing GVB calculations is widely available, for example, in the GAMESS [15, 16] and GAUSSIAN [17] packages.

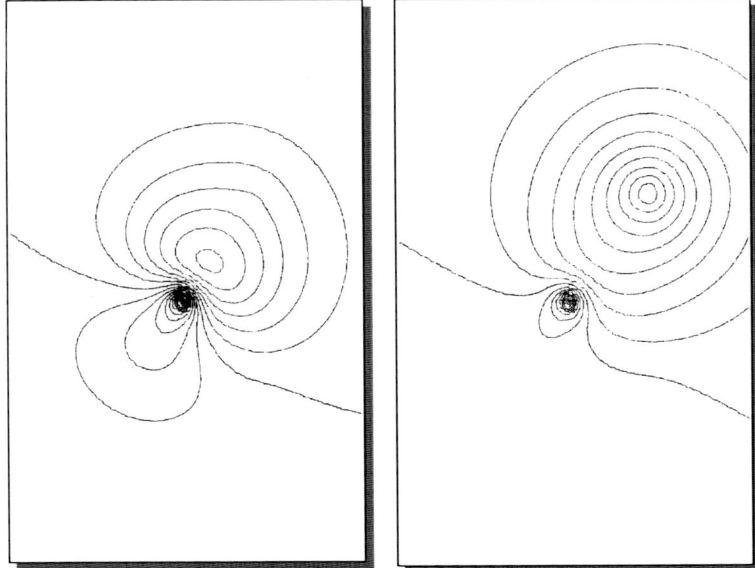

Fig. 1 Amplitude of the generalized Coulson-Fischer orbitals for the methane molecule. See text for details (Taken from [79].)

6 The Multireference Correlation Problem

Equation (4) defines the multireference correlation problem based on a molecular orbital reference function for the ground state of the hydrogen molecule. This approach to the molecular electronic structure problem provides the 'mainstream' theoretical and computational apparatus in use today.

It is not possible in the limited space available here to give a detailed account of the many facets of modern multireference electron correlation theory. This would deviate too far from our current purpose. However, two aspects deserve special emphasis.

The first aspect is the choice of reference function. This is to some extent arbitrary, but is certainly influenced by the physics and chemistry of the problem being studied. Various schemes have been proposed to remove (or at least reduce) the degree of arbitrariness in choosing a reference function and thus make possible a routine approach to the multireference correlation problem comparable to that employed in the single reference function case. Amongst early schemes, the *optimized double configuration* (ODC) [90] and *optimized valence configuration* (OVC) [91] approaches of Wahl and Das should be mentioned. The ODC method employs a two-configuration function for each bonding pair and thus can be equivalent to the pair function model of Hurley et al. [87]. More recent schemes include the *full orbital reaction space* reference function of Ruedenberg and Sundbarg [92, 93] and the *complete active space* reference function described by Roos, Taylor and Siegbahn [94].

The second aspect warranting special mention is the so-called *intruder state problem*. This problem has been found to plague practical many-body multireference formalisms. The presence of intruder states, which may be unphysical, can impair or even destroy the convergence of many-body, multireference expansions. Hubač and Wilson [95] explain:

> In spite of the success of single reference many-body methods and, in particular, (single reference) many-body perturbation theory, multireference formulations of the many-body problem have been beset by problems for more than twenty years. Multireference Rayleigh-Schrödinger perturbation theory is plagued by the so-called intruder state problem. Multireference Rayleigh-Schrödinger perturbation theory is applied to a manifold of states simultaneously and as the perturbation is switched on states within the reference space may move above some of the lower states in the complementary space. These intruder states can degrade or even destroy the convergence of the perturbation expansion.

Brillouin–Wigner methods offer a promising solution to these difficulties. (See the volume entitled *Brillouin–Wigner methods for many-body systems* [95] for details.)

7 The Coulson–Fischer Function: A Third Way in Quantum Chemistry

Equation (13) defines the Coulson–Fischer function for the ground state of the hydrogen molecule. As we have seen in Section 3, the Coulson–Fischer function is fully equivalent to both the valence bond approximation, Eq. (1), and to the molecular orbital approximation, Eq. (4). The Coulson–Fischer wave function for the hydrogen molecule combines the advantages of the valence bond and molecular formalisms in a function based on a single orbital product. In particular, the Coulson–Fischer wave function provides a description of the hydrogen molecule ground state for all nuclear geometries which is qualitatively correct. It collapses to the Hartree–Fock wave function for the ground state of the helium atom in the united atom limit.

The original Coulson–Fischer wave function for the H_2 molecule can be generalized [77] by approximating each of the space orbitals by an expansion in terms of finite analytical basis functions $\{\chi_k; k = 1, 2, \ldots, n\}$. The algebraic approximation is implemented by means of the linear expansions

$$\varphi_1(1) = \sum_{k=1}^{n} \chi_k c_{k1} \tag{22}$$

and

$$\varphi_2(2) = \sum_{k=1}^{n} \chi_k c_{k2}. \tag{23}$$

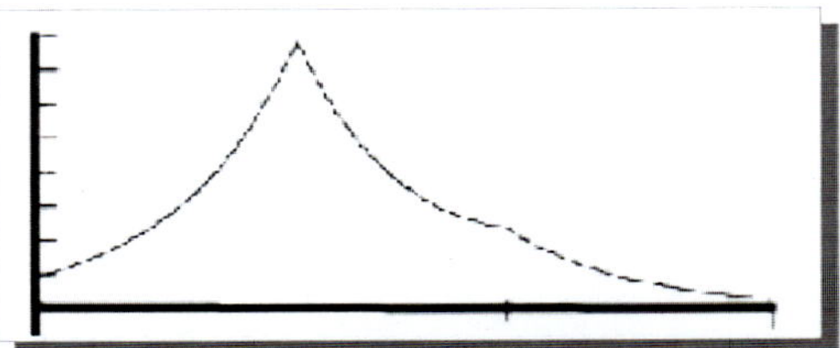

Fig. 2 Amplitude of the Coulson–Fischer orbital for the ground state of the hydrogen molecule at the experimental equilibrium nuclear separation (1.4 bohr) (Taken from [79].)

The molecular wave function is then written as

$$\Psi_{SM} = \sqrt{2!}\,\mathcal{A}\left(\varphi_1 \varphi_2 \Theta\right), \tag{24}$$

where $\mathcal{A}$ is the idempotent antisymmetrizer and $\Theta = \frac{1}{\sqrt{2}}\left(\alpha\left(1\right)\beta\left(2\right) - \alpha\left(2\right)\beta\left(1\right)\right)$ is the two-electron singlet spin function. In practice, the basis functions χ_k will most often be taken to be Gaussian-type functions. The orbital expansion coefficients $c_{k1}, k = 1, 2, \ldots, n$ and $c_{k2}, k = 1, 2, \ldots, n$ are determined by invoking the variation principle within the algebraic approximation [77] which leads to a set of orbital equations. The orbitals obtained by solutions of these equations are shown in Figs. 2 and 3. They can be described a 'distorted hydrogen $1s$ functions.'

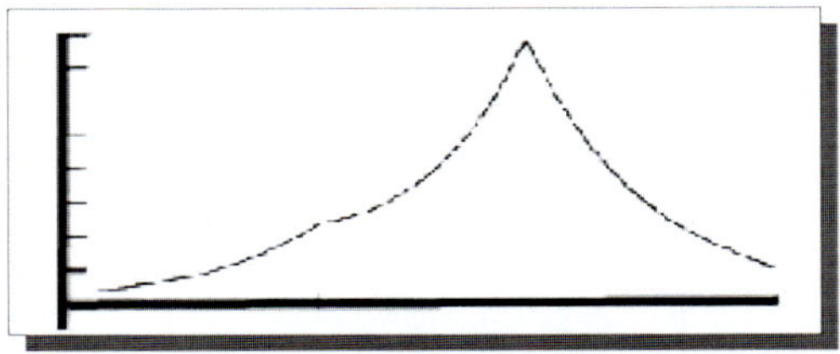

Fig. 3 Amplitude of the second Coulson–Fischer orbital for the ground state of the hydrogen molecule at the equilibrium nuclear separation. This orbital is obtained from the orbital shown in Fig. 1 by reflection in the plane perpendicular to the internuclear axis and passing through its mid-point (Taken from [79].)

The 'generalized' Coulson–Fischer wave function combines "conceptual simplicity" [77] with "results of remarkable accuracy" [77]. It recovers over $\sim 87.1\%$ of the binding energy of the hydrogen molecule ground state which should be compared with the $\sim 76.6\%$ of the binding energy supported by the single configuration molecular orbital function [96]. In Fig. 3, potential energy curves for the hydrogen molecule obtained from (a) the Hartree–Fock function, (b) the ODC multiconfiguration self-consistent field function, (c) the generalized Coulson–Fischer function [77] and (d) the extended James–Coolidge function of Kołos and Roothaan [96] are shown. The 'generalized' Coulson–Fischer approach yields a fundamental vibrational wave number of 4374.2 cm^{-1} which differs by 26.2 cm^{-1} from the

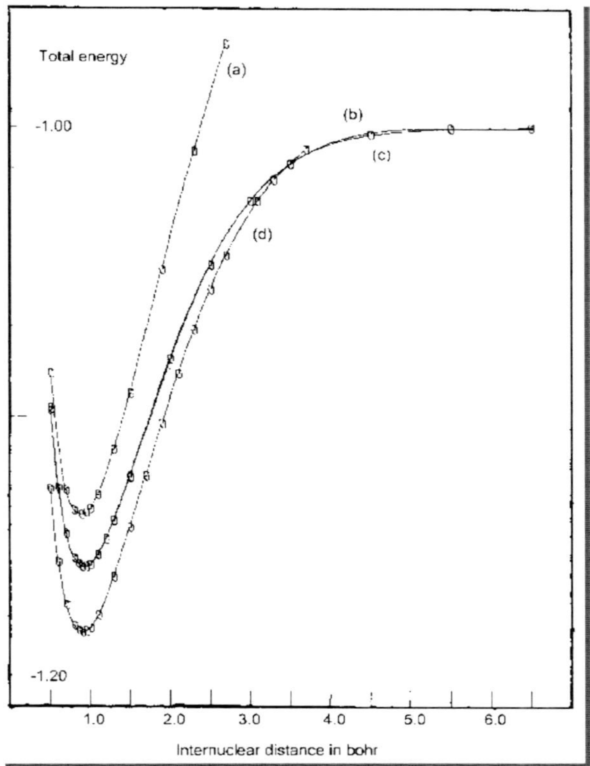

Fig. 4 Potential energy curves for the hydrogen molecule obtained from (**a**) the Hartree–Fock function [96], (**b**) the ODC multiconfiguration self-consistent field function, (**c**) the generalized Coulson–Fischer function [77] and (**d**) the extended James-Coolidge function of Kołos and Roothaan [96] (Taken from [79].)

experimental value (4400.4 cm^{-1}). The analytical molecular orbital function of Kolos and Roothaan [96] supports a fundamental vibrational wave number of 4585 cm^{-1} which differs from the experimental value by 185 cm^{-1}. The amplitude of the generalized Coulson–Fischer orbital for the hydrogen molecule at various nuclear separations is shown in Fig. 5.

We now turn our attention to the problem of constructing approximate molecular wave functions based on a single product of (spatial) orbitals for which orthogonality is not assumed. We shall require that the coupling scheme for the spin functions can be specified at will.

Taking the nonrelativistic Born–Oppenheimer Hamiltonian in which the nuclei are in fixed positions and in which there are no spin operators, we can write the exact stationary state wave function in the form [97]

$$\Psi\left(\mathbf{x}_1, \mathbf{x}_2, \ldots, \mathbf{x}_N\right) = \left(f_S^N\right)^{\frac{1}{2}} \sum_{\kappa=1}^{f_S^N} \Phi_\kappa\left(\mathbf{r}_1, \mathbf{r}_2, \ldots, \mathbf{r}_N\right) \Theta_\kappa\left(\sigma_1, \sigma_2, \ldots, \sigma_N\right), \quad (25)$$

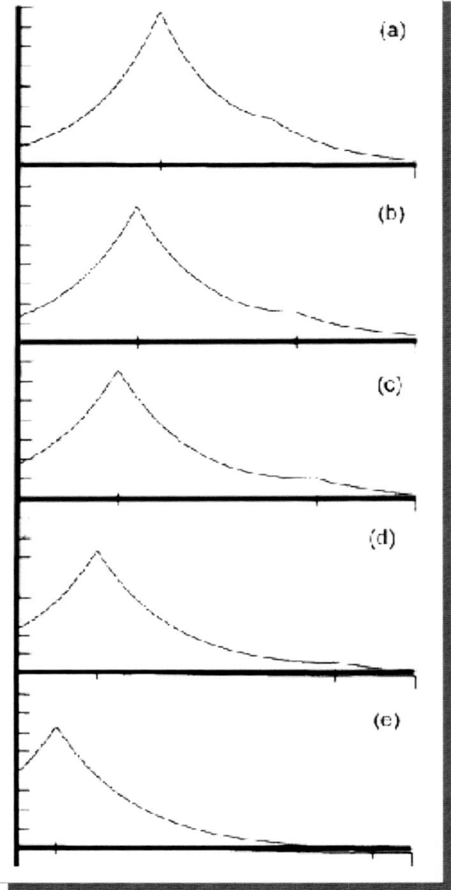

Fig. 5 Amplitude of the generalized Coulson–Fischer orbital for the hydrogen molecule at various nuclear separations: (**a**) $R = 1.4\ a_0$; (**b**) $R = 2.0\ a_0$; (**c**) $R = 2.5\ a_0$; (**d**) $R = 3.0\ a_0$; (**e**) $R = 4.0\ a_0$ (Taken from [79].)

where $\mathbf{x}_1, \mathbf{x}_2, \ldots, \mathbf{x}_N$ are the space–spin coordinates of the N electrons in the molecule. $\mathbf{r}_1, \mathbf{r}_2, \ldots, \mathbf{r}_N$ and $\sigma_1, \sigma_2, \ldots, \sigma_N$ are the corresponding space and spin coordinates, respectively. The space functions Φ_κ are degenerate eigenfunctions of the Hamiltonian operator, $\mathcal{H}$. The spin functions Θ_κ have the same spin eigenvalues, S and M.

Both the space functions, Φ_κ, and spin functions, Θ_κ, provide an irreducible representation of the group of $N!$ permutations of the indices labelling the electrons.

Applying the permutation operator P^σ, which acts on the spin variables, to the spin function Θ_κ, we obtain a linear combination of spin functions, that is

$$P^\sigma \Theta_\kappa = \sum_{\ell=1}^{f_S^N} \Theta_\ell V^S (P^\sigma)_{\ell,\kappa} . \qquad (26)$$

The matrices $\mathbf{V}^S(P)$ with elements $V^S(P)_{\ell,\kappa}$ form a matrix representation of the symmetric group characterized by the total spin and independent of its z-component.[5]

In a similar fashion, applying the permutation operator P^r to the space variables, the space functions Φ_κ yield a linear combination of space functions, that is

$$P^r \Phi_\kappa = \sum_{\ell=1}^{f_S^N} \Phi_\ell U^S \left(P^r\right)_{\ell,\kappa}. \tag{27}$$

Thus the space functions Φ_κ also form a basis for an irreducible representation of the symmetric group. The matrices $\mathbf{U}^S(P)$ with elements $U^S(P)_{\ell,\kappa}$ form a representation of the symmetric group.[6]

The space functions and the spin functions form bases for mutually dual irreducible representations of the symmetric group. The Pauli principle[7] requires that the total electronic wave function is antisymmetric with respect to the simultaneous permututation of space and spin coordinates of the electrons, that is

$$P\Psi = \epsilon_P \Psi, \tag{28}$$

where $\epsilon = +1$ for even permutations and $\epsilon = -1$ for odd permutations. We can write

$$P = P^r P^\sigma = \epsilon_P = \pm 1 \tag{29}$$

and

$$\mathbf{U}^S\left(P^r\right)\mathbf{V}^S(P^\sigma) = \pm\mathbf{I}, \tag{30}$$

where $\mathbf{I}$ is the identity matrix of dimension f_S^N. The dimension of the irreducible representations of the symmetric group arising in the above equations is

$$f_S^N = \frac{(2S+1)\,N!}{\left(\frac{1}{2}N + S + 1\right)!\left(\frac{1}{2}N - S\right)}. \tag{31}$$

---

[5] For a detailed discussion of the properties of spin functions see Chapter 3 in the *Handbook of Molecular Physics and Quantum Chemistry*, volume 2, *Molecular Electronic Structure* by Karadakov [98].

[6] For a detailed discussion of the properties of spatial functions see Chapter 4 in the *Handbook of Molecular Physics and Quantum Chemistry*, volume 2, *Molecular Electronic Structure* by Karadakov [99].

[7] For a detailed discussion of the Pauli principle see Chapter 2 in the *Handbook of Molecular Physics and Quantum Chemistry*, volume 2, *Molecular Electronic Structure* by Kaplan [100].

This is Wigner's number. It is given by the number of linearly independent spin eigenfunctions, i.e. the number of paths on a branching diagram leading to an N-electron total spin S.

We now consider the problem of constructing an acceptable approximate wave function in the form of Eq. (25) from some arbitrary spatial function $\Phi_0 (\mathbf{r}_1, \mathbf{r}_2, \ldots, \mathbf{r}_N)$. In general, Φ_0 has no particular permutational symmetry. A set of functions that forms a basis of an irreducible representation of the permutation group can be generated from Φ_0, can be obtained by applying the Wigner projection operators[8]

$$\omega_{\ell,\kappa}^S = \left(\frac{f_S^N}{N!} \right)^{\frac{1}{2}} \sum_P U_{\ell,\kappa}^S (P) \, P^r, \ \ell, \kappa = 1, 2, \ldots, N, \tag{32}$$

where the summation is over all permutations of the N electrons. For any permutation P^r

$$P^r \left(\omega_{\ell,\kappa}^S \Phi_0 \right) = \sum_{n=1}^{f_S^N} U_{n,\ell}^S (P) \left(\omega_{n,\ell}^S \Phi_0 \right) \tag{33}$$

and the set of functions $\left(\omega_{\ell,\kappa}^S \Phi_0 \right)$, $\ell = 1, 2, \ldots, f_S^N$ form a basis for an irreducible representation of the symmetric group for each value of κ. An approximate molecular wave function corresponding to the spatial function Φ_0 can be written as

$$\Psi_{S,M;\kappa}^0 = \left(\frac{1}{f_S^N} \right)^{\frac{1}{2}} \sum_{\ell=1}^{f_S^N} \left(\omega_{\ell,\kappa}^S \Phi_0 \right) \Theta_{S,M;\ell}^N, \tag{34}$$

where the spin functions $\Theta_{S,M;\ell}^N$ transform according to the representation $\epsilon_P \mathbf{U}^S (P)$ which is the dual of that used to construct the Wigner operators.

After some manipulation[9], Eq. (34) can be put in the form

$$\Psi_{S,M;\kappa}^0 = (N!)^{\frac{1}{2}} \, \mathcal{A} \left(\Phi_0 \Theta_{S,M;\kappa}^N \right), \tag{35}$$

where $\mathcal{A}$ is the antisymmetrizing operator.

In order to develop a Coulson-Fischer approach for an N-electron systems, we assume that the spatial molecular wave function is approximated by a single product of N orbitals

$$\Phi_{0(k_1,k_2,\ldots,k_N)}^N = \varphi_{k_1} (\mathbf{r}_1) \varphi_{k_2} (\mathbf{r}_2) \ldots \varphi_{k_N} (\mathbf{r}_N). \tag{36}$$

---

[8] Gerratt [72] terms the operators given here the "Young operators," after Jahn [101]. The Wigner and Young operators differ by a trivial normalization factor.

[9] For details see Gerratt [72].

This leads to an approximate molecular wave function of the form

$$\Psi_{S,M;\kappa} = (N!)^{\frac{1}{2}} \, \mathcal{A} \left(\Phi^{N}_{0(k_1,k_2,\ldots,k_N)} \Theta^{N}_{S,M;\kappa} \right) \tag{37}$$

which Gerratt [72] termed a "spin-coupled wave function", since the orbitals are coupled according to the particular scheme κ.[10]

Because the orbitals in the approximate spatial wave function (36) are non-orthogonal, the number of terms in expressions for matrix elements increases factorially with the number of orbitals, which, in its simplest form, is equal to the number of electrons in the system under investigation. This is the principal drawback of approximations to the spatial wave function based on Eq. (36).

The imposition of orthogonality restrictions on the orbitals can simplify calculations using Eqs. (36) and (37) considerably. For example, inactive or core orbitals can be taken to be doubly occupied to a good approximation and then, without loss of generality, required to be mutually orthogonal and orthogonal to the valence orbitals [73, 106]. Orbitals may be divided into mutually exclusive groups such that orbitals belonging to different groups are restricted to be orthogonal [79]. The simplest model of this type is the pair function model of Hurley et al. [87] in which the spatial wave function has the form

$$(\varphi_1\varphi_2)(\varphi_3\varphi_4)\ldots(\varphi_{\mu-1}\varphi_\mu)\ldots(\varphi_{N-1}\varphi_N), \tag{38}$$

where only orbitals within the same parentheses are nonorthogonal. The orbitals are said to satisfy the $(\mu - 1, \mu)$-orthogonality condition [76, 78]. In Fig. 5, we illustrate the effects of orthogonality restrictions in the LiH ground state, On the left-hand side of this figure, the description of the LiH ground state, based on a spatial wave function written as a product of four nonorthogonal orbitals, is shown. Two orbitals are associated with the core of the Li atom. One of these extends further from the nucleus than the other. Of the two valence orbitals, one is sp-hybrid-*like*, the other is $1s_H$-*like*. On the right-hand side of Fig. 5, the orbitals corresponding to a wave function, based on a spatial function of the form $\varphi^2_{1s}\,(\varphi_1\varphi_2)$, are displayed. The core is now described by a single doubly occupied orbital. The two valence orbitals are similar to the valence orbitals displayed on the left-hand side of the figure in the chemically important valence region; one is sp-hybrid-*like* and the other is $1s_H$-*like*. Both orbitals contain a node in the chemically less important core region. The introduction of orthogonality constraints can reduce the computational complexity of practical calculations, whilst retaining the form of the valence orbitals in chemically significant regions of space. The equivalent orbital, or localized molecular orbital, associated with the valence region is also shown in the right-hand side of Fig. 5 as a *dashed line*.

---

[10] Gallup and Goddard and their respective collaborators developed methodologies for handling spatial wave functions constructed from a product of nonorthogonal orbitals. The interested reader is referred to their original publications [102–105].

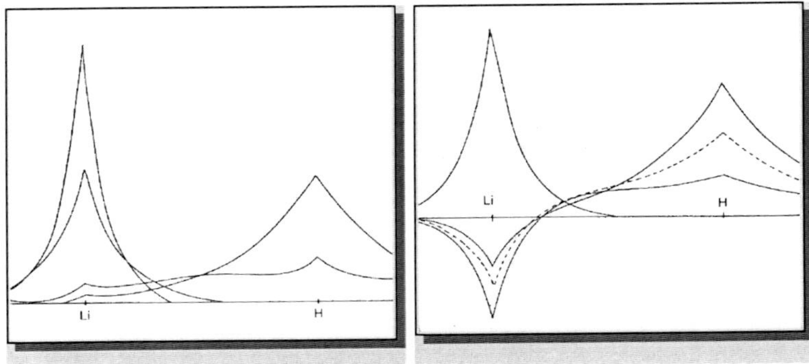

Fig. 6 Amplitude of the generalized Coulson–Fischer orbitals for the lithium hydride molecule. See text for details (Taken from [79].)

In 1999, the author wrote [107]

Recent years have witnessed a growing interest in the development of hierarchical tree methods in describing many-body systems. Hierarchical tree structures provide a systematic scheme for determining the 'closeness' of different particles without explicitly calculating the interaction between them. Barnes and Hut [108] explained that such methods work 'in the same way as humans interact with neighboring individuals, more distant villages and larger states and countries'.

and continued

The introduction of hierarchical orthogonality restrictions seeks to exploit the structure of the molecule under study.

A sequence of models and associated orthogonality restrictions based on the structural formula of the target molecule was introduced [107]. These models interpolate between the pair function model, which involves the most severe orthogonality, restrictions, and the spin-coupled wave function, in which all orbitals are nonorthogonal. For example, the pair function description of the ground state of the BeH_2 molecule with doubly occupied core orbitals, is based on a spatial function with the form

$$\varphi_c^2 \left(\varphi_1 \varphi_2\right) \left(\varphi_3 \varphi_4\right). \tag{39}$$

(Orbitals within the same parentheses are nonorthogonal. Orbitals in different parentheses are orthogonal.) The valence orbitals φ_1 and φ_2 and the core orbital φ_c are shown in Fig. 6. The orbitals φ_3 and φ_4 are related to φ_1 and φ_2 by a reflection operation. One of the orbitals in a given pair is sp-hybrid-*like*, the other is $1s_H$-*like*. In order to develop a hierarchy of models, we find it convenient to write the valence pair functions in the form

$$(\varphi_1 \varphi_2)$$
$$(\varphi_3 \varphi_4) \tag{40}$$

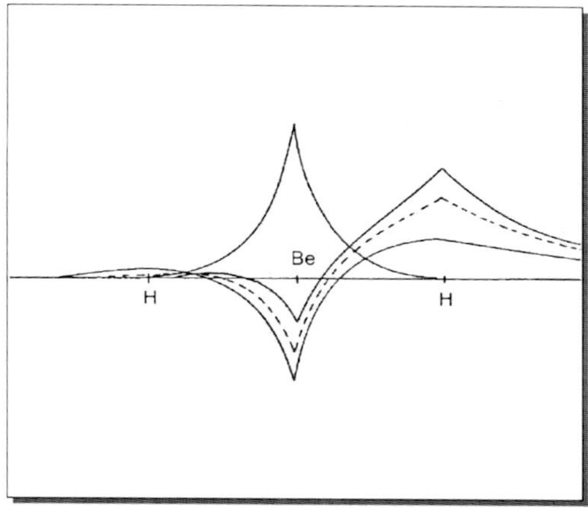

Fig. 7 Amplitude of the generalized Coulson–Fischer orbitals for the BeH$_2$ molecule. See text for details (Taken from [79].)

so that each pair function is defined in a distinct row. The spatial wave function in which all valence orbitals are nonorthogonal has the form

$$\varphi_c^2 \left(\varphi_1\varphi_2\varphi_3\varphi_4\right). \tag{41}$$

A model in which the orthogonality restrictions imposed on the pair function orbitals are relaxed, can be represented as follows:

$$\varphi_3 \; (\varphi_1\varphi_2)$$
$$\varphi_1 \; (\varphi_3\varphi_4), \tag{42}$$

where we have used the convention that in each row the two orbitals comprising the pair function on which the hierarchy is based are given in parentheses. Orbitals to the left (right) of the parenthetic term are allowed to overlap with the left (right)-hand orbital in parenthesis. In the first pair function $(\varphi_1\varphi_2)$, we take φ_1 to be the sp-hybrid-*like* orbital. This is allowed to have a nonzero overlap with φ_3, the sp-hybrid-*like* orbital in the second pair function. In this intermediate model, overlap is allowed between the sp-hybrid-*like* orbitals associated with different bonds, but the $1s_H$-*like* orbitals are required to be orthogonal to the orbitals associated with other pair functions.

We have restricted our attention to a very simple example here. A more complete description is given elsewhere [107], together with an application to a more complicated molecular system.

8 A Collaborative Virtual Environment: Future Prospects

In a recent paper [2] entitled "A Collaborative Virtual Environment for Molecular Electronic Structure Theory," Wilson and Hubač have described how

> Advances in information and communication technology are facilitating widespread cooperation between groups and individuals, who may be physically located at geographically distributed sites (- sites in different laboratories, perhaps in different countries or even different continents), in a way that may disrupt and challenge the traditional structures and institutions of science (as well as bringing change to society as a whole). Collaborative virtual environments ... have the potential to transform the 'scientific method' itself by fuelling the genesis, dissemination and accumulation of new ideas and concepts, and the exchange of alternative perspectives on current problems and strategies for their solution. Because of their openness and their global reach, as well as their emergent and thus agile nature, such environments may transform the practice of science over the next decades.

The paper by Wilson and Hubač described a *European Metalaboratory for multireference quantum chemical methods* carried out under the auspices of the EU COST programme. Specifically, a collaborative virtual environment was created in order to develop many-body methods based on Brillouin–Wigner theory. However, this environment was not open. Eight scientists from six countries (the Czech Republic, Germany, Greece, Poland, Slovakia, and the United Kingdom) participated in the project. Their results were presented in the literature in the usual way. At the time of writing, a volume[11] which summarizes the main results of the project is *in press*.

Here we are proposing the creation of an *open* collaborative virtual environment for the development of the Coulson–Fischer method for molecular wave functions. To this end we have created web pages at http://quantumsystems.googlepages.com/cve:theCoulson-Fischertheory, which is intended to form an element of a collaborative virtual environment for the further advancement of the Coulson–Fischer theory.

In his recently published report, Karadakov [52] expresses his view that

> in quantitative terms current-day VB theory is just as far from catching up with MO theory as twenty years ago.

He points to the relatively small number of researchers developing VB methodology, coupled with the lack of coordination of their efforts. The proposed open collaborative virtual environment should facilitate a new level of interaction and coordination but it envisaged that rather than focussing on VB theory, the emphasis should be on the Coulson–Fischer approach – a third way in quantum chemistry.

References

1. S. Wilson, P. J. Grout, J. Maruani, G. Delgado-Barrio, P. Piecuch, eds., *Frontiers in Quantum Systems in Chemistry and Physics*, Progress in Theoretical Chemistry & Physics, **18**, Springer (2008)
2. S. Wilson, I. Hubač. In: *Frontiers in Quantum Systems in Chemistry and Physics*, S. Wilson, P. J. Grout, J. Maruani, G. Delgado-Barrio, P. Piecuch, eds., p. 561, Springer (2008)
3. R. S. Mulliken, C. C. J. Roothaan, Proc. Nat. Acad. Sci. USA **45**, 394 (1959)
4. P. A. M. Dirac, Proc. Roy. Soc. **A123**, 714 (1929)
5. S. F. Boys, Proc. Roy. Soc. **A200**, 542 (1950)

---

[11] The volume is entitled *Brillouin-Wigner methods for many-body systems* [95] by I. Hubač and S. Wilson and will appear in the *Progress in Theoretical Chemistry & Physics* series.

6. R. McWeeny, Nature. **166**, 21 (1950)

7. S. Wilson, *Chemistry by Computer*, Plenum Press, New York (1986)

8. G. A. Moore, Electronics **38**, No. 8, April 19 (1965)

9. J. M. Roberts, *Twentieth Century – A History of the World 1901 to present*, p. 562, Allen Lane, London (1999)

10. W. J. Hehre, L. Radom, P. von Schleyer, J. A. Pople, *Ab initio Molecular Orbital Theory*, John Wiley, Chichester (1986)

11. L. A. Curtiss, K. Raghavachari, P. C. Redfern, V. Rassolov, J. A. Pople, J. Chem. Phys. **109**, 7764 (1998)

12. L. A. Curtiss, P. C. Redfern, K. Raghavachari, V. Rassolov, J. A. Pople, J. Chem. Phys. **110**, 4703 (1999)

13. J. A. Pople, Rev. Mod. Phys. **71**, 1267 (1999)

14. S. Wilson. In: *Chemical Modelling – Applications and Theory*, Specialist Periodical Report, A. Hinchliffe, ed., p. 208, Royal Society of Chemistry, London (2008)

15. M. W. Schmidt, K. K. Baldridge, J. A. Boatz, S. T. Elbert, M. S. Gordon, J. H. Jensen, S. Koseki, N. Matsunaga, K. A. Nguyen, S. Su, T. L. Windus, M. Dupuis, J. A. Montgomery, J. Comput. Chem. **14**, 1347 (1993)

16. M. S. Gordon, M. W. Schmidt. In: *Theory and Applications of Computational Chemistry: the first forty years*, C. E. Dykstra, G. Frenking, K. S. Kim, G. E. Scuseria, eds., p. 1167, Elsevier, Amsterdam (2005)

17. M. J. Frisch, G. W. Trucks, H. B. Schlegel, G. E. Scuseria, M. A. Robb, J. R. Cheeseman, J. A. Montgomery, Jr., T. Vreven, K. N. Kudin, J. C. Burant, J. M. Millam, S. S. Iyengar, J. Tomasi, V. Barone, B. Mennucci, M. Cossi, G. Scalmani, N. Rega, G. A. Petersson, H. Nakatsuji, M. Hada, M. Ehara, K. Toyota, R. Fukuda, J. Hasegawa, M. Ishida, T. Nakajima, Y. Honda, O. Kitao, H. Nakai, M. Klene, X. Li, J. E. Knox, H. P. Hratchian, J. B. Cross, V. Bakken, C. Adamo, J. Jaramillo, R. Gomperts, R. E. Stratmann, O. Yazyev, A. J. Austin, R. Cammi, C. Pomelli, J. W. Ochterski, P. Y. Ayala, K. Morokuma, G. A. Voth, P. Salvador, J. J. Dannenberg, V. G. Zakrzewski, S. Dapprich, A. D. Daniels, M. C. Strain, O. Farkas, D. K. Malick, A. D. Rabuck, K. Raghavachari, J. B. Foresman, J. V. Ortiz, Q. Cui, A. G. Baboul, S. Clifford, J. Cioslowski, B. B. Stefanov, G. Liu, A. Liashenko, P. Piskorz, I. Komaromi, R. L. Martin, D. J. Fox, T. Keith, M. A. Al-Laham, C. Y. Peng, A. Nanayakkara, M. Challacombe, P. M. W. Gill, B. Johnson, W. Chen, M. W. Wong, C. Gonzalez, J. A. Pople, Gaussian 03, Gaussian, Inc., Wallingford CT (2004)

18. C. A. Coulson, I. Fischer, Philos. Mag. **40**. 386 (1949)

19. W. Heitler, F. London, Zeits. für Physik **44**, 455 (1927)

20. F. London, Zeits. für Physik **46**, 455 (1928)

21. F. Hund, Zeits. für Physik **51**, 759 (1928)

22. F. Hund, Zeits. für Physik **73**, 1 (1931)

23. F. Hund, Zeits. für Physik **74**, 1 (1932)

24. R. S. Mulliken, Phys. Rev. **32**, 186 (1928)

25. R. S. Mulliken, Phys. Rev. **32**, 761 (1928)

26. R. S. Mulliken, Phys. Rev. **33**, 730 (1929)

27. R. S. Mulliken, Phys. Rev. **41**, 49 (1932)

28. L. Pauling, Proc. Nat. Acad. Sci. USA **14**, 359 (1928)

29. L. Pauling, J. Am. Chem. Soc. **53**, 1367 (1931)

30. L. Pauling, J. Am. Chem. Soc. **53**, 3225 (1931)

31. L. Pauling, *The Nature of the Chemical Bond*, Cornell University Press, Ithaca, New York (1939)

32. J. C. Slater, Phys. Rev. **37**, 481 (1931)

33. J. C. Slater, Phys. Rev. **38**, 1109 (1931)

34. J. C. Slater, Phys. Rev. **41**, 255 (1931)

35. G. N. Lewis, J. Am. Chem. Soc. **38**, 762 (1916)

36. J. H. Van Vleck, A. Sherman, Rev. Mod. Phys. **7**, 167 (1935)

37. R. Hoffmann, S. Shaik, P. C. Hiberty, Accounts of Chemical Research **36**, 750 (2003)
38. S. S. Shaik, P. C. Hiberty, Helv. Chim. Acta. **86**, 1063 (2003)
39. P. Watson: *A Terrible Beauty. The People and Ideas that Shaped the Modern Mind. A History*, Weidenfeld & Nicolson, London (2000)
40. T. Hager: *Force of Nature: The Life of Linus Pauling*, p. 217, Simon & Schuster, New York (1995)
41. C. C. J. Roothaan, Rev. Mod. Phys. **23**, 69 (1951)
42. G. G. Hall, Proc. Roy. Soc. A **205**, 541 (1951)
43. J. M. Schulman, D. N. Kaufman, J. Chem. Phys. **53**, 477 (1970)
44. U. Kaldor, Phys. Rev. A **7**, 427 (1973)
45. S. Wilson, D. M. Silver, Phys. Rev. A **14**, 1949 (1976)
46. S. Wilson, *Electron Correlation in Molecules*, Dover, New York (2007)
47. S. Wang, Phys. Rev. **31**, 579 (1928)
48. S. Weinbaum, J. Chem. Phys. **1**, 593 (1933)
49. C. A. Coulson, Trans. Faraday Soc. **33**, 1479 (1937)
50. C. A. Coulson, Z. Luz, Monatshefte für Chemie **98**, 62 (1967)
51. R. McWeeny, Intern. J. Quantum Chem. **74**, 87 (1999)
52. P. B. Karadakov. In: *Chemical Modelling Application and Theory*, Specialist Periodical Report, A. Hinchliffe, ed., p. 312, Royal Society of Chemistry, London (2008)
53. G. A. Gallup, *Valence Bond Methods – Theory and Applications*, Cambridge University Press, Cambridge (2002)
54. F. Weinhold, C. R. Landis, *Valency and Bonding: A Natural Bond Orbital Donor-Acceptor Perspective*, Cambridge University Press, Cambridge (2005)
55. S. S. Shaik, P. C. Hiberty, *A Chemist's Guide to Valence Bond Theory*, Wiley, New York (2007)
56. S. Wilson, M. Raimondi, D. L. Cooper, eds., *Quantum Theory of Chemical Bonding, Special Issue in Memory of Joseph Gerratt*, International Journal of Quantum Chemistry, **74** (2) (1999)
57. D. L. Cooper, ed., *Valence Bond Methods*, Theoretical and Computational Chemistry, volume 10, Elsevier, Amsterdam (2002)
58. S. Wilson, P. F. Bernath, R. McWeeny, eds., *Handbook of Molecular Physics and Quantum Chemistry*, volume 2: *Molecular Electronic Structure*, John Wiley, Chichester (2003)
59. D. L. Cooper, P. B. Karadakov, T. Thorsteinsson, Modern Valence-Bond Description of Gas-Phase Pericyclic Reactions. In: *Theoretical and Computational Chemistry*, volume 10, p. 41, D. L. Cooper, ed., Elsevier, Amsterdam (2002)
60. H. Nakano, K. Sorakubo, K. Nakayama, K. Hirao, Complete Active Space Valence Bond (CASVB) Method and its Application to Chemical Reactions. In: *Theoretical and Computational Chemistry*, volume 10, p. 55, D. L. Cooper, ed., Elsevier, Amsterdam (2002)
61. J. H. Van Lenthe, F. Dijkstra, R. W. A. Havenith, TURTLE – A Gradient VBSCF Program. Theory and Studies of Aromaticity. In: *Theoretical and Computational Chemistry*, volume 10, p. 79, D. L. Cooper, ed., Elsevier, Amsterdam (2002)
62. W. Hiberty, Y. Mo, Z. Cao, Q. Zhang, A Spin-Free Approach for Valence Bond Theory and its Applications. In: *Theoretical and Computational Chemistry*, volume 10, p. 143, D. L. Cooper, ed., Elsevier, Amsterdam (2002)
63. P. C. Hiberty, S. Shaik, BOVB – A Valence Bond Method Incorporating Static and Dynamic Electon Correlation Effects. In: *Theoretical and Computational Chemistry*, volume 10, p. 187, D. L. Cooper, ed., Elsevier, Amsterdam (2002)
64. J. J. W. McDouall, The Biorthogonal Valence Bond Method. In: *Theoretical and Computational Chemistry*, volume 10, p. 227, Ed: D. L. Cooper, Elsevier, Amsterdam (2002)
65. M. Sironi, M. Raimondi, R. Martinazzo, F. Gianturco, D. L. Cooper, Recent Developments of the SCVB Method. In: *Theoretical and Computational Chemistry*, volume 10, p. 261, D. L. Cooper, ed., Elsevier, Amsterdam (2002)
66. R. McWeeny, J. Li, Valence Bond Theory Using Symmetric Group Methods. In: Handbook of Molecular Physics and Quantum Chemistry, volume 2: Molecular Electronic Structure, chapter 10, p. 122, S. Wilson, P. F. Bernath, R. McWeeny (eds.) Wiley, Chichester (2003)

67. M. Sironi, D. L. Cooper, M. Raimondi, Valence Bond Theory: Determinantal Methods. In: *Handbook of Molecular Physics and Quantum Chemistry*, volume 2: *Molecular Electronic Structure*, chapter 11, p. 140, S. Wilson, P. F. Bernath, R. McWeeny, eds., John Wiley, Chichester (2003)

68. J. Gerratt, D. L. Cooper, P. Karaddakov, M. Raimondi, Spin-coupled Theory. In: *Handbook of Molecular Physics and Quantum Chemistry*, volume 2: *Molecular Electronic Structure*, chapter 12, p. 148, S. Wilson, P. F. Bernath, R. McWeeny, ed., John Wiley, Chichester (2003)

69. D. L. Cooper, Valence Bond Theory: Other Methods. In: *Handbook of Molecular Physics and Quantum Chemistry*, volume 2: *Molecular Electronic Structure*, chapter 13, p. 169, S. Wilson, P. F. Bernath, R. McWeeny, eds., John Wiley, Chichester (2003)

70. J. Gerratt, W. N. Lipscomb, Proc. Nat. Acad. Sci. **59**, 332 (1968)

71. J. Gerratt, W. N. Lipscomb, Intern. J. Quantum Chem. **74**, 83 (1999)

72. J. Gerratt, Adv. At. Mol. Phys. **7**, 141 (1971)

73. J. Gerratt, Specialist Periodical Reports: Theoretical Chemistry **1**, 60 (1974)

74. S. Wilson, J. Gerratt, In: *Proceedings of the* SRC *Atlas Symposium: Quantum Chemistry – The State of the Art*, V. R. Saunders, J. Brown, eds., St. Catherine's College, Oxford, p. 109, Atlas Computer Laboratory, Science Research Council, Chilton, Oxfordshire (1974)

75. N. C. Pyper, J. Gerratt. In: *Proceedings of the* SRC *Atlas Symposium: Quantum Chemistry – The State of the Art*, V. R. Saunders, J. Brown, eds., St. Catherine's College, Oxford, p. 93, Atlas Computer Laboratory, Science Research Council, Chilton, Oxfordshire (1974)

76. S. Wilson, J. Gerratt, Molec. Phys. **30**, 765 (1975)

77. S. Wilson, J. Gerratt, Molec. Phys. **30**, 777 (1975)

78. S. Wilson, J. Gerratt, Molec. Phys. **30**, 789 (1975)

79. S. Wilson, *A Self-Consistent Group Function Model for Molecular Wave Functions*, doctoral dissertation, University of Bristol (1975)

80. N. C. Pyper, J. Gerratt, Proc. Roy. Soc. (London) A **355**, 407 (1977)

81. D. L. Cooper, J. Gerratt, M. Raimondi, Adv. Chem. Phys. **69**, 319 (1987)

82. D. L. Cooper, J. Gerratt, M. Raimondi, Int. Rev. Phys. Chem. **7**, 59 (1988)

83. J. Gerratt, D. L. Cooper, M. Raimondi. In: *Valence Bond and Chemical Structure*, D. J. Klein, N. Trinajstic, eds., Elsevier, Amsterdam (1990)

84. D. L. Cooper, J. Gerratt, M. Raimondi, Chem. Rev. **91**, 929 (1991)

85. J. Gerratt, D. L. Cooper, P. B. Karadakov, M. Raimondi, Chem. Soc. Rev. **26**, 87 (1997)

86. P. B. Karadakov, Progr. Phys. Chem. **94**, 2 (1998)

87. A. C. Hurley, J. E. Lennard-Jones, J. A. Pople, Proc. Roy. Soc. (London) **A220**, 446 (1953)

88. A. C. Hurley, Proc. Roy. Soc. (London) **A235**, 224 (1956)

89. W. J. Hunt, P. J. Hay, W. A. Goddard, J. Chem. Phys. **57**, 738 (1972)

90. A. C. Wahl, G. Das. In: *Methods of electronic structure theory*, H. F. Schaefer III, ed., Plenum, New York (1977)

91. A. C. Wahl, G. Das, Adv. Quantum Chem. **5**, 261 (1970)

92. K. Ruedenberg, K. R. Sundbarg. In: *Quantum Science*, J. L. Calais, O. Goscinski, J. Linderberg, Y. Ohrn, eds., Plenum, New York (1976)

93. K. Ruedenberg, M. W. Schmidt, M. M. Gilbert, S. T. Elbert, Chem. Phys. **71**, 41 (1982)

94. B. O. Roos, P. R. Taylor, P. E. M. Siegbahn, Chem. Phys. **48**, 157 (1980)

95. I. Hubač, S. Wilson, *Brillouin-Wigner methods for many-body systems*, Progress in Theoretical Chemistry and Physics, Springer (*in press*)

96. W. Kolos, C. C. J. Roothaan, Rev. Mod. Phys. **32**, 219 (1960)

97. E. P. Wigner, *Group Theory and its Application to the Quantum Mechanics of Atomic Spectra*, Academic Press, New York (1959)

98. P. B. Karadakov, Spin Functions. In: *Handbook of Molecular Physics and Quantum Chemistry*, volume 2: *Molecular Electronic Structure*, chapter 3, p. 25, S. Wilson, P. F. Bernath, R. McWeeny, eds., John Wiley, Chichester (2003)

99. P. B. Karadakov, Spatial Wavefunctions, In: *Handbook of Molecular Physics and Quantum Chemistry*, volume 2: *Molecular Electronic Structure*, chapter 4, p. 33, S. Wilson, P. F. Bernath, R. McWeeny, eds., John Wiley, Chichester (2003)

100. I. G. Kaplan, The Pauli Principle, In: *Handbook of Molecular Physics and Quantum Chemistry*, volume 2: *Molecular Electronic Structure*, chapter 2, p. 15, S. Wilson, P. F. Bernath, R. McWeeny, eds., John Wiley, Chichester (2003)
101. H. A. Jahn, Phys. Rev. **96**, 989 (1954)
102. G. A. Gallup, Adv. Quantum Chem. **7**, 113 (1973)
103. G. A. Gallup, Adv. Quantum Chem. **16**, 229 (1982)
104. W. A. Goddard III, Phys. Rev. **157**, 73 (1967); *ibid.* **157**, 81 (1967)
105. W. A. Goddard III, J. Chem. Phys. **48**, 450 (1968); *ibid.* **48**, 5377 (1968)
106. S. J. McNicholas, F. R. Manby, Intern. J. Quantum Chem. **74**, 97 (1999)
107. S. Wilson, Intern. J. Quantum Chem. **74**, 135 (1999)
108. J. Barnes, P. Hut, Nature **324**, 446 (1986)

Part V

Advances in Density Functional Theory

Energy Densities of Exchange and Correlation in the Slowly Varying Region of the Airy Gas

John P. Perdew, Lucian A. Constantin, and Adrienn Ruzsinszky

Abstract In Kohn–Sham theory, where the integrated exchange and correlation energies are functionals of the electron density, several recent approximate functionals make use of the energy densities of exchange and correlation for nonuniform reference systems or of exchange for the real system of interest. A relevant case in which to examine these energy densities is the Airy gas. Here we evaluate the conventional exact-exchange energy density from the occupied orbitals, and the conventional correlation energy density within the random phase approximation using the occupied and unoccupied orbitals. We find, as expected, that the exchange energy density in the region of slowly varying electron density demonstrates *a mostly negative* small correction to the local density approximation (LDA) but has no second-order density-gradient expansion (although the integrated exchange energy of a finite system has one). We also find, as expected, that the exchange-*correlation* energy density demonstrates a *positive* small correction to the LDA in the region of high and slowly varying electron density. It too appears to have no second-order density-gradient expansion. An appendix shows that a slowly varying density, scaled uniformly to the high-density limit in which exchange dominates, remains slowly varying only for exchange and not for correlation.

Keywords: Exchange · Correlation · Energy density · Airy gas · Gradient expansion · Density functional theory

1 Introduction: Why Energy Densities Matter Now

In density functional theory [1, 2], the exchange-correlation energy $E_{xc} = E_x + E_c$ is evaluated as an integral over ordinary three-dimensional space:

$$E_{xc} = \int d^3r \, n(\mathbf{r}) \, \varepsilon_{xc}(\mathbf{r}). \tag{1}$$

J.P. Perdew (✉), L.A. Constantin, A. Ruzsinszky
Department of Physics and Engineering Physics, Tulane University, New Orleans, Louisiana 70118, USA, e-mail: perdew@tulane.edu, lconstan@tulane.edu, aruzsinszky@gmail.com

P. Piecuch et al. (eds.), *Advances in the Theory of Atomic and Molecular Systems*, Progress in Theoretical Chemistry and Physics 19, DOI 10.1007/978-90-481-2596-8_14, © Springer Science+Business Media B.V. 2009

Here $n(\mathbf{r})$ is the electron density, ε_{xc} is the exchange-correlation energy per particle, and $n\varepsilon_{xc}$ is the energy density. The earliest density functional, the local density approximation (LDA) [1], simply replaced the exact energy density by that of an electron gas of uniform density equal to the local density. But, unlike the electron density and integrated energy, the energy density is not an observable, and only for the uniform gas is it uniquely defined. In a nonuniform system, one can add to it any functional of the electron density that integrates to zero over all space, without changing E_{xc} (a sort of gauge transformation). Thus standard approximations beyond LDA [2], such as the generalized gradient approximation (GGA), the meta-GGA, and the exact-exchange hybrid, were developed without reference to the energy density, by satisfying exact constraints on E_{xc} or by fitting integrated energies to data, or both.

The conventional exchange-correlation energy density is generated by the choice [3, 4]

$$\varepsilon_{xc}(\mathbf{r}) = \frac{1}{2} \int d^3 r' \, n_{xc}(\mathbf{r}, \mathbf{r}')/|\mathbf{r}' - \mathbf{r}|. \tag{2}$$

Here $n_{xc}(\mathbf{r}, \mathbf{r}')$ is the density at $\mathbf{r}'$ of the coupling constant-averaged exchange-correlation hole around an electron at $\mathbf{r}$. At the lower limit of the coupling constant integration, we have the exact exchange hole density n_x, constructed from the occupied Kohn–Sham orbitals and leading to a standard Fock integral for E_x. But the unoccupied Kohn–Sham orbitals are needed for the correlation hole density n_c in the random phase approximation (RPA) [4]. More accurate than RPA is Quantum Monte Carlo (QMC). QMC has been used to generate the xc energy density for a few systems like solid silicon or the silicon atom, and to fit a simple density functional to it [5]. However, this expression does not [6] seem to carry over so well to a jellium slab within RPA.

Some time ago, Folland [7] compared exact and LDA exchange energy densities in atoms and found substantial similarity. Nevertheless, it has always been clear [8] that the standard gradient corrections to the LDA in the integrated exchange energy do not model the corresponding corrections in the conventional exchange energy density.

Kohn and Mattsson [9] proposed the Airy gas as a second reference system for density functional theory, to supplement the uniform gas. Vitos [10] constructed a GGA (using only the ingredients n and $|\nabla n|$) to fit the conventional exchange energy density of the Airy gas. Armiento and Mattsson [11] combined a Vitos-like GGA for exchange with a GGA for correlation constructed differently, and produced a successful GGA for solids that anticipated other work [12–14]. Recently we have constructed the correlation energy density of the Airy gas within the RPA, which we have combined with a short-range correction to RPA to make an exchange-correlation GGA [15] by fitting to conventional energy densities.

An independent line of thought also revives our interest in energy densities. The standard global hybrid functionals [2] mix a large constant fraction of a semilocal (LDA, GGA, or meta-GGA) exchange energy with a complementary smaller constant fraction of the exact exchange energy, achieving improved accuracy. But finding the right fraction of exact exchange for every region requires a local hybrid or hyper-GGA [16], in which the fraction of each energy density varies with position $\mathbf{r}$ in a way determined by the electron density distribution. For consistency, one needs to know the exact exchange energy density, not in the conventional gauge but in the gauge of the semilocal functional, which requires some modeling [17]. Alternatively and preferably, but with greater effort, one can transform the semilocal and exact exchange energy densities to the same unambiguous gauge [18].

In this work, we will use the Airy gas to confirm an old conclusion [19, 20]: Although a second-order gradient expansion exists for the integrated exchange energy, no such expansion exists for the conventional exchange energy density. This conclusion to some extent limits the efficacy of fitting to a conventional exchange energy density or modeling the exact exchange energy density in the gauge of a semilocal functional. The absence of a second-order gradient expansion for the energy density in the conventional gauge need not imply its absence in other gauges, as discussed again at the end of Section 6, nor does it limit the accuracy of density functionals constructed in the more standard way described at the beginning of this section.

2 Airy Gas Model (with Uniform Scaling)

The Airy gas [9] is a system of noninteracting electrons which provides the Kohn–Sham orbitals for an evaluation of, say, the kinetic energy density in a given gauge [21, 22], the conventional exchange energy density (L.A. Constantin, A. Ruzsinszky, and J.P. Perdew, unpublished manuscript) [10, 11], or the RPA conventional correlation energy density [15]. The noninteracting electrons in their spin-unpolarized ground state see a linear Kohn–Sham or effective potential

$$v_{\text{eff}}(z) = \begin{cases} -Fz, & (-\infty < z < L) \\ \infty, & (z \geq L). \end{cases} \tag{3}$$

The constant force on each electron is positive, and the infinite potential barrier (needed to normalize the orbitals) is far away: $L \rightarrow \infty$. The chemical potential is $\mu = 0$. Classically, electrons are confined to the half space $z > 0$, but in our quantum treatment they tunnel into the classically forbidden region $z < 0$. The linear variation of the effective potential in the region close to $z = 0$ can model the Kohn–Sham potential in the surface or edge region of jellium [23], and the evanescent decay of the electron density for large negative z models the tail of the electron density (which is not in principle described by any gradient expansion).

In practice, as in [15], we replace the true Airy gas described above by an Airy gas model in which $L/l = 20$ with l defined below. This system has 19 occupied orbitals for the motion along the z axis. For the RPA calculation, we keep the lowest 50 of the unoccupied orbitals. For $z \ll 20l$ this is a rather faithful model of the true Airy gas – a striking example of the nearsightedness of the electron density [24].

The true Airy gas and its solution in terms of the Airy functions are discussed in detail in [9] and [15]. An important feature of the solution is the existence of a scaling length

$$l = (2F)^{-1/3}. \tag{4}$$

This means that the density $n(z)$ for any force F can be found from the density $n_0(z)$ for $F = 1/2$ by scaling [9]:

$$n(z) = l^{-3}n_0(z/l). \tag{5}$$

As the force F increases, the scaling length l decreases and the electron density at positive z increases because of both factors in Eq. (5). Figure 1 shows the electron density for $F = 0.10$. Note that the electronic edge region is $-l < z < l$. (We use atomic units, with energies in hartrees and distances in bohrs.) Equation (5) is an

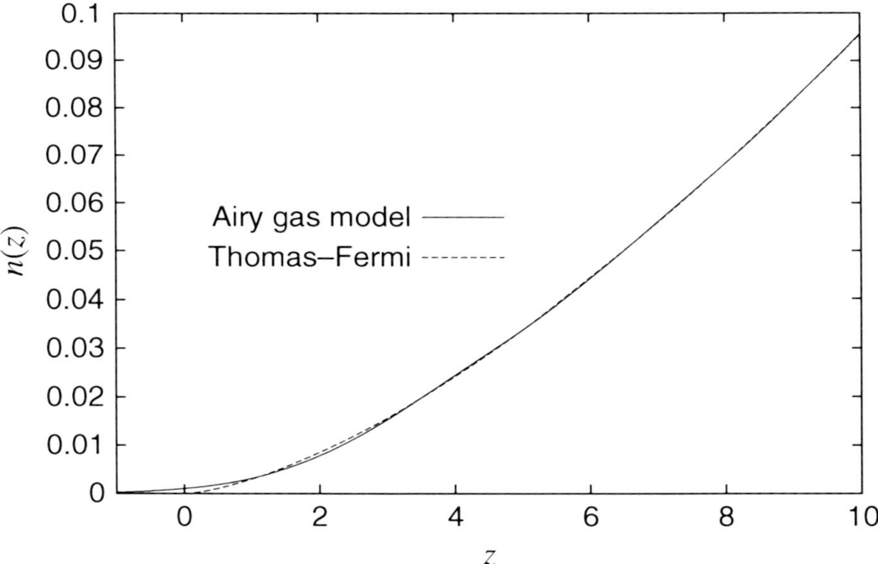

Fig. 1 Exact and Thomas–Fermi electron density n as a function of position z for the Airy gas model with force $F = 0.10$. The scaling length is $l = 1.71$. The edge region is $-l < z < l$ and the Thomas–Fermi density is reasonably accurate for $z > l$. The infinite barrier is at $z = 20l = 34.2$. The magnitudes of the densities in this figure are valence-electron-like; the density parameter r_s (the radius of a sphere containing on average one electron) is about 3.3 at $z = l$ and about 1.3 at $z = 10$ (atomic units)

example of a uniform scaling of the electron density [25]:

$$n(\mathbf{r}) \rightarrow n_\lambda(\mathbf{r}) = \lambda^3 n(\lambda \mathbf{r}), \tag{6}$$

which leaves the electron number of a finite system unchanged. Under uniform density scaling, the exact or approximate exchange energy of a finite system scales as [25]

$$E_x[n_\lambda] = \lambda E_x[n]. \tag{7}$$

The exchange energy density $f_x = n\varepsilon_x$ for the Airy gas then scales as

$$f_x(z) = l^{-4} f_{x0}(z/l). \tag{8}$$

Like the electron density, it has to be computed only once, for one value of F. The correlation energy density has no exact scaling equality [25] and must be computed separately for each F. In the low-density limit ($\lambda \rightarrow 0$), correlation scales like exchange, but in the high-density limit ($\lambda \rightarrow \infty$) correlation varies much more weakly with λ.

3 Region of Slowly Varying Electron Density

For the kinetic and exchange energies, a slowly varying density is one that varies slowly over the local Fermi wavelength $\lambda_F = 2\pi/k_F$. This condition is reasonably satisfied in the valence regions of many bulk solids, and even in the energetically important portions of many solid surfaces [13]. For any electron density $n(\mathbf{r})$, we define the local Fermi wavevector $k_F = [3\pi^2 n(\mathbf{r})]^{1/3}$, the reduced density gradient

$$s = |\nabla n|/[2k_F n], \tag{9}$$

its square

$$p = s^2, \tag{10}$$

and the reduced density Laplacian

$$q = \nabla^2 n/[(2k_F)^2 n]. \tag{11}$$

In a slowly varying region, p and q are small in magnitude compared to unity, the LDA is nearly correct, and one can hope that the small corrections to LDA can be gradient-expanded in powers of ∇. Under uniform scaling, $s(\mathbf{r}) \rightarrow s(\lambda \mathbf{r})$ and $q(\mathbf{r}) \rightarrow q(\lambda \mathbf{r})$.

The Airy gas for $x \gg l$ is (apart from small Friedel oscillations) an example of a region of slowly varying density. To get simple analytic expressions that will show this, let us find the Thomas–Fermi (TF) approximation [26] to the density, which follows from making the LDA for the kinetic energy and solving the TF Euler equation for $z > 0$:

$$(1/2)k_F^2 - Fz = 0. \tag{12}$$

This equation says that the total energy of a Fermi-level electron at position z equals the chemical potential. We immediately find

$$n_{TF}(z) = (2Fz)^{3/2}/(3\pi^2), \tag{13}$$

which has the expected behaviors: It grows with z, and scales as in Eq. (5). Like the classical density, the TF density vanishes for all negative z. Figure 1 shows how the exact electron density approaches its TF limit with increasing z. Now

$$p_{TF} = 9/[32Fz^3], \tag{14}$$

$$q_{TF} = 3/[32Fz^3] = p_{TF}/3. \tag{15}$$

Clearly then $z >> l$ is a region of slowly varying density, in which the reduced gradient and Laplacian tend to zero. Figure 2 shows how the exact s approaches

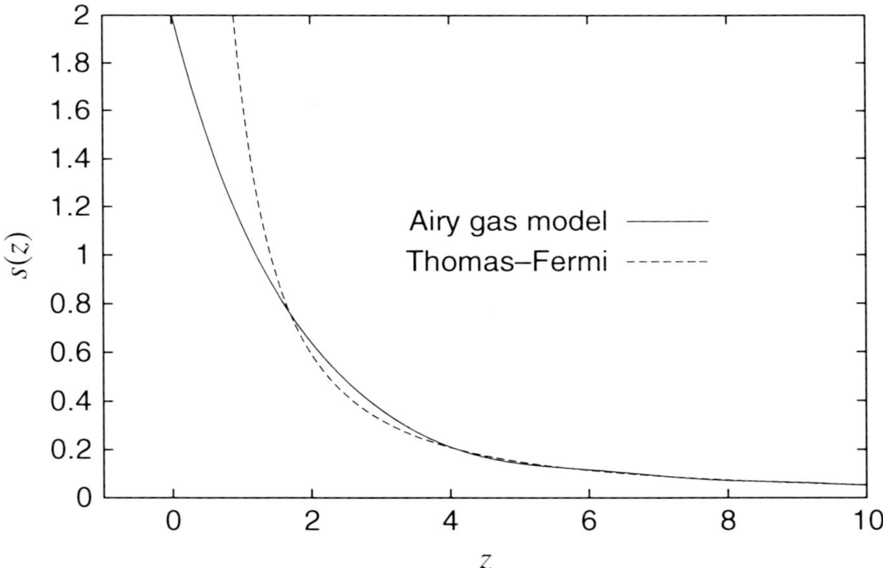

Fig. 2 Exact and Thomas–Fermi reduced density gradient s of Eq. (9) as a function of position z for the Airy gas model with force $F = 0.10$ (atomic units)

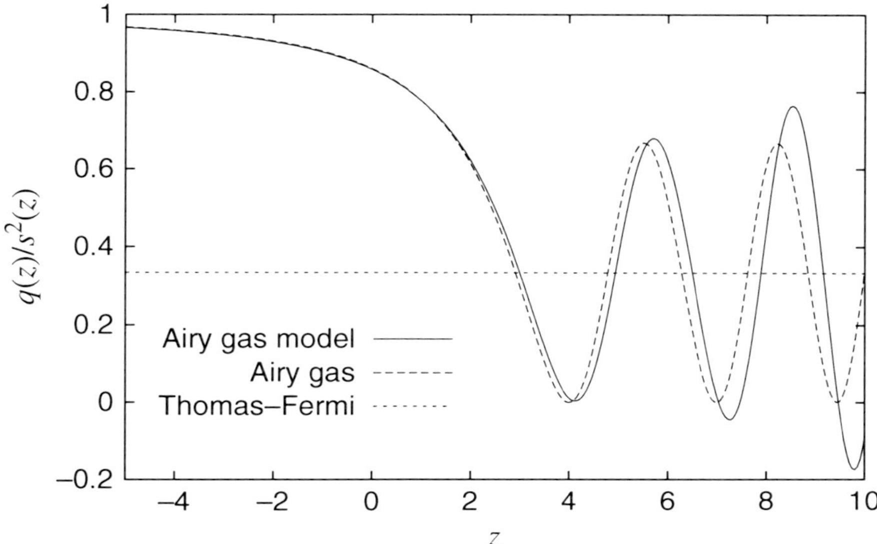

Fig. 3 Exact and Thomas–Fermi ratio of the reduced Laplacian q of Eq. (11) to the square of the reduced gradient s of Eq. (9), as a function of position z, for the Airy gas model with force $F = 0.10$. Also shown for comparison is the same ratio in the true Airy gas, where the Friedel oscillations [9] appear to be damped with increasing z. The visible difference between the true Airy gas and the Airy gas model (with an infinite barrier at $z = 20l$) suggests that the ratio plotted here is less nearsighted than the density itself (atomic units)

its TF limit with increasing z, and Fig. 3 shows how the exact ratio q/p displays Friedel oscillations around its TF value of 1/3, within our Airy gas model.

For a density that is everywhere slowly varying, the exchange energy has a known second-order gradient expansion [27, 28], which discards the $O(\nabla^4)$ terms in

$$E_x[n] = \int d^3r \, n \, \varepsilon_x^{unif}[1 + \mu_x p + O(\nabla^4)]. \tag{16}$$

Here

$$\varepsilon_x^{unif} = -3k_F/[4\pi] \tag{17}$$

is the exchange energy per electron of a gas of uniform density n. The LDA exchange energy is negative, and the second-order gradient expansion for the exchange energy (with gradient coefficient $\mu_x = 10/81 = 0.1235$) is even more negative. For the TF density of the Airy gas,

$$n\varepsilon_x^{unif} = -(2Fz)^2/[4\pi^3]. \tag{18}$$

Note that $n\varepsilon_x^{unif} p$ goes to zero like $1/z$ as $z \to \infty$ in the true Airy gas, and thus is not quite integrable over z.

4 Exchange Energy Density of the Airy Gas: Is There a Second-Order Gradient Expansion for the Slowly Varying Region?

If there was a second-order gradient expansion for the exchange energy density, it would have to take the form [20] of the zero-th and second-order terms in

$$n \, \varepsilon_x = n \, \varepsilon_x^{unif} [1 + Ap + 3(\mu_x - A)q + O(\nabla^4)], \tag{19}$$

where A is a constant. Integration by parts on Eq. (19) recovers Eq. (16). No form other than Eq. (19) is consistent with that condition and with symmetry and scaling requirements. Note that, if the TF approximation for the Airy gas were perfect, we could use Eq. (15) to reduce the square bracket of Eq. (19) to that of Eq. (16), independent of A.

Early analytic work [19] suggested that the expansion (19) does not exist. This work replaced the Coulomb interaction $1/|\mathbf{r}' - \mathbf{r}| = 1/R$ by the screened Coulomb interaction $\exp(-\alpha R)/R$ and evaluated the coefficients of p and q as functions of α/k_F analytically. In the limit $\alpha \to 0$, the coefficient of q diverged, although Eq. (16) was recovered (with a slightly different gradient coefficient 7/81 appropriate to this limit). More recently, Armiento and Mattsson [20] examined the exact exchange energy density of the Matthieu gas (in which the noninteracting electrons

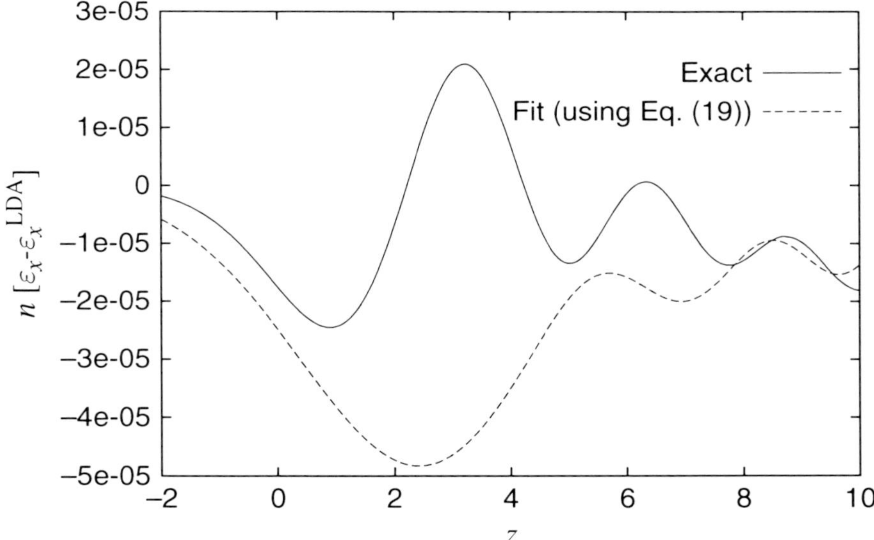

Fig. 4 Exact and fitted (using Eq. (19)) deviation of the conventional exchange energy density from LDA exchange, as a function of position z for the Airy gas model with force $F = 0.10$ (atomic units)

experience a sinusoidal effective potential) for the $1/R$ interaction, and in the slowly varying limit found no second-order gradient expansion for this energy density.

Here we will similarly examine the exact exchange energy density in the slowly varying region of the Airy gas with $F = 0.10$. We can compute the exchange energy density accurately for $z < 10$ noting that p is small compared to unity (less than or about equal to 0.4) for $z > 2$. We vary A in Eq. (19) to minimize the integral over z from 2 to 10 of the absolute value of the difference between the second-order gradient expansion of Eq. (19) and the exact exchange energy density, finding $A = 0.15$ and a very poor fit (Fig. 4). Thus we confirm that there is no second-order gradient expansion for the exchange energy density.

As a proof-of-principle for our fitting, we show in Appendix 1 that we can fit the exact positive noninteracting kinetic energy density of the Airy gas to a known second-order gradient expansion.

5 Exchange and Correlation Together in the Random Phase Approximation

In a uniform gas, correlation has two length scales: the Fermi wavelength $\lambda_F = 2\pi/k_F$ and the screening length $\lambda_s = 1/k_s$, where $k_s = (4k_F/\pi)^{1/2}$. The validity of the second-order gradient expansion for the exchange-correlation energy [29], which discards the $O(\nabla^4)$ terms in

$$E_{xc}[n] = \int d^3r\, n\, \varepsilon_x^{unif}[1 + \mu_{xc}(n)p + O(\nabla^4)] + \int d^3r\, n\, \varepsilon_c^{unif}, \qquad (20)$$

requires that the density vary slowly on both length scales [31]. A further condition is that the second-order contribution $n\varepsilon_x^{unif}\mu_c(n)p$ to correlation should be small in magnitude compared to the local contribution $n\varepsilon_c^{unif}$. In real systems, the additional conditions are seldom if ever satisfied [30], but it is easy to see that they are satisfied in the large-z limit for the Airy gas, and in particular they are satisfied for $z > 2$ when $F = 0.10$. In Appendix 2, we show that a slowly varying density, scaled uniformly to the high-density limit ($\lambda \to \infty$) in which exchange fully dominates correlation, remains slowly varying only for exchange and not for correlation. This explains how it can be that the high-density limit of the gradient expansion for E_{xc} differs in second order from the high-density limit of the gradient expansion for E_x.

The high-density limit of $\mu_{xc}(n)$ was derived by Ma and Brueckner [29]. Within or beyond RPA, it is a *negative* number -0.1331. Since ε_x^{unif} is negative, the second-order gradient term makes a positive contribution to the negative E_{xc}, unlike the negative contribution of the second-order term to the negative E_x. Within the RPA, the density dependence of $\mu_{xc}(n) = \mu_x + \mu_c(n)$ is very weak [31], and we shall henceforth ignore it.

Within the RPA with $F = 0.10$, we can evaluate the correlation energy density accurately only for $z < 7$ (Fig. 5). Note that, while the correction to LDA exchange

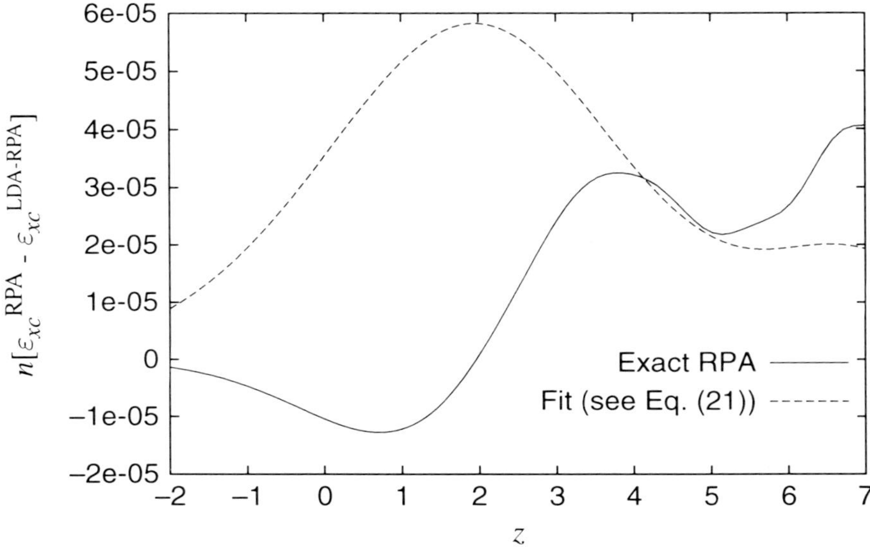

Fig. 5 Exact and fitted (using Eq. (21)) deviation of the RPA conventional exchange-correlation energy density from RPA-LDA, as a function of position z for the Airy gas model with force $F = 0.10$ (atomic units)

in Fig. 4 is mostly negative, the correction to LDA exchange-correlation in Fig. 5 is mostly positive. These results for the conventional energy densities might have been anticipated from the integrated gradient expansions.

As we move toward large z in the Airy gas, the electron density increases without limit. In the local term, which becomes increasingly dominant, exchange overwhelms correlation as z increases, but correlation actually contributes more than exchange to the relatively small correction to LDA.

To see if there is a second-order gradient expansion for the conventional exchange-correlation energy density in RPA, we have fitted B in

$$n\varepsilon_x^{unif}[Bp + 3(\mu_{xc} - B)q], \tag{21}$$

where $\mu_{xc} = -0.1331$, to $n[\varepsilon_{xc}^{RPA} - \varepsilon_{xc}^{unif\,RPA}]$ over $2 < z < 7$, much as we did for exchange in Section 4. The best fit ($B = -0.14$) is poor, as shown in Fig. 5. This leads us to suspect (although with less confidence than for exchange) that there is no second-order gradient expansion for the conventional RPA exchange-correlation energy density, or at least none with our assumed density-independent gradient coefficients.

For completeness, we have also looked for a second-order gradient expansion of the conventional correlation energy density, by fitting C in

$$n\varepsilon_x^{unif}[Cp + 3(\mu_c - C)q], \tag{22}$$

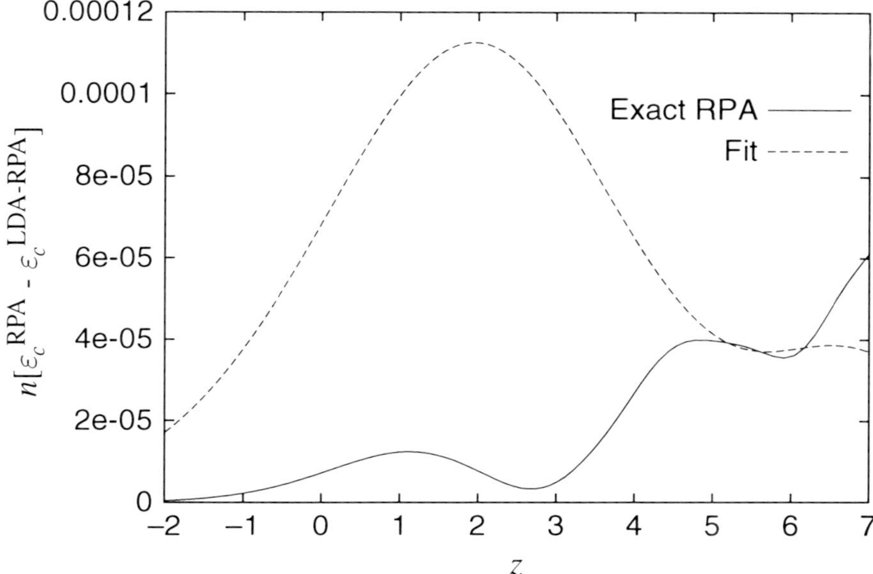

Fig. 6 Exact and fitted (using Eq. (22)) deviation of the RPA conventional correlation energy density from RPA-LDA, as a function of position z, for the Airy gas model with force $F = 0.10$ (atomic units)

where $\mu_c = -0.2566$, to $n[\varepsilon_c^{RPA} - \varepsilon_c^{unif\,RPA}]$ over $2 < z < 7$, with the result $C = -0.277$ and the poor fit of Fig. 6.

6 Conclusions

Previous work [19, 20] strongly suggests that there is no second-order gradient expansion for the conventional exchange energy density. We confirm this by examining the slowly varying region of the Airy gas. This to some extent limits the efficacy of GGAs (and meta-GGAs) that are fitted to a conventional exchange energy density: They cannot be right for the relatively small integrated correction to LDA in the slowly varying limit, except for slow density variations very similar to those to which they were fitted. Even if there was a gradient expansion of the form of Eq. (19), with a finite coefficient of q different from zero, this difficulty would persist for a GGA but not for a meta-GGA. (There need not be any similar difficulty for GGAs and meta-GGAs constructed in more standard ways, such as the PBEsol [13] GGA for exchange.) The absence of a gradient expansion for the conventional exchange energy density also to some extent limits the efficacy of modeling the exact exchange energy density in the gauge of a semilocal functional; this modeling will also fail for the relatively small integrated correction to LDA for slowly varying densities.

Nevertheless, we do find a relatively small and *mostly negative* correction to the LDA conventional exchange energy density in the slowly varying regions, as the second-order gradient expansion for the integrated exchange energy might suggest.

For the conventional exchange-correlation energy density, we find a relatively small *positive* correction to the LDA conventional exchange-correlation energy density in the slowly varying regions, as the second-order gradient expansion for the integrated exchange-correlation energy might suggest. We also find evidence (weaker than for the case of exchange alone) that a second-order gradient expansion of the conventional exchange-correlation energy density in RPA does not exist, at least with our assumed density-independent gradient coefficients.

We have also shown that a slowly varying electron density, scaled uniformly to the high-density limit in which exchange dominates, remains slowly varying for exchange but not for correlation. This explains why the high-density limit of the gradient expansion for the integrated exchange-correlation energy differs in second-order from the gradient expansion for the integrated exchange energy.

For the construction of improved density functionals, it could be useful to find an energy-density gauge that has a well-behaved gradient expansion. Coordinate transformations under the double integration of Eqs. (1)–(2) are one way [32, 33] to change the gauge of the energy density, and might lead to such a useful gauge.

Acknowledgement This work was supported in part by the National Science Foundation under grant DMR-0501588. We thank Piotr Piecuch for constant encouragement and technical support.

Appendix 1 Proof-of-Principle for the Fitting: The Positive Noninteracting Kinetic Energy Density

The positive kinetic energy density of the occupied Kohn–Sham orbitals is

$$nt_s = (1/2) \sum_i |\nabla \psi_i|^2. \tag{A1}$$

The second-order gradient approximation to $n[t_s - t_s^{unif}]$, where $t_s^{unif} = 3k_F^2/10$ is [34].

$$nt_s[5p/27 + Dq], \tag{A2}$$

where $D = (40/3)(1/6) = 2.222$. We have fitted D in Eq. (A2) to $n[t_s - t_s^{unif}]$ over our Airy gas model with $2 < z < 10$. The best fit, with $D = 2.64 \approx (40/3)(1/5)$, is shown in Fig. 7. The fit is good. If we had chosen not to fit the kinetic energy density itself but its integral over one wavelength of the Friedel oscillation, we would presumably have found $D = (40/3)(1/6)$, as discussed around Fig. 3 of Ref. [22].

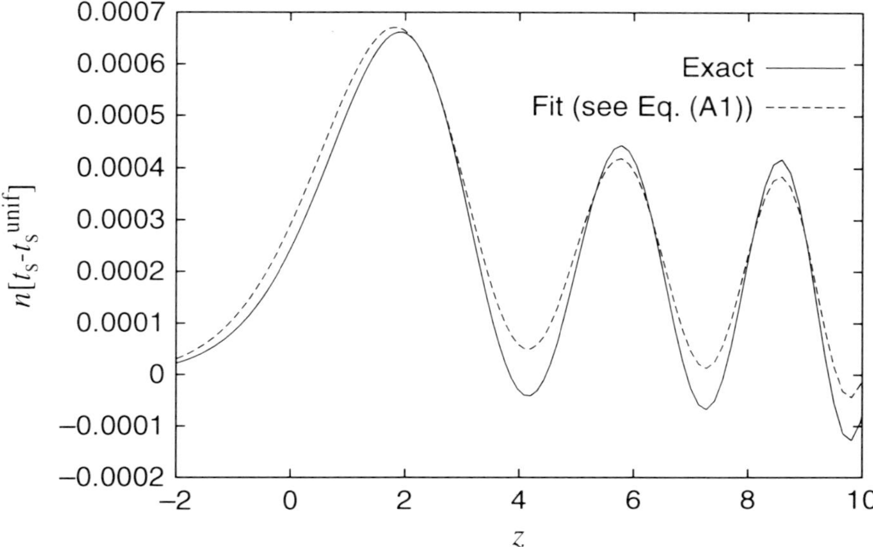

Fig. 7 Exact and fitted (using Eq. (A2)) deviation of the positive noninteracting kinetic energy density from its LDA or TF approximation, as a function of position z, for the Airy gas model with force $F = 0.10$ (atomic units)

Appendix 2 A Slowly Varying Electron Density, Scaled Uniformly to the High-Density Limit, Remains Slowly Varying for Exchange but not for Correlation

Start with a density that is so slowly varying that the second-order gradient expansions are valid for E_x (Eq. (16)) and for E_{xc} (Eq. (20)). This requires that the reduced density gradients on both length scales of Section 5 must be small. Thus p and q of Eqs. (10) and (11) have magnitudes much less than 1, and so do $p_c = (k_F/k_s)^2 p$ and $q_c = (k_F/k_s)^2 q$. Moreover, the second-order contribution to the correlation energy per particle, $\varepsilon_x^{unif} \mu_c p \sim k_F p$ must be small compared to the local part ε_c^{unif}.

Now make the uniform density scaling of Eq. (6) and let the scale parameter $\lambda \to \infty$. One easily finds $p \sim \lambda^0$, $q \sim \lambda^0$, $p_c \sim \lambda$, $q_c \sim \lambda$, $\varepsilon_x^{unif} \mu_c p \sim \lambda$, and $\varepsilon_c^{unif} \to -\ln \lambda$. Thus all the conditions for convergence of gradient expansion for exchange hold up, while all conditions for convergence of the gradient expansion for correlation break down, in this high-density limit.

References

1. W. Kohn, L. J. Sham, Phys. Rev. **140**, A1133 (1965)
2. *A Primer in Density Functional Theory*, ed. by C. Fiolhais, F. Nogueira, M. Marques (Springer, Berlin, 2003)

3. O. Gunnarsson, B. I. Lundqvist, Phys. Rev. B **13**, 4274 (1976)
4. D. C. Langreth, J. P. Perdew, Phys. Rev. B **15**, 2884 (1977)
5. A. C. Cancio, M. Y. Chou, Phys. Rev. B **74**, 08102 (2006)
6. L. A. Constantin, J. M. Pitarke, J. Chem. Theory Comput. **5**, 895 (2009)
7. N. O. Folland, Phys. Rev. A **3**, 1535 (1971)
8. J. P. Perdew, Y. Wang, Phys. Rev. B **33**, 8800 (1986)
9. W. Kohn, A. E. Mattsson, Phys. Rev. Lett. **81**, 3487 (1998)
10. L. Vitos, B. Johansson, J. Kollar, H. L. Skriver, Phys. Rev. B **62**, 10046 (2000)
11. R. Armiento, A. E. Mattsson, Phys. Rev. B **72**, 085108 (2005)
12. Z. G. Wu, R. E. Cohen, Phys. Rev. B **73**, 235116 (2006)
13. J. P. Perdew, A. Ruzsinszky, G. I. Csonka, O. A. Vydrov, G. E. Scuseria, L. A. Constantin, X. Zhou, K. Burke, Phys. Rev. Lett. **100**, 136406 (2008); ibid. **102**, 039902 (2009)
14. Y. Zhao, D. G. Truhlar, J. Chem. Phys. **128**, 184109 (2008)
15. L. A. Constantin, A. Ruzsinszky, J. P. Perdew, Phys. Rev. B, to appear (2009)
16. J. P. Perdew, V. N. Staroverov, J. Tao, G. E. Scuseria, Phys. Rev. A **78**, 052513 (2008)
17. J. Tao, J. P. Perdew, V. N. Staroverov, G. E. Scuseria, Phys. Rev. A **77**, 012509 (2008
18. K. Burke, F. G. Cruz, K. C. Lam, J. Chem. Phys. **109**, 8161 (1998)
19. J. P. Perdew, Y. Wang, in *Mathematics Applied to Science*, ed. by J.A. Goldstein, S. Rosencrans, G. Sod (Academic, New York, 1988)
20. R. Armiento, A. E. Mattsson, Phys. Rev. B **66**, 165117 (2002)
21. L. Vitos, B. Johansson, J. Kollar, H. L. Skriver, Phys. Rev. A **61**, 052511 (2000)
22. L. A. Constantin, A. Ruzsinszky, Phys. Rev. B **79**, 115117 (2009).
23. V. Sahni, J. B. Krieger, J. Gruenebaum, Phys. Rev. B **15**, 1941 (1977)
24. W. Kohn, E. Prodan, Proc. Nat. Acad. Sci. (USA) **102**, 11635 (2005)
25. M. Levy, J. P. Perdew, Phys. Rev. B **31**, 6264 (1985)
26. L. H. Thomas, Proc. Camb. Phil. Soc. **23**, 542 (1927); E. Fermi Z. Phys. **48**, 73 (1928)
27. P. R. Antoniewicz, L. Kleinman, Phys. Rev. B **31**, 6779 (1985)
28. P. S. Svendsen, U. von Barth, Phys. Rev. B **54**, 17392 (1996)
29. S.-K. Ma, K. A. Brueckner, Phys. Rev. **165**, 18 (1968)
30. J. P. Perdew, L. A. Constantin, E. Sagvolden, K. Burke, Phys. Rev. Lett. **97**, 223002 (2006)
31. D. C. Langreth, J. P. Perdew, Phys. Rev. B **21**, 5469 (1980)
32. J. Tao, J. Chem. Phys. **115**, 3519 (2001)
33. J. Tao, M. Springborg, J. P. Perdew, J. Chem. Phys. **119**, 6457 (2003)
34. M. Brack, B. K. Jennings, Y. H. Chu, Phys. Lett. **65B**, 1 (1976)

Orbital-Free Embedding Effective Potential in Analytically Solvable Cases

Andreas Savin and Tomasz A. Wesolowski

Abstract The effective embedding potential introduced by Wesolowski and Warshel [*J.* Phys. Chem., **97** (1993) 8050] depends on two electron densities: that of the environment (n_B) and that of the investigated embedded subsystem (n_A). In this work, we analyze this potential for pairs n_A and n_B, for which it can be obtained analytically. The obtained potentials are used to illustrate the challenges in taking into account the Pauli exclusion principle.

Keywords: Embedding potential · Density functional theory · Kinetic energy functional · Orbital-free embedding

1 Introduction

Computer simulation methods based on the idea of embedding are commonly used in numerical simulations of condensed matter: solids, liquids, interfaces, macromolecules (especially biomolecues), etc. The underlying concept behind the embedding strategy is very simple – one part of a total system is selected to be described by means of quantum mechanical descriptors such as orbitals whereas the remaining part of the whole system is considered as a source of some additional potential. Many strategies to construct embedding potentials are known in various areas of computational chemistry and computational material sciences. Usually, the embedding potential is postulated taking into account system-dependent parameters. Wesolowski and Warshel used the basic concepts of the Hohenberg–Kohn–Sham density functional theory [1, 2]: the reference system of non-interacting electrons, the functional of the kinetic energy in such a system

T.A. Wesolowski (✉)
Department of Physical Chemistry, University of Geneva, Geneva, Switzerland,
e-mail: tomasz.wesolowski@unige.ch

A. Savin
Laboratoire de Chimie Theorique, CNRS and Universite Pierre et Marie Curie (Paris VI), Paris, France, e-mail: andreas.savin@lct.jussieu.fr

P. Piecuch et al. (eds.), *Advances in the Theory of Atomic and Molecular Systems*, Progress in Theoretical Chemistry and Physics 19, DOI 10.1007/978-90-481-2596-8_15, © Springer Science+Business Media B.V. 2009

$(T_s[n])$, and exchange-correlation functional $(E_{xc}[n])$, the external potential $v_{ext}(\mathbf{r})$, and the exchange-correlation potential $v_{xc}[n](\mathbf{r}) = \frac{\delta E_{xc}[n]}{\delta n(\mathbf{r})}$ to obtain one-electron equations for embedded orbitals (Eqs. (20) and (21) in Ref. [3]):

$$\left[-\frac{1}{2}\nabla^2 + v_{ext}^A(\mathbf{r}) + \int \frac{n_A(\mathbf{r}')}{|\mathbf{r}' - \mathbf{r}|} d\mathbf{r}' + v_{xc}[n_A](\mathbf{r}) + v_{KSCED}^{emb}[n_A, n_B](\mathbf{r}) \right] \varphi_i^A = \varepsilon_i \varphi_i^A, \tag{1}$$

where $v_{KSCED}^{emb}[n_A, n_B](\mathbf{r})$ is a system-independent expression for the embedding potential:

$$v_{KSCED}^{emb}[n_A, n_B](\mathbf{r}) = v_{ext}^B(\mathbf{r}) + \int \frac{n_B(\mathbf{r}')}{|\mathbf{r}' - \mathbf{r}|} d\mathbf{r}' + \tag{2}$$

$$\left. \frac{\delta E_{xc}[n]}{\delta n(\mathbf{r})} \right|_{n=n_A+n_B} - \left. \frac{\delta E_{xc}[n]}{\delta n(\mathbf{r})} \right|_{n=n_A} +$$

$$\left. \frac{\delta T_s[n]}{\delta n(\mathbf{r})} \right|_{n=n_A+n_B} - \left. \frac{\delta T_s[n]}{\delta n(\mathbf{r})} \right|_{n=n_A}$$

and where n_A denotes the density constructed from the embedded orbitals φ_i

$$\sum_i^{Nocc} 2|\varphi_i|^2 = n_A(r) . \tag{3}$$

Throughout this chapter, equations are written in atomic units. For the sake of simplicity, equations are given for spin-compensated electron densities: hence the factor 2 in Eq. (3). The acronym KSCED stands for the Kohn–Sham equations with constrained electron density and is used to distinguish the two effective potentials expressed as density functionals: the one in the considered one-electron equations, which involves an additional constraint (see Eq. (5) below), from that in the Kohn–Sham equations.

The above expression for the embedding potential, which was given explicitly in Eq. (3) of Ref. [4], shows clearly that except for $v_{ext}^B(\mathbf{r})$ the position dependency of every other term in $v_{KSCED}^{emb}[n_A, n_B](\mathbf{r})$ is determined by the position dependency of n_A and n_B. The symbol $v_t^{nadd}(\mathbf{r})$ will be used throughout this chapter the last two terms in Eq. (2), i.e., for the difference:

$$v_t^{nadd}[n_A, n_B](\mathbf{r}) = \left. \frac{\delta T_s[n]}{\delta n(\mathbf{r})} \right|_{n=n_A+n_B} - \left. \frac{\delta T_s[n]}{\delta n(\mathbf{r})} \right|_{n=n_A}, \tag{4}$$

arising from the fact that the functional $T_s[n]$ is not additive (see also the subsequent sections).

The embedding potential given in Eq. (2) was obtained by requiring that the ground-state energy of the total system including both the investigated subsystem described by means of Kohn–Sham orbitals and the environment is stationary for a given choice for n_B which is not optimized (frozen). For this reason, we refer colloquially to methods using Eq. (2) as frozen density functional theory (FDFT) [3], frozen density embedding [5], etc. The derivation of Eq. (1) given in Ref. [3] (see also Ref. [6]) provided a new interpretation of embedding methodology in numerical simulations. Embedding calculations can be seen as the constrained optimization problem with the following weak constraint imposed on the total density n:

$$C[n] \geq 0, \tag{5}$$

where $C[n] = \min(n - n_B)$ and n_B is the component of the electron total electron density which is not subject to optimization.

Note that the most common constraints in Euler–Lagrange equation take the form $C[n] = 0$, where $C[n]$ is some density functional. For instance, the constraint $C[n] = \int n(\mathbf{r})d\mathbf{r} - N = 0$ is used in the derivation of the Kohn–Sham equations. Additional constraints expressed as $C[n] = 0$ are also used in some computational schemes such as the procedure to generate diabatic electronic states for the evaluation of the rate of the electron-transfer reaction [7].

The results of a partial optimization of the total electron density in which $n_B(\mathbf{r})$ is frozen depend on the choice made for $n_B(\mathbf{r})$. Unless also n_B is included in the optimization process (see for instance the *freeze-and-thaw* procedure of Ref. [4]), such partial minimization might lead to the total density which differs from the true ground-state electron density of the whole system, n_o. If n_B is chosen to be such that $n_0(\mathbf{r}) - n_B(\mathbf{r}) < 0$ in some domains, the density, n_A, obtained from Eqs. (1) and (3) cannot be equal to the complementary density $n_o - n_B$. Note that non-negativity of the complementary density for any given n_B cannot be verified a priori. Moreover, the densities n_A obtained from Eqs. (1) and (3) are *pure*-state non-interacting v-representable by construction. It is also not possible a priori to tell whether the density $n_o - n_B$ belongs to this class. These concerns can be avoided here because $v_t^{nadd}[n_A, n_B]$ is a functional of two electron densities and it is a well-defined quantity regardless $n_A + n_B$, n_A, and n_B are ground-state densities of some non-interacting systems or not. Note that $v_t^{nadd}[n_A, n_B]$ is defined using the functional $T_s[n]$ for which extension exists (see Definitions and Notations section).

The embedding potential of Eq. (2) has been used as the basis for various computational methods. In most of our own numerical simulations (see for instance Refs. [8–11]), we use Eq. (1), and the partitioning of the total effective potential into its environment-free and embedding components is rather a technical issue. The orbital-free effective embedding potential given in Eq. (2) has been also used outside the domain for which it was derived, i.e., in combination with wave-function-based methods [12–14]. A detailed analysis of such a pragmatic combination of different treatment of the electron–electron interactions in the embedding potential and in the embedded component of the total electron density given in Ref. [15] revealed that it leads to double counting of significant energy contributions if Hartree–Fock method is used in combination with embedding potential given in Eq. (2). The same analyses showed that the magnitude of this double counting can be reduced or even entirely eliminated if the embedded system is described by means of a multi-determinantal "wavefunction." More recently, it was demonstrated that Eq. (2) provides the exact form of the potential to be used if the embedded object is described by means of one-particle reduced density matrices and the corresponding functional [16].

It is worthwhile to recall here that the position dependency of $v_t^{nadd}(\mathbf{r})$ is not explicit but it originates from the inhomogeneity of the densities n_A and n_B. For the same potential, understood as a functional of n_A and n_B, the symbol $v_t^{nadd}[n_A, n_B](\mathbf{r})$ or just $(v_t^{nadd}[n_A, n_B])$ will be used. Such a distinction is of key importance for practical applications of Eq. (1) as they require the use of approximants to the functional $v_t^{nadd}[n_A, n_B]$. Moreover, the use of Eq. (1) in the more general framework of linear-response strategy for excited states involves functional derivatives of $v_t^{nadd}[n_A, n_B]$ with respect to n_A [10, 17].

In order to derive analytic forms of approximants to $v_t^{nadd}[n_A, n_B]$, various strategies are possible. A straightforward one relies on some known approximants to $T_s[n]$ which are used to derive analytic expression for $v_t^{nadd}[n_A, n_B]$ [18, 12, 19]. Such a strategy is based on the assumption that a reasonable approximant for $T_s[n]$ leads to a reasonable approximant to $v_t^{nadd}[n_A, n_B]$. Our dedicated studies on the relation between such approximants [18, 20–22] showed that this assumption is not founded at least for the most common approximants. A more refined strategy relies on the direct analysis of the quality of electron density obtained from a given approximant to $v_t^{nadd}[n_A, n_B]$ [18, 20–22] rather than on the performance of the parent approximant to $T_s[n]$. Note that in calculations based on the Wesolowski–Warshel embedding formalism the absolute values of $T_s[n]$ obtained from an approximant to this functional are not needed. This strategy lead us to the GGA97 approximant [21] which was chosen as the remedy for erratic results obtained using the second-order gradient expression for $T_s[n]$ [18]. Finally, approximants to $v_t^{nadd}[n_A, n_B]$ can be derived using exact properties of this functional as guidelines or imposed conditions. This strategy uses such pairs of n_A and n_B for which the exact dependence of v_t^{nadd} on position is available. For instance, the analysis of the behavior of $v_t^{nadd}[n_A, n_B]$ for $n_A \rightarrow 0$ and $\int n_B d\mathbf{r} = 2$ led us to a new approximant referred to as NDSD (non-decomposable approximant using second derivatives) [23] superior in accuracy to the GGA97 one. It is useful to recall here that a "shortcut strategy" can be applied also in practice. It consists of using some analytic expression for $v_t^{nadd}(\mathbf{r})$ which depends explicitly on position [24]. This way of overcoming flaws of existing approximants to $v_t^{nadd}[n_A, n_B]$ proceeds without constructing any new approximant to $v_t^{nadd}[n_A, n_B]$.

The principal objective of this work is to obtain the exact form of $v_t^{nadd}(\mathbf{r})$ for some cases (choices n_A and n_B). The considered cases make it possible not only to obtain $v_t^{nadd}(\mathbf{r})$ by means of the analytic inversion but also to illustrate the challenges to account for the Pauli exclusion principle by means of a multiplicative potential which is the functional of two densities, $v_t^{nadd}[n_A, n_B]$. In this work, no approximant to $v_t^{nadd}[n_A, n_B]$ is constructed, but the obtained shapes of $v_t^{nadd}(\mathbf{r})$ are to be used as guidelines for constructing approximants to $v_t^{nadd}[n_A, n_B]$ in the future works.

It is worthwhile to point out at this point that this work uses extensively the inversion technique to obtain the external potential in a system of non-interacting electrons (Kohn–Sham system) which yield a given arbitrarily chosen target electron density. Finding numerically the potential associated with a given arbitrarily chosen target electron density is a well-known issue in density functional theory [25–27] and was even used recently [28] for obtaining the embedding potential in an alternative way to that given in Eq. (1). In this work, we analyze specific systems for which the inversion can be made analytically. The analytical inversion applied here cannot be seen, however, as an alternative to the numerical inversion techniques.

2 Definitions and Notations

The key quantity analyzed in this work is the potential $v_t^{nadd}(\mathbf{r})$ which is defined as the difference between the functional derivatives of the functional $T_s[n]$ at two different densities, n. The kinetic energy obtained in Kohn–Sham calculations provides the numerical value of the functional $T_s[n]$ defined in the Levy's constrained search [29]:

$$T_s[n] = \min_{\Psi \to n} \langle \Psi | \hat{T} | \Psi \rangle \ . \tag{6}$$

Such a definition requires that n is pure-state non-interacting v-representable, i.e., it is a ground-state density of some non-interacting system. In the particular cases analyzed later in this work, the density $n_A + n_B$ belongs to this narrower class. Unfortunately, pure-state non-interacting v-representability of n_A cannot be assured. Therefore, interpreting $v_t(\mathbf{r})$ as the functional $v_t[n_A, n_B]$ is justified only if either both the densities $n_A + n_B$ and n_A are pure-state non-interacting v-representable or if the definition of $T_s[n]$ is extendable to a wider class. Owing to the Levy–Lieb [30, 31] extension of $T_s[n]$ for ensembles

$$T_s[n] = \min_{\omega_i, \Psi_i \sum_i \omega_i |\Psi_i|^2 \to n} \sum_i \omega_i \langle \Psi_i | T | \Psi_i \rangle , \tag{7}$$

we shall not be concerned with the restriction of pure-state non-interacting v-representability of the densities considered in this work. Throughout this work, the following convention concerning notation for electron density is used:

n_B is this component of the electron density which is not represented by means of orbitals and which is not optimized,

n_A denotes the density obtained from Eqs. (1) and (3),

n is just the sum of n_A and n_B,

n_o is the exact total ground-state electron density.

3 Analytical Results

3.1 Choice of the System for Obtaining $v^{nadd}[n_A, n_B]$

A fictitious four-electron spherically symmetric system for which the exact Kohn–Sham potential would read $v_{KS} = -\frac{1}{r}$ is considered in this work. The identification of the external potential $v_{ext}(r)$ which would correspond to such simple form of the Kohn–Sham potential is of no concern for the present considerations. For the purposes of the present analyses, it is crucial that the two doubly occupied Kohn–Sham orbitals have the known exact analytic form of the hydrogenic wavefunctions $1s$ and $2s$. Note that the considered Kohn–Sham potential is neither the exact nor a reasonable approximant to the Kohn–Sham potential for a beryllium atom. The model bears some resemblance to a model used for a different purpose (the excited state of a two-electron system) in Ref. [32], Eq. (9). Although the analytical form of the dependence of $v_t^{nadd}[n_A, n_B]$ on n_A and n_B is not obtained, the exact form of $v_t^{nadd}(\mathbf{r})$ can be constructed analytically for various choices made for n_A and n_B in the considered fictitious system. Below, the construction of $v_t^{nadd}(\mathbf{r})$ is outlined.

In this case, the total density of the Kohn–Sham system reads

$$n_o(\mathbf{r}) = 2\left(\phi_{1s}^2 + \phi_{2s}^2\right) , \tag{8}$$

and the Kohn–Sham potential is

$$v_{KS} = -\frac{1}{r} . \tag{9}$$

Let us consider the following decomposition of n_o into the n_A and n_B components which are obtained as the combinations of orbital densities:

$$n_A(\mathbf{r}) = 2\left((1-w)\phi_{1s}^2 + w\phi_{2s}^2\right) , \tag{10}$$

$$n_B(\mathbf{r}) = 2\left(w\phi_{1s}^2 + (1-w)\phi_{2s}^2\right) . \tag{11}$$

At $w = 0$, the frozen density – n_B – is that of the valence, the doubly occupied $2s$ orbital.

At $w = 1$, the frozen density – n_B – is that of the core, i.e., that of the doubly occupied $1s$ orbital.

As w increases from 0 to 1, the density n_B evolves from that localized in the valence to that localized in the core.

In this work, the effective potential which leads to the complementary density $n_A(\mathbf{r})$ is constructed for several choices made for w and consequently for n_B. Since the potential obtained in this way has also the form given in Eqs. (1) and (2),

$v_t^{nadd}(\mathbf{r})$ can be subsequently obtained. The subsequent sections concern the evolution of $v_t^{nadd}(\mathbf{r})$ as w changes.

3.2 Case I: n_B Taken as Valence Electron Density

Let us first choose $w = 0$, i.e., $n_B = 2\varphi_{2s}^2$. The optimal n_A obtained from Eq. (1) with such a choice for n_B equals the complementary density $n_A = 2\varphi_{1s}^2$ only if the potential $v_t^{nadd}(\mathbf{r})$ disappears (or it is constant). Indeed, if $v_t^{nadd} = 0$, the effective potentials in either the Kohn–Sham equations for the total system or in Eq. (1) (for the chosen n_B) are the same. As a consequence, φ_{1s} and φ_{2s} are eigenvectors of either equations and their eigenvalues are ordered in the same way. The lowest eigenvalue corresponds to φ_{1s}, whereas the φ_{2s} is the solution which corresponds to the excited state of the reference system of non-interacting electrons. Therefore, the ground-state orbital obtained from Eq. (1) is also the square root of the target density $n_o - n_B$ (modulo the phase factor). The Aufbau principle is not violated. It is worthwhile to notice that despite the fact that the overlap between two orbital densities $n_A = 2\varphi_{1s}^2$ and $n_B = 2\varphi_{2s}^2$ is non-zero, $v_t^{nadd} = 0$ (or constant). The simplest approximant to $v_t^{nadd}[n_A, n_B]$ derived from local density approximation (i.e., Thomas-Fermi approximant for $T_s[n]$) leads to a non-negative $v_t^{nadd}[n_A, n_B]$ if the densities do overlap. This indicates a systematic flaw of this approximant.

3.3 Case II: n_B Taken as Valence Electron Density with Small Admixture of Core Electron Density

Let us now consider the case when w is very small, but non-zero, i.e., transfer a very small amount of the previous n_B density from the valence shell to the core.

On the one hand, the density to be determined from Eq. (1) is now asymptotically determined by the $2s$ density. As a consequence, its eigenvalue must be equal to that of the $2s$ orbital, i.e., $-1/8$, in order to have a potential which asymptotically goes to zero.

On the other hand, the potential cannot be changed in an important way to yield essentially the same density as for $w = 0$.

The situation reminds of that of going from the system with N electrons to that of $N+1$ electrons: in order to satisfy both requirements, the potential will be essentially shifted by a "constant", $(-1/2)-(-1/8))$, thus shifting the eigenvalue [33–36]. This happens over all space, except for the asymptotic region, where the φ_{2s}^2 dominates. Coming into that region, the potential falls down, finally approaching 0.

The potential can be explicitly calculated, by inverting Eq. (1). Let $v_s(r)$ denote the whole potential in this equation. This potential can be expressed as $v_{KS}[n](r) + v_t^{nadd}[n_A, n_B](r)$ (see Eq. (1)). For the considered system, $v_s(r)$ can be obtained as

$$v_s(r) = \frac{1}{2}\frac{\nabla^2\sqrt{n - n_B}}{\sqrt{n - n_B}} + constant \,, \tag{12}$$

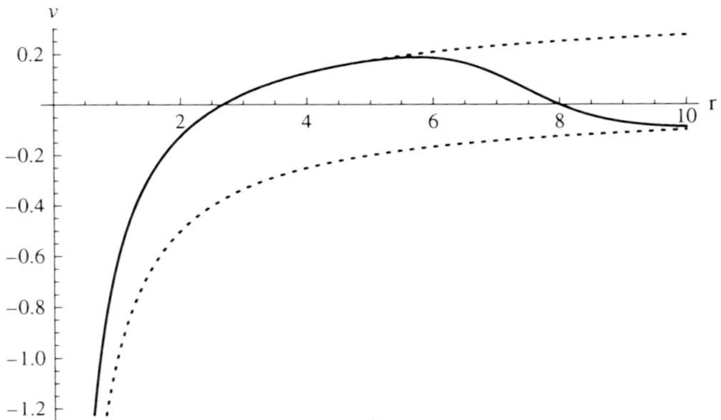

Fig. 1 The potential ($v(r)$ – bold line) obtained from Eq. (12) together with the Kohn–Sham potential ($-1/r$ – lower dotted line) and the shifted Kohn–Sham potential ($-1/r + 1/2 - 1/8$ – upper dotted line) obtained for $w = 0.001$, i.e., $n_A = 0.999n_{1s} + 0.001n_{2s}$ and $n_B = 0.001n_{1s} + 0.999n_{2s}$

choosing the constant such that the potential goes to 0 when $r \to \infty$. Note that both $v_s(r)$ and $v_{KS}(r)$ are external potentials in a reference system of non-interacting electrons which are associated with different densities – $v_s(r)$ with $n - n_B$ whereas $v_{KS}(r)$ with n. In the chosen example, $v_{KS}(r) = -1/r$. The difference between the two potentials obtained for $w = 0.001$ is shown in Fig. 1

$v_t^{nadd}(\mathbf{r})$ is shown in Fig. 2 for the same $w = 0.001$, together with $4\pi r^2 n_A(r)$. On the scale of the plot, the contribution of the $2s$ density is not noticeable. For higher values of w, it appears as a shoulder or for even higher values of w yields a second maximum in $4\pi r^2 n_A(r)$.

As the contribution of the $2s$ density increases, the jump in $v_t^{nadd}(r)$ is further displaced toward the origin, as can be seen by comparing Figs. 1 and 3 which show $v(r)$ for $w = 0.001$ and $w = 0.01$, respectively.

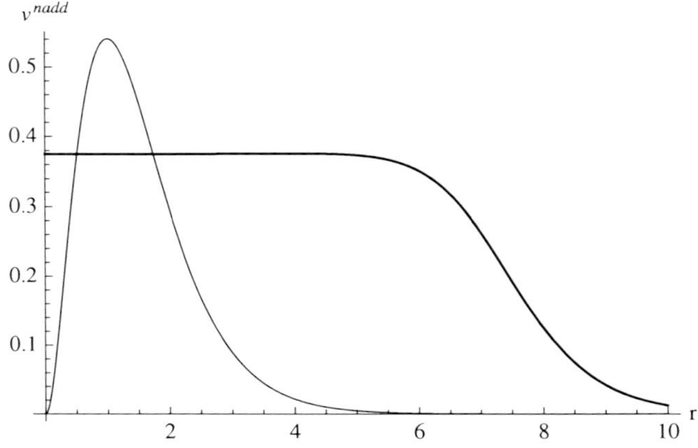

Fig. 2 $v_t^{nadd}(r)$ (bold line) $-1/r$, obtained for $w = 0.001$, i.e., $n_A = 0.999n_{1s} + 0.001n_{2s}$ and $n_B = 0.001n_{1s} + 0.999n_{2s}$. The arbitrarily normalized plot of $4\pi r^2 n_A$ is also shown (thin line)

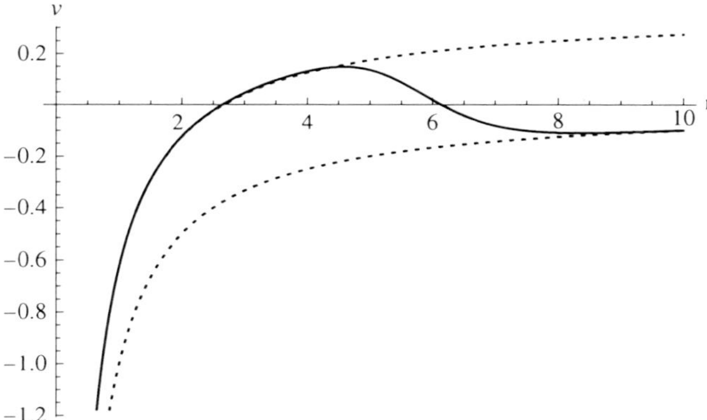

Fig. 3 The potential ($v(r)$ – bold line) obtained from Eq. (12) together with the Kohn–Sham potential ($-1/r$ – lower dotted line) and the shifted Kohn–Sham potential ($-1/r + 1/2 - 1/8$ – upper dotted line) obtained for $w = 0.01$, i.e., $n_A = 0.99n_{1s} + 0.01n_{2s}$ and $n_B = 0.01n_{1s} + 0.99n_{2s}$

3.4 Case III: n_B Taken as Valence Electron Density with Noticeable Admixture of Core Electron Density

As the contribution of the $2s$ density to n_A (i.e., the density to be determined from Eq. (1)) increases, a bump starts to be apparent in the plot of the inverted potential $v(r)$ (compare Figs. 3, 4, and 5).

The bump in $v(r)$ originates from the rapid variation of v_t^{nadd} component (compare Figs. 4 and 5 with Figs. 6 and 7, respectively).

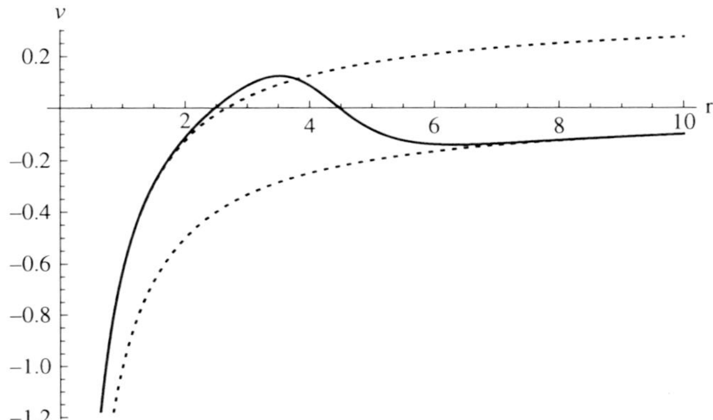

Fig. 4 The potential ($v(r)$ – bold line) obtained from Eq. (12) together with the Kohn–Sham potential ($-1/r$ – lower dotted line) and the shifted Kohn–Sham potential ($-1/r + 1/2 - 1/8$ – upper dotted line) obtained for $w = 0.1$, i.e., $n_A = 0.9n_{1s} + 0.01n_{2s}$ and $n_B = 0.1n_{1s} + 0.9n_{2s}$

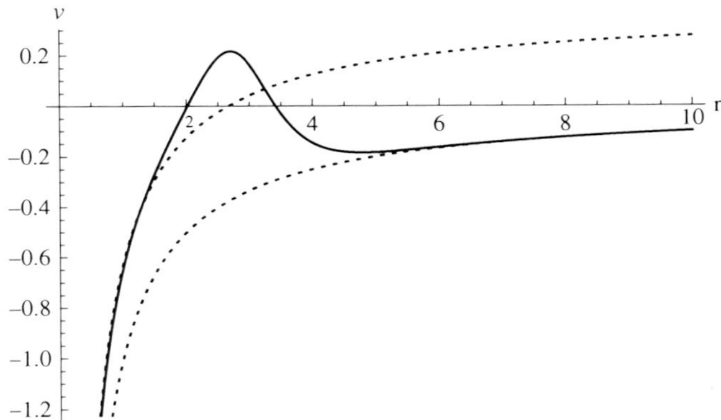

Fig. 5 The potential ($v(r)$ – bold line) obtained from Eq. (12) together with the Kohn–Sham potential ($-1/r$ – lower dotted line) and the shifted Kohn–Sham potential ($-1/r + 1/2 - 1/8$ – upper dotted line) obtained for $w = 0.5$, i.e., $n_A = 0.5n_{1s} + 0.5n_{2s}$ and $n_B = 0.5n_{1s} + 0.5n_{2s}$

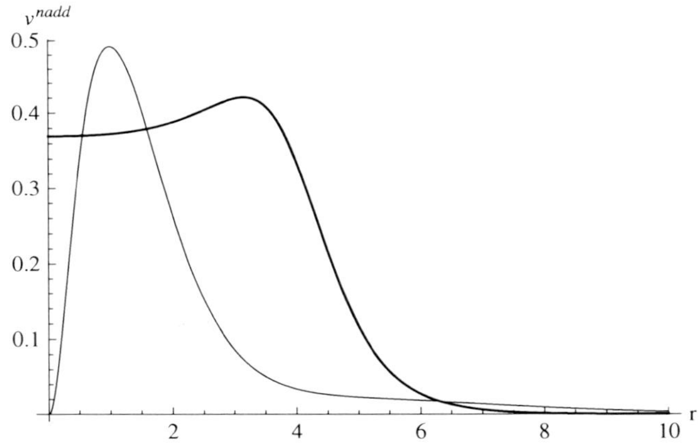

Fig. 6 $v_t^{nadd}(r)$ (bold line) $-1/r$, obtained for $w = 0.9$, i.e., $n_A = 0.9n_{1s} + 0.1n_{2s}$ and $n_B = 0.1n_{1s} + 0.9n_{2s}$. The arbitrarily normalized plot of $4\pi r^2 n_A$ is also shown (thin line)

3.5 Case IV: n_B Taken as Core Electron Density with Small Admixture of Valence Electron Density

The bump in $v(r)$ (or in v_t^{nadd}) gets more and more pronounced as w increases. Figs. 8 and 9 show $v(r)$ for $w = 0.9$ and for $w = 0.99$, respectively.

The origin of the dip in the density (and the bump in the potential needed to produce it) is clear: the contribution of the 2s density to n_A increases with w; at 100 per cent 2s density, one even has $n_A(r = 2) = 0$.

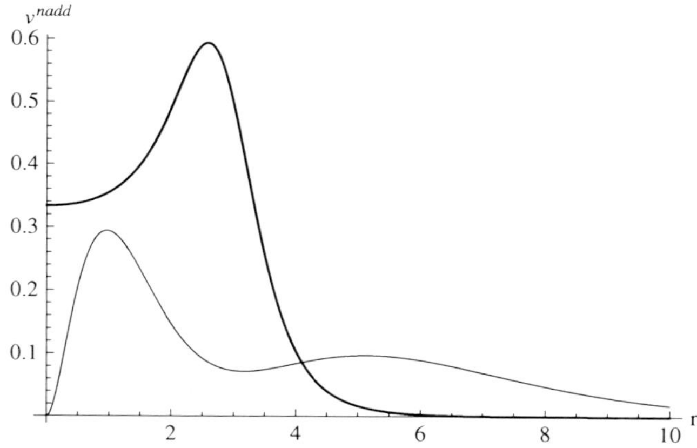

Fig. 7 $v_t^{nadd}(r)$ (*bold line*) $-1/r$, obtained for $w = 0.5$, i.e., $n_A = 0.5n_{1s} + 0.5n_{2s}$ and $n_B = 0.5n_{1s} + 0.5n_{2s}$. The arbitrarily normalized plot of $4\pi r^2 n_A$ is also shown (*thin line*)

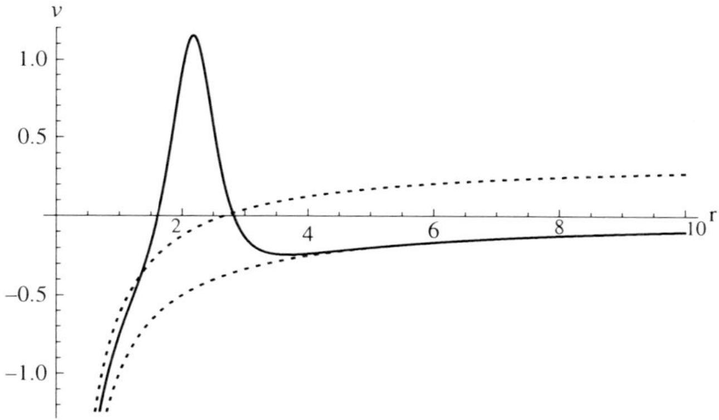

Fig. 8 The potential ($v(r)$ – *bold line*) obtained from Eq. (12) together with the Kohn–Sham potential ($-1/r$ – *lower dotted line*) and the shifted Kohn–Sham potential ($-1/r + 1/2 - 1/8$ – *upper dotted line*) obtained for $w = 0.9$, i.e., $n_A = 0.1n_{1s} + 0.9n_{2s}$ and $n_B = 0.9n_{1s} + 0.1n_{2s}$

3.6 Case V: n_B Taken as Core Electron Density

At the first sight, it seems that when n_B is the core electron density ($n_B = 2 \cdot 1s^2$), the density $n_o - n_B$ can be obtained from Eqs. (1) and (3) by putting $v_t^{nadd}(r) = 0$, i.e., for $v(r) + v_t^{nadd}(r) = -1/r$. Indeed, the orbital $2s$ is one of the eigenfunctions in the equation $\left[-\frac{1}{2}\nabla^2 - 1/r\right]\varphi = \varepsilon\varphi$. It is, however, not *the* lowest eigenvalue. For $v_t^{nadd}(r) = 0$, Eqs. (1) and (3) would lead, therefore, to the total electron density $n_A + n_B = 4 \cdot 1s^2$ which would violate the Pauli exclusion principle. Satisfying the requirement that the ground-state orbital obtained from Eq. (1) yields such density n_A that $n_A = n_o - n_B$ for n_B being the core electron density must be reflected in the

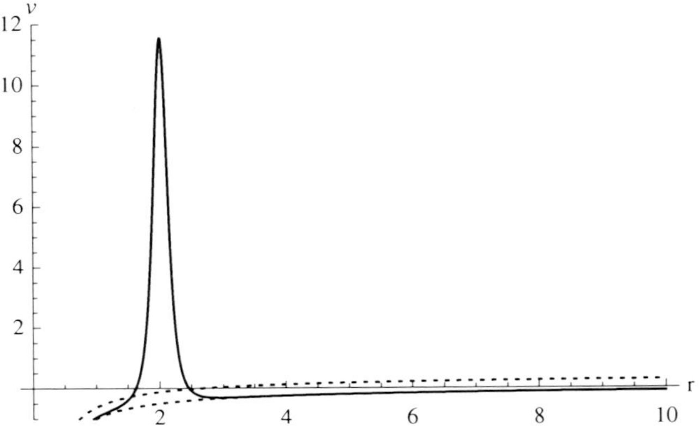

Fig. 9 The potential ($v(r)$ – bold line) obtained from Eq. (12) together with the Kohn–Sham potential ($-1/r$ – lower dotted line) and the shifted Kohn–Sham potential ($-1/r + 1/2 - 1/8$ – upper dotted line) obtained for $w = 0.99$, i.e., $n_A = 0.01n_{1s} + 0.99n_{2s}$ and $n_B = 0.99n_{1s} + 0.01n_{2s}$

$v_t^{nadd}(r)$ component of the total potential. The solution to this paradox is provided by the existence of the bump which is transformed into a barrier that is infinitely high and vanishingly thin. This barrier can be seen as a change in the boundary conditions on the Schrödinger equation.

Figure 11 shows the appearance of the barrier-like character of $v_t^{nadd}(r)$ for $w = 0.999$.

A final question remains to be clarified. The integral of n_A on each of the segments (r between 0 and 2 and from 2 to infinity, respectively) is a non-integer number. What is the wave function in such a case? In fact, in this case we have to deal with two spatial regions: the inner sphere and the outer spherical shell. The $2s$ orbital is an eigenfunction of the Hamiltonian on both segments and the eigenvalue is, of course, the same ($-1/8$). To obtain the ground state with the density given, we use the nodeless functions on each of the segments; next, we produce ensembles with weights equal to the integral of the $2s$ orbital from 0 to 2 and from 2 to infinity (0.053 and 0.947, respectively) and get the correct density.

$v_t^{nadd}(r)$ evolves thus from $v_t^{nadd}(r) = 0$ at $w = 0$ to a delta-like potential as w approaches 1. It is worthwhile to look more closely at the $w = 0.5$ case. If $w = 0.5$, the not-optimized component of the electron density, n_B, is exactly the same as the target electron density $n_A = n_o - n_B = 1/2n_o$. The equation $[-\frac{1}{2}\nabla^2 + v(r)]\varphi = \varepsilon\varphi$, where $v(r)$ is the potential shown in Fig. 5, leads to such eigenfunction that $2\varphi^2 = n_o/2$. It is worthwhile to notice that no orbital representation of the density n_B has been used so far in our considerations. In the $w = 0.5$ case, one can trivially represent also n_B by means of an orbital given by the square root of $n_B = 1/2n_o$. Since $n_A = n_B$ the two "orbitals" are the same (module phase factor) and are obviously strongly non-orthogonal. Nevertheless, Eq. (1) still leads to the exact ground-state electron density for the total system. This example illustrates that the embedded orbitals obtained from Eq. (1) and orbitals obtained from some

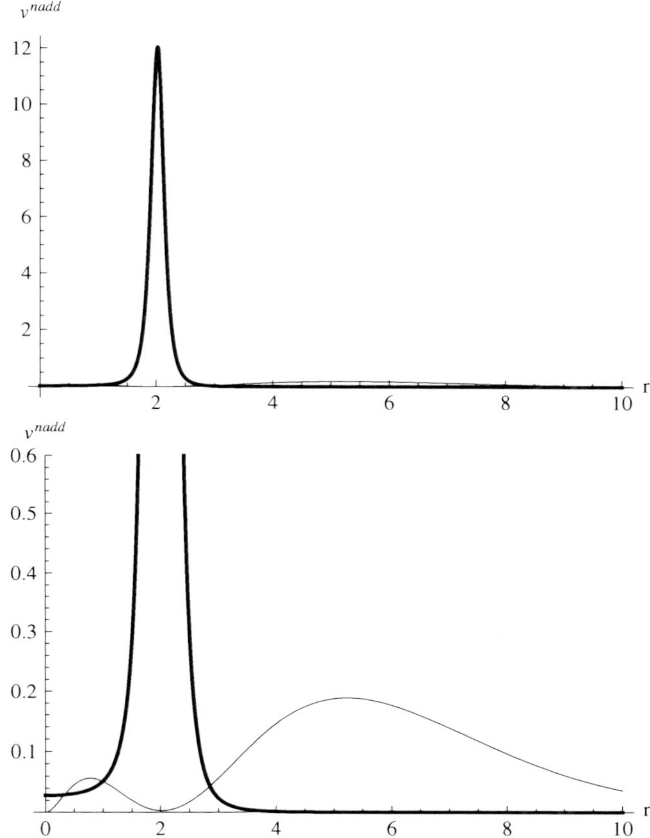

Fig. 10 $v_t^{nadd}(r)$ (*bold* line) $-1/r$, obtained for $w = 0.9$, i.e., $n_A = 0.9n_{1s} + 0.1n_{2s}$ and $n_B = 0.1n_{1s} + 0.9n_{2s}$. The arbitrarily normalized plot of the radial density, $4\pi r^2 n_A$, is also shown (*thin* line). The *upper* and *lower* figures show the same potential at different ranges

reconstruction of the assumed density n_B represent different objects which should not be confused with any wave function representation of the total system.

4 Conclusions

In the embedding formalism introduced by Wesolowski and Warshel [3], the total electron density is partitioned into two components. One of them is not optimized (frozen) and the other is subject to optimization. The optimized component is treated in a Kohn–Sham-like way, i.e., by means of a reference system of non-interacting electrons. The multiplicative potential in one-electron equations for embedded orbitals, Eq. (1) or Eqs. (20) and (21) of Ref. [3], differs from the Kohn–Sham

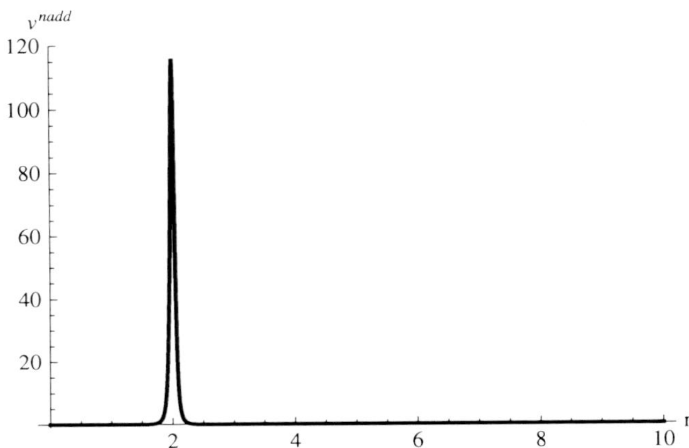

Fig. 11 $v_t^{nadd}(r)$ (*bold line*) $-1/r$, obtained for $w = 0.999$, i.e., $n_A = 0.999n_{1s} + 0.001n_{2s}$ and $n_B = 0.001n_{1s} + 0.999n_{2s}$. The arbitrarily normalized plot of the radial density, $4\pi r^2 n_A$, is also shown (*thin line*)

potential by an additional term arising from non-additivity of the kinetic energy, $v_t^{nadd}(r)$. The question arises how such multiplicative potential takes into account the Pauli exclusion principle. In particular, if the non-optimized component of the total density coincides with the orbital density of the lowest lying Kohn–Sham orbital of the whole system, how to avoid that the complementary density obtained from Eq. (1) collapses on the not-optimized one?

A simple example is analyzed of the case when the *exact* effective potential in one-electron equations for embedded orbitals (Eqs. (20) and (21) of Ref. [3]) can be obtained analytically. The Kohn–Sham orbitals for the considered total system are the hydrogenic functions 1s and 2s. For particularly partitioned total density of such system, in which the not-optimized component of the electron density is a mixture of 1s and 2s orbital densities, the exact potential in Eq. (1) can be obtained analytically. In the considered examples, the density n_A obtained from Eq. (1) complements perfectly the chosen not-optimized one, n_B, so they add to the exact ground-state density. Depending on the choice for n_B and n_A resulting from it the following situations are observed:

- If n_A is the density that is given by the lowest-lying orbital the Kohn–Sham potential for the whole system coincides with the effective potential in Eq. (1). No additional potential is needed ($v_t^{nadd}(r) = 0$).
- If n_A is essentially given by the lowest-lying orbital, with only a small admixture of the higher-lying orbtials, the additional potential is mainly shifted in the region of interest; this shift aligns the orbital energy of the lower-lying orbital to that of the higher-lying one, allowing the mixing of the densities of different states.

- As the contribution to n_A due to the higher-lying level increases, a bump shows up in the nodal region of the corresponding orbital (density). The bump forbids electrons to enter the region close to the node of the high-lying orbital.
- If n_A is the density of the high-lying orbital, the bump is transformed into an infinite barrier and an ensemble description has to be used for n_A. In this case, $v_t^{nadd}(r)$ becomes an infinitely high barrier of vanishing thickness.

Turning back to the issue of developing approximants to $v_t^{nadd}[n_A, n_B]$, i.e., representing the potential $v_t^{nadd}(\mathbf{r})$ as functional of the two electron densities without introducing any explicit position-dependency, the analytic results obtained in this work exposed rather serious difficulties if one would aim at some universal, system- and also partitioning-independent approximant to $v_t^{nadd}[n_A, n_B]$. For instance, the obtained shapes of the $v_t^{nadd}(\mathbf{r})$ indicate that this potential changes abruptly with minor modifications of n_A and n_B. Moreover, examples were given where $v_t^{nadd}[n_A, n_B] \neq v_t^{nadd}[n_B, n_A] = const$ and the densities n_A and n_B do overlap and are far from uniform.

Acknowledgements The authors thank Prof. Pietro Cortona for helpful discussions. T. A. W. and A. S. acknowledge the support from the grants by Swiss National Research Foundation (Project 200020-116760) and ANR (Project 07-BLAN-0272), respectively.

References

1. P. Hohenberg, W. Kohn, Phys. Rev. B **136**, 864 (1964)
2. W. Kohn, L. J. Sham, Phys. Rev. **140**, A1133 (1965)
3. T. A. Wesołowski, A. Warshel, J. Phys. Chem. **97**, 8050 (1993)
4. T. A. Wesolowski, J. Weber, Chem. Phys. Lett. **248**, 71 (1996)
5. ADF 2003.01, SCM, Theoretical Chemistry, Vrije Universiteit, Amsterdam, The Netherlands, http://www.scm.com.
6. T. A. Wesolowski, in *Computational Chemistry: Reviews of Current Trends*, vol. X, ed. by J. Leszczynski (World Scientific, Singapore, 2006), p. 1
7. Q. Wu, T. Van Voorhis, J. Chem. Phys. **125**, 164105 (2006)
8. T. A. Wesołowski, Chem. Phys. Lett. **311**, 87 (1999)
9. M. Zbiri, M. Atanasov, C. Daul, J.-M. Garcia Lastra, T. A. Wesolowski, Chem. Phys. Lett. **397**, 441 (2004)
10. T. A. Wesolowski, J. Am. Chem. Soc. **126**, 11444 (2004)
11. M. Leopoldini, N. Russo, M. Toscano, M. Dulak, T. A. Wesolowski, Chem. Eur. J. **12**, 2532 (2006)
12. T. Klüner, N. Govind, Y. A. Wang, E. A. Carter, J. Chem. Phys. **116**, 42 (2002)
13. D. Lahav, T. Kluner, J. Phys. – Cond. Matt. **19**, 226001 (2007)
14. C. R. Jacob, L. Visscher, J. Chem. Phys. **128**, 155102 (2008)
15. T. A. Wesolowski, Phys. Rev. A **77**, 012504 (2008)
16. K. Pernal, T. A. Wesolowski, Int. J. Quantum. Chem. **109**, 2520 (2009)
17. M. Casida, T. A. Wesolowski, Int. J. Quantun Chem. **96**, 577 (2004)
18. T. A. Wesolowski, J. Weber, Int. J. Quantum Chem. **61**, 303 (1997)
19. M. Hodak, W. Lu, J. Bernholc, J. Chem. Phys. **128**, 014101 (2008)
20. T. A. Wesolowski, H. Chermette, J. Weber, J. Chem. Phys. **105**, 9182 (1996)

21. T. A. Wesolowski, J. Chem. Phys. **106**, 8516 (1997)
22. Y. A. Bernard, M. Dulak, J. W. Kaminski, T. A. Wesolowski, J. Phys. A **41**, 055302 (2008)
23. J.-M. Garcia Lastra, J. W. Kaminski, T.A. Wesolowski, J. Chem. Phys. **129**, 074107 (2008)
24. C. R. Jacob, S. M. Beyhan, L. Visscher, J. Chem. Phys. **126**, 234116 (2007)
25. Q. Zhao, R. C. Morrison, R. G. Parr, Phys. Rev. A **50**, 2138 (1994)
26. E. J. Baerends, Phys. Rev. Lett. **87**, 133004 (2001)
27. W. Yang, Q. Wu, Phys Rev. Lett. **89**, 143002 (2002)
28. O. Roncero, M. P. de Lara-Castells, P. Villarreal, F. Flores, J. Ortega, M. Paniagua, A. Aguado, J. Chem. Phys. **129**, 184104 (2008)
29. M. Levy, Proc. Natl. Acad. Sci. USA **76**, 6062 (1979)
30. M. Levy, Phys. Rev. A **26**, 1200 (1982)
31. E. H. Lieb, Int. J. Quantum Chem. **24**, 243 (1983)
32. J. P. Perdew, M. Levy, Phys. Rev. B **31**, 6264 (1985)
33. J. P. Perdew, R. G Parr, M. Levy, J. L. Balduz Jr., Phys. Rev. Lett. **49**, 1691 (1982)
34. D. J. Tozer, N. C. Handy, J. Chem. Phys. **109**, 10180 (1998)
35. F. Colonna, A. Savin, J. Chem. Phys. **110**, 2828 (1999)
36. F. Della Sala, A. Görling, Andreas, Phys. Rev. Lett. **89**, 033003 (2002)

A Simple Analytical Density Model for Atoms and Ions Based on a Semiexplicit Density Functional

Ignacio Porras and Francisco Cordobés-Aguilar

Abstract In this work a semiexplicit density energy functional is minimized with an analytical closed form for the density of an atom or ion. From this approach, an asymptotic formula for the energy in terms of Z and N is obtained. The resulting atomic density approximates the Hartree–Fock result, averaging over shell effects.

Keywords: Density functionals · Density models · Thomas–Fermi–Dirac

1 Introduction

For some computational techniques in quantum chemistry a simple zero-th order approximation of the electron density of any atom of the system can be useful as the starting point of an iterative procedure. A very simple description of the electron density and binding energy of any atom or ion allows a rapid evaluation of very complex structures. This is the spirit of the orbital-free, explicit density functional approaches, usually based on the Thomas–Fermi–Dirac model and its extensions [1].

In addition to this, the dependence of some atomic density-dependent properties with respect to the nuclear charge and the number of electrons are only shown up by means of simple density models like the semiclassical Thomas–Fermi one and its modifications [2]. As an example, from this approach the following simple formula estimates reasonably well the total binding energy of neutral atoms:

$$E = -0.7687Z^{7/3} + \frac{1}{2}Z^2 - 0.294Z^{5/3}\text{a.u.} \tag{1}$$

However, this formula is found by including the Scott correction [3] to the energy of the Thomas–Fermi–Dirac model, and this approach does not provide a correspondingly corrected density.

I. Porras (✉), F. Cordobés-Aguilar
Departamento de Física Atómica, Molecular y Nuclear, Facultad de Ciencias,
Universidad de Granada, E-18071 Granada, Spain,
e-mail:porras@ugr.es, co.cordobes@gmail.com

P. Piecuch et al. (eds.), *Advances in the Theory of Atomic and Molecular Systems*,
Progress in Theoretical Chemistry and Physics 19, DOI 10.1007/978-90-481-2596-8_16,
© Springer Science+Business Media B.V. 2009

Furthermore, analytical models of the electron density are useful for testing density functional approaches. These models are usually restricted to few electron atoms.

However, the application of semiclassical models derived from the Thomas–Fermi approach has some drawbacks, including an overestimation of the electron density near the nucleus, which affects the dependence of low-order radial expectation values with Z, and the impossibility of describing atomic anions [1].

In the past few years, we have proposed a model for correcting the wrong description of the electron density near the nucleus of the semiclassical approaches [4]. This provides values of average atomic properties at the accuracy of the Hartree–Fock procedure with a very simple approach [5], allows relativistic extensions [6], and can describe negative ions [7].

The goal of this work is to apply this model to find a simple analytical form for the electron density which provides reasonable energy estimates. By expressing this semiclassical approach as the minimization of a semiexplicit density functional, we will minimize variationally a given form for the density which approximates very well the numerical solution of the model. This density will be useful for further applications, as will be discussed in the conclusions.

2 Density Functional Model of Atoms

The energy functional of our model, which approximates the ground state energy of an atom or ion, is obtained as a density functional by using single-particle orbitals for describing electrons in a near-nucleus region $\mathcal{R}_1 = \{r \leq r_0\}$, and local plane waves in the outer region $\mathcal{R}_2 = \{r > r_0\}$. Outside the near-nucleus region, the potential acting on any electron can be assumed as slowly varying, and therefore it is justified to use local plane waves, as in the Thomas–Fermi–Dirac-based approaches, for a basic shell-averaged description. In this region exchange can be approximately included by means of the Dirac form. In the near-nucleus region, where the nuclear attraction in dominant and the previous assumption fails, we assume that we can describe the system by single-particle orbitals under an effective potential that can be expanded in a power series of the radial coordinate r. Assuming this latter region is small, we neglect its contribution to the exchange energy, which depends on integrals over the region of products of different orbitals which can be assumed to overlap slightly for small r. With these assumptions [8], the following energy functional is obtained[1]:

$$E[\rho] = E_1[\{\phi_i(\rho)\}] + E_2[\rho] + U_{12}[\{\phi_i(\rho)\}, \rho], \tag{2}$$

where

$$E_1 = T_1[\{\phi_i(\rho)\}] + V_1[\{\phi_i(\rho)\}] + U_{11}[\{\phi_i(\rho)\}], \tag{3}$$

---

[1] Atomic units are used throughout this chapter.

$$E_2 = T_2[\rho] + V_2[\rho] + U_{22}[\rho] + K_2[\rho],\tag{4}$$

and the particular terms are

$$T_1 = \sum_{i=1}^{N} \int_{\mathcal{R}_1} d\mathbf{r}\, \phi_i^*(\mathbf{r}) \left(-\frac{1}{2}\nabla^2\right) \phi_i(\mathbf{r}),\tag{5}$$

$$V_1 = \sum_{i=1}^{N} \int_{\mathcal{R}_1} d\mathbf{r}\, \phi_i^*(\mathbf{r}) \left(-\frac{Z}{r}\right) \phi_i(\mathbf{r}),\tag{6}$$

$$U_{11} = \frac{1}{2} \sum_{i,j=1}^{N} \int_{\mathcal{R}_1} d\mathbf{r} \int_{\mathcal{R}_1} d\mathbf{r}'\, \phi_i^*(\mathbf{r})\phi_j^*(\mathbf{r}')\frac{1}{|\mathbf{r}-\mathbf{r}'|}\phi_i(\mathbf{r})\phi_j(\mathbf{r}'),\tag{7}$$

$$U_{12} = \sum_{i=1}^{N} \int_{\mathcal{R}_1} d\mathbf{r}\, \phi_i^*(\mathbf{r})\phi_i(\mathbf{r}) \int_{\mathcal{R}_2} d\mathbf{r}'\, \frac{\rho(\mathbf{r}')}{|\mathbf{r}-\mathbf{r}'|},\tag{8}$$

$$T_2 = C_k \int_{\mathcal{R}_2} d\mathbf{r}\, [\rho(\mathbf{r})]^{5/3}, \qquad C_k = \frac{3}{10}\left(3\pi^2\right)^{2/3},\tag{9}$$

$$V_2 = \int_{\mathcal{R}_2} d\mathbf{r}\, v(\mathbf{r})\rho(\mathbf{r}),\tag{10}$$

$$U_{22} = \frac{1}{2} \int_{\mathcal{R}_2} d\mathbf{r} \int_{\mathcal{R}_2} d\mathbf{r}'\, \frac{\rho(\mathbf{r})\rho(\mathbf{r}')}{|\mathbf{r}-\mathbf{r}'|},\tag{11}$$

and

$$K_2 = C_e \int_{\mathcal{R}_2} d\mathbf{r}\, [\rho(\mathbf{r})]^{4/3}, \qquad C_e = -\frac{3}{4}\left(\frac{3}{\pi}\right)^{1/3}.\tag{12}$$

The terms of E_1 and U_{12} depend implicitly on ρ by means of the relationship

$$\rho(\mathbf{r}) = \sum_{i=1}^{N} \phi_i^*(\mathbf{r})\phi_i(\mathbf{r}).\tag{13}$$

We propose a spherical density for minimizing $E[\rho]$ of the form

$$\rho(r) = \begin{cases} \rho_1(r) & r \le r_0 \\ \rho_2(r) & r > r_0 \end{cases},\tag{14}$$

with the condition that the average radial density $\rho(r)$ and its first derivative $\rho'(r)$ be continuous at $r = r_0$. This has two different expressions which are obtained below.

The energy will be evaluated separating contributions from both regions in the way:

$$E[\rho] = \left(E_1 + \frac{1}{2}U_{12} \right) + \left(E_2 + \frac{1}{2}U_{12} \right) = \tilde{E}_1 + \tilde{E}_2, \tag{15}$$

where $\tilde{E}_1$ and $\tilde{E}_2$ denote respectively the first and second terms in parenthesis.

2.1 Density and Energy from $\mathcal{R}_1$

For $r < r_0$ a minimum is found when the density is constructed by means of single-particle orbitals which satisfy

$$\left[-\frac{1}{2}\nabla^2 + V(\mathbf{r}) \right] \phi_i(\mathbf{r}) = \varepsilon_i \phi_i(\mathbf{r}), \qquad \mathbf{r} \in \mathcal{R}_1, \tag{16}$$

where we have introduced $V(\mathbf{r})$ (the effective potential), which denotes

$$V(\mathbf{r}) = -\frac{Z}{r} + \int \frac{\rho(\mathbf{r}')}{|\mathbf{r} - \mathbf{r}'|}\, d\mathbf{r}'. \tag{17}$$

An approximate solution to Eq. (16) can be found assuming that, for small r_0, the effective potential can be approximated by a small-r expansion in which we will only retain the two first terms:

$$V(r) = -\frac{Z}{r} + V_0 + O(r^2), \tag{18}$$

where $V_0 = \langle r^{-1} \rangle$. For an isolated atom or ion, the term of order r is zero. In the case of external fields, this could not be the case and then it should be taken into account.

Then we solve, for s-states, the previous equation up to order r^3 by using the expression

$$\phi_i(r) = a_i \left[1 + b_i r + c_i r^2 + d_i r^3 + O(r^4) \right], \tag{19}$$

which gives

$$b_i = -Z, \tag{20}$$

$$V_0 - \varepsilon_i = 3c_i - Z^2, \tag{21}$$

$$d_i = \frac{Z}{6}(Z^2 - 4c_i). \tag{22}$$

The total density can be written then in terms of the constants

$$A = \sum_i a_i^2, \qquad C = \frac{\sum_i c_i a_i^2}{\sum_i a_i^2}, \tag{23}$$

with the result

$$\rho_1(r) = A\left[1 - 2Zr + (2C + Z^2)r^2 + \frac{Z}{3}(Z^2 - 10C)r^3 + O(r^4)\right]. \tag{24}$$

The number of electrons in region $\mathcal{R}_1$, denoted by N_1, is given by

$$N_1 = \int_{\mathcal{R}_1} \rho_1(r)\, d\mathbf{r} = \frac{4}{3}\pi A r_0^3 \left[1 - \frac{3}{2}Zr_0 + \frac{3}{5}(2C + Z^2)r_0^2 + \frac{1}{6}Z(Z^2 - 10C)r_0^3\right]. \tag{25}$$

Using Eqs. (16) and (18), the energy term $\tilde{E}_1$ can be approximated by

$$\begin{aligned}
\tilde{E}_1 &= \int_{\mathcal{R}_1} d\mathbf{r} \sum_i \phi_i^*(r)\left[\varepsilon_i - \frac{1}{2}\left(V(r) + \frac{Z}{r}\right)\right]\phi_i(r) \\
&= \int_{\mathcal{R}_1} d\mathbf{r} \sum_i |\phi_i|^2 \left[\frac{1}{2}V_0 - 3c_i + Z^2 + O(r^2)\right] \\
&\approx \left(\frac{1}{2}\langle r^{-1}\rangle - 3C + Z^2\right) N_1.
\end{aligned} \tag{26}$$

In the previous expression the radial expectation value $\langle r^{-1}\rangle$ appears. The contribution to this quantity from this region $\mathcal{R}_1$ is given by

$$\begin{aligned}
\langle r^{-1}\rangle_1 &= \int_{\mathcal{R}_1} \frac{1}{r}\rho_1(r)\, d\mathbf{r} \\
&= 2\pi A r_0^2 \left[1 - \frac{4}{3}Zr_0 + \frac{1}{2}(2C + Z^2)r_0^2 + \frac{2}{15}Z(Z^2 - 10C)r_0^3\right]. \tag{27}
\end{aligned}$$

2.2 Density and Energy from $\mathcal{R}_2$

The minimization of the energy functional with respect to the density for $r > r_0$ leads to the integral equation of the Thomas–Fermi–Dirac model restricted to this region. This was solved numerically with some constraints that must be imposed because of a wrong asymptotic behavior of the exact solution when $r \rightarrow \infty$, which is usually circumvented by cutting the electron distribution at a finite radius [7, 8]. But in this work we will adopt a different strategy which will give us a more physical density and with an analytical closed form. As the aim of our approach is to find tractable description of atoms, we will minimize the energy function with a variational procedure for a simple density form:

$$\rho_2(r) = \frac{B}{r^{3/2}} e^{-\sigma(r-r_0)}. \tag{28}$$

This form is obtained from the small r behavior of the Thomas–Fermi–Dirac density and an exponential decreasing behavior. This is a very simple form that approximates reasonably well the values of $\rho_2(r)$ obtained from the numerical resolution mentioned above.

The contributions to the norm and to $\langle r^{-1} \rangle$ from this region are obtained by means of a change of variable $u = \sigma(r - r_0)$ with the result

$$\begin{aligned}
N_2 &= \int_{\mathcal{R}_2} \rho_2(r)\, d\mathbf{r} = 4\pi B\sigma^{-3/2} \int_0^\infty (u + \sigma r_0)^{1/2} e^{-u}\, du \\
&= 4\pi B\sigma^{-3/2} \frac{\sqrt{\pi}}{2} \left\{ e^{\sigma r_0} \left[1 - \mathrm{Erf}(\sqrt{\sigma r_0}) \right] + \frac{2}{\sqrt{\pi}} \sqrt{\sigma r_0} \right\},
\end{aligned} \tag{29}$$

and

$$\begin{aligned}
\langle r^{-1} \rangle_2 &= \int_{\mathcal{R}_2} \frac{1}{r} \rho_2(r)\, d\mathbf{r} = 4\pi B\sigma^{-1/2} \int_0^\infty (u + \sigma r_0)^{-1/2} e^{-u}\, du \\
&= 4\pi B\sigma^{-1/2} \sqrt{\pi}\, e^{\sigma r_0} \left[1 - \mathrm{Erf}(\sqrt{\sigma r_0}) \right],
\end{aligned} \tag{30}$$

where $\mathrm{Erf}(z)$ denotes the error function, defined by

$$\mathrm{Erf}(z) = \frac{2}{\sqrt{\pi}} \int_0^z e^{-t^2}\, dt. \tag{31}$$

Now we evaluate the contributions to the energy:

$$\tilde{E}_2 = T_2 + V_2 + \frac{1}{2} U_{12} + U_{22} + K_2 : \tag{32}$$

$$\begin{aligned}
T_2 &= C_k \int_{\mathcal{R}_2} [\rho_2(r)]^{5/3}\, d\mathbf{r} \\
&= 4\pi C_k B^{5/3} \sigma^{-1/2} \int_0^\infty (u + \sigma r_0)^{-1/2} e^{-\frac{5}{3}u}\, du \\
&= 4\pi C_k B^{5/3} \sigma^{-1/2} \sqrt{\frac{3\pi}{5}}\, e^{\frac{5}{3}\sigma r_0} \left[1 - \mathrm{Erf}\left(\sqrt{\frac{5\sigma r_0}{3}} \right) \right],
\end{aligned}$$

$$V_2 = -Z \langle r^{-1} \rangle_2, \tag{33}$$

$$\frac{1}{2} U_{12} = \frac{1}{2} N_1 \langle r^{-1} \rangle_2, \tag{34}$$

$$
\begin{aligned}
U_{22} &= (4\pi)^2 \int_{r_0}^{\infty} r^2 \rho_2(r)\, dr \int_{r}^{\infty} r' \rho_2(r')\, dr' \\
&= (4\pi)^2 B^2 \sigma^{-2} \int_{0}^{\infty} (u + \sigma r_0)^{1/2} e^{-u}\, du \int_{u}^{\infty} (v + \sigma r_0)^{-1/2} e^{-v}\, dv \\
&= (4\pi)^2 \sqrt{\pi}\, B^2 \sigma^{-2} e^{\sigma r_0} \int_{0}^{\infty} (u + \sigma r_0)^{1/2} e^{-u} \left[1 - \mathrm{Erf}(\sqrt{u + \sigma r_0}) \right],
\end{aligned} \tag{35}
$$

$$K_2 = C_e \int_{\mathcal{R}_2} [\rho_2(r)]^{4/3}\, d\mathbf{r} = 3\pi C_e B^{4/3} \sigma^{-1}. \tag{36}$$

Therefore, the energy is written in terms of the error function and the integral in Eq. (35), which can be done by numerical quadrature for any value of the product σr_0.

2.3 Matching Conditions and Parameter Scaling

In the previous equations, the parameters which appear are r_0, A, C, B, and σ. The role of the matching radius r_0, which is the only input parameter of this model has been discussed in previous work [8]. Different conditions for fixing r_0 for all atoms, i.e., the continuity of the energy density, the obtention of the Scott correction for the energy of the non-interacting electrom atom, or the best fit of the total atomic binding energy, lead to similar values: r_0 scales with Z, being the leading term approximately equal to $r_0 = 1/(2Z)$. We will adopt this value from the start.

From this assumption we can find the scaling of the other parameters. As it will be shown below, all the results are consistent with the following assumptions for the main dependence with Z:

$$C = \overline{C} Z, \tag{37}$$

$$A = \overline{A}\, Z^3, \tag{38}$$

$$B = \overline{B}\, Z^{3/2}, \tag{39}$$

$$\sigma = \overline{\sigma} Z N^{-2/3}, \tag{40}$$

where the last condition is required by the correct normalization.

The value of A and C are related to the other parameters by means of the continuity of $\rho(r)$ and its derivative at $r = r_0$. If we match the ratio

$$-\frac{\rho_2'(r_0)}{Z\rho_2(r_0)} = 3 + \frac{\sigma}{Z} \tag{41}$$

to $-\rho_1'(r_0)/(Z\rho_1(r_0))$, we obtain

$$\overline{C} = \frac{3 + 7x}{6 - 2x}, \tag{42}$$

where we have introduced

$$x \equiv \frac{\sigma}{Z} = \overline{\sigma} N^{-2/3}. \tag{43}$$

Thus, x does not depend on Z at first order.

The value of $\overline{A}$ is related to $\overline{B}$ by matching $\rho_1(r_0)$ to $\rho_2(r_0)$), which leads to

$$\overline{A} = 3\sqrt{2\overline{B}}\left(1 - \frac{x}{3}\right). \tag{44}$$

Finally, the value of B must be fixed from the normalization condition. The contribution to the norm from both regions are

$$N_1 = \pi\overline{B}\,\frac{14 - x}{15\sqrt{2}}, \tag{45}$$

and N_2 given by Eq. (29). This means that N_1 is independent of Z and N at first order.

The value of $\overline{B}$ can be related to σ from the condition $N = N_1 + N_2$, which leads to

$$\overline{B} = \frac{30x^{3/2}N}{\pi^{3/2}60e^{x/2}\left[1 - \mathrm{Erf}(\sqrt{x/2})\right] + \sqrt{2x}\pi(60 + 14x - x^2)}, \tag{46}$$

which is independent of N provided that x scales with $N^{-2/3}$, which agrees with Eq. (43).

Then, all parameters are expressed as a function of the only variational parameter $\overline{\sigma}$, by means of $x = \overline{\sigma} N^{-2/3}$. $\overline{\sigma}$ is assumed to be independent of Z and N. In the large N-limit, x can be treated as a perturbation, which leads to simple asymptotic expressions for the energy, as will be described below.

3 Asymptotic Expressions for the Energy in the Large Z and N Limit

We will assume, as in the standard applications of the statistical method, that we are describing an atom or ion with both large values of Z and N. Then, we will expand the results for the different energy terms in power series of $x^{1/2}$, which is proportional to $N^{-1/3}$.

From Eq. (26), using Eqs. (25), (37), (38), (42), (43), and (44), we find the contribution $\tilde{E}_1$:

$$
\begin{aligned}
\tilde{E}_1 &= -Z^2 N \left[\frac{7}{30\sqrt{2\pi}} x^{3/2} + O\left(x^{5/2}\right) \right] \\
&= -\frac{7}{30\sqrt{2\pi}} \bar{\sigma}^{3/2} Z^2 + O\left(Z^3 N^{-2/3}\right),
\end{aligned}
\tag{47}
$$

and from Eqs. (33), (34), (35), and (36), we obtain the contributions to $\tilde{E}_2$, by expanding the error function and the expression for $\overline{B}$ (Eq. (46)), and retaining consistently the leading terms:

$$
\begin{aligned}
T_2 &= Z^2 N^{5/3} \left[\frac{3^{13/6}\pi^{1/3}}{2^{2/3}5^{3/2}} x^2 - \frac{3^{5/3}\pi^{1/3}}{5(2\pi)^{1/6}} x^{5/2} + O\left(x^{7/2}\right) \right] \\
&= \frac{3^{13/6}\pi^{1/3}}{2^{2/3}5^{3/2}} \bar{\sigma}^2 Z^2 N^{1/3} - \frac{3^{5/3}\pi^{1/3}}{5(2\pi)^{1/6}} \bar{\sigma}^{5/2} Z^2 + O\left(Z^2 N^{-2/3}\right),
\end{aligned}
\tag{48}
$$

$$
\begin{aligned}
V_2 &= N Z^2 \left[-2x + 2\sqrt{\frac{2}{\pi}} x^{3/2} + O\left(x^{5/2}\right) \right] \\
&= -2\bar{\sigma} Z^2 N^{1/3} + 2\sqrt{\frac{2}{\pi}} \bar{\sigma}^{3/2} Z^2 + O\left(Z^2 N^{-2/3}\right),
\end{aligned}
\tag{49}
$$

$$
\begin{aligned}
\frac{1}{2} U_{12} &= Z N^2 \left[\frac{7}{15\sqrt{2\pi}} x^{5/2} + O\left(x^3\right) \right] \\
&= \frac{7}{15\sqrt{2\pi}} \bar{\sigma}^{5/2} Z N^{1/3} + O(Z),
\end{aligned}
\tag{50}
$$

$$
\begin{aligned}
K_2 &= Z N^{4/3} \left[\frac{3^{7/3}}{4(2\pi)^{4/3}} x + O\left(x^2\right) \right] \\
&= \frac{3^{7/3}}{4(2\pi)^{4/3}} \bar{\sigma} Z N^{2/3} + O(Z).
\end{aligned}
\tag{51}
$$

For U_{22}, given by Eq. (35), with $\sigma r_0 = x/2$, we make use of:

$$\int_0^\infty (u + x/2)^{1/2} e^{-u} \left[1 - \mathrm{Erf}(\sqrt{u + x/2}) \right] = \frac{\sqrt{\pi}}{2} e^{x/2} \left[1 - \mathrm{Erf}(\sqrt{x/2}) \right] + \sqrt{\frac{x}{2}} +$$
$$+ \int_0^\infty (u + x/2)^{1/2} e^{-u} \mathrm{Erf}(\sqrt{u + x/2}),$$

$$(52)$$

and work with the last integral as follows, relating the error function to the Kummer hypergeometric function $M(a; c; z)$:

$$\int_0^\infty du \left(u + \frac{x}{2} \right)^{1/2} e^{-u} \mathrm{Erf}\left(\sqrt{u + \frac{x}{2}} \right) = \frac{2}{\sqrt{\pi}} \int_0^\infty du \left(u + \frac{x}{2} \right) e^{-2u - \frac{x}{2}}$$
$$M\left(1; \frac{3}{2}; u + \frac{x}{2} \right)$$
$$= \frac{2}{\sqrt{\pi}} e^{x/2} \int_{x/2}^\infty dt\, t\, e^{-2t} M(1; 3/2; t)$$
$$= \frac{2}{\sqrt{\pi}} e^{x/2} \int_0^\infty dt\, t\, e^{-2t} M(1; 3/2; t) +$$
$$- \frac{2}{\sqrt{\pi}} e^{x/2} \int_0^{x/2} dt\, t\, e^{-2t} M(1; 3/2; t)$$
$$\sim \frac{\pi + 2}{8} \frac{2}{\sqrt{\pi}} e^{x/2} + O(x^2).$$

$$(53)$$

Therefore, we find

$$U_{22} = ZN^2 \left[\frac{\pi - 2}{\pi} x + \frac{\sqrt{2}(3\pi - 10\sqrt{\pi} - 6)}{15\pi^{(3/2)}} x^{5/2} + O\left(x^3 \right) \right]$$
$$= \frac{\pi - 2}{\pi} \overline{\sigma} Z N^{4/3} + \frac{\sqrt{2}(3\pi - 10\sqrt{\pi} - 6)}{15\pi^{(3/2)}} \overline{\sigma}^{5/2} Z N^{1/3} + O(Z).$$

$$(54)$$

In the previous expressions we have been consistent and retained the following terms in powers of $N^{1/3}$: $Z^2 N^{1/3}$, $Z^2 N^0$, and $Z^2 N^{-1/3}$ as well as from $ZN^{4/3}$ to $ZN^{2/3}$. For neutral atoms, this mean that we are describing the terms with $Z^{7/3}$, Z^2, and $Z^{5/3}$, which includes exchange effects and with a remainder ($Z^{4/3}$) of the order of the correlation energy (not considered in this work).

Joining all terms, and using numerical values for the coefficients, we find the final expression for the energy in terms of the variational parameter $\overline{\sigma}$:

$$E(Z, N) = (0.891944\overline{\sigma}^2 - 2\overline{\sigma})Z^2 N^{1/3} + (1.50268\overline{\sigma}^{3/2} - 0.91876\overline{\sigma}^{5/2})Z^2 +$$
$$+ 0.36338\overline{\sigma} Z N^{4/3} - 0.279887\overline{\sigma} Z N^{2/3}.$$

$$(55)$$

It can be noticed how the electron–electron interaction in region $\mathcal{R}_1$ contributes to terms of smaller order $(ZN^{1/3})$ than those retained in this equation. This means that some of the approximations performed (exclusion of exchange in this region and therefore inclusion of nonphysical self interaction) do not play any role in the final results.

The minimization for any atom of given Z and N can be performed numerically very easily. We particularize the results for neutral atoms. For $Z = N$, this formula gives

$$E(Z, Z) = (-1.63662\overline{\sigma} + 0.891944\overline{\sigma}^2)Z^{7/3} + (1.50268\overline{\sigma}^{3/2} - 0.91876\overline{\sigma}^{5/2})Z^2 +$$
$$- (0.279887\overline{\sigma} + 0.0559453\overline{\sigma}^{5/2})Z^{5/3}.$$

$$(56)$$

The leading term has a minimum for $\overline{\sigma} = 0.917445$. With this value this term is equal to $-0.750754Z^{7/3}$, close to the first term of Eq. (1).

The minimization of this quantity leads to energy estimations which compare reasonably well to the values obtained from restricted Hartree–Fock calculations [9], taking into account the simplicity of the model. This is illustrated in Table 1, where the values of $\overline{\sigma}$ which minimize Eqs. (55) and (56) are also displayed.

Table 1 Total energies for some neutral atoms and cations obtained from the present approach compared to Hartree–Fock values [9]

Z	N	$\overline{\sigma}$	$-E$	$-E(\mathrm{HF})$
10	9	0.954	116.32	127.82
10	10	0.947	117.85	128.55
20	19	0.920	623.76	676.57
20	20	0.920	627.72	676.75
30	29	0.913	1657.5	1777.5
30	30	0.914	1665.2	1777.9
40	39	0.910	3309.4	3538.8
40	40	0.912	3321.8	3539.0
50	50	0.910	5670.2	6023.0
60	60	0.910	8771.8	9284.0
70	70	0.909	12680	13392
80	80	0.909	17443	18409
90	90	0.909	23103	24360

It is remarkable to check how $\overline{\sigma}$ is approximately constant for different values of Z, a result which agrees with the scaling assumptions.

The errors of the energy prediction by this formula decrease from 7 to 3%. We know from previous work that the numerical solution of the model provides average

Fig. 1 Electron density of
Krypton given by Eqs. (14),
(24), and (28) (*solid line*),
with the parameters obtained
from the minimization of Eq.
(56), compared to
Hartree–Fock values [9]

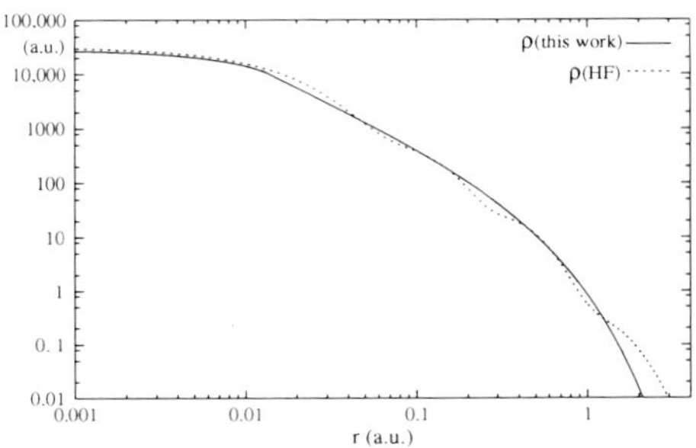

properties with errors less than 1%. The present model can be adjusted by two ways:
(i) a fine tuning of Zr_0 (including a dependence in N) or (ii) inclusion of factors
depending on more variational parameters in the model density for $r \geq r_0$, both at
the cost of losing simplicity in the model.

The comparison of the density provided by this model, given by Eqs. (14), (24),
and (28), with Hartree–Fock values of $\rho(r)$ is illustrated in Fig. 1 for the particular
case of Krypton ($Z = 36$). For this atom, the value of the parameters are $\overline{\sigma} =
0.9056, \overline{A} = 0.6137, \overline{C} = 0.6139$, and $\overline{B} = 0.07438$. The resulting density is very
accurate near the nucleus and averages shell effects for larger distances. However, it
has a steeper asymptotic decrease at large r.

4 Conclusions

In this work a simple analytical atomic density model is obtained from the expression of a modified Thomas–Fermi–Dirac model with quantum corrections near the
nucleus as the minimization of a semiexplicit density functional. The use of a simple
exponential analytical form for the density outside the near-nucleus region and the
resolution of a single-particle Schrödinger equation with an effective potential near
the origin allows us to solve easily the problem and obtain an asymptotic expression
for the energy of an atom or ion in terms of the nuclear charge Z and the number of
electrons N.

This provides a simple description that can be useful for different problems, for
example, to study polarization effects in atoms. Because of its simplicity this model
can be applied to electron-atom scattering problems in which the polarization of the
electron cloud from the incoming electron can be taken into account by including
angular dependence in the variational density. Another line of application is the
study of Rydberg atoms beyond the frozen-core approximation.

Acknowledgements This work has been supported in part by the Spanish Ministerio de Ciencia e
Innovación (FPA2008-03488) and from the Junta de Andalucía (FQM-220).

References

1. R. G. Parr, W. Yang, *Density Functional Theory of Atoms and Molecules* (Oxford University Press, New York, 1989)
2. B.-G. Englert, *Semiclassical Theory of Atoms* (Springer-Verlag, Berlin, 1988)
3. J. Scott, Phil. Mag **43**, 859 (1952)
4. I. Porras, A. Moya, Phys. Rev. A **59**, 1859 (1999)
5. A. Moya, I. Porras, in *Quantum Systems in Chemistry and Physics*, Vol. 1, *Basic Problems and Model Systems*, ed. by S. Wilson, J. Maruani, A. Hernández-Laguna, Vol. 2 of *Progress in Theoretical Chemistry and Physics* (Kluwer, Dordrecht, 2000), pp. 215–225
6. I. Porras, A. Moya, in *Quantum Systems in Chemistry and Physics*, Vol. 1, *Basic Problems and Model Systems*, ed. by S. Wilson, J. Maruani, A. Hernández-Laguna, Vol. 2 of *Progress in Theoretical Chemistry and Physics* (Kluwer, Dordrecht, 2000), pp. 195–213
7. I. Porras, A. Moya, Int. J. Quantum Chem. **99**, 288 (2004)
8. I. Porras, J. Math. Chem. (2009) DOI 10.1007/s10910-009-9554-0
9. T. Koga, K. Kanayama, S. Watanabe, A. J. Thakkar, Int. J. Quantum Chem. **71**, 491 (1999)

Part VI
Advances in Concepts and Models

The Jahn–Teller Effect: Implications in Electronic Structure Calculations

Isaac B. Bersuker

Abstract An updated and extended formulation of the generalized Jahn–Teller effect (JTE) is given, which includes the proof of the general validity of this effect as *the only source of instabilities and distortions of high-symmetry configurations* of polyatomic systems. The immediate implication in electronic structure calculations is that when unstable states are considered, the excited electronic states that produce the instability of the ground state should be involved in the calculations. It is shown that when molecular systems are distorted, but there are no apparent degeneracies or close in energy states, *the JTE are hidden in the excited states* of the undistorted configuration, even when the energy gap to these states is very large. For molecular systems with half-closed-shell electronic configurations e^2 and t^3, which produce totally symmetric charge distribution and are not subject to the JTE, distortions were shown to occur due to the *strong pseudo-JT mixing of two excited states*, and the distortion is accompanied by *orbital disproportionation*. In some systems, this produces two coexisting states, low-spin (LS) distorted and high-spin (HS) undistorted, and a novel phenomenon: *JT-induced spin-crossover*.

Keywords: Jahn–Teller Effect · Configurational instability · Electronic structure Calculations · Orbital disproportionation · Spin crossover

1 Updated Formulation of the Generalized Jahn–Teller Effect

The generalized Jahn–Teller effect (JTE) [including proper JTE, pseudo-JTE (PJTE), and Renner–Teller effects (RTE)] in its present understanding is a local feature of *any* polyatomic system that describes its properties in high-symmetry configurations

I.B. Bersuker (✉)
Department of Chemistry and Biochemistry, The University of Texas at Austin, Austin, TX 78712, USA, e-mail: bersuker@cm.utexas.edu

P. Piecuch et al. (eds.), *Advances in the Theory of Atomic and Molecular Systems*, Progress in Theoretical Chemistry and Physics 19, DOI 10.1007/978-90-481-2596-8_17, © Springer Science+Business Media B.V. 2009

with respect to nuclear displacements from this configuration [1]. This understanding is essentially enlarged and much different from that introduced by E. Teller [2] based on a discussion with L. Landau [3] and widely used in textbooks. The latter continue to treat the JTE as a small effect of instability and spontaneous distortion relevant to specific situations of electronic degeneracy in nonlinear molecules, which is not entirely true. In the modern formulation (see below) the JTE is possible, in principle, in any polyatomic systems without a priori exceptions.

If not restricted to the special case of electronic degeneracy, interactions of electronic states with nuclear displacements that lie in the basis of the JTE look like the well-known general electron-vibrational (in molecules) and electron–phonon (in crystals) interactions. In fact, however, JT vibronic couplings are different from the general cases, and the difference is due to the different number of electronic states involved in the interaction with vibrations. In the usual approach, the interaction of one electronic nondegenerate (usually ground) state or band with vibrations is considered, and therefore it is nonzero for totally symmetric vibrations only. Distinguished from this general case, the JTE involves necessarily *two or more electronic states (bands)*, degenerate or with a limited energy gap between them (pseudo-degenerate), which allow for interaction also with low-symmetry nuclear displacements. The latter may produce peculiar (unusual) adiabatic potential energy surfaces (APES) with conical intersections, instabilities, distortions, and pseudo-rotations, and a variety of important observable properties, jointly termed JTE.

Since two or more electronic states and low-symmetry nuclear displacements are present in any quantum polyatomic system with more than two atoms, there are no a priori exceptions from possible JTE in such systems. The question is only that, dependent on the system parameters, the JTE may be small and may be unobservable directly. For nuclear configurations with zero energy gaps between the interacting electronic states (exact degeneracy), the APES has no minimum due to the JTE (see below for some limited exceptions), but if the vibronic coupling constants are small, there is only splitting of vibrational frequencies and no structural instability. This is true also for weak RTE. The weak PJTE just softens (lowers the vibrational frequency of) the state under consideration in the direction of the active coordinate, but in many cases this softening cannot be observed directly as we do not know the primary frequency without the PJT interaction (still there are indirect indications of the PJTE in this case too). The strong PJTE results in instability and distortions which can be observed directly via a variety of consequences for observable properties [1]. The latter may be qualitatively different for JT, PJT, and RT effects, respectively.

This modern understanding of the JTE is based on the latest achievements of the theory. In the primary ("primitive") formulation based on the JT theorem [2] in which only exact degeneracy and the interaction with only linear terms of vibronic coupling were taken into account, the JTE states that *in electronic degenerate states of nonlinear molecular systems the nuclear configuration is unstable with respect to low-symmetry distortions that remove the degeneracy*, the twofold spin degeneracy being an exception. The limitation of linear vibronic coupling resulted in the exclusion of linear molecules making them an exception from the JTE; with the inclusion of quadratic terms of vibronic coupling linear molecules in degenerate states may become unstable; this is the RTE.

The limitation of exact degeneracy was first removed by Opik and Pryce [4], but they considered that the degeneracy is lifted by a small perturbation transforming the point of degeneracy into an avoiding crossing one, for which the JTE remains, albeit slightly modified. The idea was essentially extended much later to include interactions with any excited states (via large energy gaps) and to show that this interaction is of fundamental importance as it is the only source of instabilities and distortions in polyatomic systems in nondegenerate states. Because of its extreme importance and to introduce some denotations used below, we bring here a simple formulation of the PJTE.

Consider the APES of a two-level system with the ground state 1 and excited state 2 and an energy gap between them Δ which interacts (mix) under the symmetrized nuclear displacement Q_Γ. Using perturbation theory with respect to the linear vibronic coupling term $(\partial H/\partial Q_\Gamma)_0 Q_\Gamma$ we easily obtain [1] that the primary curvature (the curvature without vibronic coupling) of the ground state K_0^Γ,

$$K_0^\Gamma = \langle 1 \left| (\partial^2 H/\partial Q_\Gamma^2)_0 \right| 1 \rangle, \tag{1}$$

is lowered by the amount $(F_\Gamma^{12})^2/\Delta_{12}$,

$$K^\Gamma = K_0^\Gamma - (F_\Gamma^{12})^2/\Delta_{12}, \tag{2}$$

where F_Γ^{12} is the PJT vibronic coupling constant,

$$F_\Gamma^{12} = \langle 1 \left| (\partial H/\partial Q_\Gamma)_0 \right| 2 \rangle. \tag{3}$$

Similarly, for a multilevel problem in the linear approximation,

$$K^\Gamma = K_0^\Gamma - \sum_j (F_\Gamma^{1j})^2/\Delta_{1j}. \tag{4}$$

At the point of extremum of the APES in the Q_Γ direction,

$$\langle 1 \left| (\partial H/\partial Q_\Gamma)_0 \right| 1 \rangle = 0 \tag{5}$$

(we call this point high-symmetry configuration), the curvature K^Γ coincides with the force constant, the latter is thus the sum of two terms:

$$K^\Gamma = K_0^\Gamma + K_v^\Gamma, \tag{6}$$

where K_0^Γ after Eq. (1) is the rigidity of the system with regard to Q_Γ displacements of the nuclei in the fixed electron distribution, while the negative PJT vibronic coupling contribution K_v^Γ stands for the softening of the system in this direction due to electrons partly following the nuclei (Eqs. (2) and (4) are well-known second-order perturbation theory expression but they remain the same in a two-level or multi-level problem; by no means should they be considered as just small perturbations). If

$$|K_v^\Gamma| > K_0^\Gamma \tag{7}$$

(or for a two-level system $\Delta < F_\Gamma^2/K_0^\Gamma$), the force constant (5) is negative and the system is unstable in the direction Q_Γ.

Thus, the condition (7) is *sufficient* to make the system unstable. But is it a *necessary* condition? In other words, can the system become unstable beyond the condition (7), that is, can the inequality $K_0^\Gamma < 0$ be realized? We succeeded to show [5] that at the extremum point (5)

$$K_0^\Gamma > 0 \tag{8}$$

always, meaning that the PJT coupling to the appropriate excited states is the only possible source of instability of the ground state high-symmetry configuration (a similar statement can be formulated for the instability of excited states). This means also that the condition $|K_v^\Gamma| > K_0^\Gamma$ is both *necessary and sufficient* for instability of the systems.

For atoms, the condition (8) is trivial. Indeed, since the charge distribution around the nucleus obeys the condition of minimum energy, any displacement of the nucleus in the fixed electron cloud (equivalent to the displacement of the latter with respect to the fixed nucleus) will increase the energy. This argumentation (of some authors) does not hold for molecules because when there are two or more nuclei, the energy minimum of charge distribution for fixed nuclei does not mean energy minimum with regard to nuclear displacements; the later may decrease the nuclear repulsion. Nevertheless, it was shown both analytically and by ab initio calculations [5, 6] that the condition (8) at the point (5) is valid *always*.

Thus with these proofs, two important additions to the previous traditional understanding of the JTE emerged: (1) any polyatomic system may be subject to the JTE (in its extended formulation including the PJTE) and (2) if there are instabilities and distortions of high-symmetry configurations, they are due to and only due to the JTE. Together with the role of quadratic terms of the vibronic coupling, the extended formulation of the JTE that includes the latest achievements in this field is as follows [1, 7]:

> The necessary and sufficient condition of instability (lack of minimum of the APES) of high-symmetry configurations of any polyatomic system is the presence of two or more electronic states, degenerate or nondegenerate, that are interacting sufficiently strong under the nuclear displacements in the direction of instability, the twofold spin degeneracy being an exception.

As compared with the previous formulation, this one does not restrict the JTE to exact degeneracies and excludes other mechanism of instability. The only restriction is the requirement of "high-symmetry configuration" in the sense of symmetry-induced degenerate states and Eq. (5) for nondegenerate states. The meaning of the latter requirement is that the system should be force-equilibrated; if there is no extremum of the APES in nondegenerate states there is no equilibrium configuration and no problem of instability. The twofold spin degeneracy is an obvious exception from the JTE since in accordance with the Kramers theorem only magnetic interactions can remove this degeneracy whereas the vibronic coupling is pure electrostatic.

The consequences of the extended formulation of the JTE are vast for both fundamental understanding of the origin of molecular and solid-state structure and applications [1]. In particular, it leads directly to the conclusion that all the structural symmetry breakings in molecular systems and condensed matter are triggered by the JTE [8]. Together with the statement in particle physics that "symmetry breaking is always associated with a degeneracy" [9] (this statement includes also a particle analog of pseudo-degeneracy), we may speculate that *nature tends to avoid degeneracies by means of symmetry breaking.* In molecular systems and condensed matter, this tendency is conveyed via the JTE.

The statement "nature tends to avoid degeneracies" should be understood in the sense that the degeneracy will be removed provided there are degrees of freedom to do it. In the absence of such appropriate degrees of freedom, the degeneracy remains. For instance, in an isolated system in a degenerate electronic E state the degeneracy will be removed by the JTE, while the double degeneracy of the ground vibronic level will be removed by the Coriolis interaction. Another example is the above-mentioned Kramers twofold spin degeneracy which can be lifted only in the presence of magnetic fields, e.g., the magnetic field of the nuclei.

As for practical applications of the generalized JTE in its extended formulation, they are numerous and continuously increasing involving such important fields as molecular shapes, stereochemistry, chemical activation and mechanism of chemical reactions, all-range spectroscopy, electron-conformational changes in biology, impurity physics, lattice formation and phase transitions, etc. [1].

In this chapter, we limit the presentation by some recent achievements in this field showing how the JTE directly affects electronic structure calculations and serves as a tool for rationalization of the results. Presenting a most general analytical model to which the computational results should be related the JTE conveys the computer experiments of ab initio calculations to the theory of electronic and vibronic structure.

2 Influence of Excited States on Ground State Geometry: The PJTE

The extended formulation of the generalized JTE above states that the necessary and sufficient condition of instability of the high-symmetry configuration of any polyatomic system is the presence of two or more electronic states that interact sufficiently strongly under the nuclear displacements in the direction of instability. Configurational instabilities are present in a vast majority of processes in chemistry, physics, and biology, including, e.g., transition states of chemical reactions, conformational changes in biology, and phase transitions in physics (for examples see [1]). In all these situations, the possible instabilities (their trigger mechanisms) are controlled by the *two or more electronic states;* there is no instability within just one electronic state.

If the ground state in the high-symmetry configuration is degenerate (meaning it has the two or more electronic states), it may produce the instability by itself. More often the ground electronic state is nondegenerate, and then the instability is possible *only* if there are appropriate excited states that produce the necessary PJT interaction. In this way we get a general approach to (a tool for) solving molecular and solid-state problems in which the excited states acquire a key role: they determine both the possibility and the direction of instability (e.g., the mechanism of the elementary act of the chemical reaction starting at its transition state).

The role of excited states in the formation of the instability of the ground state comes out clearly from the practice of ab initio calculations. Indeed, it is well known that in general, one cannot get instability and energy barriers without including some representation of the corresponding excited states in the basis set or in the singles of configuration interaction. The negative PJT contribution of the excited states to the curvature of the APES of the ground state resulting in its instability was confirmed also directly by means of ab initio calculations for specific molecular systems (see, e.g., in [1, 6, 10]). A direct probe of the role of excited atomic states (in the basis sets) in getting the instability was performed recently (P. Garcia-Fernandez, I. B. Bersuker, unpublished). The CaF_2 molecule was shown by ab initio calculations to have a bent geometry in its ground state due to the PJT instability of the linear configuration with the main contribution to the PJTE due to the excited states formed mainly by the excited atomic d states of Ca [11]. Based on this information one can predict that by excluding the atomic d states from the basis set there will be no instability. Indeed, ab initio CCSD(T) calculations with a F–cc-pvtz basis set yield a bent configuration for CaF_2 in the ground state with an angle $\alpha = 154°$ and interatomic distance $R = 2.006\,Å$ when the full basis set is involved, and a linear configuration with $\alpha = 180°$ and $R = 2.059\,Å$ if the excited atomic d states are excluded from the basis set (P. Garcia-Fernandez, I. B. Bersuker, unpublished).

The role of the JTE is most important in interpretation (rationalization) of experimental results including results of ab initio calculations. With regard to the latter the JTE may serve as a general (based on first principles) analytical model for understanding and generalizations. In the majority of cases the results of ab initio calculations are published "as they are" with discussion of methods used and accuracies achieved in comparison with other similar calculations, which is an important problem by itself, also because they yield necessary numbers good for comparison with the experimental data. But very rarely the question is raised why the results are "as they are," meaning what is the origin of the molecular characteristics obtained from the calculations. As an illustrative example we mention the results of a recent paper entitled *Why are some* ML_2 *molecules* (M=Ca, Sr, Ba; L=H, F, Br) *bent while others are linear?* [11]. While the ab initio calculations only yield that some of these molecules are linear and others are bent, the analysis of the results from the point of view of the PJTE shows convincingly what is the difference in the electronic structure of the atoms that makes the molecular geometry different. Indeed, Fig. 1

shows the MO scheme for such ML_2 molecules from which it is seen that the PJT mixing of the HOMO σ_u formed mainly by the ligand orbitals with the unoccupied π_g orbitals formed by mainly central atomic d orbitals may produce the odd (bending) nuclear displacements. Obviously, both the energy gap between these states Δ and the vibronic coupling constant F are specific for the atoms M and L, and only some of them obey the condition of instability (7) [11].

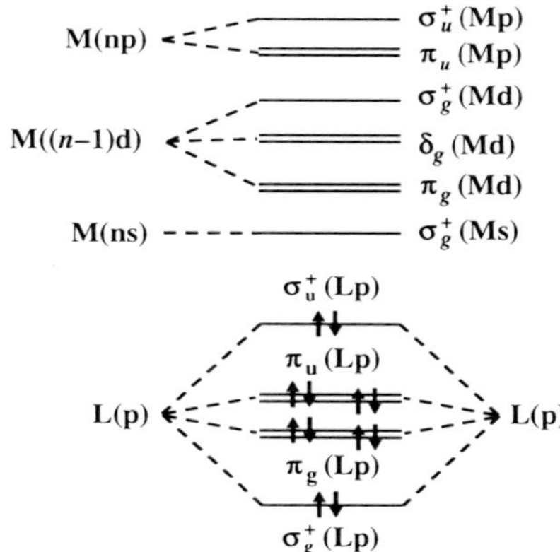

Fig. 1 Molecular orbital (MO) scheme of the valence states in the ML_2 molecules. Shown in parenthesis are the main atomic orbital contributions to the MOs

Another example is prediction of possible noncentrosymmetric linear configuration of XYX molecules as a result of the PJT mixing of the electronic state under consideration with a higher-in-energy one of opposite parity. Such a configuration with two nonequivalent Y–X bonds to two equivalent atoms X seems to be unreasonable, but the bonding interpretation of the PJTE [12] suggests that under certain conditions the additional covalency on p bonding gained by the shortening of one of the Y–X bond is larger than the loss on distorted (stressed) σ bond, and two such bonds cannot be formed simultaneously. The two minima of the APES with nonequivalent Y–X bonds in each of them and a dipole moment of the molecule was found in the lowest excited $^2\Pi_u$ state of $CuCl_2$ [12], and there is a reasonable confidence that this is not the only case of such PJT-predicted distorted configurations. A similar noncentrosymmetric linear configuration was found by ab initio calculations of the BNB (bor-nitrogen-bor) molecule in the ground state [13]. Although the barrier between the minima in this case is small and hardly observable directly (the dipole moment in the minima is more significant), it is a matter of principle that the PJTE may be present even in very small molecules. On the other hand very small barriers (and even flat potentials with zero barriers) in such noncentrosymmetric systems may influence some of the observable properties including lower odd-stretching vibrational frequencies, higher polarizabilities along the molecular axis, and higher dispersion of molecular beams in perpendicular electric fields.

3 Distorted Configurations of Nondegenerate Ground States Produced by JTE in Excited States

The generalized formulation of the JTE may raise some questions that require special explanation. If the instability of any polyatomic system is of JT origin, why are there systems with no apparent electronic degeneracy or pseudo-degeneracy which are unstable in the high-symmetry configuration and stable in configurations of lower symmetry? In other words, there are stable molecular systems in low-symmetry configurations for which the nearest high-symmetry configuration has no degenerate ground state and no low-lying excited states, and hence no apparent JTE. Recent developments in the JTE theory casting light on this question elucidate also some peculiarities of electronic structure calculations with geometry optimization.

As shown below, it comes out that in all cases when the JT origin of distortions is not seen explicitly, the instability is still due to the JTE, but the latter is "hidden" in the excited states of the high-symmetry configurations.

Usually, exploring molecular shapes, the nuclear configuration at the global minimum of the APES is sought for, but not much attention is paid to the problem of *origin* of this configuration. If the geometry of the system in the global minimum has lower symmetry than the nearest possible higher symmetry configuration, the latter should be unstable *due to JTE in the ground or excited states*.

The distortion produced by excited state JTE in the high-symmetry configuration, if strong enough, may offset the energy gap to the ground state, cross the latter, and produce the distorted global minimum. Figure 2 shows a general picture on how this may happen. In the next section another case of hidden JTE is discussed.

For an excited state $E \otimes e$ problem the condition that its distortion will produce a global minimum with a distorted configuration is $(F_E^2/2K_E) > \Delta$, where F_E and K_E are the vibronic coupling and primary force constants, respectively. For a $T \otimes (e + t_2)$ problem the corresponding conditions are either $(F_E^2/2K_E) > \Delta$ when e distortions are advantageous or $(2F_T^2/3K_T) > \Delta$ in case of t_2 distortions.

A straightforward example of such hidden JTE, the ozone molecule O_3, was considered recently [14]. Ab initio calculations of the electronic structure of this molecule were performed multiply. Figure 3 shows some of the results obtained by means of high-level ab initio calculations for the ground state with geometry optimization [15, 16].

The APES of O_3 has three equivalent minima (Fig. 3(a)) in which the molecule was shown to have a distorted (obtuse) triangular configuration and a central minimum at higher energy for the undistorted regular triangular geometry. Figure 3(b) shows the cross-section of the surface along one of the minima. The electronic ground state of this molecule is nondegenerate, neither in the undistorted nor the distorted configurations, so there is no JTE in the ground state, nor are there low lying excited states to justify an assumption of a PJTE. Nevertheless we explicitly see the distortions. So where is the JT origin of these distortions?

To answer this question ab initio electronic structure calculations including excited states were performed [14]. The results for the cross-section along one of the minima are shown in Fig. 4. In comparison with Fig. 3(b) we see that there is an excited state which for the undistorted configuration is a E term, and the global minimum for the distorted configuration is just a component of this degenerate term in the $E \otimes e$ problem that produces the three minima of the APES (the interaction with the ground A term at the crossing is very weak). In this picture the JT origin of the three equivalent distorted configurations is seen explicitly as originating from the strong JTE in the excited state with essential contribution of quadratic terms of the vibronic coupling.

Note that the energy gap from the ground A state to the excited E state in the undistorted configuration is relatively large, $\sim$8.5 eV, so the "classical" thinking of the JTE as a small structural deviation from the configuration of the degenerate state could not apprehend such an effect of distortion with $Q_\vartheta = 0.69$ Å and with a stabilization energy more than 9 eV (in our early ab initio calculations [6] we encountered cases of strong PJTE between states with energy gaps of 10–15 eV). The paradigm of the JTE as resulting in small distortions should be eliminated ("forget it"). The JT distortions may be of any size as *all* the distortions are of JT nature.

To reveal that the JT origin of the distortions is not the end of the story, the authors of the above electronic structure calculations of O_3 (or any other ab initio

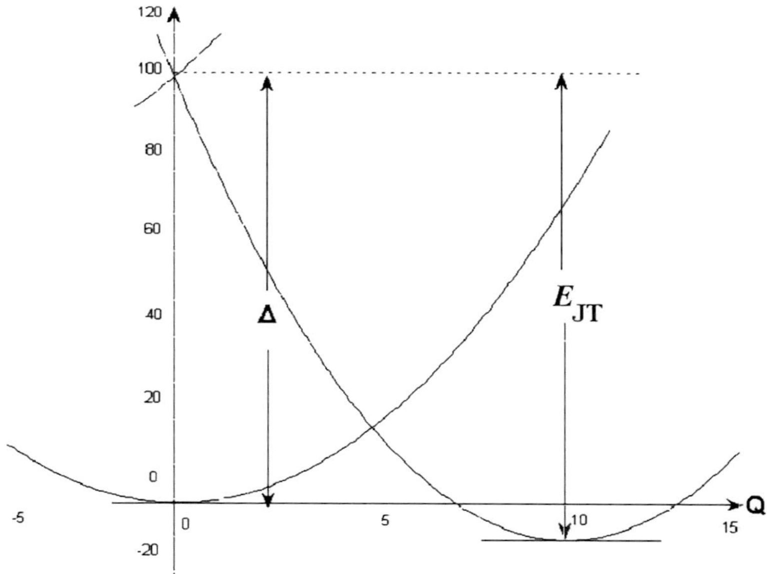

Fig. 2 Illustration of the origin of the distortion in the global minimum due to the JTE hidden in the excited state; it takes place when the JT stabilization energy E_{JT} is larger then the energy gap Δ to the ground state of the high-symmetry configuration and in the direction of the JT-active coordinate

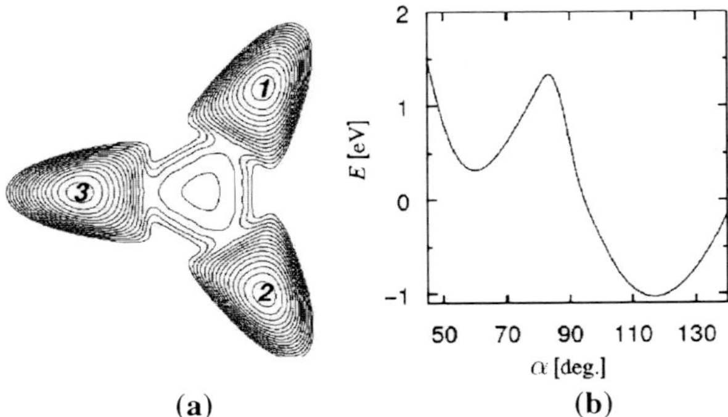

(a)   **(b)**

Fig. 3 Ab initio calculations for the ground state APES of the ozone molecule [15, 16]: (**a**) Equipotential contours showing three minima of three equivalent obtuse-triangular distortions and a shallow minimum (in the centre) of the undistorted regular-triangular configuration [15]; (**b**) Cross-section of the APES along one of the minima [16] (α is the angle at the distinguished oxygen atom in the isosceles configuration)

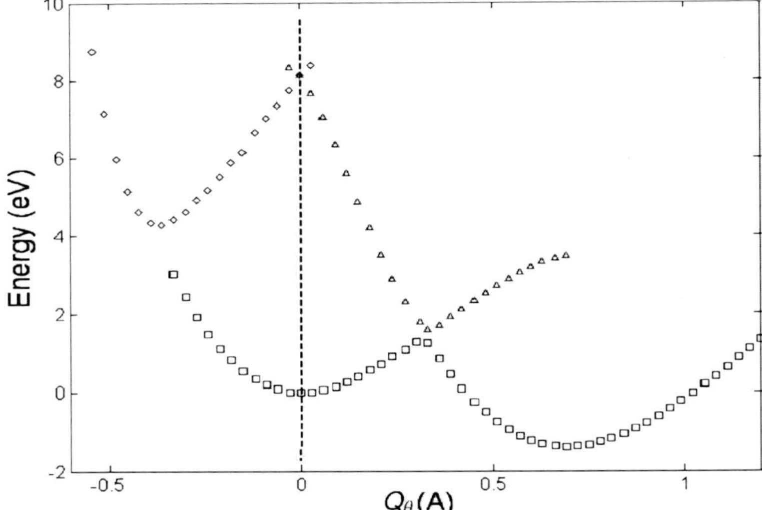

Fig. 4 Cross-section of the APES of the ozone molecule along the Q_θ component of the double-degenerate e mode obtained by numerical ab initio calculations including the highly excited E state, explicitly demonstrating that the ground state distorted configurations are due to the JTE in the excited state [14]. The global minimum is at $Q_\theta = 0.69$Å and the E–A avoided crossing takes place at $Q_\theta \sim 0.35$Å

calculations with geometry optimization that result in distorted configurations) may argue that it is nice to know the origin of the minima, but this does not change the validity of their results on the geometry of the system (the global minimum) and vibrational frequencies (the curvature at the minimum). Indeed, the disclosure of the JT origin of the distorted configurations does not influence the validity of the numerical results obtained by direct calculations. But calculations are not performed just for the sake of calculations: the JTE essentially influences the *physical interpretation* of the numerical results. If the minima are of JT origin, the properties of the system should bear all the features of the JTE that produced them. In particular, in the case of the ozone molecule the minima emerge as a result of the JT $E \otimes e$ problem for which the wavefunctions and energy levels should be subject to the topological (Berry) phase which may essentially change the results. The differences include first of all the ordering and spacing of the vibronic energy levels, their ground state degeneracy, and fractional (semi-integer) quantum numbers of the vibrations when the Berry phase is included [1]; in turn this may change the spectroscopic and thermodynamic properties. With the Berry phase included the ordering of the vibronic energy levels is as follows:

$$E, A_{1(2)}, A_{2(1)}, E, E, A_{1(2)}, A_{2(1)}, E, E, A_{1(2)}, A_{2(1)}, \ldots$$

and their quantum numbers are fractional, whereas if the Berry phase is ignored we have

$$A_1, E, E, A_{1(2)}, A_{2(1)}, E, E, A_{1(2)}, A_{2(1)}, E, E, \ldots$$

and the quantum numbers are integers. Thus by revealing the hidden JTE, the JT origin of the distorted global minimum configuration, we get the correct observable spectroscopic and thermodynamic properties of the system which are essentially different from those obtained by electronic structure calculations of the ground state.

Of particular interest is the fractional (half-integer) quantum numbers of the vibronic energy levels as they influence directly the spectroscopic properties, e.g., the Coriolis splitting of the ground state. For a triangular X_3 (symmetric top) molecule the rotational energy is given by the following approximate expression [17]:

$$E = BJ(J+1) - (B-C)K_c^2 \pm 2C\zeta K_c, \tag{9}$$

where B and C are the rotational constants (the C axis is perpendicular to the X_3 plan), J and K_c are the rotational quantum numbers of a symmetric top, and the last term describes the Coriolis interactions with the Coriolis constant ζ. For strong JTE or PJTE the effective Coriolis constant can be taken equal to the quantum number m of the vibronic level [1, 17]. It emerges from Eq. (9) that the Coriolis splitting equals $4mCK_c$, and for integer m values will differ essentially from those for half-integer m. Moreover, the ground vibronic state with $m = 0$ should not be split by the Coriolis interaction, whereas it should be split in the state with fractional quantization where $m = \pm 1/2$. This example shows that by rationalizing the results of electronic structure calculations by means of the JTE theory one may reveal new physically observable effects.

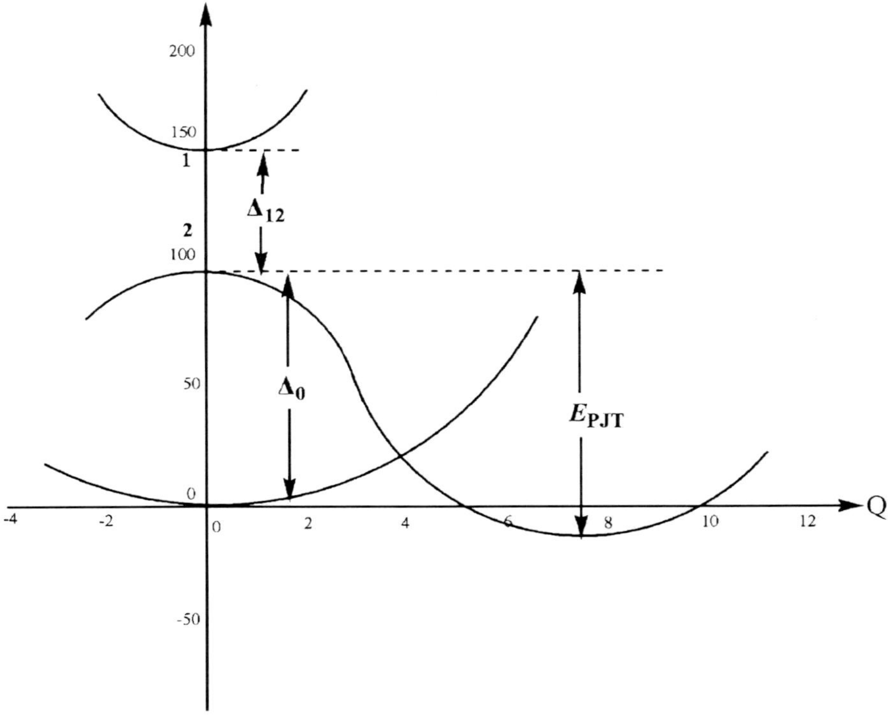

Fig. 5 A strong PJTE between two excited states with $E_{PJT} > \Delta_0$ produces a global minimum with a configuration distorted in the direction of the PJT active coordinate Q_Γ

4 Distorted Configurations of Nondegenerate Ground States Produced by PJT Coupling Between Two Excited States

Another case of hidden JTE is similar to the previous one in the sense that it involves excited states, but it is different in mechanism and consequences. In this case the distorted ground state configuration is produced by a strong PJT mixing of two excited states (of the high-symmetry configuration) with an energy gap Δ_{12} and a stabilization energy E_{PJT} larger than Δ_0 (Fig. 5).

The condition that the excited-state PJT-distorted configuration produces a global minimum of the APES is thus [1]

$$\frac{\left(F_\Gamma^{(12)}\right)^2}{2K_0} - \Delta_{12} + \frac{\Delta_{12}K_0}{2\left(F_\Gamma^{(12)}\right)^2} > \Delta_0, \tag{10}$$

where F_Γ^{12} and K_0 are defined above.

Distinguished from the JT case where the possible distortion is restricted by the JT-active modes only, the PJT-induced distortion may be of any kind dependent on the symmetries of the mixing states. Another distinguished feature of the excited-state PJT-induced distortion is that it leads to orbital disproportionation discussed in Section 5.

This second type of hidden JTE is even more "hidden" than the first type in the previous section. The best examples of this kind of JTE induced by PJT coupling between two excited states are in systems with half-filled closed shells of degenerate e and t orbitals, meaning electronic e^2 and t^3 configurations. Indeed in the ground state, according to Hund's rule, the electronic configurations have the highest possible spin, ^3A in e^2 and ^4A in t^3, as in $(e_\vartheta \uparrow; e_\varepsilon \uparrow)$ and $(t_x \uparrow; t_y \uparrow; t_z \uparrow)$, respectively. Since the charge distribution in these configurations is totally symmetric with respect to the geometry of the system and the electronic state is nondegenerate, no JTE is expected in these ground states. Other distributions of the electrons on these orbitals result in excited terms with lower spin, ^1E and ^1A in e^2, and ^2E, $^2T_1$, and $^2T_2$ in t^3. In accordance with the earlier (primitive) formulation of the JT theorem the nuclear configuration (geometry) of the system in these excited degenerate states should be unstable. Unexpectedly, it was shown [18, 19] that *in violation of the earlier formulation of the JTE* all these states are non-JT, meaning that the totally symmetric charge distribution of the e^2 and t^3 electron configurations is not violated by the electron interactions in the excited states. Since the spin of the latter is different from that in the ground state, there is no PJT interaction between them either. Nevertheless many of these systems are distorted in the ground state. So where is the JTE in these systems?

Analyzing this situation it was found that in systems with electronic e^2 configurations there is a strong PJTE between the two excited states ^1E and ^1A, approximately twice as strong as the expected JTE in the same system with just one e electron [12]. The possibility of such a PJTE, in general, was indicated earlier [19], but it was not apprehended that it may produce a global minimum with a distorted configuration. Calculations including the PJT mixing of excited states E–A of Na$_3$ were performed to explain its two-photon ionization spectra [20].

Consider, for example, the triangular molecule Si$_3$ with D$_{3h}$ symmetry. Experimental spectroscopic data indicate that, similar to O$_3$, this molecule in the ground state has a distorted (obtuse triangular) configuration with C$_{2v}$ symmetry. Figure 6 illustrates some of the results of ab initio MRCI/cc-pqtz calculations of the electronic structure of this molecule (including excited states) and APES in the cross-section along the mode of distortion (Q_θ coordinate) [18]. We see that the ground electronic state in the undistorted geometry is a spin triplet $^3A_2'$, while the excited states are singlets $^1E'$ and $^1A'$ with a very small JTE in the $^1E'$ state (which cannot overcome the energy gap to the ground state to produce the global distortion as in the O$_3$ case), but a strong PJTE $(^1E' + {}^1A') \otimes e'$. In the direction of the distortion one of the components of the $^1E'$ term is stabilized by the strong PJT coupling with the excited $^1A_1'$ state and crosses the ground triplet state of the undistorted configuration to produce the global minimum with a distorted geometry. The latter is in agreement with the experimental data on infrared spectra [21]. The small JTE in the $^1E'$

state is due to the "contamination" of the non-JT pure e^2 configuration with other
(non-e^2) configurations in the process of ab initio calculations with configuration
interaction.

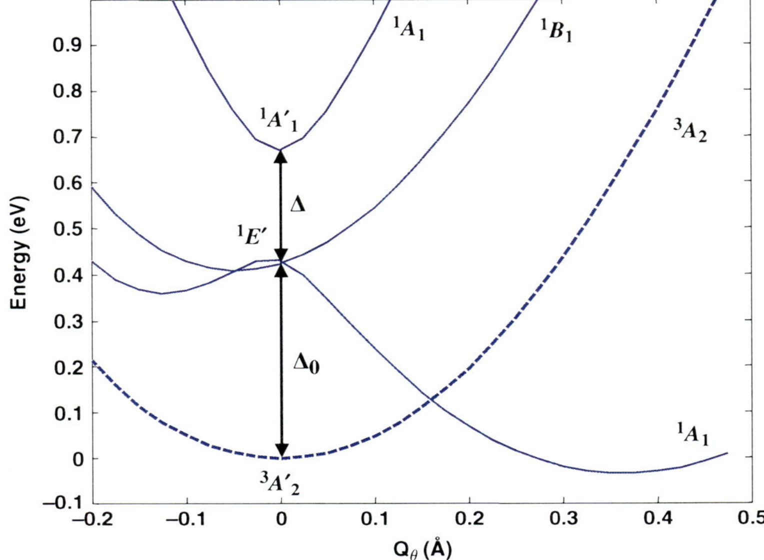

Fig. 6 Cross-section of the APES along the Q_θ coordinate for the terms arising from the electronic
e^2 configuration of Si_3 [18]. Its main features are (as predicted by the theory) a very weak JTE in
the excited E state, a strong PJTE between the A component of this state and the higher A state
producing the global minimum with a distorted configuration, and a second conical intersection
along Q_θ (with two more, equivalent, in the full e space). The spin-triplet state is shown by *dashed
line*

Figure 6 shows also one of the additional conical intersection in the Q_θ direction,
and there are two other equivalent in the e space of the distortions in accordance with
the JTE theory for the $E \otimes e$ problem [1]. Because of these additional conical intersections there are no Berry phase implications in this case: the transition between
the minima along the lowest barriers goes around four conical intersections instead
of one [22].

The PJTE in excited states of systems, with electronic e^2 configurations which
produce global minima with distorted geometries and orbital disproportionation (see
below) in addition to Si_3, was confirmed also by ab initio calculations of a series of
molecular system from different classes including Si_3C, Si_4, $Na_4{}^-$, and CuF_3 [18].

Moving to systems with half-closed-shell electronic t^3 configurations, we find
a similar totally symmetric charge distribution in all its states, ground and excited
(including degenerate states), which makes all of them non-JT, in violation of the
primitive formulation of the JTE. Again, in these cases there is a strong PJTE that
mixes two excited states with the result that the lower one is pushed down to overcome the energy gap to the ground state and to produce a global minimum with a
distorted configuration. For the electronic $t_2{}^3$ configuration the energy terms are $^4A_2$

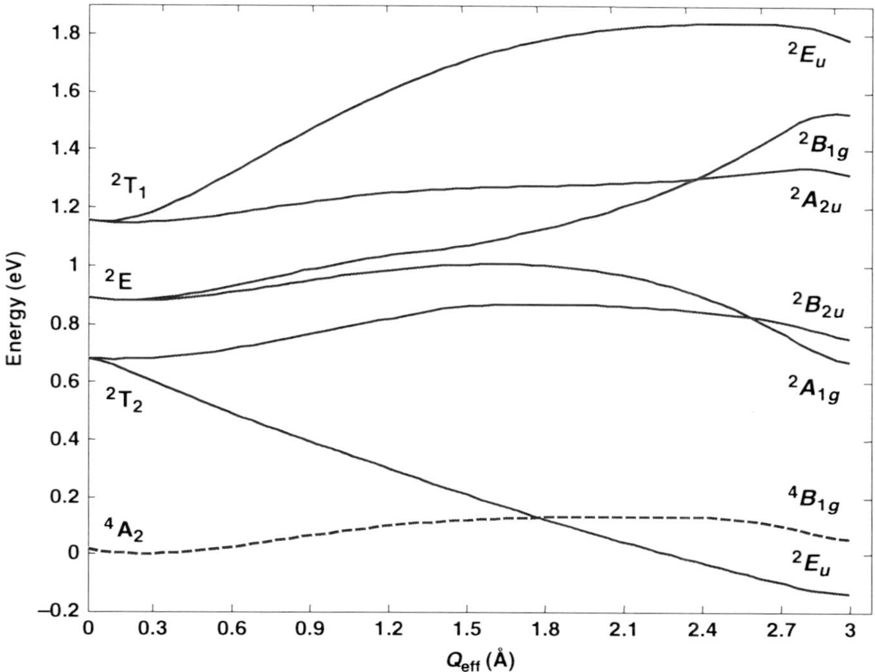

Fig. 7 Cross-section of the APES of Na_4^- along the e-mode distortion transforming the system from tetrahedral ($Q_{eff} = 0$) to square-planar geometry due to the $\left(T_1 + T_2\right) \otimes e$ PJT coupling

(usually the ground one), ^2E, $^2T_1$, and $^2T_2$ (the results for t_1^3 are similar), and the strong PJT problem under consideration is $\left(^2T_1 + {}^2T_2\right) \otimes e$.

Consider the example of the Na_4^- cluster [18]. In the high-symmetry configuration the four sodium atoms are arranged in a tetrahedron. The four $3s$ valence orbitals in this conformation form a_1 and t_2 symmetrized orbitals. In the Na_4^- system the valence electronic configuration is $a_1^2 t_2^3$ producing electronic terms $^4A_2$, $^2T_1$, ^2E, and $^2T_2$ from the t_2^3 configuration. CASSCF calculations of the electronic structure of this system in the ground and excited states as a function of the tetragonal e displacements using the cc-$pvtz$ basis set and the s valence orbitals of Na as the active space are illustrated in Fig. 7.

As expected from the general theory [18], there is no significant JT distortions in any of the states formed by the t_2^3 configuration, but there is a strong PJTE of the type $\left(^2T_1 + {}^2T_2\right) \otimes e$ that pushes down one of the components of the $^2T_2$ term making it the absolute minimum in which the tetrahedron is distorted in the e direction. We have thus a spin-quadruplet ground state in the undistorted tetrahedral configuration and a spin-doublet state in the distorted global minimum with the shape of a rhombus.

The t_{1u}^3 configuration was also explored in the fullerene anions C_{60}^{3-}. For this system the orbital disproportionation (see Section 5) was first revealed by Ceulemans, Chibotaru, and Cimpoesu [23] by direct estimation of the electron interactions in the distorted configuration in order to explain the origin of conductivity in the alkaline-doped fullerides A_3C_{60}.

5 PJT-Induced Orbital Disproportionation and Spin Crossover

Analyzing the wavefunctions in the distorted configurations in the general case of electronic e^2 configurations it was shown [18] that the distortion induced by the PJT mixing of two excited states is accompanied by *orbital disproportionation* of the type $(|\varepsilon \uparrow; \varepsilon \downarrow\rangle - |\theta \uparrow; \theta \downarrow\rangle) \rightarrow |\theta \uparrow; \theta \downarrow\rangle$ or $(|\varepsilon \uparrow; \varepsilon \downarrow\rangle - |\theta \uparrow; \theta \downarrow\rangle) \rightarrow |\varepsilon \uparrow; \varepsilon \downarrow\rangle$, meaning that in the ground state of the distorted geometry the two electrons occupy one e orbital with opposite spins instead of the proportionate totally symmetric distribution of the two electrons on the two orbitals in all the electronic states of the undistorted configuration. The ab initio calculations for Si_3 fully confirm this prediction [18]. The orbital disproportionation provides for a transparent physical picture on why and how the distortion takes place. The wavefunctions of the excited singlet terms $^1A_1$ and ^1E before PJT mixing are as follows:

$$^1A_1 = \frac{1}{\sqrt{2}} (|\varepsilon \uparrow; \varepsilon \downarrow\rangle + |\theta \uparrow; \theta \downarrow\rangle)$$

$$^1E_\theta = \frac{1}{\sqrt{2}} (|\varepsilon \uparrow; \varepsilon \downarrow\rangle - |\theta \uparrow; \theta \downarrow\rangle).$$

$$^1E_\varepsilon = \frac{1}{\sqrt{2}} (|\theta \uparrow; \varepsilon \downarrow\rangle + |\theta \downarrow; \varepsilon \uparrow\rangle). \tag{11}$$

In all these states the charge distribution is symmetrical with respect to the θ and ε components. Due to the PJTE the $^1E_\theta$ component mixes with the $^1A_1$ function to result in their linear combination which in the case of sufficiently strong vibronic coupling produces a disproportionate distribution of either $|\varepsilon \uparrow; \varepsilon \downarrow\rangle$ or $|\theta \uparrow; \theta \downarrow\rangle$ [18]. In any of these cases the charge distribution is nontotally symmetric and distorts the high-symmetry configuration. In other words, if the hidden PJTE conditions are met, it is more energetically convenient for the system to pair its electrons on the same orbital and distort the nuclear framework than to remain symmetrical in either the ground or the excited state of the undistorted configuration.

A quite similar effect takes place in the case of electron configuration t^3. In this case the PJT strong vibronic mixing of two excited states $^2T_1$ and $^2T_2$ results in a lower orbitally disproportionate component of the type $|t_x \uparrow; t_z \downarrow; t_z \uparrow\rangle$, while the ground quadruplet state $^4A_2$ corresponds to the Hund's rule distribution $|t_x \uparrow; t_y \uparrow; t_z \uparrow\rangle$.

As follows from these results, orbital disproportionation in systems with half-closed-shell electronic configurations is necessarily accompanied by lowering the spin of the electronic ground state. For the e^2 configuration it means transition from the high-spin (HS) triplet ^3A state to the low-spin (LS) singlet state ^1A, while for t^3 this transition is from the quadruplet (HS = 3/2) to the doublet (LS = 1/2) state. Since this transition is induced by the PJT distortion originating from an excited

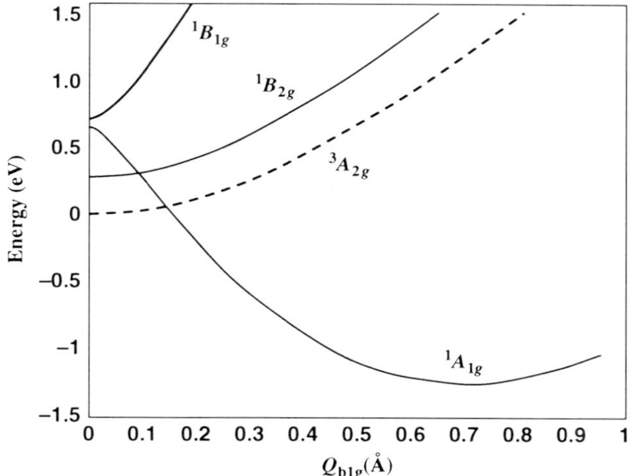

Fig. 8 Cross-section of the APES of Si_4 along the b_1-mode that distorts the system from square-planar to rhombic geometry due to the $(^1A_1 +^1 B_1) \otimes e$ PJT coupling between two excited states

electronic state, the two states, HS undistorted and LS distorted, coexist in two minima of the APES which may have close energies. In between these two minima there may be a crossing between the two states of different spin, a *spin crossover*. The results of ab initio calculations in Ref. [18], some of which are presented in Figs. 6, 7, 8, and 9, show explicitly the spin crossover that takes place in the specific molecules under consideration. It may take place in any molecular system with electronic e^2 or t^3 configurations, meaning molecules with at least one threefold axis of symmetry and appropriate number of electrons.

The spin crossover phenomenon is known to take place in cubic coordination systems of transition metal compounds (TMC) with electronic configurations d^4–d^7 that may produce either HS or LS complexes, subject to the strength of the ligand field [24, 25]. For some values of the latter the two electronic configurations, HS

Table 1 The parameters of the JT spin-crossover in several systems. Δ_{ex} is the energy difference between the ground states of the high-spin and low-spin configurations, and δ_{HS} and δ_{LS} are the respective energy barriers, the energy difference between the minima and the crossing point between the two spin states (Fig. 9). All the energies are read off the zero-point vibrations

	Method	Δ_{ex} (eV)	δ_{HS} (eV)	δ_{LS} (eV)
Si_3C	CASPT2/cc-pvtz	2.180	−0.148	2.031
Na_3[a]	MRCI/cc-pvtz	0.367	−0.020	0.364
Si_4	CASPT2/cc-pvtz	1.341	0.004	1.241
Na_4^-	CASPT2/cc-pvtz	0.141	0.107	0.251
Si_3	MRCI/cc-pqtz	0.132	0.062	0.194
CuF_3	CASPT2/Roos	0.190	0.541	0.712
C_{60}^{3-}	DFT (LDA) [21]	0.157	−0.004	−0.001

[a] Excited state.

and LS, may be close in energy so they can cross over as a function of the breathing mode of the system (metal–ligand distance). This spin crossover is known for a long time and has been subjected to more intensive studies during more than two decades because, in principle, systems with two spin states may serve as molecular materials for electronics [25]. However, the observation of the two states and transitions between them under perturbations (required for such materials) encounters essential difficulties because of fast radiationless transitions between them (very short lifetime of the higher in energy state due to its fast relaxation to the lower one). The two spin states in TMC cannot be observed beyond low temperatures because of their poor separation in space and fast relaxation due to the relatively high spin-orbital interaction in the metal [25]. So far they have been observed only for some compounds in optical LS $\rightarrow$ HS excitation at low temperatures <50 K and mostly as a cooperative effect in solids [25].

The crossover induced by the PJTE and orbital disproportionation is essentially different from the spin crossover in TMC produced by the strength of the ligand field. Indeed, (1) the PJT-induced spin crossover takes place in a variety of molecular systems, small to moderate, organic and inorganic, as well as in metal-containing molecules, as illustrated on a series of molecular systems taken as examples [18]; (2) the HS—LS intersystem relaxation rate in the PJT case is expected to be much lower than in TMC because the two spin states have different nuclear configurations, distorted and undistorted, producing a significant barrier between them and a small Franck–Condon factor, while the spin-orbital interaction in light-atom molecules

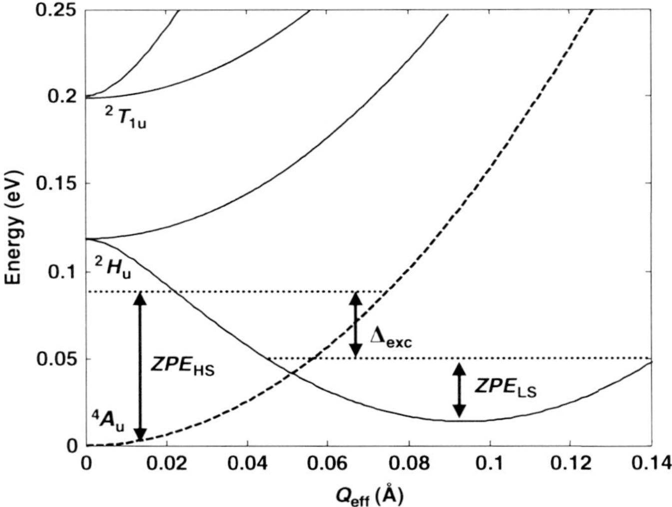

Fig. 9 Cross-section of the APES of C_{60}^{3-} along the *effective mode* that takes the system from the undistorted high-spin minimum to the distorted low-spin minimum due to the multimode ($H_u + T_{1u}$) $\otimes h_g$ PJT coupling [26]. The zero-point energies (ZPE) along *this mode* (which are different from the global ZPE) show that the lowest vibronic state associated with the electronic spin-doublet state is lower than that for the spin-quadruplet

is smaller by orders of magnitude than in TMC; (3) based on these considerations (followed by numerical estimates) it can be assumed that in the PJT-induced spin-crossover the switch between the two states (in both directions) under perturbations can be observed as a single-molecule phenomenon and at relatively high temperatures. The molecule of Si_4 (Fig. 8) seems to be one of appropriate candidates for testing this effect: in this case the two spin states $^3A_{2g}$ and $^1A_{1g}$ are separated also by a structural barrier which includes distortions along three coordinates; b_{1g}, a_{2u}, and a_{1g}. For the molecular systems mentioned above, for which numerical calculations were carried out, numerical estimates for the positions of the two minima on the APES, their energies, and the point of crossover of the terms with different spin are given in Table 1 with the denotations shown in Fig. 9. The values of the energy barriers are read off the zero-point vibrations. We see that the spin-crossover parameter values vary in considerably large ranges. Preliminary estimates show that the relaxation rate in some of these systems is by several orders of magnitude lower than in TMC. Since the number and the variety of molecules with e^2 and t^3 electronic configurations are practically unlimited, we may hope that systems with required combinations of these parameter values in specific limits and low relaxation rates are feasible. This *JT single-molecule spin crossover* is a new phenomenon that may also have applications as novel materials for electronics and yet another example on how JTE interpretation of results of electronic structure calculations lead to novel physical effects.

6 Conclusions

An outline of the generalized JTE implications in electronic structure calculations is given showing the importance of this effect in both choosing the method and basis set of ab initio computation and rationalization of the results. The latter aspect is of special importance in transforming computer experiments in theoretical explanation and prediction of molecular properties. As the only source of instability and distortions of any polyatomic system the JTE serves as a general tool of (approach to) problem solving which in electronic structure calculations allow one to make conclusions based on first principles.

Using a series of molecular systems from different classes it is shown that if electronic structure calculations with geometry optimization yield a global minimum with a distorted configuration, meaning a geometric symmetry lower than the nearby higher symmetry one, the distortion is produced by the JTE in the undistorted systems. If there is no apparent JTE in the undistorted configuration (no degenerate ground state and no low-lying excited states to assume PJT mixing), the JTE is still there, hidden in the excited states. Two qualitatively different cases of hidden JTE can be distinguished: (1) the distortion is due to a strong JTE in one of the excited states of the undistorted configuration and (2) it is created by a strong PJTE mixing two excited states of the latter. Examples with ab initio calculations illustrate both cases. An interesting consequence of the hidden PJTE is that it results in orbital disproportionation and the spin-crossover phenomenon. In general, excited electronic states cannot be ignored in any full analysis of molecular properties (even in the ground state), and the JTE is the unique model that allows for a reasonable rationalization of experimental and computational results.

References

1. I. B. Bersuker, *The Jahn Teller Effect* (Cambridge University Press, Cambridge, UK, 2006)
2. H. A. Jahn, E. Teller, Proc. Roy. Soc. London, A **161**, 220 (1937)
3. E. Teller, in *The Jahn-Teller Effect in Molecules and Crystals*, ed. by R. Englman (Wiley, London, 1972), foreword
4. U. Öpik, M. H. L. Pryce, Proc. Roy. Soc. London, A **238**, 425 (1957)
5. I. B. Bersuker, Nouv. J. Chim. **4**, 139 (1980); Teor. Eksp. Khim. **16**, 291 (1980); Pure Appl. Chem. **60**, 1167 (1988); Fiz. Tverdogo Tela, **30**, 1738 (1988)
6. I. B. Bersuker, N. N Gorinchoi, V. Z. Polinger, Theor. Chim. Acta **66**, 161 (1984)
7. I. B. Bersuker, in *Fundamental World of Quantum Chemistry*, ed. by E. J. Brändas, E. S. Kryachko (Kluwer, Dordrecht, 2004), Vol. 3, p. 257
8. I. B. Bersuker, Adv. Quant. Chem. **44**, 1 (2003)
9. S. Weinberg, *Quantum Theory of Fields* (Cambridge University Press, Cambridge, 1995), Chap. 11
10. I. B. Bersuker, N. B. Balabanov, D. Pekker, J. E. Boggs, J. Chem. Phys. **117**, 10478 (2002); I. B. Bersuker, V. Z. Polinger, N. N. Gorinchoi, J. Struct. Chem.: THEOCHEM **5**, 369 (1992)
11. P. Garcia-Fernandez, I. B. Bersuker, J. E. Boggs, J. Phys. Chem. A **111**, 10409 (2007)
12. W. Zou, I. B. Bersuker, J. E. Boggs, J. Chem. Phys. **129**, 114107 (2008)
13. Y. Liu, W. Zou, I. B. Bersuker, J. E. Boggs, J. Chem. Phys., **130**, 184305 (2009)
14. P. Garcia-Fernandez, I. B. Bersuker, J. E. Boggs, Phys. Rev. Lett. **96** 163005 (2006)
15. D. Babikov, B. K. Kendrick, R. B. Walker, R. T. Pack, P. Fleurat-Lesard, R. Schinke, J. Chem. Phys. **118**, 6298 (2003)
16. R. Siebert, P. Fleurat-Lessard, R. Schinke, M. Bittererová, S. C. Farantos, J. Chem. Phys. **116**, 9749 (2002); R. Schinke, P. Fleurat-Lessard, J. Chem. Phys. **121**, 5789 (2004)
17. G. Herzberg, *Electronic Spectra and Electronic Structure of Polyatomic Molecules* (Van Nostrand, Toronto, 1966)
18. P. Garcia-Fernandez, I. B. Bersuker, J. E. Boggs, J. Chem. Phys. **125**, 104102 (2006)
19. A. Ceulemans, Chem. Phys. **66**, 169 (1982); A. Ceulemans, D. Beyens, L. G. Vanquickenborne, J. Am. Chem. Soc. **104**, 2988 (1982); A. Ceulemans, Top. Curr. Chem. **171**, 27 (1994)
20. R. Meiswinkel, H. Koppel, Chem. Phys. **144**, 117 (1990)
21. S. Li, R. J. Van Zee, W. Weltner, Jr., K. Raghavachari, Chem. Phys. Lett. **243**, 275 (1995); J. Fulara, P. Freivogel, M. Grutter, J. P. Maier, J. Phys. Chem. **100**, 18042 (1996)
22. W. Zwanziger, E. R. Grant, J. Chem. Phys. **87**, 2954 (1987); H. Koizumi, I. B. Bersuker, Phys. Rev. Lett. **83**, 3009 (1999)
23. L. F. Chibotaru, A. Ceulemans, Phys. Rev. B **53**, 15522 (1996); A. Ceulemans, L. F. Chibotaru, F. Cimpoesu, Phys. Rev. Lett. **78**, 3725 (1997)
24. H. A. Goodwin, Coord. Chem. Rev. **18**, 293 (1976); E. K. Barefield, D. H. Busch, S. M. Nelson, Q. Rev. Chem. Soc. **22**, 457 (1968)
25. A. Hauser, Top. Curr. Chem. **234**, 155 (2004); O. Kahn, C. J. Martinez, Science **279**, 5347 (1998)
26. M. Lüders, A. Bordon, N. Manini, A. Dal Corso, M. Fabrizio, E. Tossatti, Phil. Mag. B **82**, 1611 (2002); N. Manini, A. Dal Corso, M. Fabrizio, E. Tossatti, Phil. Mag. B **81**, 793 (2001); M. Lüders, N. Manini, P. Gattari, E. Tossatti, Eur. Phys. J. B **35**, 57 (2003)

Rules for Excited States of Degenerate Systems: Interpretation by Frozen Orbital Analysis

Hiromi Nakai

Abstract This review addresses the rules for the ordering and the splitting of the excited states for the transitions between degenerate orbitals. First, the generality of the rules for the degenerate excitations is examined numerically by highly correlated methods for various types of systems, having $D_{\infty h}$, $C_{\infty v}$, T_d, O_h, and I_h symmetries. Next, the qualitative interpretation of the rules for degenerate excitations is demonstrated by adopting the frozen-orbital approximation/analysis.

Keywords: Frozen orbital analysis · Excited state · Degenerate system · Ab initio calculation

1 Introduction and Rules for Degenerate Excitations

The characterization of electronic excited states has attracted much attention in connection with photochemistry. For example, transition metal complexes are characterized by a variety of absorption spectra in the visible and ultraviolet (UV) regions. The absorption spectra essentially give us information about the electronic excited states corresponding to dipole-allowed transitions due to their high symmetries, while some of the data in crystalline fields indicate the existence of several excited states to which dipole transitions are forbidden in the absence of perturbation. Most photochemical reactions of metal complexes, which are occasionally important as homogeneous photocatalytic reactions, involve both allowed and forbidden excited states. Thus, the systematic understanding of the nature of these excited states is essential in designing photochemical reactions.

Theoretical studies have played important roles in clarifying the excited states, in particular, of high-symmetry systems. The excited-state calculations with the configuration interaction based on the symmetry-adapted cluster expansion (SAC-CI) method [1, 2] have given reasonable and reliable assignments for experimental absorption spectra [3]. The SAC-CI method is a linear-response theory using the

H. Nakai (✉)
Department of Chemistry and Biochemistry, School of Advanced Science and Engineering, Waseda University, Tokyo 169-8555, Japan, e-mail: nakai@waseda.jp

P. Piecuch et al. (eds.), *Advances in the Theory of Atomic and Molecular Systems*, 363
Progress in Theoretical Chemistry and Physics 19, DOI 10.1007/978-90-481-2596-8_18,
© Springer Science+Business Media B.V. 2009

coupled-cluster ground-state wave function. For example, the strong peak at around 7.12 eV in the UV spectrum of an MoF_6 octahedral complex was assigned to the dipole-allowed $^1T_{1u}$ state having the t_{1u} to t_{2g} excitation nature [4, 5]. The relevant orbitals, t_{1u} and t_{2g}, are characterized by the π bonding and antibonding molecular orbitals (MOs) of the Mo—F bonds, respectively. Strictly, the t_{1u} and t_{2g} MOs are dominantly composed of the $2p$ atomic orbitals (AOs) of F and the $4d$ AOs of Mo, respectively.

The SAC-CI calculations of MoF_6 also demonstrated the existence of three other states, i.e., $^1A_{2u}$, ^1E_u, and $^1T_{2u}$, with the same t_{1u} to t_{2g} excitation character. It is natural to determine the existence of these four states according to the group theory, namely, the direct product of t_{1u} and t_{2g} leads to four irreducible representations, a_{2u}, e_u, t_{1u}, and t_{2u}. The calculated excitation energies were 6.67 ($^1A_{2u}$), 6.73 (^1E_u), 6.81 ($^1T_{2u}$), and 7.24 ($^1T_{2u}$) eV, respectively. These results led us to the following question: what determines the ordering and the splitting of the excited states for the transitions between degenerate orbitals. We have intensively examined the excited states of high-symmetry systems, e.g., linear molecules ($D_{\infty h}$, $C_{\infty v}$), tetrahedral (T_d) and octahedral (O_h) metal complexes, $(B_{12}H_{12})^{2-}$ and $C_{60}(I_h)$. Finally, we have discovered the following rules:

In singlet excitations between occupied and virtual degenerate orbitals with valence character,

(i) *the highest transition is dipole-allowed,*
(ii) *the splitting between the dipole-allowed and dipole-forbidden states is larger than those among the dipole-forbidden states.*

While the above rules of degenerate excitations are restricted to the valence excitations, the other cases, such as core and Rydberg excitations, bring about considerably smaller splittings as easily expected.

The composition of this review is as follows: Section 2 describes the numerical examples of the rules for degenerate excitations. The data in the next section are obtained by highly correlated methods, since the effects of electron correlations are essential for accurate descriptions of the excited states. Section 3 demonstrates the interpretation of the rules by using the simplified model that corresponds to the frozen-orbital approximation (FZOA) [4]. In the excitation energy formulas to which the FZOA leads, the splitting schemes are related to the specific two-electron integrals, whose values are qualitatively analyzed by the relevant orbital characters. Finally, the summary is addressed in Section 4.

2 Numerical Examples of Rules for Degenerate Excitations

This section demonstrates the generality of the rules for degenerate excitations numerically. The data are partially taken from the literature. Some data are newly obtained by the SAC-CI calculations [1, 2] with Huzinaga–Dunning double-zeta plus polarization (DZP) basis sets [6, 7] and/or by the time-dependent density functional theory (TDDFT) calculations [8, 9] with Dunning's correlation-consistent polarized valence double-zeta (cc-pVDZ) basis sets [10]. For metal atoms such as Cu and Ag, the effective core potential plus valence double-zeta (CEP-31G) basis sets [11] are adopted. The TDDFT calculations are performed with the hybrid functional including Becke's three parameters for Becke88 exchange [12] and Lee–Yang–Parr correlation [13] functionals (B3LYP) [14]; thus, we use the abbreviation of TD-B3LYP. For comparison, we also tabulate the data calculated by the frozen-orbital approximation/analysis (FZOA) [4], where the MOs are frozen at the HF ground state. The excited states obtained by the FZOA correspond to the results of the configuration interaction with singles (CIS) within the minimum active space. The detailed treatment will be explained in Section 3.

2.1 Doubly-Degenerate Case

The first examples are doubly-degenerate excitations. The simplest ones are doubly-degenerate π–π^* excitations in diatomic molecules such as N_2 ($D_{\infty h}$) and CO ($C_{\infty v}$). The doubly-degenerate highest-occupied MOs (HOMOs), as denoted by π, belong to the irreducible representations of π_u in $D_{\infty h}$ and π in $C_{\infty v}$, while the doubly-degenerate lowest-unoccupied MOs (LUMOs), as denoted by π^*, belong to those of π_g in $D_{\infty h}$ and π in $C_{\infty v}$, respectively. The direct products between π_u and π_g and between π and π can be decomposed as follows:

$$\pi_u \otimes \pi_g = \sigma_u^+ + \sigma_u^- + \delta_u (D_{\infty h}), \tag{1}$$

$$\pi \otimes \pi = \sigma^+ + \sigma^- + \delta (C_{\infty v}). \tag{2}$$

As a result, the π–π^* excitations bring about three singlet states, i.e., $\{^1\Sigma_u^+, {}^1\Sigma_u^-, {}^1\Delta_u\}$ and $\{^1\Sigma^+, {}^1\Sigma^-, {}^1\Delta\}$. Here, $^1\Sigma_u^+$ and $^1\Sigma^+$ are dipole-allowed.

Table 1 shows the π–π^* excited states of diatomic molecules $\{N_2, P_2, As_2\}$ ($D_{\infty h}$) and $\{CO, CS, CSe, SiO, GeO\}$ ($C_{\infty v}$), which formally possess triple bonds, calculated at the SAC-CI/DZP and FZOA/DZP levels. Experimental data [15–18] are also shown in Table 1. Since the dipole-allowed $^1\Sigma_u^+/{}^1\Sigma^+$ states in these systems strongly mix between the π–π^* and Rydberg transitions and their excitation energies exceed ionization thresholds, some assignments have not been confirmed. The dipole-allowed $^1\Sigma_u^+/{}^1\Sigma^+$ states, whose data are written in italics, appear in the highest level in any system. Furthermore, the energy splittings between $^1\Sigma_u^+$ and $^1\Delta_u$ in the $D_{\infty h}$ systems are considerably larger than those between $^1\Delta_u$ and $^1\Sigma_u^-$. For example, the energy splittings between $^1\Sigma_u^+$ and $^1\Delta_u$ and between $^1\Delta_u$ and $^1\Sigma_u^-$ in N_2 are calculated to be 5.37 and 0.43 eV by the SAC-CI/DZP method, respectively. The former is one order of magnitude larger than the latter. Similar trends are seen in the $C_{\infty v}$ systems. Thus, the rules (i) and (ii) are established in these systems. The experimental data for N_2, CO, and SiO support the rules.

Note that the crude treatment, FZOA, can reproduce these trends, while it over-estimates the energy gaps between $^1\Sigma_u^+$ and $^1\Delta_u$ in $D_{\infty h}$ and between $^1\Sigma^+$ and

Table 1 π–π^* Excited states of diatomic molecules, N_2, P_2, As_2, CO, CS, CSe, SiO, and GeO, calculated at the SAC-CI/DZP and FZOA/DZP levels. Excitation energies (E) and energy splittings (ΔE) are given in eV

System	Point group	Main conf.	FZOA/DZP			SAC-CI/DZP			Exptl.
			State	E	ΔE	State	E	ΔE	
$N\equiv N$	$D_{\infty h}$	$\pi_u \rightarrow \pi_g$	$^1\Sigma_u^+$	19.79		$^1\Sigma_u^+$	16.02		14.48/14.25[a]
					] 10.17			] 5.37	
			$^1\Delta_u$	9.62		$^1\Delta_u$	10.65		10.27[b]
					] 0.46			] 0.43	
			$^1\Sigma_u^-$	9.16		$^1\Sigma_u^-$	10.22		9.92[b]
$P\equiv P$	$D_{\infty h}$	$\pi_u \rightarrow \pi_g$	$^1\Sigma_u^+$	9.26		$^1\Sigma_u^+$	7.00		5.82/8.22[c]
					] 5.50			] 2.59	
			$^1\Delta_u$	3.76		$^1\Delta_u$	4.41		
					] 0.27			] 0.22	
			$^1\Sigma_u^-$	3.49		$^1\Sigma_u^-$	4.19		
$As\equiv As$	$D_{\infty h}$	$\pi_u \rightarrow \pi_g$	$^1\Sigma_u^+$	8.46		$^1\Sigma_u^+$	6.13		
					] 5.35			] 2.50	
			$^1\Delta_u$	3.11		$^1\Delta_u$	3.63		
					] 0.29			] 0.28	
			$^1\Sigma_u^-$	2.82		$^1\Sigma_u^-$	3.35		
$C\equiv O$	$C_{\infty v}$	$\pi \rightarrow \pi$	$^1\Sigma^+$	18.89		$^1\Sigma^+$	14.51		$(>13)^c$
					] 8.14			] 3.88	
			$^1\Delta$	10.75		$^1\Delta$	10.63		
					] 0.33			] 0.14	10.23[c,d]
			$^1\Sigma^-$	10.42		$^1\Sigma^-$	10.49		
									9.88[c,d]
$C\equiv S$	$C_{\infty v}$	$\pi \rightarrow \pi$	$^1\Sigma^+$	12.57		$^1\Sigma^+$	8.91		8.91/8.04[c]
					] 6.76			] 3.88	
			$^1\Delta$	5.81		$^1\Delta$	6.01		
					] 0.29			] 0.14	
			$^1\Sigma^-$	5.52		$^1\Sigma^-$	5.86		
$C\equiv Se$	$C_{\infty v}$	$\pi \rightarrow \pi$	$^1\Sigma^+$	11.56		$^1\Sigma^+$	8.09		
					] 6.62			] 2.68	
			$^1\Delta$	4.94		$^1\Delta$	5.41		
					] 0.30			] 0.22	
			$^1\Sigma^-$	4.64		$^1\Sigma^-$	5.19		
$Si\equiv O$	$C_{\infty v}$	$\pi \rightarrow \pi$	$^1\Sigma^+$	10.95		$^1\Sigma^+$	7.67		6.55[c]
					] 3.96			] 1.74	
			$^1\Delta$	6.99		$^1\Delta$	5.93		4.81[c]
					] 0.11			] 0.08	
			$^1\Sigma^-$	6.88		$^1\Sigma^-$	5.85		4.79[c]

Table 1 (continued)

System	Point group	Main conf.	FZOA/DZP			SAC-CI/DZP			Exptl.
			State	E	ΔE	State	E	ΔE	
$Ge\equiv O$	$C_{\infty v}$	$\pi \to \pi$	$^1\Sigma^+$	*10.38*		$^1\Sigma^+$	*7.05*		
					] *4.48*			] *1.87*	
			$^1\Delta$	5.90		$^1\Delta$	5.18		
					] 0.15			] 0.13	
			$^1\Sigma^-$	5.75		$^1\Sigma^-$	5.05		

[a] From [15].
[b] From [16].
[c] From [17].
[d] From [18].

$^1\Delta$ in $C_{\infty v}$ by a factor of ~ 2 relative to the accurate SAC-CI values. That is because the dipole-allowed $^1\Sigma_u^+/^1\Sigma^+$ states have strong configuration interactions between the $\pi-\pi^*$ and the Rydberg transitions in these systems. In the N_2 analogs, $\{N_2, P_2, As_2\}$, as well as CO analogs, $\{CO, CS, CSe\}$, the splittings between the dipole-allowed and dipole-forbidden states monotonically decrease as the systems get heavier. Among CO, SiO, and GeO, such monotonic decrease cannot be seen.

Table 2 shows the $\pi-\pi^*$ excited states of acetylene derivatives, C_2H_2, C_2F_2, C_2Cl_2, C_2Cu_2, and C_2Ag_2 ($D_{\infty h}$), which formally possess triple bonds, calculated at the TD-B3LYP/cc-pVDZ+CEP-31G level. Only the experimental datum for C_2H_2 [19] is given in Table 2. The data of the dipole-allowed $^1\Sigma_u^+$ states are written in italics. The occupied π and unoccupied π^* MOs are delocalized from the $C\equiv C$ region to the edge atoms except for acetylene, since the halogen and metal possess valence p orbitals. The splittings between $^1\Sigma_u^+$ and $^1\Delta_u$ in acetylide are smaller than those of acetylene and acetyl halides, but still one order of magnitude larger than those between $^1\Delta_u$ and $^1\Sigma_u^-$. As a result, the rules (i) and (ii) are established in these systems as in the diatomic case.

Table 3 shows the $n-\pi^*$ excited states of CO_2 analogs, that is, CO_2 and CS_2 ($D_{\infty h}$) and N_2O ($C_{\infty v}$), calculated at the TD-B3LYP/cc-pVDZ level. Experimental data [20] are also shown in Table 3. The molecules formally possess two double bonds. From another point of view, a three-center four-electron (3c-4e) bond exists in each π direction, i.e., π_x or π_y plane. Thus, the doubly-degenerate HOMOs have non-bonding character. The HOMOs, as denoted by n, belong to the irreducible representation of π_g, while the LUMOs, as denoted by π^*, to that of π_u in $D_{\infty h}$. While this ordering is the opposite of that for the diatomic molecules, the three states resulting from the $n-\pi^*$ excitations are in the same order: namely, $^1\Sigma_u^+$, $^1\Sigma_u^-$, and $^1\Delta_u$. In spite of the different orbital characters, we can find that the rules (i) and (ii) hold in these systems as in the diatomic case. Note that the experimental assignments for the lowest $^1\Sigma_u^-$ state of CO_2 and N_2O might be incorrect. The large gaps between the $^1\Sigma_u^-$ and $^1\Delta_u$ states might be assumed in the assignment. In such a situation, the rule (ii) is of assistance in obtaining a correct assignment.

Table 4 shows the $\pi-\pi^*$ excitation energies of benzene derivatives, i.e., C_6H_6, C_6F_6, and C_6Cl_6 (D_{6h}), by the TD-B3LYP calculations. The experimental data for

Table 2 π–π^* Excited states of acetylene derivatives, C_2H_2, C_2F_2, C_2Cl_2, C_2Cu_2, and C_2Ag_2, calculated at the TD-B3LYP/cc-pVDZ+CEP-31G level. Excitation energies (E) and energy splittings (ΔE) are given in eV

System	Point group	Main conf.	State	E	ΔE	Exptl.
HC≡CH	$D_{\infty h}$	$\pi_u \to \pi_g$	$^1\Sigma_u^+$	11.23		9.28
					] 4.17	
			$^1\Delta_u$	7.06		
					] 0.23	
			$^1\Sigma_u^-$	6.83		
FC≡CF	$D_{\infty h}$	$\pi_u \to \pi_g$	$^1\Sigma_u^+$	12.44		
					] 4.43	
			$^1\Delta_u$	8.01		
					] 0.18	
			$^1\Sigma_u^-$	7.83		
ClC≡CCl	$D_{\infty h}$	$\pi_u \to \pi_g$	$^1\Sigma_u^+$	9.54		
					] 2.99	
			$^1\Delta_u$	6.55		
					] 0.00	
			$^1\Sigma_u^-$	6.55		
CuC≡CCu	$D_{\infty h}$	$\pi_u \to \pi_g$	$^1\Sigma_u^+$	5.13		
					] 0.87	
			$^1\Delta_u$	4.26		
					] 0.03	
			$^1\Sigma_u^-$	4.23		
AgC≡CAg	$D_{\infty h}$	$\pi_u \to \pi_g$	$^1\Sigma_u^+$	5.19		
					] 0.79	
			$^1\Delta_u$	4.40		
					] 0.03	
			$^1\Sigma_u^-$	4.37		

C_6H_6 [21] and C_6Cl_6 [22] are shown in Table 4. Since these systems belong to a different point group, D_{6h}, than linear systems, the decomposition of the reducible representation varies as follows:

$$e_{1g} \otimes e_{1u} = b_{1u} + b_{2u} + e_{1u}(D_{6h}). \tag{3}$$

Among the three states, $^1B_{1u}$, $^1B_{2u}$, and $^1E_{1u}$, the degenerate $^1E_{1u}$ state is dipole-allowed. Since the dipole-allowed states written in italics appear in the highest level, the rule (i) holds in these systems. Because the two splittings are comparable, the rule (ii) does not work for the benzene derivatives. The energy splitting between $^1E_{1u}$ and $^1B_{1u}$ decreases as heavier systems are examined, namely, 1.05, 0.92, and 0.53 eV in C_6H_6, C_6F_6, and C_6Cl_6, respectively. The π and π^* orbitals of the

Table 3 n–π^* Excited states of CO_2 analog, CO_2, CS_2, and N_2O, calculated at the TD-B3LYP/cc-pVDZ level. Excitation energies (E) and energy splittings (ΔE) are given in eV

System	Point group	Main conf.	State	E	ΔE		Exptl.
O=C=O	$D_{\infty h}$	$\pi_u \rightarrow \pi_g$	$^1\Sigma_u^+$	12.89			11.08
					3.93	]	
			$^1\Delta_u$	8.96			8.41
					0.32	]	
			$^1\Sigma_u^-$	8.64			(6.53)
S=C=S	$D_{\infty h}$	$\pi_u \rightarrow \pi_g$	$^1\Sigma_u^+$	7.19			6.29
					2.91	]	
			$^1\Delta_u$	4.28			3.89
					0.28	]	
			$^1\Sigma_u^-$	4.00			3.49
N=N=O	$C_{\infty v}$	$\pi \rightarrow \pi$	$^1\Sigma^+$	11.61			9.66
					4.64	]	
			$^1\Delta$	6.97			6.81
					0.38	]	
			$^1\Sigma^-$	6.59			(4.54)

Table 4 π–π^* Excited states of benzene derivatives, C_6H_6, C_6F_6, and C_6Cl_6, calculated at the TD-B3LYP/cc-pVDZ level. Excitation energies (E) and energy splittings (ΔE) are given in eV

System	Point group	Main conf.	State	E	ΔE		Exptl.
C_6H_6	D_{6h}	$e_{1g} \rightarrow e_{1u}$	$^1E_{1u}$	7.25			6.94
					1.05	]	
			$^1B_{1u}$	6.20			6.20
					0.73	]	
			$^1B_{2u}$	5.47			4.90
C_6F_6	D_{6h}	$e_{1g} \rightarrow e_{1u}$	$^1E_{1u}$	7.01			
					0.92	]	
			$^1B_{1u}$	6.09			
					0.81	]	
			$^1B_{2u}$	5.28			
C_6Cl_6	D_{6h}	$e_{1g} \rightarrow e_{1u}$	$^1E_{1u}$	5.67			5.75
					0.53	]	
			$^1B_{1u}$	5.14			
					0.62	]	
			$^1B_{2u}$	4.52			

benzene ring are delocalized to the halogen atoms. The delocalization is expected to be greater in C_6Cl_6 than in C_6F_6. While the splittings between $^1E_{1u}$ and $^1B_{1u}$ are slightly larger than those between $^1B_{1u}$ and $^1B_{2u}$ in C_6H_6 and C_6F_6, the opposite is true in C_6Cl_6.

2.2 Triply-Degenerate Case

This section gives the numerical examples for triply-degenerate excitations. Triply-degenerate irreducible representations, $(t; t_1, t_2; t_{1g}, t_{2g}, t_{1u}, t_{2u})$, appear in tetrahedral $(T; T_h; T_d)$, octahedral $(O; O_h)$, and higher-symmetry point groups. The triply-degenerate MOs are important in transition metal complexes, since five d orbitals are split into triply- and doubly-degenerate MOs under the ligand fields, e.g., t_{2g} and e_g in O_h.

Table 5 shows the triply-degenerate excitations in MoF_6 and $Mo(CO)_6$ (O_h), whose calculated and experimental data are taken from Refs. [5] and [23], respectively. The main configurations are given in Table 5, namely, $1t_{2u} \rightarrow 3t_{2g}$ and $6t_{1u} \rightarrow 3t_{2g}$ in MoF_6 and $3t_{2g} \rightarrow 12t_{1u}$ and $3t_{2g} \rightarrow 3t_{2u}$ in $Mo(CO)_6$. In these complexes, several other excited states having the gerade-to-gerade transitions in the lower energy region, in which all states are dipole-forbidden, exist. The Mo atom in MoF_6 formally possesses a $(4d)^0$ configuration, while that in $Mo(CO)_6$ a $(4d)^{10}$ configuration. Thus, the lower excited states in MoF_6 and $Mo(CO)_6$ correspond to ligand-to-metal charge transfer (LMCT) and metal-to-ligand charge transfer (MLCT) transitions, respectively. To be precise, both occupied and unoccupied MOs are delocalized within metal and ligand orbitals, and show slightly π-type bonding and antibonding characters, respectively. The details are provided in the original papers [4, 5, 23].

The direct products of $t_{1u} \otimes t_{2g}$ and $t_{2u} \otimes t_{2g}$ can be decomposed as follows:

$$t_{1u} \otimes t_{2g} = a_{2u} + e_u + t_{1u} + t_{2u}, \tag{4}$$

$$t_{2u} \otimes t_{2g} = a_{1u} + e_u + t_{1u} + t_{2u}. \tag{5}$$

As a result, the $\{t_{1u} \rightarrow t_{2g}, t_{2g} \rightarrow t_{1u}\}$ and $\{t_{2u} \rightarrow t_{2g}, t_{2g} \rightarrow t_{2u}\}$ excitations bring about four singlet states, i.e., $\{^1A_{2u}, {}^1E_u, {}^1T_{1u}, {}^1T_{2u}\}$ and $\{^1A_{1u}, {}^1E_u, {}^1T_{1u}, {}^1T_{2u}\}$, respectively. Here, the $^1T_{1u}$ state is dipole-allowed. The dipole-allowed $^1T_{1u}$ states, whose data are written in italics in Table 5, appear in the highest level in each system. Furthermore, the energy splittings between $^1T_{1u}$ and $^1T_{2u}$ in MoF_6 and $^1T_{1u}$ and $^1A_{2u}$ in $Mo(CO)_6$ are larger than the other splittings. For example, for the $6t_{1u} \rightarrow 3t_{2g}$ transition in MoF_6, the energy splittings of $^1T_{1u}-^1T_{2u}$, $^1T_{2u}-^1E_u$, and $^1E_u-^1A_{2u}$ are 0.43, 0.08, and 0.06 eV, respectively. Thus, the rules (i) and (ii) are established in these systems with the triply-degenerate excitations.

It is notable that the FZOA treatment can reproduce the orderings of the four states, except for the $3t_{2g} \rightarrow 12t_{1u}$ transition in $Mo(CO)_6$, and the qualitative trends for the splittings, although it tends to overestimate the excitation energies

Table 5 t–t Excited states of octahedral metal complexes, MoF_6 and $Mo(CO)_6$, calculated at the SAC-CI/DZP and FZOA/DZP levels. Excitation energies (E) and energy splittings (ΔE) are given in eV

System	Point group	Main conf.	FZOA/DZP State	E	ΔE	SAC-CI/DZP State	E	ΔE	Exptl.
MoF_6	O_h	$1t_{2u} \rightarrow 3t_{2g}$	$^1T_{1u}$	10.05		$^1T_{1u}$	6.62		6.54
					] 1.69			] 0.71	
			$^1T_{2u}$	8.36		$^1T_{2u}$	5.92		
					] 0.56			] 0.23	
			^1E_u	7.80		^1E_u	5.69		
					] 0.04			] 0.00	
			$^1A_{2u}$	7.76		$^1A_{2u}$	5.69		
MoF_6	O_h	$6t_{1u} \rightarrow 3t_{2g}$	$^1T_{1u}$	9.84		$^1T_{1u}$	7.24		7.12
					] 1.08			] 0.43	
			$^1T_{2u}$	8.76		$^1T_{2u}$	6.81		
					] 0.06			] 0.08	
			^1E_u	8.70		^1E_u	6.73		
					] 0.07			] 0.06	
			$^1A_{1u}$	8.63		$^1A_{1u}$	6.67		
$Mo(CO)_6$	O_h	$3t_{2g} \rightarrow 12t_{1u}$	$^1T_{1u}$	6.83		$^1T_{1u}$	5.29		5.45
					] 1.38			] 0.69	
			$^1T_{2u}$	5.46		$^1T_{2u}$	4.59		
					] 0.08			] 0.03	
			$^1A_{2u}$	5.38		$^1A_{2u}$	4.57		~4.66
					] 0.12			] 0.06	
			^1E_u	5.26		^1E_u	4.51		4.33
$Mo(CO)_6$	O_h	$3t_{2g} \rightarrow 3t_{2u}$	$^1T_{1u}$	8.10		$^1T_{1u}$	6.90		
					] 1.03			] 1.01	
			$^1T_{2u}$	7.07		$^1T_{2u}$	5.89		~5.89
					] 0.25			] 0.23	
			$^1A_{1u}$	6.82		$^1A_{1u}$	5.66		
					] 0.04			] 0.06	
			^1E_u	6.78		^1E_u	5.59		

and the energy gaps between the highest dipole-allowed state and the second dipole-forbidden state.

The next examples for triply-degenerate excitations concern the tetrahedral transition metal oxides, CrO_4^{2-}, MoO_4^{2-}, MnO_4^-, TcO_4^-, RuO_4, and OsO_4 (T_d). Table 6 shows the excited states with the $t_1 \rightarrow e$ excitations in these complexes, whose calculated and experimental data are taken from Refs. [24–28]. Except for MoO_4^{2-}, the $t_1 \rightarrow e$ excitations lead to the lowest and second lowest states, which concern the colors of solutions of these complexes, namely, CrO_4^{2-} (yellow), MoO_4^{2-} (colorless), MnO_4^- (purple), RuO_4 (orange), and OsO_4 (colorless). Since the metal atom in these complexes formally possesses a $(nd)^0$ configuration ($n = 3$,

Table 6 t-e Excited states of tetrahedral metal oxides, CrO_4^{2-}, MoO_4^{2-}, MnO_4^-, TcO_4^-, RuO_4, and OsO_4, calculated at the SAC-CI/DZP level. Excitation energies (E) and energy splittings (ΔE) are given in eV

System	Point group	Main conf.	SAC-CI/DZP			Exptl.
			State	E	ΔE	
CrO_4^{2-}	T_d	$t_1 \rightarrow e$	$^1T_2$	3.41		3.38
					] 0.46	
			$^1T_1$	2.95		2.95
MoO_4^{2-}	T_d	$t_1 \rightarrow e$	$^1T_2$	5.52		5.29
					] 0.41	
			$^1T_1$	5.11		
MnO_4^-	T_d	$t_1 \rightarrow e$	$^1T_2$	2.57		2.27
					] 0.39	
			$^1T_1$	2.18		
TcO_4^-	T_d	$t_1 \rightarrow e$	$^1T_2$	4.28		4.27
					] 0.45	
			$^1T_1$	3.83		
RuO_4	T_d	$t_1 \rightarrow e$	$^1T_2$	3.22		3.22
					] 0.52	
			$^1T_1$	2.70		
OsO_4	T_d	$t_1 \rightarrow e$	$^1T_2$	3.90		4.34
					] 0.68	
			$^1T_1$	3.22		

4, 5), the lower excited states correspond to LMCT transitions. To be precise, both occupied t_1 and unoccupied e MOs are delocalized within metal and oxygen orbitals, and show slightly π-type bonding and antibonding characters, respectively. Since the direct products of $t_1 \otimes e$ can be decomposed as

$$t_1 \otimes e = t_1 + t_2, \tag{6}$$

the $t_1 \rightarrow e$ excitation brings about two singlet states, i.e., $^1T_1$ and $^1T_2$. Here, the $^1T_2$ state is dipole-allowed. In all systems, the dipole-allowed $^1T_2$ states appear in the higher level. Thus, the rule (i) holds here. Furthermore, the energy splittings between $^1T_2$ and $^1T_1$ are moderately large, that is, in the range of 0.39–0.68 eV.

2.3 Quadruply-Degenerate Case

The next example is a quadruply-degenerate excitation. Quadruply-degenerate irreducible representations (g; g_g, g_u) are seen in icosahedral (I; I_h) and higher-symmetry point groups. In general, it is difficult to find the quadruply-degenerate MOs in

Table 7 g–g Excited states of icosahedral $(B_{12}H_{12})^{2-}$ compound calculated at the SAC-CI/cc-pVDZ and FZOA/cc-pVDZ levels. Excitation energies (E) and energy splittings (ΔE) are given in eV

System	Point group	Main conf.	FZOA/cc-pVDZ			SAC-CI/cc-pVDZ		
			State	E	ΔE	State	E	ΔE
$(B_{12}H_{12})^{2-}$	I_h	$g_u \to g_g$	$^1T_{1u}$	12.06		$^1T_{1u}$	9.76	
					$]\ 2.80$			$]\ 0.97$
			$^1T_{2u}$	9.26		$^1T_{2u}$	8.79	
					$]\ 0.11$			$]\ 0.26$
			^1G_u	9.15		^1G_u	8.53	
					$]\ 0.31$			$]\ 0.10$
			^1H_u	8.84		^1H_u	8.43	
					$]\ 0.07$			$]\ 0.16$
			^1A_u	8.77		A_u	8.27	

the valence region. The HOMOs and LUMOs of $(B_{12}H_{12})^{2-}$ (I_h) correspond to quadruply-degenerate g_u and g_g symmetries, respectively. Those MOs, which mainly consist of $2p$ orbitals of B, exhibit π bonding and antibonding characters.

Table 7 shows the HOMO–LUMO excitations in $(B_{12}H_{12})^{2-}$, whose data are taken from Ref. [29]. Since the direct products of $g_u \otimes g_g$ can be decomposed as

$$g_u \otimes g_g = a_u + t_{1u} + t_{2u} + g_u + h_u, \tag{7}$$

the $g_u \to g_g$ excitation leads to five singlet states, i.e., ^1A_u, $^1T_{1u}$, $^1T_{2u}$, ^1G_u, and ^1H_u. Here, the $^1T_{1u}$ state is dipole-allowed. The dipole-allowed $^1T_{1u}$ state, whose data are written in italics in Table 7, appears in the highest level. Furthermore, the energy splitting between $^1T_{1u}$ and $^1T_{2u}$ is one order of magnitude larger than the other splittings. Therefore, it can be confirmed that the rules (i) and (ii) hold even in this complicated system with the quadruply-degenerate excitations.

Although the FZOA treatment tends to overestimate the excitation energies and the energy gaps between the highest dipole-allowed state and the second dipole-forbidden state, it can reproduce the orderings of the five states and the qualitative trends for the splittings.

2.4 Quintuply-Degenerate Case

The final case is the quintuply-degenerate excitation. As in the quadruply-degenerate case, the quintuply-degenerate irreducible representations (h; h_g, h_u) are seen in icosahedral (I; I_h) and higher-symmetry point groups. The HOMOs and next-HOMOs of $C_{60}(I_h)$ correspond to quintuply-degenerate h_u and h_g symmetries, respectively, whereas the LUMOs and next-LUMOs to triply-degenerate t_{1u} and t_{1g} symmetries, respectively. Those MOs, which mainly consist of $2p$ orbitals of C, exhibit π bonding and antibonding characters. It should be noted that the h_g

Table 8 h–t Excited states of icosahedral C_{60} molecule calculated at the SAC-CI/cc-pVDZ and FZOA/6-31G levels. Excitation energies (E) and energy splittings (ΔE) are given in eV

System	Point group	Main conf.	FZOA/6-31G			TD-B3LYP/cc-pVDZ			Exptl.
			State	E	ΔE	State	E	ΔE	
C_{60}	I_h	$h_g \rightarrow t_{1u}$	$^1T_{1u}$	*4.94*		$^1T_{1u}$	*3.45*		*3.04*
					] *0.80*			] *0.18*	
			$^1T_{2u}$	4.14		^1G_u	3.27		
					] 0.10			] 0.05	
			^1G_u	4.04		$^1T_{2u}$	3.22		
					] 0.05			] 0.02	
			^1H_u	3.99		^1H_u	3.20		
C_{60}	I_h	$h_u \rightarrow t_{1g}$	$^1T_{1u}$	*5.43*		$^1T_{1u}$	*3.89*		*3.78*
					] *1.39*			] *0.36*	
			^1G_u	4.04		$^1T_{2u}$	3.53		
					] 0.01			] 0.01	
			^1H_u	4.03		^1G_u	3.52		
					] 0.04			] 0.05	
			$^1T_{2u}$	3.99		^1H_u	3.47		

MOs become lower than the next-HOMOs by varying the used basis sets and/or exchange-correlation (XC) functional because the energy levels between the h_g and the g_g MOs are significantly close to each other.

Since the gerade-to-gerade and ungerade-to-ungerade transitions lead to no dipole-allowed excitations, we investigate the excitations from the next-HOMOs to the LUMOs and from the HOMOs to the next-LUMOs: i.e., $h_g \rightarrow t_{1u}$ and $h_u \rightarrow t_{1g}$, as shown in Table 8. The FZOA/6-31G results are taken from Ref. [29]; the TD-B3LYP/cc-pVDZ results are newly obtained in this review. The experimental data are taken from Ref. [30], which observed an absorption spectrum for C_{60} in hexane solution. The experimental study predicted the existence of at least eight dipole-allowed states below 7 eV.

The direct products of $h_g \otimes t_{1u}$ and $h_u \otimes t_{1g}$ can be decomposed as follows:

$$h_g \otimes t_{1u} = t_{1u} + t_{2u} + g_u + h_u, \tag{8}$$

$$h_u \otimes t_{1g} = t_{1u} + t_{2u} + g_u + h_u. \tag{9}$$

Thus, both $h_g \rightarrow t_{1u}$ and $h_u \rightarrow t_{1g}$ excitations lead to four singlet states, i.e., $^1T_{1u}$, $^1T_{2u}$, ^1G_u, and ^1H_u. Here, the $^1T_{1u}$ state is dipole-allowed. The dipole-allowed $^1T_{1u}$ states, whose data are written in italics in Table 8, are situated at the highest level. Furthermore, the energy splittings between the highest and the second highest states are one order of magnitude larger than the other splittings. As in the previous cases, it can be confirmed that the rules (i) and (ii) hold in this complicated system with the quintuply-degenerate excitations.

The FZOA treatment tends to overestimate the excitation energies and the energy gaps between the highest dipole-allowed state and the second dipole-forbidden state.

It does not reproduce the ordering among the dipole-forbidden states, of which splittings are considerably small for both TD-B3LYP and FZOA calculations. However, it is notable that the FZOA treatment can reproduce important trends, namely, the highest states and the splittings. This ensures the reliability of the interpretation based on the FZOA treatment.

3 Interpretation of Rules for Degenerate Excitations by Frozen Orbital Analysis

This section gives the qualitative interpretation of the rules for degenerate excitations. We adopt the FZOA treatment, which has succeeded in giving a qualitative but clear description in many cases, e.g., Koopmans' theorem for estimating the ionization potential (IP) and the electron affinity (EA), and Hund's rule of maximum multiplicity concerning the singlet–triplet separation.

3.1 Ionization Potential and Electron Affinity

One of the simplest treatments for estimating the IP is Koopmans' theorem [31], which indicates that the occupied orbital energy with an opposite sign is a reasonable approximation for the IP. Here, the IP is estimated by the difference between the Hartree–Fock (HF) ground-state and ionized-state energies.

$$\Delta E_i = E_i - E_0 = \langle \Phi_i | \hat{H} | \Phi_i \rangle - \langle \Phi_0 | \hat{H} | \Phi_0 \rangle = -\varepsilon_i. \tag{10}$$

For simplicity, we consider the closed-shell ground state. The HF ground-state wave function Φ_0 is expressed by the antisymmetrized product, i.e., Slater determinant, of the one-electron MOs $\{\phi_1, \overline{\phi}_1, \phi_2, \overline{\phi}_2, \cdots\}$:

$$\Phi_0 = \left\| \phi_1 \overline{\phi}_1 \phi_2 \overline{\phi}_2 \cdots \phi_i \overline{\phi}_i \right\|. \tag{11}$$

In the FZOA, the ionized-state wave function obtained by removing an electron from the MO $\overline{\phi}_i$, Φ_i, can be expressed by

$$\Phi_i = \left\| \phi_1 \overline{\phi}_1 \phi_2 \overline{\phi}_2 \cdots \phi_i \right\|. \tag{12}$$

This treatment is missing two effects, namely, the orbital relaxation and the electron correlation. Therefore, it cannot necessarily provide accurate results. However, it is qualitatively valid in many cases due to the cancellation of the two effects. A similar treatment is applicable to the EA. The electron-attached state obtaining by adding an electron to ϕ_a is described by

$$\Phi^a = \left\| \phi_1 \overline{\phi}_1 \phi_2 \overline{\phi}_2 \cdots \phi_i \overline{\phi}_i \phi_a \right\|. \tag{13}$$

Then, the EA becomes

$$\Delta E^a = E_0 - E^a = \left\langle \Phi_0 \left| \hat{H} \right| \Phi_0 \right\rangle - \left\langle \Phi^a \left| \hat{H} \right| \Phi^a \right\rangle = -\varepsilon_a. \tag{14}$$

The Koopmans' theorem for the EA tends to give less favorable results than that for the IP because the cancellation between the orbital relaxation and the electron correlation is not expected.

There is something similar to the Koopmans' theorem in the Kohn–Sham (KS) DFT [32]. It is Janak's theorem [33], which states that the derivative of the total energy with respect to the occupation number of the KS orbital $\overline{\phi}_i$ is exactly the KS orbital energy:

$$\Delta E_i = -\frac{\partial E}{\partial n_i} = -\varepsilon_i^{KS}. \tag{15}$$

This is true not only for the (unknown) exact XC potential but for all commonly used approximate XC potentials.

3.2 Singlet and Triplet Excitation Energies

This section describes the FZOA treatment for the singlet and triplet excited states. In the FZOA, the configuration state functions (CSFs) for the singlet- and triplet-type excitations from $\{\phi_i, \overline{\phi}_i\}$ to $\{\phi_a, \overline{\phi}_a\}$ are expressed by

$$^1\Phi_i^a = \frac{1}{\sqrt{2}} \left\| \phi_1 \overline{\phi}_1 \phi_2 \overline{\phi}_2 \cdots (\phi_i \overline{\phi}_a + \phi_a \overline{\phi}_i) \right\|, \tag{16}$$

$$^3\Phi_i^a = \frac{1}{\sqrt{2}} \left\| \phi_1 \overline{\phi}_1 \phi_2 \overline{\phi}_2 \cdots (\phi_i \overline{\phi}_a - \phi_a \overline{\phi}_i) \right\|. \tag{17}$$

Here, both CSFs possess a zero quantum number for the z-component of the spin angular momentum operator ($M_s = 0$). The corresponding excitation energies are derived as follows:

$$\begin{aligned} ^1\Delta E_i^a = {}^1 E_i^a - E_0 &= \left\langle {}^1\Phi_i^a \left| \hat{H} \right| {}^1\Phi_i^a \right\rangle - \left\langle \Phi_0 \left| \hat{H} \right| \Phi_0 \right\rangle \\ &= \varepsilon_a - \varepsilon_i - J_{ia} + 2K_{ia}, \end{aligned} \tag{18}$$

$$\begin{aligned} ^3\Delta E_i^a = {}^3 E_i^a - E_0 &= \left\langle {}^3\Phi_i^a \left| \hat{H} \right| {}^3\Phi_i^a \right\rangle - \left\langle \Phi_0 \left| \hat{H} \right| \Phi_0 \right\rangle \\ &= \varepsilon_a - \varepsilon_i - J_{ia}, \end{aligned} \tag{19}$$

where J and K are Coulomb and exchange integrals, respectively. As a result, the singlet–triplet separation is estimated by

$$\Delta = {}^{1}E_{i}^{a} - {}^{3}E_{i}^{a} = {}^{1}\Delta E_{i}^{a} - {}^{3}\Delta E_{i}^{a} = 2K_{ia}. \tag{20}$$

Since the exchange integral is a positive definite, one can immediately obtain the correct sign for the energy splitting. This explains Hund's rule of maximum multiplicity in which, for a given electronic configuration, the state with the highest multiplicity has the lowest energy [34]. This explanation for singlet–triplet separation is generally accepted in many texts on quantum mechanics. However, in fact, the difference is negative when it is evaluated with accurate wave functions. Many theoretical studies have attempted to clarify the reasons for this discrepancy, for example, see Refs. [35–37].

According to the study on He by Kohl [35], the electron repulsion energy for the same pair of spatial orbitals, which is smaller for the triplet than for the singlet, is not the dominant contribution to the total energy. In the transition from the FZOA treatment to the exact solution, the redistribution of charge results in a substantial lowering of the nuclear attraction term in the triplet, which determines the energy levels of the singlet and triplet states. Nevertheless, we think that a discussion of the FZOA treatment is important.

3.3 Excitation Energies for Degenerate Excitations

This section presents formulas for the excitation energies between degenerate MOs based on the FZOA treatment. We first deal with the simplest case, that is, the π–π^* excitation in $N_2(D_{\infty h})$. Figure 1 illustrates the π and π^* MOs and the excitations between these degenerate MOs. When we adopt the Bethe's symmetry-lowering method from $D_{\infty h}$ to D_{2h}, the occupied π_u MOs, $\{\phi_i, \overline{\phi_i}\}$ and $\{\phi_j, \overline{\phi_j}\}$, are assigned to b_{3u} and b_{2u}, that is, π_x and π_y, respectively. Similarly, the unoccupied π_g MOs, $\{\phi_a, \overline{\phi_a}\}$ and $\{\phi_b, \overline{\phi_b}\}$, are assigned to b_{2g} and b_{3g}, that is, π_x^* and π_y^*, respectively.

There are four CSFs for the singlet-type excitations among these degenerate MOs, namely,

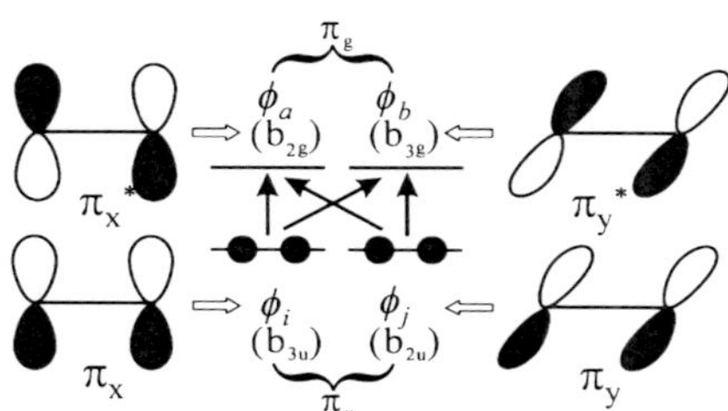

Fig. 1 Excitation from doubly-degenerate π_u to π_g MOs in $D_{\infty h}$. The occupied ϕ_i and ϕ_j MOs are assigned to b_{3u} and b_{2u} in D_{2h} subgroup symmetry, respectively. The unoccupied ϕ_a and ϕ_b MOs are assigned to b_{2g} and b_{3g}, respectively

$$ {}^1\Phi_i^a = \frac{1}{\sqrt{2}} \left\| \phi_1\overline{\phi}_1\phi_2\overline{\phi}_2 \cdots \left(\phi_i\overline{\phi}_a + \phi_a\overline{\phi}_i\right)\phi_j\overline{\phi}_j \right\| \quad ({}^1B_{1u} \text{ in } D_{2h}), \tag{21} $$

$$ {}^1\Phi_i^b = \frac{1}{\sqrt{2}} \left\| \phi_1\overline{\phi}_1\phi_2\overline{\phi}_2 \cdots \left(\phi_i\overline{\phi}_b + \phi_b\overline{\phi}_i\right)\phi_j\overline{\phi}_j \right\| \quad ({}^1A_u \text{ in } D_{2h}), \tag{22} $$

$$ {}^1\Phi_j^a = \frac{1}{\sqrt{2}} \left\| \phi_1\overline{\phi}_1\phi_2\overline{\phi}_2 \cdots \phi_i\overline{\phi}_i \left(\phi_j\overline{\phi}_a + \phi_a\overline{\phi}_j\right) \right\| \quad ({}^1A_u \text{ in } D_{2h}), \tag{23} $$

$$ {}^1\Phi_j^b = \frac{1}{\sqrt{2}} \left\| \phi_1\overline{\phi}_1\phi_2\overline{\phi}_2 \cdots \phi_i\overline{\phi}_i \left(\phi_j\overline{\phi}_b + \phi_b\overline{\phi}_j\right) \right\| \quad ({}^1B_{1u} \text{ in } D_{2h}). \tag{24} $$

Here, while these CSFs naturally satisfy the spin symmetry, they do not adapt the space symmetry. Spin- and space-symmetry-adapted wave functions are given by

$$ \Psi({}^1\Sigma_u^+) = \frac{1}{\sqrt{2}} \left({}^1\Phi_i^a + {}^1\Phi_j^b\right) ({}^1B_{1u} \text{ in } D_{2h}), \tag{25} $$

$$ \Psi({}^1\Delta_u) = \frac{1}{\sqrt{2}} \left({}^1\Phi_i^a - {}^1\Phi_j^b\right) ({}^1B_{1u} \text{ in } D_{2h}), \tag{26} $$

$$ \Psi({}^1\Delta_u) = \frac{1}{\sqrt{2}} \left({}^1\Phi_i^b + {}^1\Phi_j^a\right) ({}^1A_u \text{ in } D_{2h}), \tag{27} $$

$$ \Psi({}^1\Sigma_u^-) = \frac{1}{\sqrt{2}} \left({}^1\Phi_i^b - {}^1\Phi_j^a\right) ({}^1A_u \text{ in } D_{2h}). \tag{28} $$

As a result, the excitation energies for the four states are derived as follows:

$$ \Delta E\left({}^1\Sigma_u^+\right) = E\left({}^1\Sigma_u^+\right) - E_0 = \left\langle \Psi({}^1\Sigma_u^+) \left| \hat{H} \right| \Psi({}^1\Sigma_u^+) \right\rangle - \left\langle \Phi_0 \left| \hat{H} \right| \Phi_0 \right\rangle $$
$$ = (\varepsilon_a - \varepsilon_i) + (-J_{ia} + 2K_{ia}) + \{2(ai|jb) - (ab|ij)\}, \tag{29} $$

$$ \Delta E\left({}^1\Delta_u\right) = (\varepsilon_a - \varepsilon_i) + (-J_{ia} + 2K_{ia}) - \{2(ai|jb) - (ab|ij)\}, \tag{30} $$

$$ \Delta E\left({}^1\Delta_u\right) = (\varepsilon_a - \varepsilon_i) + \left(-J_{ja} + 2K_{ja}\right) + \{2(aj|ib) - (ab|ij)\}, \tag{31} $$

$$ \Delta E\left({}^1\Sigma_u^-\right) = (\varepsilon_a - \varepsilon_i) + \left(-J_{ja} + 2K_{ja}\right) - \{2(aj|ib) - (ab|ij)\}, \tag{32} $$

where $(ai|jb)$, $(aj|ib)$, and $(ab|ij)$ are electron repulsion integrals in the MO basis. To derive Eqs. (29) (30), (31), and (32), we use the following relations based on the degeneracy

$$ \varepsilon_i = \varepsilon_j, \varepsilon_a = \varepsilon_b, J_{ia} = J_{jb}, $$
$$ J_{ja} = J_{ib}, K_{ia} = K_{jb}, K_{ja} = K_{ib}. \tag{33} $$

Note that the values of J_{ia} and J_{ja} and also those of K_{ia} and K_{ja} are different, despite the use of the symmetry rule. The energy expressions in Eqs. (29), (30), (31), and (32) correspond to the CIS results within the minimum active space, i.e., occupied π_u and unoccupied π_g MOs.

The right-hand sides of Eqs. (29) (30), (31), and (32) can be divided into three parts:

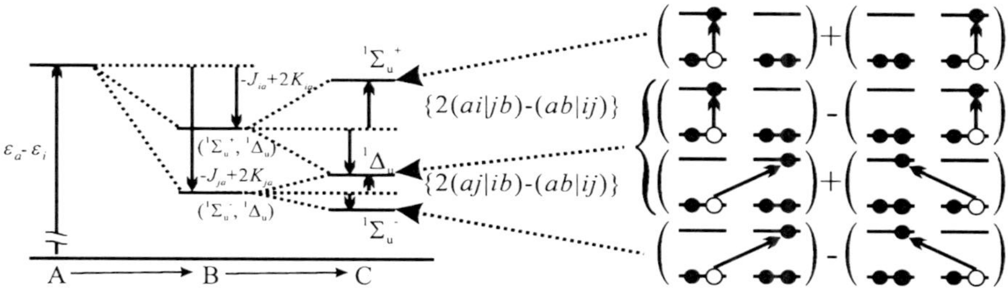

Fig. 2 Splitting scheme of the singlet excited states, namely, $^1\Sigma_u^+$, $^1\Delta_u$, and $^1\Sigma_u^-$, for the $\pi_u \to \pi_g$ transitions. The splitting between ($^1\Sigma_u^+$, $^1\Delta_u$) and ($^1\Delta_u$, $^1\Sigma_u^-$) is due to the B term. Further splittings between $^1\Sigma_u^+$ and $^1\Delta_u$ and between $^1\Delta_u$ and $^1\Sigma_u^-$ are due to the C term

$$\Delta E = A + B + C, \tag{34}$$

where A is the orbital energy difference, B consists of the Coulomb and exchange integrals, and C includes the remaining integrals. Figure 2 schematically illustrates the energy splittings of the four (or three) singlet excited states for the $\pi_u \to \pi_g$ transitions, that is, $^1\Sigma_u^+$, $^1\Delta_u$, $^1\Delta_u$, and $^1\Sigma_u^-$, of which two are degenerate. The energy level on the left-hand side in Fig. 2 is determined by the A term, which does not produce energy splittings due to the degenerate MOs. The energy levels in the middle in Fig. 2 are split into two by the B term, which produces energy splittings between ($^1\Sigma_u^+$, $^1\Delta_u$) and ($^1\Delta_u$, $^1\Sigma_u^-$). The four (or three) energy levels on the right-hand side in Fig. 2 are due to the C term, which leads to further energy splittings between $^1\Sigma_u^+$ and $^1\Delta_u$ and between $^1\Delta_u$ and $^1\Sigma_u^-$.

Analyses of two-electron integrals such as J_{ia}, J_{ja}, K_{ia}, K_{ja}, $(ai|jb)$, $(aj|ib)$, and $(ab|ij)$ are important for an understanding of the above energy splittings. The spatial distributions of the electron densities, $\phi_i^*(\mathbf{r})\phi_i(\mathbf{r})$ and $\phi_a^*(\mathbf{r})\phi_a(\mathbf{r})$, are associated with the Coulomb integral J_{ia}, namely, the integral decreases as the distance between the electron densities increases. Since we are considering the closed-shell ground-state case, namely, the restricted HF (RHF) treatment, the spatial distribution of $\bar{\phi}_i^*(\mathbf{r})\bar{\phi}_i(\mathbf{r})$ is equivalent to that of $\phi_i^*(\mathbf{r})\phi_i(\mathbf{r})$ and $\bar{\phi}_a^*(\mathbf{r})\bar{\phi}_a(\mathbf{r})$ to $\phi_a^*(\mathbf{r})\phi_a(\mathbf{r})$. For J_{ja}, the spatial distributions of the electron densities, $\phi_j^*(\mathbf{r})\phi_j(\mathbf{r})$ and $\phi_a^*(\mathbf{r})\phi_a(\mathbf{r})$, are associated.

For the exchange integrals K_{ia} and K_{ja}, the overlap distributions (or the transition densities), $\phi_a^*(\mathbf{r})\phi_i(\mathbf{r})$ and $\phi_a^*(\mathbf{r})\phi_j(\mathbf{r})$, play the key roles, respectively. Of course, both the integrals $\int \phi_a^*(\mathbf{r})\phi_i(\mathbf{r})d\mathbf{r}$ and $\int \phi_a^*(\mathbf{r})\phi_j(\mathbf{r})d\mathbf{r}$ vanish due to the orthogonality of the MOs. Furthermore, both overlap distributions of $\phi_a^*(\mathbf{r})\phi_i(\mathbf{r})$ and $\phi_j^*(\mathbf{r})\phi_b(\mathbf{r})$ are important for $(ai|jb)$, whereas those of $\phi_a^*(\mathbf{r})\phi_j(\mathbf{r})$ and $\phi_i^*(\mathbf{r})\phi_b(\mathbf{r})$ for $(aj|ib)$. As mentioned above, $\phi_a^*(\mathbf{r})\phi_i(\mathbf{r})$, $\phi_a^*(\mathbf{r})\phi_j(\mathbf{r})$, $\phi_i^*(\mathbf{r})\phi_b(\mathbf{r})$, and $\phi_j^*(\mathbf{r})\phi_b(\mathbf{r})$ are equivalent to $\bar{\phi}_a^*(\mathbf{r})\bar{\phi}_i(\mathbf{r})$, $\bar{\phi}_a^*(\mathbf{r})\bar{\phi}_j(\mathbf{r})$, $\bar{\phi}_i^*(\mathbf{r})\bar{\phi}_b(\mathbf{r})$, and $\bar{\phi}_j^*(\mathbf{r})\bar{\phi}_b(\mathbf{r})$ in the RHF treatment, respectively.

The intensities of the absorption bands are calculated by using the square of the transition dipole moment. The transition dipole moments for the four states for the $\pi_u \to \pi_g$ transitions, namely, $^1\Sigma_u^+$, $^1\Delta_u$, $^1\Delta_u$, and $^1\Sigma_u^-$, are estimated as follows:

$$\langle \Psi(^1\Sigma_u^+) |\mathbf{r}| \Phi_0\rangle = \frac{1}{\sqrt{2}} \left(\langle ^1\Phi_i^a |\mathbf{r}| \Phi_0\rangle + \langle ^1\Phi_j^b |\mathbf{r}| \Phi_0\rangle \right) = \langle \phi_a |\mathbf{r}| \phi_i\rangle + \langle \phi_b |\mathbf{r}| \phi_j\rangle$$
$$= 2 \langle \phi_a |\mathbf{r}| \phi_i\rangle , \tag{35}$$

$$\langle \Psi(^1\Delta_u) |\mathbf{r}| \Phi_0\rangle = \frac{1}{\sqrt{2}} \left(\langle ^1\Phi_i^a |\mathbf{r}| \Phi_0\rangle - \langle ^1\Phi_j^b |\mathbf{r}| \Phi_0\rangle \right) = \langle \phi_a |\mathbf{r}| \phi_i\rangle - \langle \phi_b |\mathbf{r}| \phi_j\rangle = 0, \tag{36}$$

$$\langle \Psi(^1\Delta_u) |\mathbf{r}| \Phi_0\rangle = \frac{1}{\sqrt{2}} \left(\langle ^1\Phi_j^a |\mathbf{r}| \Phi_0\rangle + \langle ^1\Phi_i^b |\mathbf{r}| \Phi_0\rangle \right) = \langle \phi_a |\mathbf{r}| \phi_j\rangle + \langle \phi_b |\mathbf{r}| \phi_i\rangle = 0, \tag{37}$$

$$\langle \Psi(^1\Sigma_u^-) |\mathbf{r}| \Phi_0\rangle = \frac{1}{\sqrt{2}} \left(\langle ^1\Phi_j^a |\mathbf{r}| \Phi_0\rangle - \langle ^1\Phi_i^b |\mathbf{r}| \Phi_0\rangle \right) = \langle \phi_a |\mathbf{r}| \phi_j\rangle - \langle \phi_b |\mathbf{r}| \phi_i\rangle = 0. \tag{38}$$

Here, we use the equivalency of the integrals for α and β MOs in the arithmetic manipulation. Because the irreducible representations of ϕ_i, ϕ_j, ϕ_a, and ϕ_b in the D_{2h} symmetry are b_{3u}, b_{2u}, b_{2g}, and b_{3g}, respectively, the direct products of $\phi_a \otimes \phi_i$, $\phi_b \otimes \phi_j$, $\phi_a \otimes \phi_j$, and $\phi_b \otimes \phi_i$ are b_{1u}, b_{1u}, a_u, and a_u, respectively. The dipole operator x, y, z belong to b_{3u}, b_{2u}, and b_{1u}, respectively. As a result, the individual terms in the middle of Eqs. (35) and (36) become nonzero, while those in Eqs. (37) and (38) become zero. To derive the final expressions of Eqs. (35) and (36), we use the following relations owing to the degeneracy,

$$\langle \phi_a |\mathbf{r}| \phi_i\rangle = \langle \phi_b |\mathbf{r}| \phi_j\rangle . \tag{39}$$

It should be noted that the integration for the transition dipole $\langle \phi_a |\mathbf{r}| \phi_i\rangle$ includes the overlap distribution (or transition density), $\phi_a^*(\mathbf{r})\phi_i(\mathbf{r})$. Therefore, the integral $\langle \phi_a |\mathbf{r}| \phi_i\rangle$ should be closely related to the two-electron integrals, K_{ia} and $(ai|jb)$.

The energy expressions in Eqs. (29), (30), (31), and (32) and the transition dipole moments in Eqs. (35), (36), (37), and (38) are general to π–π^* excitations in $D_{\infty h}$ systems, since they are derived only using the group theory. It is true not only for diatomic cases but also for polyatomic cases. For example, the excitations in CO_2 occur from π_g to π_u, not from π_u to π_g, the excitation energy expressions are the same as Eqs. (29) (30), (31), and (32). Similar expressions can be derived for the π–π^* excitations in $C_{\infty v}$ systems, where the expressions for the $^1\Sigma_u^+$, $^1\Delta_u$, and $^1\Sigma_u^-$ states in $D_{\infty h}$, that is, Eqs. (29) (30), (31), and (32) and (35), (36), (37), and (38), are assigned to those for the $^1\Sigma^+$, $^1\Delta$, and $^1\Sigma^-$ states, respectively.

In what follows, we discuss the triply-degenerate excitations, namely, t–t excitations in O_h and T_d systems. Figure 3 illustrates the excitations between the t_{1u} and the t_{2g} MOs. With the use of the Bethe's symmetry-lowering method from O_h to D_{2h}, the occupied t_{1u} MOs, $\{\phi_i, \overline{\phi}_i\}$, $\{\phi_j, \overline{\phi}_j\}$, and $\{\phi_k, \overline{\phi}_k\}$, are assigned to b_{3u} (x), b_{2u} (y), and b_{1u} (z), respectively. Similarly, the unoccupied t_{2g} MOs, $\{\phi_a, \overline{\phi}_a\}$, $\{\phi_b, \overline{\phi}_b\}$, and $\{\phi_c, \overline{\phi}_c\}$, are assigned to b_{3g} (yz), b_{2g} (zx), and b_{1g} (xy), respectively.

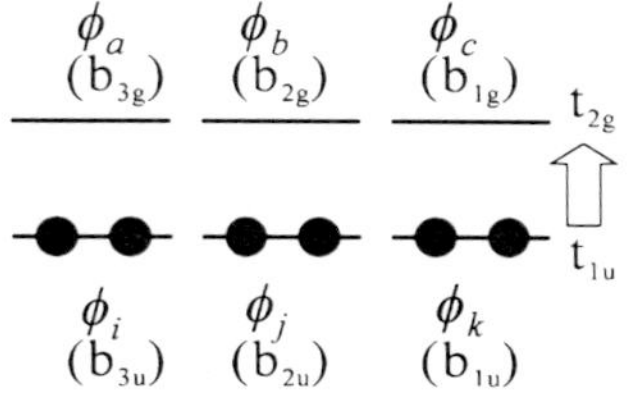

Fig. 3 Excitation from triply-degenerate t_{1u} to t_{2g} MOs in O_h. The occupied ϕ_i, ϕ_j and ϕ_k MOs are assigned to b_{3u}, b_{2u}, and b_{1u} in D_{2h} subgroup symmetry, respectively. The unoccupied ϕ_a, ϕ_b, and ϕ_c MOs are assigned to b_{3g}, b_{2g}, and b_{1g}, respectively

Spin- and space-symmetry-adapted wave functions for the singlet-type excitations among these degenerate MOs are derived by

$$\Psi(^1T_{1u}) = \frac{1}{\sqrt{2}}\left(^1\Phi_j^c + {}^1\Phi_k^b\right)(^1B_{3u} \text{ in } D_{2h}), \tag{40}$$

$$\Psi(^1T_{1u}) = \frac{1}{\sqrt{2}}\left(^1\Phi_k^a + {}^1\Phi_i^c\right)(^1B_{2u} \text{ in } D_{2h}), \tag{41}$$

$$\Psi(^1T_{1u}) = \frac{1}{\sqrt{2}}\left(^1\Phi_i^b + {}^1\Phi_j^a\right)(^1B_{1u} \text{ in } D_{2h}), \tag{42}$$

$$\Psi(^1T_{2u}) = \frac{1}{\sqrt{2}}\left(^1\Phi_j^c - {}^1\Phi_k^b\right)(^1B_{3u} \text{ in } D_{2h}), \tag{43}$$

$$\Psi(^1T_{2u}) = \frac{1}{\sqrt{2}}\left(^1\Phi_k^a - {}^1\Phi_i^c\right)(^1B_{2u} \text{ in } D_{2h}), \tag{44}$$

$$\Psi(^1T_{2u}) = \frac{1}{\sqrt{2}}\left(^1\Phi_i^b - {}^1\Phi_j^a\right)(^1B_{1u} \text{ in } D_{2h}), \tag{45}$$

$$\Psi(^1E_u) = \frac{1}{\sqrt{2}}\left(^1\Phi_i^a - {}^1\Phi_j^b\right)(^1A_u \text{ in } D_{2h}), \tag{46}$$

$$\Psi(^1E_u) = \frac{1}{\sqrt{6}}\left(^1\Phi_i^a + {}^1\Phi_j^b - 2\,{}^1\Phi_k^c\right)(^1A_u \text{ in } D_{2h}), \tag{47}$$

$$\Psi(^1A_{2u}) = \frac{1}{\sqrt{3}}\left(^1\Phi_i^a + {}^1\Phi_j^b + {}^1\Phi_k^c\right)(^1A_u \text{ in } D_{2h}). \tag{48}$$

Finally, the excitation energies for the four states are derived as follows:

$$\Delta E\left(^1T_{1u}\right) = (\varepsilon_a - \varepsilon_i) + \left(-J_{ja} + 2K_{ja}\right) + \{2(aj|ib) - (ab|ij)\}, \tag{49}$$

$$\Delta E\left(^1T_{2u}\right) = (\varepsilon_a - \varepsilon_i) + \left(-J_{ja} + 2K_{ja}\right) - \{2(aj|ib) - (ab|ij)\}, \tag{50}$$

$$\Delta E\left(^1E_u\right) = (\varepsilon_a - \varepsilon_i) + (-J_{ia} + 2K_{ia}) - \{2(ai|jb) - (ab|ij)\}, \tag{51}$$

$$\Delta E\left(^1A_{2u}\right) = (\varepsilon_a - \varepsilon_i) + (-J_{ia} + 2K_{ia}) + \{2(ai|jb) - (ab|ij)\}. \tag{52}$$

The energy expressions in Eqs. (49), (50), (51), and (52) correspond to the CIS results within the minimum active space, i.e., occupied t_{1u} and unoccupied t_{2g} MOs.

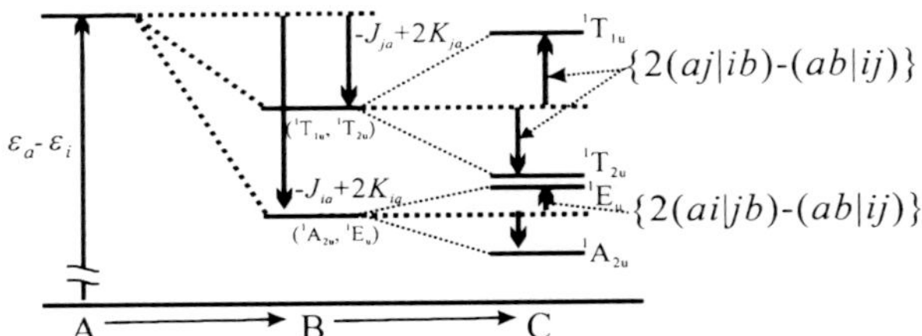

Fig. 4 Splitting scheme of the singlet excited states, namely, $^1T_{1u}$, $^1T_{2u}$, ^1E_u, and $^1A_{2u}$, for the $t_{1u} \rightarrow t_{2g}$ transitions. The splitting between ($^1T_{1u}$, $^1T_{2u}$) and (^1E_u, $^1A_{2u}$) is due to the B term. Further splittings between $^1T_{1u}$ and $^1T_{2u}$ and between ^1E_u and $^1A_{2u}$ are due to the C term

To derive Eqs. (49), (50), (51), and (52), we use the relations due to the degeneracy, e.g.,

$$\varepsilon_i = \varepsilon_j = \varepsilon_k, \quad \varepsilon_a = \varepsilon_b = \varepsilon_c,$$

$$J_{ia} = J_{jb} = J_{kc}, \quad J_{ja} = J_{kb} = J_{ic} = J_{ka} = J_{ib} = J_{jc},$$

$$K_{ia} = K_{jb} = K_{kc}, \quad K_{ja} = K_{kb} = K_{ic} = K_{ka} = K_{ib} = K_{jc},$$

$$(ai\,|jb) = (bj\,|kc) = (ck\,|ia), \quad (ab\,|ij) = (bc\,|jk) = (ca\,|ki). \tag{53}$$

The categorization in Eq. (34) is also applicable to Eqs. (49), (50), (51), and (52). Figure 4 schematically illustrates the energy splittings of the four singlet excited states for the $t_{1u} \rightarrow t_{2g}$ transitions, that is, $^1T_{1u}$, $^1T_{2u}$, ^1E_u, and $^1A_{2u}$. The A term leads to no splittings in Fig. 4. The B term splits the energy levels into two, that is, ($^1T_{1u}$, $^1T_{2u}$) and (^1E_u, $^1A_{2u}$). The complete splittings are accomplished by taking the C term into account.

As in the doubly-degenerate case, the analyses of two-electron integrals such as J_{ia}, J_{ja}, K_{ia}, K_{ja}, $(ai|jb)$, $(aj|ib)$, and $(ab|ij)$ are essential for the interpretation of the energy splittings. While the spatial distributions of the electron densities are related to the Coulomb integrals, those of the transition densities such as $\phi_a^*(\mathbf{r})\phi_i(\mathbf{r})$ and $\phi_j^*(\mathbf{r})\phi_b(\mathbf{r})$ play the key roles for the exchange integrals like K_{ia} and the two-electron integrals in the C term such as $(ai|jb)$.

The transition dipole moments for the four states for the $t_{1u} \rightarrow t_{2g}$ transitions are calculated as follows:

$$\langle \Psi(^1T_{1u})\,|\mathbf{r}|\,\Phi_0\rangle = \frac{1}{\sqrt{2}}\left(\langle {}^1\Phi_i^b\,|\mathbf{r}|\,\Phi_0\rangle + \langle {}^1\Phi_j^a\,|\mathbf{r}|\,\Phi_0\rangle\right) = \langle \phi_b\,|\mathbf{r}|\,\phi_i\rangle + \langle \phi_a\,|\mathbf{r}|\,\phi_j\rangle$$

$$= 2\langle \phi_a\,|\mathbf{r}|\,\phi_j\rangle, \tag{54}$$

$$\langle \Psi(^1T_{2u})\,|\mathbf{r}|\,\Phi_0\rangle = \frac{1}{\sqrt{2}}\left(\langle {}^1\Phi_i^b\,|\mathbf{r}|\,\Phi_0\rangle - \langle {}^1\Phi_j^a\,|\mathbf{r}|\,\Phi_0\rangle\right) = \langle \phi_b\,|\mathbf{r}|\,\phi_i\rangle - \langle \phi_a\,|\mathbf{r}|\,\phi_j\rangle = 0, \tag{55}$$

$$\langle \Psi(^1E_u)\,|\mathbf{r}|\,\Phi_0\rangle = \frac{1}{\sqrt{2}}\left(\langle {}^1\Phi_i^a\,|\mathbf{r}|\,\Phi_0\rangle - \langle {}^1\Phi_j^b\,|\mathbf{r}|\,\Phi_0\rangle\right) = \langle \phi_a\,|\mathbf{r}|\,\phi_i\rangle - \langle \phi_b\,|\mathbf{r}|\,\phi_j\rangle = 0, \tag{56}$$

$$\langle \Psi(^1A_{2u}) |\mathbf{r}| \Phi_0\rangle = \frac{1}{\sqrt{2}} \left(\langle ^1\Phi_i^a |\mathbf{r}| \Phi_0\rangle + \langle ^1\Phi_j^b |\mathbf{r}| \Phi_0\rangle\right) = \langle \phi_a |\mathbf{r}| \phi_i\rangle + \langle \phi_b |\mathbf{r}| \phi_j\rangle = 0.$$

(57)

Here, the individual terms in the middle of Eqs. (54) and (57) are nonzero, while those in Eqs. (56) and (57) are zero. Since the integration for the transition dipole $\langle \phi_a |\mathbf{r}| \phi_j\rangle$ includes the overlap distribution (or transition density), $\phi_a^*(\mathbf{r})\,\phi_j(\mathbf{r})$, this integral is closely related to the two-electron integrals, K_{ja} and $(aj|ib)$.

The energy expressions in Eqs. (49), (50), (51), and (52) and the transition dipole moments in Eqs. (54), (55), (56), and (57) are general to $t_{1u} \rightarrow t_{2g}$, $t_{2u} \rightarrow t_{1g}$, $t_{1g} \rightarrow t_{2u}$, and $t_{2g} \rightarrow t_{1u}$ transitions in O_h. For $t_{1u} \rightarrow t_{1g}$, $t_{2u} \rightarrow t_{2g}$, $t_{1g} \rightarrow t_{1u}$, and $t_{2g} \rightarrow t_{2u}$ transitions in O_h, the formulas for the $^1A_{2u}$ state are assigned to those for the $^1A_{1u}$ state. Furthermore, the excitation energy formulas for the $^1T_{1u}$ and $^1T_{2u}$ states are interchanged, namely, Eqs. (49) and (50). Details are discussed in Refs. [4] and [5]. Furthermore, similar formulas can be given for the $t_1 \rightarrow t_1$, $t_2 \rightarrow t_2$, $t_1 \rightarrow t_2$, and $t_2 \rightarrow t_1$ transitions in T_d.

Next, we discuss the quadruply-degenerate excitations in I_h systems. Figure 5 illustrates the excitations between the g_u and g_g MOs. With the use of the Bethe's symmetry-lowering method from I_h to D_{2h}, the occupied g_u MOs, $\{\phi_i, \overline{\phi}_i\}$, $\{\phi_j, \overline{\phi}_j\}$, $\{\phi_k, \overline{\phi}_k\}$, and $\{\phi_l, \overline{\phi}_l\}$, are assigned to a_u, b_{1u}, b_{2u}, and b_{3u}, respectively. Similarly, the unoccupied g_g MOs, $\{\phi_a, \overline{\phi}_a\}$, $\{\phi_b, \overline{\phi}_b\}$, $\{\phi_c, \overline{\phi}_c\}$, and $\{\phi_d, \overline{\phi}_d\}$, are assigned to a_g, b_{1g}, b_{2g}, and b_{3g}, respectively. Spin- and space-symmetry-adapted wave functions for the singlet-type excitations among these degenerate MOs are given by

$$\Psi(^1T_{1u}) = \frac{1}{2}\left(^1\Phi_i^b + {}^1\Phi_j^a + {}^1\Phi_k^d + {}^1\Phi_l^c\right)\;(^1B_{1u}\text{ in }D_{2h}),$$

(58)

$$\Psi(^1T_{1u}) = \frac{1}{2}\left(^1\Phi_i^c + {}^1\Phi_j^d + {}^1\Phi_k^a + {}^1\Phi_l^b\right)\;(^1B_{2u}\text{ in }D_{2h}),$$

(59)

$$\Psi(^1T_{1u}) = \frac{1}{2}\left(^1\Phi_i^d + {}^1\Phi_j^c + {}^1\Phi_k^b + {}^1\Phi_l^a\right)\;(^1B_{3u}\text{ in }D_{2h}),$$

(60)

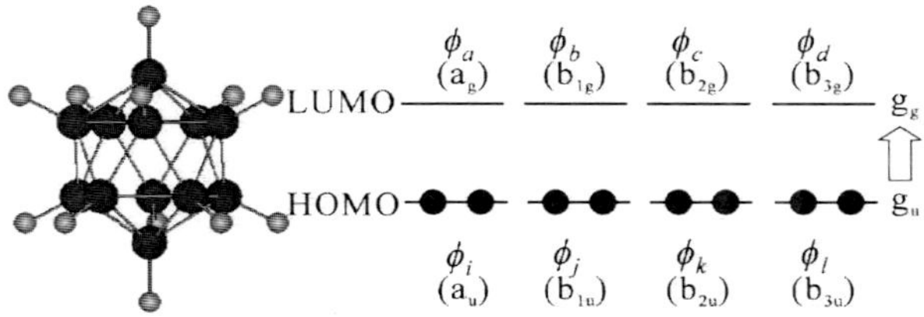

Fig. 5 Structure of $(B_{12}H_{12})^{2-}$ and excitation from quadruply-degenerate g_u to g_g MOs in I_h. The occupied ϕ_i, ϕ_j, ϕ_k, and ϕ_l MOs are assigned to a_u, b_{1u}, b_{2u}, and b_{3u} and in D_{2h} subgroup symmetry, respectively. The unoccupied ϕ_a, ϕ_b, ϕ_c, and ϕ_d MOs are assigned to a_g, b_{1g}, b_{2g}, and b_{3g}, respectively

$$\Psi(^1T_{2u}) = \frac{1}{2}\left(^1\Phi_i^b + {}^1\Phi_j^a - {}^1\Phi_k^d - {}^1\Phi_l^c\right)(^1B_{1u}\text{ in }D_{2h}), \tag{61}$$

$$\Psi(^1T_{2u}) = \frac{1}{2}\left(^1\Phi_i^c + {}^1\Phi_j^d - {}^1\Phi_k^a - {}^1\Phi_l^b\right)(^1B_{2u}\text{ in }D_{2h}), \tag{62}$$

$$\Psi(^1T_{2u}) = \frac{1}{2}\left(^1\Phi_i^d + {}^1\Phi_j^c - {}^1\Phi_k^b - {}^1\Phi_l^a\right)(^1B_{3u}\text{ in }D_{2h}), \tag{63}$$

$$\Psi(^1G_u) = \frac{1}{2\sqrt{3}}\left(3^1\Phi_i^a - {}^1\Phi_j^b - {}^1\Phi_k^c - {}^1\Phi_l^d\right)(^1A_u\text{ in }D_{2h}), \tag{64}$$

$$\Psi(^1G_u) = \frac{1}{2\sqrt{3}}\left(^1\Phi_i^b - {}^1\Phi_j^a - \sqrt{5}\,{}^1\Phi_k^d + \sqrt{5}\,{}^1\Phi_l^c\right)(^1B_{1u}\text{ in }D_{2h}), \tag{65}$$

$$\Psi(^1G_u) = \frac{1}{2\sqrt{3}}\left(^1\Phi_i^c - \sqrt{5}\,{}^1\Phi_j^d - {}^1\Phi_k^a + \sqrt{5}\,{}^1\Phi_l^b\right)(^1B_{2u}\text{ in }D_{2h}), \tag{66}$$

$$\Psi(^1G_u) = \frac{1}{2\sqrt{3}}\left(^1\Phi_i^d + \sqrt{5}\,{}^1\Phi_j^c - \sqrt{5}\,{}^1\Phi_k^b + {}^1\Phi_l^a\right)(^1B_{3u}\text{ in }D_{2h}), \tag{67}$$

$$\Psi(^1H_u) = \frac{1}{\sqrt{2}}\left(^1\Phi_k^c - {}^1\Phi_l^d\right)(^1A_u\text{ in }D_{2h}), \tag{68}$$

$$\Psi(^1H_u) = \frac{1}{\sqrt{6}}\left(2^1\Phi_j^b - {}^1\Phi_k^c - {}^1\Phi_l^d\right)(^1A_u\text{ in }D_{2h}), \tag{69}$$

$$\Psi(^1H_u) = \frac{1}{2\sqrt{3}}\left(\sqrt{5}\,{}^1\Phi_i^b - \sqrt{5}\,{}^1\Phi_j^a + {}^1\Phi_k^d - {}^1\Phi_l^c\right)(^1B_{1u}\text{ in }D_{2h}), \tag{70}$$

$$\Psi(^1H_u) = \frac{1}{2\sqrt{3}}\left(\sqrt{5}\,{}^1\Phi_i^c + {}^1\Phi_j^d - \sqrt{5}\,{}^1\Phi_k^a - {}^1\Phi_l^b\right)(^1B_{2u}\text{ in }D_{2h}), \tag{71}$$

$$\Psi(^1H_u) = \frac{1}{2\sqrt{3}}\left(\sqrt{5}\,{}^1\Phi_i^d + {}^1\Phi_j^c + {}^1\Phi_k^b - \sqrt{5}\,{}^1\Phi_l^a\right)(^1B_{3u}\text{ in }D_{2h}), \tag{72}$$

$$\Psi(^1A_u) = \frac{1}{2}\left(^1\Phi_i^a + {}^1\Phi_j^b + {}^1\Phi_k^c + {}^1\Phi_l^d\right)(^1A_u\text{ in }D_{2h}). \tag{73}$$

When using the symmetry rules for the integrals of degenerate MOs, the excitation energies for the five states, i.e., ^1A_u, $^1T_{1u}$, $^1T_{2u}$, ^1G_u, and ^1H_u, are derived as follows:

$$\Delta E\left(^1T_{1u}\right) = \Delta\varepsilon + \left\{\frac{1}{2}\left(-J_{ib} + 2K_{ib}\right) + \frac{1}{2}\left(-J_{kd} + 2K_{kd}\right)\right\} + \left[\frac{1}{2}\{2(aj|ib) - (ab|ij)\}\right.$$

$$\left. + \frac{1}{2}\{2(bj|kc) - (bc|jk)\} + 2\{2(bi|kd) - (bd|ik)\}\right], \tag{74}$$

$$\Delta E\left(^1T_{2u}\right) = \Delta\varepsilon + \left\{\frac{1}{2}\left(-J_{ib} + 2K_{ib}\right) + \frac{1}{2}\left(-J_{kd} + 2K_{kd}\right)\right\} + \left[\frac{1}{2}\{2(aj|ib) - (ab|ij)\}\right.$$

$$\left. + \frac{1}{2}\{2(bj|kc) - (bc|jk)\} - 2\{2(bi|kd) - (bd|ik)\}\right], \tag{75}$$

$$\Delta E\left({}^{1}G_{u}\right) = \Delta\varepsilon + \left\{\frac{3}{4}\left(-J_{ia} + 2K_{ia}\right) + \frac{1}{4}\left(-J_{jb} + 2K_{jb}\right)\right\}$$

$$+ \left[-\frac{3}{2}\{2(ai|jb) - (ab|ij)\} + \frac{1}{2}\{2(bj|kc) - (bc|jk)\}\right], \tag{76}$$

$$\Delta E\left({}^{1}H_{u}\right) = \Delta\varepsilon + \left(-J_{jb} + 2K_{jb}\right) + \{2(bj|kc) - (bc|jk)\}, \tag{77}$$

$$\Delta E\left({}^{1}A_{u}\right) = \Delta\varepsilon + \left\{\frac{1}{4}\left(-J_{ia} + 2K_{ia}\right) + \frac{3}{4}\left(-J_{jb} + 2K_{jb}\right)\right\}$$

$$+ \left[\frac{3}{2}\{2(ai|jb) - (ab|ij)\} + \frac{3}{2}\{2(bj|kc) - (bc|jk)\}\right], \tag{78}$$

where the orbital energy difference is described by $\Delta\varepsilon$. The energy expressions in Eqs. (74), (75), (76), (77), and (78) correspond to the CIS results within the minimum active space, i.e., occupied g_u and unoccupied g_g MOs.

The categorization in Eq. (34) is also applicable to Eqs. (74), (75), (76), (77), and (78). Furthermore, the integrals in the B term are expressed by J_{ov} and K_{ov}, and those in the C term by $(oo'|vv')$ and $(ov|o'v')$. Here, o and v mean occupied and unoccupied MOs, respectively. Figure 6 schematically illustrates the energy splittings of the five singlet excited states for the $g_u \rightarrow g_g$ transitions, that is, ${}^{1}A_u$, ${}^{1}T_{1u}$, ${}^{1}T_{2u}$, ${}^{1}G_u$, and ${}^{1}H_u$. Details are discussed in Ref. [21].

Finally, we show the excitation energy formulas of the quintuply-degenerate excitations in I_h systems. Figure 7 illustrates the excitations between the h_u and the t_{1g} MOs and those between the h_g and the t_{1u} MOs. When adopting the Bethe's symmetry-lowering method from I_h to D_{2h}, spin- and space-symmetry-adapted wave functions for the singlet-type excitations among these degenerate MOs are given by

$$\Psi\left({}^{1}T_{1u}\right) = \frac{1}{\sqrt{10}}\left(\sqrt{2}\,{}^{1}\Phi_{i}^{a} + \sqrt{2}\,{}^{1}\Phi_{j}^{a} + \sqrt{3}\,{}^{1}\Phi_{l}^{c} + \sqrt{3}\,{}^{1}\Phi_{m}^{b}\right)\;({}^{1}B_{1u}\text{ in }D_{2h}), \tag{79}$$

$$\Psi\left({}^{1}T_{1u}\right) = \frac{1}{\sqrt{10}}\left(\sqrt{2}\,{}^{1}\Phi_{i}^{b} + \sqrt{2}\,{}^{1}\Phi_{j}^{b} + \sqrt{3}\,{}^{1}\Phi_{k}^{c} + \sqrt{3}\,{}^{1}\Phi_{m}^{a}\right)\;({}^{1}B_{2u}\text{ in }D_{2h}), \tag{80}$$

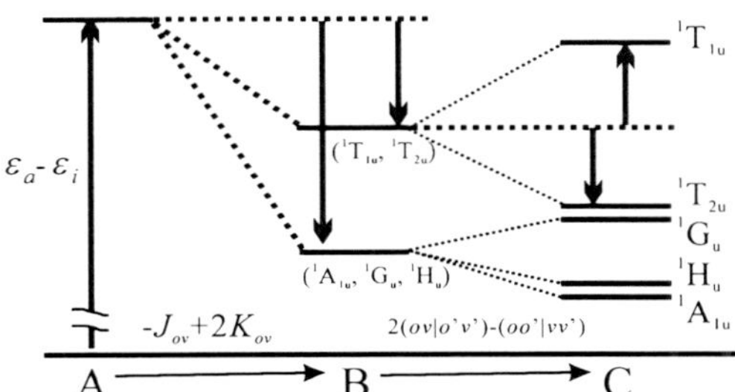

Fig. 6 Splitting scheme of the singlet excited states, namely, ${}^{1}T_{1u}$, ${}^{1}T_{2u}$, ${}^{1}G_u$, ${}^{1}H_u$, and ${}^{1}A_{1u}$, for the $g_u \rightarrow g_g$ transitions. The splitting between $({}^{1}T_{1u}, {}^{1}T_{2u})$ and $({}^{1}G_u, {}^{1}H_u, {}^{1}A_{1u})$ is due to the B term. Further splittings between ${}^{1}T_{1u}$ and ${}^{1}T_{2u}$ and among ${}^{1}G_u$, ${}^{1}H_u$, and ${}^{1}A_{1u}$ are due to the C term

                                                                H. Nakai

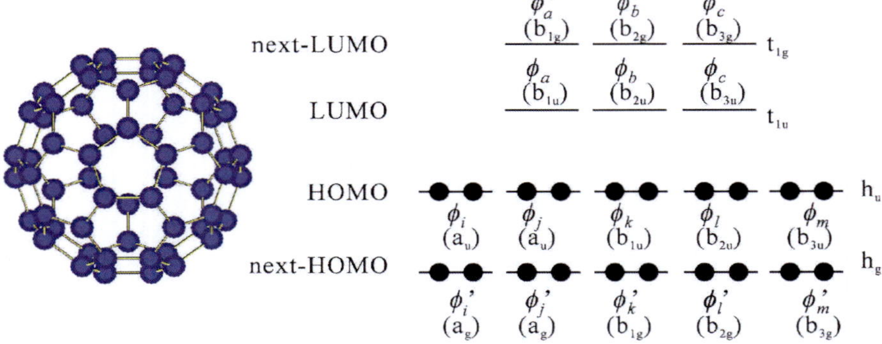

Fig. 7 Structure of C_{60} and excitation from quintuply-degenerate h_u/h_g to triply-degenerate t_{1g}/t_{1u} MOs in I_h

$$\Psi(^1T_{1u}) = \frac{1}{\sqrt{10}} \left(\sqrt{2}\,^1\Phi_i^c + \sqrt{2}\,^1\Phi_j^c + \sqrt{3}\,^1\Phi_k^b + \sqrt{3}\,^1\Phi_l^a \right) (^1B_{3u} \text{ in } D_{2h}), \quad (81)$$

$$\Psi(^1T_{2u}) = \frac{1}{2\sqrt{10}} \left\{ \left(\sqrt{5}+\sqrt{3}\right) {}^1\Phi_i^a - \left(\sqrt{5}-\sqrt{3}\right) {}^1\Phi_j^a \right.$$
$$\left. + \sqrt{2}\left(\sqrt{5}-1\right) {}^1\Phi_l^c - \sqrt{2}\left(\sqrt{5}+1\right) {}^1\Phi_m^b \right\} (^1B_{1u} \text{ in } D_{2h}), \quad (82)$$

$$\Psi(^1T_{2u}) = \frac{1}{2\sqrt{10}} \left\{ \left(\sqrt{5}+\sqrt{3}\right) {}^1\Phi_i^b - \left(\sqrt{5}-\sqrt{3}\right) {}^1\Phi_j^b \right.$$
$$\left. + \sqrt{2}\left(\sqrt{5}-1\right) {}^1\Phi_m^a - \sqrt{2}\left(\sqrt{5}+1\right) {}^1\Phi_k^c \right\} (^1B_{2u} \text{ in } D_{2h}), \quad (83)$$

$$\Psi(^1T_{2u}) = \frac{1}{2\sqrt{10}} \left\{ \left(\sqrt{5}+\sqrt{3}\right) {}^1\Phi_i^c - \left(\sqrt{5}-\sqrt{3}\right) {}^1\Phi_j^c \right.$$
$$\left. + \sqrt{2}\left(\sqrt{5}-1\right) {}^1\Phi_k^b - \sqrt{2}\left(\sqrt{5}+1\right) {}^1\Phi_l^a \right\} (^1B_{3u} \text{ in } D_{2h}), \quad (84)$$

$$\Psi(^1G_u) = \frac{1}{\sqrt{3}} \left({}^1\Phi_k^a + {}^1\Phi_l^b + {}^1\Phi_m^c \right) (^1A_u \text{ in } D_{2h}), \quad (85)$$

$$\Psi(^1G_u) = \frac{1}{4\sqrt{15}} \left\{ \sqrt{2}\left(\sqrt{5}-3\sqrt{3}\right) {}^1\Phi_i^a - \sqrt{2}\left(\sqrt{5}+3\sqrt{3}\right) {}^1\Phi_j^a \right.$$
$$\left. + 2\left(3+\sqrt{5}\right) {}^1\Phi_l^c + 2\left(3-\sqrt{5}\right) {}^1\Phi_m^b \right\} (^1B_{1u} \text{ in } D_{2h}), \quad (86)$$

$$\Psi(^1\mathrm{G_u}) = \frac{1}{4\sqrt{15}} \left\{ \sqrt{2}\left(\sqrt{5} - 3\sqrt{3}\right)\, {}^1\Phi_i^b - \sqrt{2}\left(\sqrt{5} + 3\sqrt{3}\right)\, {}^1\Phi_j^b \right.$$
$$\left. + 2\left(3 + \sqrt{5}\right)\, {}^1\Phi_m^a + 2\left(3 - \sqrt{5}\right)\, {}^1\Phi_k^c \right\} \; (^1\mathrm{B_{2u}} \text{ in } \mathrm{D_{2h}}), \tag{87}$$

$$\Psi(^1\mathrm{G_u}) = \frac{1}{4\sqrt{15}} \left\{ \sqrt{2}\left(\sqrt{5} - 3\sqrt{3}\right)\, {}^1\Phi_i^c - \sqrt{2}\left(\sqrt{5} + 3\sqrt{3}\right)\, {}^1\Phi_j^c \right.$$
$$\left. + 2\left(3 + \sqrt{5}\right)\, {}^1\Phi_k^b + 2\left(3 - \sqrt{5}\right)\, {}^1\Phi_l^a \right\} \; (^1\mathrm{B_{3u}} \text{ in } \mathrm{D_{2h}}), \tag{88}$$

$$\Psi(^1\mathrm{H_u}) = \frac{1}{\sqrt{2}} \left({}^1\Phi_k^a - {}^1\Phi_l^b \right) (^1\mathrm{A_u} \text{ in } \mathrm{D_{2h}}), \tag{89}$$

$$\Psi(^1\mathrm{E_u}) = \frac{1}{\sqrt{6}} \left({}^1\Phi_k^a + {}^1\Phi_l^b - 2\,{}^1\Phi_m^c \right) (^1\mathrm{A_u} \text{ in } \mathrm{D_{2h}}), \tag{90}$$

$$\Psi(^1\mathrm{H_u}) = \frac{1}{\sqrt{6}} \left({}^1\Phi_i^a - {}^1\Phi_j^a + \sqrt{2}\,{}^1\Phi_l^c - \sqrt{2}\,{}^1\Phi_m^b \right) (^1\mathrm{B_{1u}} \text{ in } \mathrm{D_{2h}}), \tag{91}$$

$$\Psi(^1\mathrm{H_u}) = \frac{1}{\sqrt{6}} \left({}^1\Phi_i^b - {}^1\Phi_j^b + \sqrt{2}\,{}^1\Phi_m^a - \sqrt{2}\,{}^1\Phi_k^c \right) (^1\mathrm{B_{2u}} \text{ in } \mathrm{D_{2h}}), \tag{92}$$

$$\Psi(^1\mathrm{H_u}) = \frac{1}{\sqrt{6}} \left({}^1\Phi_i^c - {}^1\Phi_j^c + \sqrt{2}\,{}^1\Phi_k^b - \sqrt{2}\,{}^1\Phi_l^a \right) (^1\mathrm{B_{3u}} \text{ in } \mathrm{D_{2h}}). \tag{93}$$

If one takes the symmetry rules for the integrals into account, the excitation energies for the four states, i.e., $^1\mathrm{T_{1u}}$, $^1\mathrm{T_{2u}}$, $^1\mathrm{G_u}$, and $^1\mathrm{H_u}$, are derived as follows:

$$\Delta E\left(^1\mathrm{T_{1u}}\right) = \Delta\varepsilon + \left\{ \frac{2}{5}(-J_{ia} + 2K_{ia}) + \frac{1}{2}(-J_{lc} + 2K_{lc}) \right\}$$
$$+ \left[\frac{2}{5}\{2(ai|ja) - (aa|ij)\} + \frac{3}{5}\{2(bm|lc) - (bc|lm)\} \right.$$
$$\left. + \frac{4}{5\sqrt{6}}\{2(ai|lc) - (ac|il)\} \right], \tag{94}$$

$$\Delta E\left(^1\mathrm{T_{2u}}\right) = \Delta\varepsilon + \left\{ \frac{2}{5}(-J_{ia} + 2K_{ia}) + \frac{3}{5}(-J_{lc} + 2K_{lc}) \right\}$$
$$+ \left[-\frac{1}{10}\{2(aj|ia) - (aa|ij)\} - \frac{2}{5}\{2(bm|lc) - (bc|lm)\} \right.$$
$$\left. - \frac{1}{5\sqrt{6}}\{2(ai|lc) - (ac|il)\} \right], \tag{95}$$

$$\Delta E \left({}^1\mathrm{G_u}\right) = \Delta\varepsilon + \left\{ \frac{8}{15}\left(-J_{ia} + 2K_{ia}\right) + \frac{7}{15}\left(-J_{lc} + 2K_{lc}\right)\right\}$$
$$+ \left[-\frac{11}{30}\left\{2(aj|ia) - (aa|ij)\right\} + \frac{2}{15}\left\{2(bm|lc) - (bc|lm)\right\}\right.$$
$$\left. -\frac{3}{5\sqrt{6}}\left\{2(ai|lc) - (ac|il)\right\}\right], \tag{96}$$

$$\Delta E \left({}^1\mathrm{H_u}\right) = \Delta\varepsilon + \left\{ \frac{1}{3}\left(-J_{ia} + 2K_{ia}\right) + \frac{2}{3}\left(-J_{lc} + 2K_{lc}\right)\right\}$$
$$+ \left[-\frac{1}{3}\left\{2(aj|ia) - (aa|ij)\right\} - \frac{1}{6}\left\{2(bm|lc) - (bc|lm)\right\}\right]. \tag{97}$$

The excitation energy formulas in Eqs. (94), (95), (96), and (97) correspond to the CIS results within the minimum active space, i.e., $(\mathrm{h_u} \times \mathrm{t_{1g}})$ or $(\mathrm{h_g} \times \mathrm{t_{1u}})$.

As seen here, the categorization in Eq. (34) is universal for the excitation energy formulas of the degenerate excitations. Even for the complicated cases such as quadruply- and quintuply-degenerate MOs, the A term is the orbital energy difference, the B term consists of the Coulomb and exchange integrals expressed by J_{ov} and K_{ov}, respectively, and the C term involves the remaining integrals given by $(oo'|vv')$ and $(ov|o'v')$.

3.4 Analysis of Splitting Scheme

This section presents the numerical analyses based on the FZOA treatment. We first investigate the doubly-degenerate case. As an example, the $\pi-\pi^*$ excitations in CO are analyzed in detail. Table 9 shows the numerical data for the orbital energy differences and two-electron integrals appearing in Eqs. (29), (30), (31), and (32), although the expressions for the ${}^1\Sigma_u^+$, ${}^1\Delta_u$, and ${}^1\Sigma_u^-$ states in $\mathrm{D_{\infty h}}$ are assigned to those for the ${}^1\Sigma^+$, ${}^1\Delta$, and ${}^1\Sigma^-$ states, respectively. The absolute values of the Coulomb integrals are greater than those of the exchange and the other two-electron integrals. On the contrary, the difference between J_{ia} and J_{ja} is smaller than that between K_{ia} and K_{ja}, namely, 0.74 and 2.49 eV, respectively. In the B term, the Coulomb and exchange terms have opposite sign. Thus, the two energy levels split by the B term are determined by the difference between K_{ia} and K_{ja}, not between J_{ia} and J_{ja}. In particular, K_{ia} is one order of magnitude larger than K_{ja}. In the two-electron integrals appearing in the C term, $(ai|jb)$ is one order of magnitude larger than $(aj|ib)$, just like the relationship between K_{ia} and K_{ja}, while $(ab|ij)$ is small and common to the four (or three) states. This difference determines the splittings among the ${}^1\Sigma^+$, ${}^1\Sigma^-$, and ${}^1\Delta$ states, as well as their ordering.

The exchange integrals, K_{ia} and K_{ja}, include the overlap distributions (or the transition densities), $\phi_a^*(\mathbf{r})\phi_i(\mathbf{r})$ and $\phi_a^*(\mathbf{r})\phi_j(\mathbf{r})$, respectively. The ϕ_i and ϕ_a MOs

Table 9 Numerical data (in eV) for the orbital energy difference, Coulomb, exchange, and remaining two-electron integrals appearing in the FZOA/DZP treatment for the $\pi-\pi^*$ excitation of CO

System	Point Group	Main Configuration	State	A	B			C			$A+B+C$
				$\Delta\varepsilon$	J_{ia} J_{ja}	K_{ia} K_{ja}	Total	$(ai|jb)$ $(aj|ib)$	$(ab|ij)$	Total	
CO	$C_{\infty v}$	$\pi \rightarrow \pi$	$^1\Sigma^+$	21.541	12.236	2.760	−6.714	2.221	0.373	4.069	18.896
			$^1\Delta$							−4.069	10.756
			$^1\Delta$							0.165	10.756
			$^1\Sigma^-$		11.489	0.269	−10.949	0.269		−0.165	10.426

with b_1 symmetry (in C_{2v}) are located in the xz plane, while the ϕ_j and ϕ_b MOs with b_2 symmetry in the yz plane. Therefore, the (maximum) amplitude of the overlap distribution of $\phi_a^*(\mathbf{r})\phi_i(\mathbf{r})$ is larger than that of $\phi_a^*(\mathbf{r})\phi_j(\mathbf{r})$, which brings about the great difference between K_{ia} and K_{ja}. Similarly, the (maximum) amplitude of the overlap distribution of $\phi_j^*(\mathbf{r})\phi_b(\mathbf{r})$ is larger than that of $\phi_i^*(\mathbf{r})\phi_b(\mathbf{r})$. The difference between $(ai|jb)$ and $(aj|ib)$ integrals attributes to the difference of these overlap distributions. Relationships between the overlap distributions and the integrals are schematically illustrated in Fig. 8. This is the qualitative understanding of the rule (ii) for the degenerate excitations.

The transition dipole moment also involves overlap distributions such as $\phi_a^*(\mathbf{r})\phi_i(\mathbf{r})$ and $\phi_a^*(\mathbf{r})\phi_j(\mathbf{r})$ in the integrand. The symmetry determines whether the integral for estimating the transition dipole moment becomes zero or nonzero. For example, the overlap distributions of $\phi_a^*(\mathbf{r})\phi_i(\mathbf{r})$ and $\phi_a^*(\mathbf{r})\phi_j(\mathbf{r})$ correspond to the a_1 and a_2 symmetries (in C_{2v}), respectively. Because the dipole operators, i.e., x, y, and z, have b_1, b_2, and a_1 symmetries, respectively, the direct product between $\phi_a^*(\mathbf{r})\phi_i(\mathbf{r})$ and z can lead to the totally symmetric representation a_1 and, therefore, the integral becomes nonzero. On the contrary, any combinations between $\phi_a^*(\mathbf{r})\phi_j(\mathbf{r})$ and dipole operators cannot yield the totally symmetric representation and the integrals yield zero. It is noted that the overlap distribution of $\phi_a^*(\mathbf{r})\phi_i(\mathbf{r})$ is common to the exchange integral K_{ia} in the B term, the two-electron integral $(ai|jb)$ in the C term, and the transition dipole moment $\langle \phi_a |\mathbf{r}| \phi_i \rangle$. This is the qualitative interpretation of the rule (i) for the degenerate excitations.

We next investigate the triply-degenerate cases: namely, $1t_{2u} \rightarrow 3t_{2g}$ and $6t_{1u} \rightarrow 3t_{2g}$ in MoF_6 and $3t_{2g} \rightarrow 12t_{1u}$ and $3t_{2g} \rightarrow 3t_{2u}$ in $Mo(CO)_6$. Table 10 shows the numerical data for the orbital energy differences and two-electron integrals

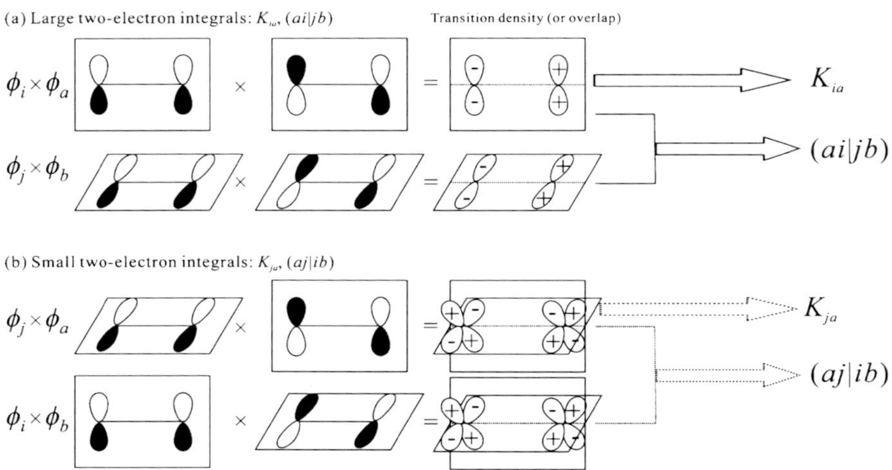

Fig. 8 Relationship between the overlap distributions $\{\phi_a^*(\mathbf{r})\phi_i(\mathbf{r}),\ \phi_j^*(\mathbf{r})\phi_b(\mathbf{r})\}$ and the two-electron integrals $\{K_{ia},\ (ai|jb)\}$ (**a**) and between $\{\phi_a^*(\mathbf{r})\phi_j(\mathbf{r}),\ \phi_i^*(\mathbf{r})\phi_b(\mathbf{r})\}$ and the two-electron integrals $\{K_{ja},\ (aj|ib)\}$

Table 10 Numerical data (in eV) for the orbital energy difference, Coulomb, exchange, and remaining two-electron integrals appearing in the FZOA/DZP treatment for the t–t excitations of MoF_6 and $Mo(CO)_6$

System	Point Group	Main Configuration	State	A $\Delta\varepsilon$	B J_{ja} / J_{ia}	B K_{ja} / K_{ia}	B Total	C $(aj\|ib)$ / $(ai\|jb)$	C $(ab\|ij)$	C Total	$A+B+C$
MoF_6	O_h	$1t_{2u} \rightarrow 3t_{2g}$	$^1T_{1u}$	15.855	7.688	0.517	−6.654	−0.448	−0.052	0.844	10.045
			$^1T_{2u}$							−0.844	8.357
			^1E_u		8.197	0.066	−8.065	−0.033		0.014	7.804
			$^1A_{2u}$							−0.028	7.762
MoF_6	O_h	$6t_{1u} \rightarrow 3t_{2g}$	$^1T_{1u}$	16.655	8.132	0.389	−7.354	0.310	0.082	0.538	9.839
			$^1T_{2u}$							−0.538	8.763
			^1E_u		8.080	0.052	−7.976	0.030		0.022	8.701
			$^1A_{1u}$							−0.044	8.635
$Mo(CO)_6$	O_h	$3t_{2g} \rightarrow 12t_{1u}$	$^1T_{1u}$	10.716	5.397	0.413	−4.571	0.371	0.053	0.689	6.834
			$^1T_{2u}$							−0.689	5.457
			$^1A_{2u}$		5.542	0.064	−5.413	0.046		0.078	5.381
			^1E_u							−0.039	5.264
$Mo(CO)_6$	O_h	$3t_{2g} \rightarrow 3t_{2u}$	$^1T_{1u}$	11.946	5.018	0.328	−4.361	−0.279	−0.043	0.514	8.099
			$^1T_{2u}$							−0.514	7.071
			$^1A_{1u}$		5.213	0.029	−5.154	−0.015		0.027	6.819
			^1E_u							−0.014	6.778

appearing in Eqs. (49), (50), (51), and (52) and some modifications. While the absolute values of the Coulomb integrals are greater than those of the exchange integrals, the differences between J_{ja} and J_{ia}, are smaller than those between K_{ja} and K_{ia} in all cases. Thus, the two energy levels split by the B term are determined by the difference between K_{ja} and K_{ia}, not between J_{ja} and J_{ia}. In the two-electron integrals appearing in the C term, $(aj|ib)$ is one order of magnitude larger than $(ai|jb)$, just like the relationship between K_{ja} and K_{ia}, while $(ab|ij)$ is small and common to the four states. This difference determines the splittings among the $^1\Sigma^+$, $^1\Sigma^-$, and $^1\Delta$ states, as well as their ordering.

The exchange integrals, K_{ja} and K_{ia}, include the overlap distributions (or the transition densities), $\phi_a^*(\mathbf{r})\phi_j(\mathbf{r})$ and $\phi_a^*(\mathbf{r})\phi_i(\mathbf{r})$, respectively. Here we discuss the case of $6t_{1u} \rightarrow 3t_{2g}$ in MoF_6 as an example. The ϕ_i MO with b_{3u} symmetry (in C_{2v}), which mainly consists of p_x AOs of F atoms, has a node on the yz plane, while the ϕ_a MO mainly consists of d_{yz} AO of Mo atom and situated on the yz plane. Therefore, the overlap is considerably small, in fact, the overlap distribution $\phi_a^*(\mathbf{r})\phi_i(\mathbf{r})$ has nodes on the xy, yz, and zx planes. On the contrary, the ϕ_j MO with b_{2u} symmetry (in C_{2v}) is mainly made from of p_y AOs of F atoms. Thus, the (maximum) amplitude of the overlap distribution $\phi_a^*(\mathbf{r})\phi_j(\mathbf{r})$ becomes larger than that of $\phi_a^*(\mathbf{r})\phi_i(\mathbf{r})$. This difference leads the great difference between K_{ja} and K_{ia}. Similarly, the (maximum) amplitude of the overlap distribution of $\phi_j^*(\mathbf{r})\phi_b(\mathbf{r})$ is larger than that of $\phi_i^*(\mathbf{r})\phi_b(\mathbf{r})$. The difference between $(aj|ib)$ and $(ai|jb)$ integrals originates from the difference of these overlap distributions. As in the case of the doubly-degenerate excitations, the above discussion presents the qualitative understanding of the rule (ii) for the degenerate excitations. Furthermore, the overlap distribution of $\phi_a^*(\mathbf{r})\phi_j(\mathbf{r})$ is common to the exchange integral K_{ja} in the B term, the two-electron integral $(aj|ib)$ in the C term, and the transition dipole moment $\langle \phi_a |\mathbf{r}| \phi_j \rangle$. This is the qualitative interpretation of the rule (i) for the degenerate excitations.

As seen here, the interpretation of the rules (i) and (ii) for the degenerate excitations are given by the analyses of the two-electron integrals appearing in the excitation energy formulas based on the FZOA treatment. The overlap distributions between the occupied and the unoccupied MOs play the key role for the analyses. Similar interpretation has been derived for the other cases such as quadruply- and quintuply-degenerate excitations. A detailed discussion can be found in the literature [29].

4 Concluding Remarks

This review explains the rules for the ordering and splitting of the excited states for the transitions between degenerate orbitals, namely,

In singlet excitations between occupied and virtual degenerate orbitals with valence character,

(i)   the highest transition is dipole-allowed,

(ii) the splitting between the dipole-allowed and forbidden states is larger than those among the dipole-forbidden states.

We first presented SAC-CI and TDDFT results for the degenerate excitations, e.g., linear molecules ($D_{\infty h}$, $C_{\infty v}$), tetrahedral and octahedral metal complexes (T_d, O_h), and $(B_{12}H_{12})^{2-}$ and $C_{60}(I_h)$. These data support the generality of the above rules. Next, we derived the formulas for the excitation energies for the degenerate excitations based on the FZOA treatment.

Since the FZOA treatment corresponds to the CIS within the minimal active space, the effects of the orbital relaxation and electron correlation are neglected. Furthermore, it is not applicable for cases in which the configuration interaction is important. However, numerical results demonstrate that the FZOA can reproduce the qualitative trends of the more accurate results obtained by the SAC-CI and TDDFT calculation in many cases.

The formulas based on the FZOA present the splitting scheme of the degenerate excitations and connect those splittings with particular molecular integrals such as K_{ia} and $(ai|jb)$. Furthermore, we analyze such molecular integrals in more detail from the viewpoint of the overlap distributions like $\phi_a^*(\mathbf{r})\phi_i(\mathbf{r})$ and $\phi_a^*(\mathbf{r})\phi_j(\mathbf{r})$. These analyses give the qualitative interpretation of the rule (ii). The formulas of the transition dipole moments are derived based on the FZOA treatment. The transition dipole moment integrals are also analyzed by using the overlap distributions. This provides an understanding of the rule (i) qualitatively.

The discussion in this review is limited to the singlet excitations. The triplet excited states of the degenerate systems also present several fascinating points. For example, the splittings become smaller than those among the singlet states and the ordering is opposite, namely, $^3\Sigma_u^- >\, ^3\Delta_u >\, ^3\Sigma_u^+$ and $^1\Sigma_u^+ >>\, ^1\Delta_u >\, ^1\Sigma_u^-$. This can also be interpreted by the FZOA treatment because the integrals such as K_{ia} and $(ai|jb)$, which play a key role for the singlet excitations, disappear in the formulas for the triplet states.

Although the FZOA treatment mentioned here is based on the HF wave function, it is possible to adopt the KS orbitals. The Tamm–Dancoff approximation (TDA) that neglects the de-excitation effect in the TDDFT gives a similar formulation to the CIS method. Consequently, the TDA/TDDFT treatment within the minimum active space presents the FZOA for the KS-DFT. The comparison between the HF- and the KS-based FZOA treatments is attractive [38]. For example, the behaviors of the A terms, which correspond to the band gaps, are significantly different. It is well-known that the pure DFT calculations underestimate band gaps by approximately 50%, whereas the HF ones overestimate them considerably. The HF/DFT hybrid calculations give reasonable band gaps in many cases. These differences are corrected by taking the B term into account. Details are addressed in Ref. [38].

Finally, we should mention some issues related to this work. Although a variety of numerical investigations have been carried out, there is still an open question of how general the rules are. As mentioned above the rules have been discovered *empirically* through the ab initio calculations. The FZOA treatment only provides the interpretation of the rules, not the proof. Therefore, the complete proof for the rules is expected to be performed in the near future. One of the most important issues is how to use the rules. There exist several rules concerning the excited states, whose number might be fewer than that of the ground-state rules. Kasha's rule [39], which states that photon emission such as fluorescence and phosphorescence, occurs only from the lowest-energy excited electronic state of the system, is relevant in understanding the emission spectrum. El-Sayed's rule [40] states that the rate of intersystem crossing from the lowest singlet state to the triplet manifold is relatively large if the radiationless transition involves an orbital-type change. For example, the intersystem crossing from the $\pi-\pi^*$ singlet state to $n-\pi^*$ triplet state occurs more likely than that from the $\pi-\pi^*$ singlet state to $\pi-\pi^*$ triplet state. Then, how about the rules for the degenerate excitations? We hope that the usage of the present rules will be intensively examined from the various aspects in the future.

References

1. H. Nakatsuji, K. Hirao, J. Chem. Phys. **68**, 2035 (1978)
2. H. Nakatsuji, Chem. Phys. Lett. **59**, 362 (1978); **67**, 329, 334 (1979)
3. H. Nakatsuji, Acta Chim. Hung. **129**, 719 (1992)
4. The frozen-orbital approximation is explained in the standard texts of quantum chemistry, such as A. Szabo, N. S. Ostlund, *Modern Quantum Cchemistry: Introduction to Advanced Electronic Structure Theory* (McGraw-Hill, New York, 1989); the concept of the frozen-orbital analysis was first proposed in the following paper: H. Nakai, H. Morita, H. Nakatsuji, J. Phys. Chem. **100**, 15753 (1996)
5. H. Nakai, H. Morita, P. Tomasello, H. Nakatsuji, J. Phys. Chem. A **102**, 2003 (1998)
6. S. Huzinaga, J. Andzelm, M. Klobukowski, E. Radzio-Andzelm, Y. Sakai, H. Tatewaki, *Gaussian Basis Sets for Molecular Calculations* (Elsevier, New York, 1984)
7. T. H. Dunning, Jr. and P. J. Hay, *Modern Theoretical Chemistry*, ed. by H. F. Schaeffer, III (Plenum, New York, 1977), Vol. 3
8. E. Runge, E. K. U. Gross, Phys. Rev. Lett. **52**, 997 (1985)
9. S. Hirata, M. Head-Gordon, Chem. Phys. Lett. **314**, 291 (1999)
10. T. H. Dunning, Jr., J. Chem. Phys. **90**, 1007 (1989)
11. W. Stevens, H. Basch, J. Krauss, J. Chem. Phys. **81**, 6026 (1984)
12. A. D. Becke, Phys. Rev. A **38**, 3098 (1997)
13. C. Lee, W. Yang, R. G. Parr, Phys. Rev. B **37**, 785 (1988)
14. A. D. Becke, J. Chem. Phys. **98**, 5648 (1993)
15. D. Stahel, M. Leoni, K. Dressler, J. Chem. Phys. **79**, 2541 (1983)
16. J. Oddershede, N. E. Grüner, G. H. F. Diercksen, Chem. Phys. **97**, 303 (1985)
17. K. P. Huber, G. Herzberg, *Constants of Diatomic Molecules* (Van Nostrand Reinhold, New York, 1979)
18. E. S. Nielsen, P. Jørgensen, J. Oddershede, J. Chem. Phys. **73**, 6238 (1980)
19. T. Pino, H. Ding, F. Güthe, J. P. Maier, J. Chem. Phys. **114**, 2208 (2001)
20. J. W. Rabalais, J. M. McDonald, V. Scherr, S. P. McGlynn, Chem. Rev. **71**, 73 (1971)
21. J. Lorentzon, P.-A. Malmquist, M. Fulscher, B. O. Roos, Theor. Chim. Acta **91**, 91 (1995)
22. A. Kaito, A. Takiri, M. Hatano, Chem. Phys. Lett. **25**, 548 (1974)
23. H. Morita, H. Nakai, H. Hanada, H. Nakatsuji, Mol. Phys. **92**, 523 (1997)
24. S. Jitsuhiro, H. Nakai, M. Hada, H. Nakatsuji, J. Chem. Phys. **101**, 1029 (1994)
25. H. Nakatsuji, S. Saito, J. Chem. Phys. **93**, 1865 (1990)
26. H. Nakai, Y. Ohmori, H. Nakatsuji, J. Chem. Phys. **95**, 8287 (1991)

27. J. Hasegawa, K. Toyota, M. Hada, H. Nakai, H. Nakatsuji, Theor. Chim. Acta **92**, 351 (1995)
28. H. Nakatsuji, S. Saito, Int. J. Quantum Chem. **39**, 93 (1991)
29. T. Baba, Y. Imamura, M. Okamoto, H. Nakai, Chem. Lett. **37**, 322 (2008)
30. E. Koudoumas, A. Ruth, S. Couris, S. Leach, Mol. Phys. **88**, 125 (1996)
31. T. Koopmans, Physica **1**, 104 (1933)
32. W. Kohn, L. J. Sham, Phys. Rev. A **140**, 1133 (1965)
33. A. F. Janak, Phys. Rev. B **18**, 7165 (1978)
34. F. Z. Hund, Phys. **33**, 345 (1925)
35. D. A. Kohl, J. Chem. Phys. **56**, 4236 (1972)
36. R. J. Boyd, Nature **310**, 480 (1984)
37. J. W. Warner, R. S. Berry, Nature **313**, 160 (1985)
38. Y. Imamura, T. Baba, H. Nakai, Chem. Lett., **38**, 258 (2009)
39. M. Kasha, Discuss. Faraday Soc. **9**, 14 (1950)
40. M. El-Sayed, Acc. Chem. Res. **1**, 8 (1968)

The Dissociation Catastrophe in Fluctuating-Charge Models and its Implications for the Concept of Atomic Electronegativity

Jiahao Chen and Todd J. Martínez

Abstract We have recently developed the QTPIE (charge transfer with polarization current equilibration) fluctuating-charge model, a new model with correct dissociation behavior for nonequilibrium geometries. The correct asymptotics originally came at the price of representing the solution in terms of charge-transfer variables instead of atomic charges. However, we have found an exact reformulation of fluctuating-charge models in terms of atomic charges again, which is made possible by the symmetries of classical electrostatics. We show how this leads to the distinction between two types of atomic electronegativities in our model. While one is a intrinsic property of individual atoms, the other takes into account the local electrical surroundings. This distinction could resolve some confusion surrounding the concept of electronegativity as to whether it is an intrinsic property of elements, or otherwise. We also use the QTPIE model to create a three-site water model and discuss simple applications.

Keywords: Fluctuating charges · Charge equilibration · Electronegativity equalization · Chemical hardness · Force fields · Molecular models · Water models

1 Introduction

Recent studies using classical molecular dynamics have found conventional additive force fields increasingly inadequate for today's system of interest, as the neglect

T.J. Martinez (✉)
Department of Chemistry, Stanford University, S.G. Mudd Bldg., Room 121, Stanford, CA 94305-5080, USA
e-mail: todd.martinez@stanford.edu

J. Chen
Department of Chemistry, Center for Advanced Theory and Molecular Simulation, Frederick Seitz Materials Research Laboratory, The Beckman Institute, University of Illinois, Urbana, Illinois 61801, USA
e-mail: chen6@uiuc.edu

P. Piecuch et al. (eds.), *Advances in the Theory of Atomic and Molecular Systems*,
Progress in Theoretical Chemistry and Physics 19, DOI 10.1007/978-90-481-2596-8_19,

of nonadditive phenomena such as polarization and charge transfer can lead to qualitative errors in simulations [1–4]. Of the two nonadditive effects, the literature on methods to incorporate polarization is more extensive. Two of the many popular types of methods for incorporating polarization are inducible dipoles [3–5], where additional variables are introduced to describe dipole moments induced by mutual polarization interactions, and Drude oscillators [6, 7], where polarization is described by the change in distance between the atomic nucleus and a fixed countercharge attached by a harmonic potential. However, neither of these methods is readily extensible to provide a description of charge transfer. This is in some sense surprising and contrary to physical intuition, as charge transfer is merely an extreme form of polarization: while polarization results in a redistribution of charge density within molecules, charge transfer is a redistribution of charge density across molecules.

In contrast, there are several classes of methods that exist for modeling both charge transfer and polarization effects: for example, fluctuating-charge models [2, 8–10], which model polarization by recomputing the charge distribution in response to changes in geometry or external perturbations; empirical valence bond (EVB) methods [11–13], which parameterize the energy contributions of individual valence bond configurations; and effective fragment potential (EFP)-type methods [14, 15], which use energy decompositions of *ab initio* data to construct parameterized effective potentials. We choose to study only fluctuating-charge models, as the other methods that treat both polarization and charge transfer are computationally far more costly. In EFPs, polarization is modeled using distributed, inducible dipoles while charge transfer is represented separately as a sum over antibonding orbitals of the electron acceptor. The latter necessitates *a priori* specification of the charge acceptors and donors, as well as the provision of parameters for every orbital being summed over. Not only is this description computationally expensive, but it also fails to provide a unified picture of polarization and charge transfer. In contrast, EVB does provide this unified treatment but suffers from the exponential growth in the number of relevant valence bond configurations with system size. In contrast, fluctuating-charge models introduce only a modest computational cost over conventional fixed-charge force fields, even for large systems. Several of these methods have been used in dynamics simulations, most notably QEq [16] in UFF [17], electronegativity equalization method (EEM) [18, 19] in ReaxFF [20], and *fluc*-q in the TIP4P-FQ water model [21, 22], thus demonstrating their utility in describing polarization effects in classical molecular dynamics.

In addition, fluctuating-charge models are theoretically appealing as they provide a unified treatment of polarization and charge transfer with only two parameters per atom. These parameters can be identified with the chemically important concepts of electronegativity [23–29] and (chemical) hardness [30–33]. These drive the redistribution of atomic charges in response to electrostatic interactions according to the principle of electronegativity equalization [25–28, 34].

2 The Dissociation Catastrophe in QEq-Type Fluctuating-Charge Models

Here, we briefly review the most common type of fluctuating-charge model and how such models are solved. The main idea of fluctuating-charge models is to assert that the electrostatic energy of a molecular system can be decomposed into two types of terms, i.e.,

$$E\left(\mathbf{q};\mathbf{R}\right) = \sum_{i=1}^{N} E_i^{at}\left(q_i\right) + \sum_{i<j} q_i q_j J_{ij}\left(\mathbf{R}\right), \tag{1}$$

where N is the number of atoms in the system, $\mathbf{q} = (q_1, \ldots, q_N)$ are the charges on each atom, each E_i^{at} is the intrinsic contribution of each individual atom, and each J_{ij} is a pairwise interaction that is dependent on the molecular geometry $\mathbf{R} = (\mathbf{R}_1, \ldots, \mathbf{R}_N)$. The atomic charges are then solved for by a minimization of the total electrostatic energy with respect to each atomic charge with a constraint on the total charge of the system, Q:

$$\sum_{i=1}^{N} q_i = Q. \tag{2}$$

In many fluctuating-charge models, the interactions J_{ij} are taken to represent some screened Coulomb interactions $J_{ij} = J_{ij}\left(\left|\mathbf{R}_i - \mathbf{R}_j\right|\right)$ that do not diverge in the small separation limit $\left|\mathbf{R}_i - \mathbf{R}_j\right| \to 0$. Screening is necessary in order to prevent numerical instabilities from occurring at small interatomic distances.

The precise method of calculating these interactions differs between the specific fluctuating-charge models in the literature: in the EEM [18, 19], the Coulomb interactions are evaluated as two-electron Coulomb integrals over spherically symmetric Gaussian-type atomic orbitals; the chemical potential equalization (CPE) [35] model uses similar integrals, but with empirical parameters for Fukui function corrections; in the QEq [16], fluc-q [21, 22], and ES+ [10] models, the Coulomb interactions are evaluated as two-electron Coulomb integrals over spherically symmetric Slater-type atomic orbitals; and in the CHARMM C22 force field [36, 37], the Coulomb interactions are screened with empirical functions. In the QTPIE model [38], we use two-electron Coulomb integrals over s-type primitive Gaussian orbitals, with orbital exponents fitted to reproduce the results from the much more expensive s-type Slater-type orbitals used in QEq. We have found that it is possible to optimize Gaussian orbitals to reproduce the Slater integrals with an accuracy of better than 10^{-3} atomic units, with exponents given in Table 1. The details of the fitting procedure are given in Appendix 1.

Table 1 Exponents of atomic orbital exponents that best reproduce the two-electron Slater integrals over the QEq orbitals. All quantities are in atomic units

Element	Slater exponent[a]	Gaussian exponent	Error[b]
H	1.0698	0.5434	0.01696
Li	0.4174	0.1668	0.00148
C	0.8563	0.2069	0.00162
N	0.9089	0.2214	0.00166
O	0.9745	0.2240	0.00167
F	0.9206	0.2313	0.00169
Na	0.4364	0.0959	0.00085
Si	0.7737	0.1052	0.00088
P	0.8257	0.1085	0.00089
S	0.8690	0.1156	0.00092
Cl	0.9154	0.1137	0.00091
K	0.4524	0.0602	0.00125
Br	1.0253	0.0701	0.00133
Rb	0.5162	0.0420	0.00121
I	1.0726	0.0686	0.00127
Cs	0.5663	0.0307	0.00114

[a] From Ref. [16].
[b] Maximum absolute error as defined in (36).

In addition, the atomic terms $E_i^{at}(q_i)$ in many fluctuating-charge models are each assumed to be a quadratic polynomial of the form

$$E_i^{at}(q_i) = E_i^0 + \chi_i q_i + \frac{1}{2}\eta_i q_i^2 + \dots, \tag{3}$$

where E_i^0 is a constant independent of charge and geometry and can thus be discarded in the energy expression for fluctuating-charge models. The other coefficients are interpreted by a formal comparison with a Taylor series expansion of $E_i^{at}(q_i)$ about $q_i = 0$ [25]

$$E_i^{at}(q_i) = E_i^{at}(0) + \left.\frac{dE_i^{at}}{dq_i}\right|_{q_i=0}(q_i - 0) + \frac{1}{2!}\left.\frac{d^2 E_i^{at}}{dq_i^2}\right|_{q_i=0}(q_i - 0)^2 + \dots. \tag{4}$$

By approximating the Taylor expansion coefficients with suitable finite difference formulas with spacing $\Delta q_i = 1$, the following well-known relationships are obtained:

$$\chi_i \equiv \left.\frac{dE_i^{at}}{dq_i}\right|_{q_i=0} \approx \frac{E_i^{at}(1) - E_i^{at}(-1)}{2} = \frac{\mathrm{IP}_i + \mathrm{EA}_i}{2}, \tag{5}$$

$$\eta_i \equiv \left.\frac{d^2 E_i^{at}}{dq_i^2}\right|_{q_i=0} \approx E_i^{at}(1) - 2E_i^{at}(0) + E_i^{at}(-1) = \mathrm{IP}_i - \mathrm{EA}_i, \tag{6}$$

where $IP_i = E_i^{at}(1) - E_i^{at}(0)$ is the ionization potential of the ith atom and $EA_i = E_i^{at}(0) - E_i^{at}(-1)$ is the electron affinity of the ith atom. In this manner, these coefficients can be identified as none other than the Mulliken electronegativity [39] and Parr–Pearson (chemical) hardness [33]. The preceding identifications allow fluctuating-charge models to be identified as rudimentary forms of density functional theory [40].

The truncation of the series expansion (4) at second order allows the solution to be found by solving a linear system of equations. The only complication is the need to enforce the constraint (2), which can be taken care of with the method of Lagrange multipliers. In this context, the Lagrange multiplier μ can be interpreted as the chemical potential, and the solution to the constrained problem is the charge distribution and chemical potential which minimizes the free energy

$$F(\mathbf{q}, \mu; Q) = E(\mathbf{q}) - \mu \left(\sum_{i=1}^{N} q_i - Q \right)$$

$$= \mu Q + \sum_{i=1}^{N} (\chi_i - \mu) q_i + \tfrac{1}{2} \sum_{ij} q_i q_j J_{ij}, \tag{7}$$

where $\eta_i = J_{ii}$. Minimizing this free energy then leads to the linear system of equations consisting of (2) and the equation

$$0 = \frac{\partial F(\mathbf{q})}{\partial q_i} = (\chi_i - \mu) + \sum_{j=1}^{N} q_j J_{ij}. \tag{8}$$

This system can be written in block-matrix notation

$$\begin{pmatrix} \mathbf{J} & \mathbf{1} \\ \mathbf{1}^T & 0 \end{pmatrix} \begin{pmatrix} \mathbf{q} \\ \mu \end{pmatrix} = \begin{pmatrix} -\chi \\ Q \end{pmatrix}, \tag{9}$$

where $\mathbf{1}$ is a column vector with entries all equal to unity. This system of equations is solved approximately in the historically important models of Del Re [41] and Gasteiger and Marsili [42]; however, all modern models solve these equations exactly for the charge distribution. It is straightforward to show (as in Appendix 2) that this linear system has the explicit solution

$$\begin{pmatrix} \mathbf{q} \\ \mu \end{pmatrix} = \begin{pmatrix} -\mathbf{J}^{-1}(\chi + \mu \mathbf{1}) \\ -\left(Q + \mathbf{1}^T \mathbf{J}^{-1} \chi\right) / \mathbf{1}^T \mathbf{J}^{-1} \mathbf{1} \end{pmatrix}. \tag{10}$$

It is instructive to solve the fluctuating-charge model above in the case of a neutral diatomic molecule. Then, (7) can be written explicitly in terms of one charge variable q_1, so that the energy is given by

$$F(q_1; \mathbf{R}) = (\chi_1 - \chi_2) q_1 + \tfrac{1}{2} (\eta_1 - 2J_{12}(|\mathbf{R}_1 - \mathbf{R}_2|) + \eta_2) q_1^2. \tag{11}$$

This is minimized by the explicit solution

$$q_1(\mathbf{R}) = \frac{\chi_2 - \chi_1}{\eta_1 - 2J_{12}(|\mathbf{R}_1 - \mathbf{R}_2|) + \eta_2}. \tag{12}$$

We therefore see that this fluctuating-charge model always predicts a nonzero charge on each atom unless they have equal electronegativities or at least one atom has infinite hardness. While this is reasonable for chemically bonded systems, it fails to describe, even qualitatively, the charge transfer behavior at infinite separation. As $|\mathbf{R}_1 - \mathbf{R}_2| \to \infty$, the Coulomb interaction vanishes, so that

$$\lim_{|\mathbf{R}_1 - \mathbf{R}_2| \to \infty} q_1(\mathbf{R}) = \frac{\chi_2 - \chi_1}{\eta_1 + \eta_2} \neq 0. \tag{13}$$

The model therefore predicts nonzero charge transfer even for dissociated systems, which is clearly unphysical for diatomic molecules in the gas phase. This leads to a dissociation catastrophe whereby intermolecular charge transfer is severely overestimated, causing electrostatic properties such as the dipole moment and the on-axis component of the polarizability to diverge. This renders such models useless for describing intermolecular charge transfers and in addition requires further constraints proscribing intermolecular charge transfer in practical simulations [22].

This unphysical prediction of nonzero charge transfer at infinity can be understood by turning off the Coulomb interaction terms in (1). Then, the noninteracting energy E^{NI} becomes the simple sum

$$E^{\mathrm{NI}}(\mathbf{q}; \mathbf{R}) = \sum_{i=1}^{N} E_i^{at}(q_i), \tag{14}$$

and where each individual term in the case of quadratic energies (3) can be written in the form

$$E_i^{at}(q_i) = \frac{1}{2}\eta_i\left(q_i + \frac{\chi_i}{\eta_i}\right)^2 - \frac{\chi_i^2}{2\eta_i} + E_i^0. \tag{15}$$

Thus, in the absence of any interatomic interactions, the charge predicted by fluctuating-charge models defaults to the solution $q_i = -\chi_i/\eta_i$, being the minimum point of the parabola (15). As both the atomic electronegativity and atomic hardness are constants, it is unclear how this problem can be solved while remaining in atom space, i.e., the solution space spanned by the vector of atomic charges $\mathbf{q}$.

The dissociation catastrophe can be interpreted as the consequence of an unrealistic assumption inherent in fluctuating-charge models, namely that pairs of atoms can exchange charge with equal facility regardless of their distance. This is true only in metallic phases, and therefore the extent to which this model fails to predict sensible charge distributions can be attributed to a fault in the underlying physics in assuming that molecular systems have metallic character. In the next section, we will discuss how to undo this assumption.

3 The QTPIE Model

In order to address this dissociation catastrophe, we had proposed the QTPIE (charge transfer with polarization current equilibration) model [38, 43], which was first formulated not in terms of atomic charges but in terms of charge-transfer variables [44, 45], sometimes called split-charge variables [46]. These new variables p_{ji} define a new solution space which we call the bond space, and they account for the amount of charge that has flowed from the jth atom to the ith atom and can be interpreted as the integral of a transient current between these two atoms. We require these variables to be antisymmetric, so that $p_{ji} = -p_{ij}$. Furthermore, we recover the atomic charges by summing over all source atoms, and thus the charge-transfer variables are related to the charge variables by the relation

$$\sum_{j=1}^{N} p_{ji} = q_i. \tag{16}$$

By applying this relation, the energy function of the QEq-type fluctuating-charge model (3) can be rewritten in terms of charge-transfer variables as

$$E\left(\mathbf{p}\right) = \sum_{i,j=1}^{N} \chi_i p_{ji} + \tfrac{1}{2} \sum_{i,j,k,l=1}^{N} p_{ki} p_{lj} J_{ij}$$
$$= \sum_{i<j} \left(\chi_i - \chi_j\right) p_{ji} + \tfrac{1}{2} \sum_{i<j,k<l} p_{ki} p_{lj} \left(J_{ij} - J_{il} - J_{kj} + J_{kl}\right), \tag{17}$$

where on the second line, we have exploited the skew-symmetry of the charge-transfer variables. We note that the skew-symmetry constraint, together with (16), already enforces overall charge neutrality, i.e. $Q = 0$, and therefore the model can be solved immediately by direct minimization of this energy with respect to the charge-transfer variables $\mathbf{p}$ without the use of the Lagrange multipliers.

We now create the QTPIE model by modifying the first term in (17) to have a pairwise and geometry-dependent electronegativity. By replacing χ_i with $\tilde{\chi}_{ji}\left(\mathbf{R}\right)$, we have the *new* energy function

$$E^{QTPIE}\left(\mathbf{p}; \mathbf{R}\right) = \sum_{i,j=1}^{N} \tilde{\chi}_{ji}\left(\mathbf{R}\right) p_{ji} + \frac{1}{2} \sum_{i,j,k,l=1}^{N} p_{ki} p_{lj} J_{ij} \left(\left|\mathbf{R}_i - \mathbf{R}_j\right|\right)$$
$$= \sum_{i<j} \left(\tilde{\chi}_{ji} - \tilde{\chi}_{ij}\right) p_{ji} + \tfrac{1}{2} \sum_{i<j,k<l} p_{ki} p_{lj} \left(J_{ij} - J_{il} - J_{kj} + J_{kl}\right),$$

$$\tag{18}$$

where on the second line the explicit distance dependence was suppressed for brevity. Equation (18) defines the QTPIE model, which is solved in bond space by the solution to the linear system of equations

$$\tilde{\mathbf{J}}\mathbf{p} = \tilde{\mathbf{v}}, \tag{19}$$

where the collection of charge-transfer variables $p_{(j,i)} = p_{ji}$ is now interpreted as a vector indexed by the multi-index (j, i), $1 \leq i < j \leq N$. This defines a new vector space, which we call the bond space, with bond hardness matrix $\tilde{J}_{(k,i),(l,j)} = J_{ij} - J_{il} - J_{kj} + J_{kl}$ and bond electronegativities $\tilde{v}_{(j,i)} = \tilde{\chi}_{ij} - \tilde{\chi}_{ji}$ that correspond to pairwise voltage differences.

For comparison purposes, we solve the QTPIE model analytically for a diatomic molecule. The model then consists of only one unknown variable p_{21}, and the model has the energy function

$$E^{QTPIE}(p_{21}; \mathbf{R}) = (\tilde{\chi}_{21}(\mathbf{R}) - \tilde{\chi}_{12}(\mathbf{R})) p_{21}$$
$$+ \frac{1}{2}(J_{11} - 2J_{12}(|\mathbf{R}_1 - \mathbf{R}_2|) + J_{22}) p_{21}^2, \tag{20}$$

which has the solution

$$q_1(\mathbf{R}) = p_{21}(\mathbf{R}) = \frac{\tilde{\chi}_{12}(\mathbf{R}) - \tilde{\chi}_{21}(\mathbf{R})}{J_{11} - 2J_{12}(|\mathbf{R}_1 - \mathbf{R}_2|) + J_{22}}. \tag{21}$$

In contrast to (12), it is possible to attenuate long-distance charge transfer as $|\mathbf{R}_1 - \mathbf{R}_2| \to \infty$ by requiring that $\tilde{\chi}_{12}(\mathbf{R}) - \tilde{\chi}_{21}(\mathbf{R}) \to 0$ at the same time. In the QTPIE model, there are several reasonable choices for the pairwise electronegativity [43], but we believe a reasonable definition of the pairwise electronegativity is

$$\tilde{\chi}_{ij}(\mathbf{R}) - \tilde{\chi}_{ji}(\mathbf{R}) = \frac{S_{ij}(|\mathbf{R}_i - \mathbf{R}_j|)}{\langle S_{ij'}(|\mathbf{R}_i - \mathbf{R}_{j'}|)\rangle_{j'}}(\chi_i - \chi_j), \tag{22}$$

which is essentially the bare atomic electronegativity χ_i weighted by $S_{ij}(|\mathbf{R}_i - \mathbf{R}_j|) = \int \phi_i(\mathbf{r}_1; \mathbf{R}_i) \phi_j(\mathbf{r}_1; \mathbf{R}_i) d\mathbf{r}_1$, the overlap integral between the atomic orbitals on the ith and jth atoms as introduced to calculate the screened Coulomb interactions (33) as described in Appendix 1, and renormalized by a system-dependent constant that rescales the weighting factor by the *average* weighting factor over all atoms,

$$\langle S_{ij'}(|\mathbf{R}_i - \mathbf{R}_{j'}|)\rangle_{j'} = \sum_{j'=1}^{N} S_{ij'}(|\mathbf{R}_i - \mathbf{R}_{j'}|)/N.$$

Indeed, this choice of pairwise electronegativity produces the correct asymptotic limit of no charge transfer at infinite separation, as the electronegativity difference vanishes due to the asymptotic property of the overlap integral that $S_{ij} \to 0$ as $|\mathbf{R}_i - \mathbf{R}_j| \to \infty$. Note that if we set all the attenuation factors to a numerical constant, say $S_{ij} = 1$, (22) reduces to just χ_i and we recover the QEq-type fluctuating-charge model of the preceding section. The dissociation catastrophe returns when the model assumes that pairs of atoms can exchange charge with equal facility regardless of their distance, thus reinforcing our earlier observation that the failures of the earlier model can be attributed to assuming that all systems have metallic electronic structure.

4 The Exact Reformulation of Models in Bond Space as Models in Atom Space

The preceding discussion shows that we have found a solution to the dissociation catastrophe and therefore have a framework for fluctuating-charge models that are useful for describing intermolecular charge transfer. However, this apparently comes at the price of representing the solution in bond space, which for an N-atom system has N times as many variables as the original representation in atom space. Perhaps surprisingly, it is possible to reformulate an arbitrary fluctuating-charge model formulated in bond space exactly as an equivalent fluctuating-charge model in atom space.

The key insight is that the bond-space hardness matrix $\tilde{\mathbf{J}}$ in (19) is rank deficient and that its nullspace is spanned by vectors describing cyclic charge transport [43]. In order to show this, we note that relationship between charges and charge-transfer variables (16) is linear. Therefore, the mapping from solutions in bond space to those in atom space can be expressed by a rectangular matrix $\mathbf{T}$ such that

$$\mathbf{T} : \mathbb{R}^{N(N-1)/2} \to \mathbb{R}^{N-1} \cong \left\{ \mathbf{q} \in \mathbb{R}^N | \mathbf{1}^T \mathbf{q} = 0 \right\}$$
$$\mathbf{Tp} = \mathbf{q}. \tag{23}$$

Then the relationship between $\mathbf{J}$ and $\tilde{\mathbf{J}}$ can be expressed as

$$\tilde{\mathbf{J}} = \mathbf{T}^T \mathbf{JT}. \tag{24}$$

We had previously introduced a directed graph G whose vertices are in one-to-one correspondence with atomic charges and edges that are in one-to-one correspondence with charge-transfer variables. For fluctuating-charge models without any proscriptions on charge transfer, G is a complete graph. Then $\mathbf{T}$ corresponds to the adjacency matrix for G, which has matrix element T_{ve} equal to 1 if the edge e points toward the vertex v, -1 if the edge e points away from the vertex v, and 0 otherwise.

While the atom-space hardness matrix $\mathbf{J}$ is of full rank, $\tilde{\mathbf{J}}$ has dimension $N(N-1)/2$ but only rank $N-1$. This is because combinations of charge-transfer variables that correspond to cyclic charge transport belong to the nullspace of $\tilde{\mathbf{J}}$. For illustrative purposes, consider a four-charge system that is described by the variables

$\{q_1, q_2, q_3, q_4\}$ in atom space and $\{p_{21}, p_{31}, p_{41}, p_{32}, p_{42}, p_{43}\}$ in bond space. For this system, $\mathbf{T}$ has the matrix representation

$$\mathbf{T} = \begin{pmatrix} 1 & 1 & 1 & 0 & 0 & 0 \\ -1 & 0 & 0 & 1 & 1 & 0 \\ 0 & -1 & 0 & -1 & 0 & 1 \\ 0 & 0 & -1 & 0 & -1 & -1 \end{pmatrix}. \tag{25}$$

Consider the combination of charge-transfer variables $p_{12} + p_{23} + p_{34} + p_{41} = -p_{21} + p_{41} - p_{32} - p_{43}$. In our present basis, this corresponds to the vector $\Gamma = (-1, 0, 1, -1, 0, -1)^T$. Then it is straightforward to verify by explicit calculation that

$$\mathbf{T}\Gamma = \begin{pmatrix} 1 & 1 & 1 & 0 & 0 & 0 \\ -1 & 0 & 0 & 1 & 1 & 0 \\ 0 & -1 & 0 & -1 & 0 & 1 \\ 0 & 0 & -1 & 0 & -1 & -1 \end{pmatrix} \begin{pmatrix} -1 \\ 0 \\ 1 \\ -1 \\ 0 \\ -1 \end{pmatrix} = \begin{pmatrix} 0 \\ 0 \\ 0 \\ 0 \end{pmatrix} = \mathbf{0}. \tag{26}$$

Hence, Γ, which represents a cyclic flow of charge, is in the nullspace of $\mathbf{T}$, and hence by the relation (24) is also in the nullspace of $\tilde{\mathbf{J}}$. We interpret this as a consequence of Kirchhoff's voltage law, which arises from the conservative nature of the electrostatic potential. By similar calculations, one can show that any cyclic combination of charge-transfer variables lies in the nullspace of $\tilde{\mathbf{J}}$, and so acyclic combinations of charge-transfer variables are the only ones that lie in the range of $\tilde{\mathbf{J}}$. An elementary result of graph theory immediately yields that the space of acyclic combinations of charge-transfer variables is spanned by $N - 1$ linearly independent vectors, and hence $\mathbf{T}$ is of rank $N - 1$. We have previously provided a rigorous proof of this fact [43].

When combined with the fact that $\mathbf{J}$ is a discretization of the Coulomb operator in a finite and localized basis and is therefore of full rank (N) for reasonable geometries that do not have degenerate or nearly coincident atoms, the composition (24) shows that $\tilde{\mathbf{J}}$ must have rank $N - 1$ and that there are always $N - 1$ physically important degrees of freedom, whether they are represented as atomic charges or charge-transfer variables. This suggests that it is possible to reformulate exactly any fluctuating-charge model represented in bond space as an equivalent model in atom space. In order to do this, we require the inverse mapping $\mathbf{T}^+$ such that $\mathbf{p} = \mathbf{T}^+\mathbf{q}$. As $\mathbf{T}$ is rectangular, the conventional inverse $\mathbf{T}^{-1}$ cannot exist, but the preceding discussion shows that the generalized inverse such as the Moore–Penrose pseudoinverse $\mathbf{T}^+$ performs the same role as the conventional inverse in that no information is lost in the inversion. Interestingly, it is possible to verify by explicit calculation that for the complete graph G, $\mathbf{T}^+ = \mathbf{T}^T/N$, so that

$$p_{ba} = \frac{q_a - q_b}{N}. \tag{27}$$

This simple relation allows the energy function of the QTPIE model (18) to be reformulated exactly as

$$E^{QTPIE}\left(\mathbf{T}^+\mathbf{q}\right) = \tilde{\mathbf{v}}\mathbf{T}^+\mathbf{q} + \tfrac{1}{2}\mathbf{q}^T\mathbf{J}\mathbf{q}$$

$$= \sum_{i=1}^{N} q_i \sum_{j=1}^{N} \frac{\tilde{\chi}_{ji}(\mathbf{R}) - \tilde{\chi}_{ij}(\mathbf{R})}{N} + \frac{1}{2}\sum_{i,j=1}^{N} q_i q_j J_{ij}. \tag{28}$$

Interestingly, this expression shows that the introduction of pairwise electronegativities results in an effective atomic electronegativity

$$\overline{\chi}_i = -\left(\tilde{\mathbf{v}}\mathbf{T}^+\right)_i = \sum_{j=1}^{N} \frac{\tilde{\chi}_{ij}(\mathbf{R}) - \tilde{\chi}_{ji}(\mathbf{R})}{N}, \tag{29}$$

which for the definition of the pairwise electronegativity (22) gives rise to

$$\overline{\chi}_i = \frac{\sum\limits_{j=1}^{N} S_{ij}\left(\left|\mathbf{R}_i - \mathbf{R}_j\right|\right)\left(\chi_i - \chi_j\right)}{\sum\limits_{j'=1}^{N} S_{ij'}\left(\left|\mathbf{R}_i - \mathbf{R}_{j'}\right|\right)} = \chi_i - \frac{\sum\limits_{j=1}^{N} S_{ij}\left(\left|\mathbf{R}_i - \mathbf{R}_j\right|\right)\chi_j}{\sum\limits_{j'=1}^{N} S_{ij'}\left(\left|\mathbf{R}_i - \mathbf{R}_{j'}\right|\right)} \tag{30}$$

that is in general different from the bare atomic electronegativity χ_i that goes into (22) due to the presence of an explicitly environment-dependent term. This is the main result of our paper.

Interestingly, the transformation from charge-transfer variables back into atomic charges leads us to identify explicit distance dependence in pairwise electronegativities with environmental effects in atomic electronegativities. It may be initially surprising that the expression for the effective atomic electronegativities contains explicit dependence on other atoms, which contradicts the long-held conventional wisdom that electronegativities are atomic properties. However, both Pauling and Mulliken had stressed in their seminal works that electronegativities must necessarily be thought of as atoms in molecules and not a property of atoms in isolation [23, 39, 47]. Indeed, if electronegativity is a measure of the ease of charge transfer, then it should also depend on the properties of the external entity that is receiving or donating charge. For this reason, we argue that the environmental effect captured in the effective atomic electronegativities is essential and correct. However, the fact that we have *two* quantities that take on the dimensions of electronegativity show that it is possible to retain the intuitive aspect of electronegativity as an intrinsic property of individual atoms and yet be consistent with the fundamental aspects of charge transfer.

Interestingly, the distinction between bare and effective atomic electronegativities arises precisely from consideration of the effects of the atomic environment in the latter. The effective atomic electronegativities contain a weighted average contribution of the electronegativity of every other atom in the system, with weighting factors that are distance dependent and vanish smoothly as the pairwise distances increase. In contrast, constant weights that were independent of molecular geometry produce a description identical to that offered by models like QEq, as the addition of a constant to every effective electronegativity in the system does not change the predicted charges. Thus, the smooth dependence of the weights on molecular geometry permits a smooth and continuous transition between the isolated atoms and the molecular setting with interacting atoms, which is difficult to achieve in other models that impose topological restrictions on the allowable flows of charge in the molecular system. Importantly, we retain the intuition afforded by an atomistic formulation while retaining also the correct dissociation behavior without the need for particular topological restrictions on charge flows (which would amount to changing $\mathbf{T}$). This avoids a potential complication with restricted-flow models in dynamical simulations whereby it may be necessary to adjust the topology of charge flow on the fly, which could lead to discontinuities in important observables over the course of the simulation.

5 Application to Model Water Systems

As a simple application of our QTPIE model, we study a series of simple water systems. As is well known, the dipole moment of a single molecule of water is 1.85 D in the gas phase [48] but increases to 2.95 ± 0.20 D in the liquid phase [49] due to cooperative polarization between the water molecules in condensed phases. The reproduction of such cooperative behavior is a useful test of polarizable water models. Here, we study whether the QTPIE model is able to reproduce the onset between gas-like behavior to bulk-like behavior in planar water chains. We chose physically reasonable internal geometries with O—H bond lengths of 1.00 Å and internal angles of $105°$, and the O—O internuclear distances of 2.87 Å is chosen to be the O—O internuclear separation in the ground state geometry of the water dimer. The water molecules are chosen to be coplanar and are aligned along their dipole moments. While such intermolecular geometries are physically unlikely to be observed, they are useful for studying the transition from gas-like to bulk-like behavior in an essentially one-dimensional system. As a further test of our charge models, we choose to parameterize the models using data only from monomer and dimer geometries and see whether these models satisfactorily reproduce the dipole moments in longer water chains, as this would be a sensitive indicator of the quality of the intermolecular electrostatic interactions.

To eliminate systematic error from improper parameterization, we reparameterized both the QEq and the QTPIE models to be applied specifically to three-site water models. About 1230 monomer geometries were generated by systematically varying the internal coordinates and bond lengths, and 890 dimer geometries were generated from fictitious high-temperature molecular dynamics runs at 30,000 K with a systematic variation in the Lennard–Jones attraction parameters to sample a wide variety of intermonomer distances. For each geometry, *ab initio* dipole moments were calculated with density-fitted local second-order Møller–Plesset perturbation theory [50] using the augmented Dunning correlation-consistent valence triple-zeta basis set [51] (DF-LMP2/aug-cc-pVTZ). We then optimized the weighted root mean square deviation between each model's predictions and the *ab initio* calculation using the derivative-free simplex algorithm [52], with a weight given by a Boltzmann weight at a temperature of 10,000K. This temperature has no physical significance and is merely chosen to generate convenient weights to penalize the contribution of geometries of higher energies that were produced in the systematic exploration of configuration space, some of which were as high as *ca.* 0.4 Hartrees above the minimum energy configurations and for all practical purposes lie in energetically inaccessible, and hence irrelevant, regions of the relevant potential energy surfaces. The resulting parameters are compared with the original QEq parameters in Table 2.

As a test of the water models obtained by this procedure, we use the models to study consecutive O—O internuclear separations of 2.87 Å and O—H bond lengths of 1.00 Å and internal angle 105°. Figure 1 shows the dipole moments as calculated from the definition as

$$d_\lambda = \sum_{i=1}^{N} q_i R_{i\lambda}, \tag{31}$$

where the Greek index λ indexes spatial components (x, y, and z), $R_{i\lambda}$ is the λth spatial component of the position of atom i, and q_i is the charge on atom i as determined above. For the purposes of comparison, we have compared the dipole moments with high-quality ab initio calculations at the DF-LMP2/aug-cc-pVTZ level of theory. In addition, we compare the results to the AMOEBA model available in the TIN-

Table 2 Parameters for the QTPIE and QEq models for a three-site water model

Parameter (eV)	QTPIE (this work)	QEq (original)[a]	QEq (reparameterized)
H electronegativity	5.366	4.528	3.678
H hardness	11.774	11.774	18.448
O electronegativity	7.651	7.651	9.591
O hardness	13.115	13.115	17.448

[a] From Ref. [16].

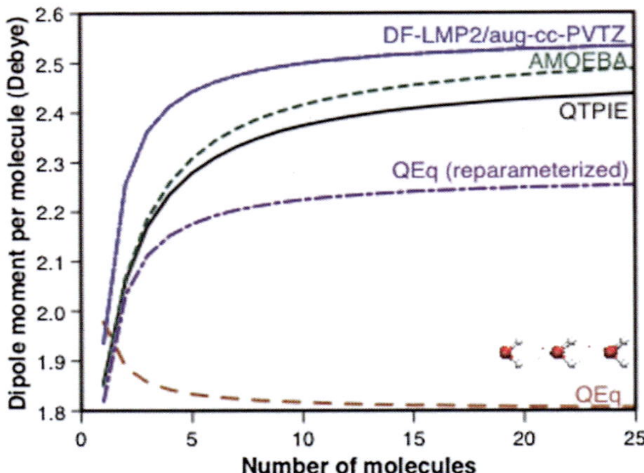

Fig. 1 Dipole moments per molecule for a sequence of planar water chains, with consecutive O—O internuclear separations of 2.87 Å and O—H bond lengths of 1.00 Å and internal angle 105°, as calculated by QTPIE (*black solid line*), DF-LMP2/aug-cc-pVTZ (*blue broken line*), AMOEBA (*green short-dashed line*), QEq (*brown dashed line*), and reparameterized QEq (*purple dash-dotted line*). The parameters used for the QTPIE and QEq models are given in Table 2

KER molecular dynamics package, which is a type of polarizable multipole model parameterized to the same level of *ab initio* theory [5].

The *ab initio* data show that dipole moment per molecule increases rapidly as a function of the chain length, and beyond approximately five water molecules gradually saturates toward a limiting value of 2.50 D per molecule. As expected, the AMOEBA model reproduces the *ab initio* data very well. By comparison, the QTPIE model is also able to reproduce the trends exhibited by the *ab initio* data and the AMOEBA model, which is especially encouraging when taking into account the much simpler description of electrostatics in QTPIE as compared to AMOEBA. Surprisingly, we see that the QEq model, using the original parameters, shows a decrease in the dipole moment per molecule with increasing chain length. This behavior is absent in the reparameterized model, but instead saturates to a value of 2.25 D per molecule, which is significantly lower than for the QTPIE and AMOEBA models.

The results suggest that QTPIE affords a qualitatively superior description of intermolecular electrostatic interactions over QEq, as even reparameterizing QEq could not produce the bulk-like dipole moments to the same level of accuracy. In contrast, the results of the QTPIE model are comparable with those of the significantly more costly AMOEBA water model, which has 14 parameters per element specifically for electrostatic interactions, as well as nonlinear, higher-order multipole interaction equations (up to the quadrupole–quadrupole level) to solve for [5]. In contrast, the QTPIE model requires only four parameters per element and solving a linear system of equations for charge–charge interactions only. Thus, the three-site water model based on QTPIE is able to reproduce satisfactorily the cooperative polarization behavior in these planar water chains with just four independent parameters and therefore shows great promise for providing a comparable level of accuracy with more computationally costly and more highly parameterized models.

6 Conclusions

The above discussion reinforces the notion that empirical atomic electronegativities must be environment dependent because the effects of the electron-accepting or electron-donating tendencies of atoms are never observed in isolation. The reason is obvious: there must be a counterpart to receive or donate charge. Even Pauling's seminal work on electronegativity and chemical bonding acknowledges that the electron-accepting or electron-donating tendencies of atoms depend on the other atoms in the molecule, and the best that can be hoped for is that this tendency for accepting or donating electrons be approximately constant over many different molecules, corresponding to many different chemical environments.

Despite this dependence on the chemical environment, electronegativities are indeed observed to vary only slightly depending on the exact molecule being considered. This strongly suggests that there must be some underlying, intrinsic atomic property that is to a large extent responsible for the observed atomic electronegativities. Here, we propose to call these quantities *bare* atomic electronegativities, as distinguished from the *effective* atomic electronegativities that are the only ones that can be directly observed.

Our work on developing fluctuating-charge models shows that, indeed, physical constraints on the qualitative behavior of these models force us to use environment-dependent electronegativities and that their dependence on the chemical environment is unavoidable and can be significant. However, this dependence is strongly local in the QTPIE model owing to the locality inherent in our definition of the distance attenuation factors. This locality is in accordance with our chemical experiences. Furthermore, it is possible to reconcile this environment dependence with the existence of an intrinsic atomic quantity; in the QTPIE model, the "bare" atomic electronegativities that are intrinsic properties of individual atoms in isolation are distinguished from the *effective* atomic electronegativities that are system-specific.

Appendix 1: Fitting of s-type Primitive Gaussian Orbitals that Best Reproduce Two-Electron Coulomb Integrals over s-Type Slater Orbitals

In this appendix, we describe how we found the exponents given in Table 1, which are of s-type primitive Gaussian orbitals such that the two-electron Coulomb integrals over them best reproduce those over s-type Slater orbitals. We construct these Gaussian orbitals by minimizing the norm of the L^2-difference between the homonuclear Coulomb integral over Slater orbitals and over Gaussian orbitals, i.e., given a Slater exponent ζ, we want the Gaussian exponent α that minimizes

$$\left\| J^G(\alpha) - J^S(\zeta) \right\|_2^2 = \left\langle J^G(\alpha), J^G(\alpha) - 2J^S(\zeta) \right\rangle_2 + \left\| J^S(\zeta) \right\|_2^2, \qquad (32)$$

where $\langle f, g \rangle_2 = \int_0^\infty f(x) g(x) dx$ is the inner product in the function space $L^2[0, \infty)$, $\|f\|_2 = \sqrt{\langle f, f \rangle_2}$ is the L^2-norm, J^G is the two-electron Coulomb integral over s-type primitive Gaussian orbitals

$$J^G(R; \alpha) = \frac{2\alpha}{\pi} \iint_{\mathbb{R}^6} \frac{e^{-\alpha|r_1 - R|^2} e^{-\alpha|r_2|^2}}{|r_1 - r_2|} dr_1 dr_2 = \frac{\mathrm{erf}\sqrt{\alpha} R}{R} \qquad (33)$$

and J^S is the two-electron Coulomb integral over s-type Slater orbitals

$$J^S(R; \zeta, n) = \frac{(2\zeta)^{4n+2}}{((2n)!)^2} \iint_{\mathbb{R}^6} \frac{|r_1 - R|^n |r_2|^n e^{-\zeta|r_1 - R|} e^{-\zeta|r_2|}}{|r_1 - r_2|} dr_1 dr_2, \qquad (34)$$

which is given in closed-form in Ref. [53]. As the Slater exponent ζ is given for each minimization, the last term in (32) can be dropped without affecting the results of the minimization, and therefore the minimization problem is solved by the Gaussian exponent α that solves the equation

$$0 = \frac{\partial}{\partial \alpha} \left\langle J^G(\alpha), J^G(\alpha) - 2J^S(\zeta) \right\rangle_2 = \left\langle 2\frac{dJ^G(\alpha)}{d\alpha}, J^G(\alpha) - J^S(\zeta) \right\rangle_2. \qquad (35)$$

We find the solution to (35) using the secant method with a trust radius of $\alpha/4$ at each iteration. The algorithm was terminated once the integral on the right-hand side of (35) was less than 10^{-16} in absolute magnitude. The results are presented in Table 1, along with the maximum absolute error as defined by

$$\mathrm{MAE} = \max_{0 \leq R < \infty} \left| J^G(R; \alpha) - J^S(R; \zeta) \right|. \qquad (36)$$

Appendix 2: Derivation of an Explicit Solution to Fluctuating-Charge Models

In this appendix, we derive the solution to the linear system of equations (9) that define a fluctuating-charge model, namely:

$$\begin{pmatrix} \mathbf{J} & \mathbf{1} \\ \mathbf{1}^T & 0 \end{pmatrix} \begin{pmatrix} \mathbf{q} \\ \mu \end{pmatrix} = \begin{pmatrix} -\chi \\ Q \end{pmatrix}. \tag{37}$$

Assume $\mathbf{J}$ is invertible. As discussed in the main text, this should be true for all reasonable geometries without degenerate or nearly coincident atoms. Then we can perform Gaussian elimination on the second row by premultiplying the first row by $-\mathbf{1}^T \mathbf{J}^{-1}$ and adding the result to the second row. This transforms the system to

$$\begin{pmatrix} \mathbf{J} & \mathbf{1} \\ \mathbf{0}^T & 0 - \mathbf{1}^T \mathbf{J}^{-1} \mathbf{1} \end{pmatrix} \begin{pmatrix} \mathbf{q} \\ \mu \end{pmatrix} = \begin{pmatrix} -\chi \\ Q + \mathbf{1}^T \mathbf{J}^{-1} \chi \end{pmatrix}, \tag{38}$$

where $0 - \mathbf{1}^T \mathbf{J}^{-1} \mathbf{1}$ is the Schur complement of $\mathbf{J}$ in this problem. Next, we solve the first row for $\mathbf{q}$, leading to

$$\begin{pmatrix} \mathbf{J} & \mathbf{0} \\ \mathbf{0}^T & -\mathbf{1}^T \mathbf{J}^{-1} \mathbf{1} \end{pmatrix} \begin{pmatrix} \mathbf{q} \\ \mu \end{pmatrix} = \begin{pmatrix} -(\chi + \mathbf{1}\mu) \\ Q + \mathbf{1}^T \mathbf{J}^{-1} \chi \end{pmatrix}. \tag{39}$$

This system of equations is now in block-diagonal form, and it is easy to write down the solution

$$\begin{pmatrix} \mathbf{q} \\ \mu \end{pmatrix} = \begin{pmatrix} -\mathbf{J}^{-1} (\chi + \mathbf{1}\mu) \\ -\left(\mathbf{1}^T \mathbf{J}^{-1} \mathbf{1}\right)^{-1} \left(Q + \mathbf{1}^T \mathbf{J}^{-1} \chi\right) \end{pmatrix}, \tag{40}$$

which is equivalent to (10). In particular, the explicit formula for the charge distribution is

$$\mathbf{q} = -\mathbf{J}^{-1} \chi - \mu \mathbf{J}^{-1} \mathbf{1} = -\mathbf{J}^{-1} \chi - \left(\frac{Q + \mathbf{1}^T \mathbf{J}^{-1} \chi}{\mathbf{1}^T \mathbf{J}^{-1} \mathbf{1}} \right) \mathbf{J}^{-1} \mathbf{1}. \tag{41}$$

It is obvious that in general, the charge distribution that solves (9) is not $\mathbf{q} = -\mathbf{J}^{-1}\chi$, which is the solution to the unconstrained problem $\mathbf{J}\mathbf{q} = -\chi$, but contains an additional term that corresponds to a correction to account for the constraint of overall charge conservation. We note that this solution has only been published previously in the ES+ model of Streitz and Mintmire [10] and was later rederived by Bultinck and Carbó-Dorca [54].

References

1. T. J. Giese, D. M. York, J. Chem. Phys. **120**, 9903–9906 (2004)
2. S. J. Stuart, B. J. Berne, J. Phys. Chem. **100**, 11934–11943 (1996)
3. A. Warshel, M. Kato, A. V. Pisliakov, J. Chem. Theory Comput. **3**, 2034–2045 (2007)
4. A. Warshel, M. Levitt, J. Mol. Bio. **103**, 227–249 (1976)
5. P. Ren, J. W. Ponder, J. Phys. Chem. B **107**, 5933–5947 (2003)
6. G. Lamoureux, B. Roux, J. Chem. Phys. **119**, 3025–3039 (2003)
7. H. Yu, W. F. van Gunsteren, Comput. Phys. Commun. **172**, 69–85 (2005)

8. S. Patel, C. L. Brooks, III, Molec. Simul. **32**, 231–249 (2006)
9. S. W. Rick, S. J. Stuart, in *Reviews in Computational Chemistry*, ed. by K. B. Lipkowitz, D. B. Boyd (Wiley, New York, 2002), Vol. 18
10. F. H. Streitz, J. W. Mintmire, Phys. Rev. B **50**, 11996 (1994)
11. U. W. Schmitt, G. A. Voth, J. Chem. Phys. **111**, 9361–9381 (1999)
12. S. M. Valone, S. R. Atlas, Phil. Mag. **86**, 2683–2711 (2006)
13. A. Warshel, Annu. Rev. Biophys. Biomol. Struct. **32**, 425–443 (2003)
14. M. S. Gordon, L. Slipchenko, H. Li, J. H. Jensen, D. C. Spellmeyer, R. Wheeler, in *Annual Reports in Computational Chemistry*, ed. by D. C. Spellmeyer, R. A. Wheeler (Elsevier, Amsterdam, 2007), Vol. 3
15. N. Gresh, G. A. Cisneros, T. A. Darden, J. P. Piquemal, J. Chem. Theory Comput. **3**, 1960–1986 (2007)
16. A. K. Rappé, W. A. Goddard, III, J. Phys. Chem. **95**, 3358–3363 (1991)
17. A. K. Rappé, C. J. Casewit, K. S. Colwell, W. A. Goddard, III, W. M. Skiff, J. Am. Chem. Soc. **114**, 10024–10035 (1992)
18. W. J. Mortier, S. K. Ghosh, S. Shankar, J. Am. Chem. Soc. **108**, 4315–4320 (1986)
19. W. J. Mortier, K. Vangenechten, J. Gasteiger, J. Am. Chem. Soc. **107**, 829–835 (1985)
20. A. C. T. Van Duin, S. Dasgupta, F. Lorant, W. A. Goddard, III, J. Phys. Chem. A **105**, 9396–9409 (2001)
21. S. W. Rick, B. J. Berne, J. Am. Chem. Soc. **118**, 672–679 (1996)
22. S. W. Rick, S. J. Stuart, B. J. Berne, J. Chem. Phys. **101**, 6141–6156 (1994)
23. L. Pauling, *The Nature of the Chemical Bond*, 2nd edition (Cornell University Press, Ithaca, NY, 1945)
24. H. O. Pritchard, H. A. Skinner, Chem. Rev. **55**, 745–786 (1955)
25. R. P. Iczkowski, J. L. Margrave, J. Am. Chem. Soc. **83**, 3547–3551 (1961)
26. J. Hinze, H. H. Jaffé, J. Am. Chem. Soc. **84**, 540–546 (1962)
27. J. Hinze, M. A. Whitehead, H. H. Jaffé, J. Am. Chem. Soc. **85**, 148–154 (1963)
28. J. Hinze, H. H. Jaffé, J. Phys. Chem. **67**, 1501–1506 (1963)
29. R. G. Parr, R. A. Donnelly, M. Levy, W. E. Palke, J. Chem. Phys. **68**, 3801–3807 (1978)
30. R. G. Pearson, J. Am. Chem. Soc. **85**, 3533–3539 (1963)
31. R. G. Pearson, Science **151**, 172–177 (1966)
32. R. G. Pearson, J. Songstad, J. Am. Chem. Soc. **89**, 1827–1836 (1967)
33. R. G. Parr, R. G. Pearson, J. Am. Chem. Soc. **105**, 7512–7516 (1983)
34. R. T. Sanderson, Science **114**, 670–672 (1951)
35. D. M. York, W. Yang, J. Chem. Phys. **104**, 159–172 (1996)
36. S. Patel, C. L. Brooks, III, J. Comp. Chem. **25**, 1–16 (2004)
37. S. Patel, A. D. Mackerell, Jr, C. L. Brooks, III, J. Comp. Chem. **25**, 1504–1514 (2004)
38. J. Chen, T. J. Martinez, Chem. Phys. Lett. **438**, 315–320 (2007)
39. R. S. Mulliken, J. Chem. Phys. **2**, 782–793 (1934)
40. R. W. Parr, W. Yang, *Density-Functional Theory of Atoms and Molecules*, 1st edition (Oxford, United Kingdom, 1989)
41. G. Del Re, J. Chem. Soc. 4031–4040 (1958)
42. J. Gasteiger, M. Marsili, Tetrahedron **36**, 3219–3228 (1980)
43. J. Chen, D. Hundertmark, T. J. Martínez, J. Chem. Phys. **129**, 214113 (2008)
44. J. L. Banks, G. A. Kaminski, R. Zhou, D. T. Mainz, B. J. Berne, R. A. Friesner, J. Chem. Phys. **110**, 741 (1999)
45. R. Chelli, P. Procacci, R. Righini, S. Califano, J. Chem. Phys. **111**, 8569–8575 (1999)
46. R. A. Nistor, J. G. Polihronov, M. H. Müser, N. J. Mosey, J. Chem. Phys. **125**, 094108 (2006)
47. L. Pauling, J. Am. Chem. Soc. **54**, 3570–3582 (1932)
48. D. R. Lide, *CRC Handbook of Chemistry and Physics*, 87 edition (CRC Press: Boca Raton, FL, 2006)
49. A. V. Gubskaya, P. G. Kusalik, J. Chem. Phys. **117**, 5290–5302 (2002)
50. H.-J. Werner, F. R. Manby, P. J. Knowles, J. Chem. Phys. **118**, 8149–8160 (2003)

51. T. H. Dunning, Jr., J. Chem. Phys. **90**, 1007–1023 (1989)
52. J. Nocedal, S. J. Wright, *Numerical Optimization* (Springer, New York, 2002)
53. N. Rosen, Phys. Rev. **38**, 255–276 (1931)
54. P. Bultinck, R. Carbo-Dorca, Chem. Phys. Lett. **364**, 357–362 (2002)

Information Planes and Complexity Measures for Atomic Systems, Ionization Processes and Isoelectronic Series

J.C. Angulo and J. Antolín

Abstract Within the present advanced review on the meaning, interpretation, and applications of the so-called 'complexity', different order-uncertainty planes emboding relevant information-theoretic magnitudes are studied in order to analyze the information content of the position and momentum electron densities of several atomic systems, including neutral atoms, singly charged ions, and isoelectronic series. The quantities substaining those planes are the exponential and the power Shannon entropies, the disequilibrium, the Fisher information, and the variance. Each plane gives rise to a measure of complexity, determined by the product of its components. In this work, the values of the so-called López-Ruiz, Mancini and Calbet (LMC), Fisher–Shannon (FS), and Cramer–Rao (CR) complexities will be provided in both conjugated spaces and interpreted from a physical point of view.

Keywords: Complexity measures · Fisher–Shannon plane · Disequilibrium · Atomic ionization · Shell–filling

1 Introduction

There have been tremendous interests in the literature to apply information theory to the electronic structure theory of atoms and molecules [1, 2]. The concepts of uncertainty, randomness, disorder, or delocalization, are basic ingredients in the study, within an information theoretic framework, of relevant structural properties for many different probability distributions appearing as descriptors of several chemical and physical systems and/or processes.

Following the usual procedures carried out within the Information Theory for quantifying the aforementioned magnitudes concerning individual distributions,

J.C. Angulo (✉)
Departamento de Física Atómica, Molecular y Nuclear and Instituto *Carlos I* de Física Teórica y Computacional, Universidad de Granada, 18071-Granada, Spain, e-mail: angulo@ugr.es

J. Antolín
Departamento de Física Aplicada, EUITIZ, Universidad de Zaragoza, 50018-Zaragoza, Spain and Instituto *Carlos I* de Física Teórica y Computacional, Universidad de Granada, 18071-Granada, Spain, e-mail: antolin@unizar.es

P. Piecuch et al. (eds.), *Advances in the Theory of Atomic and Molecular Systems*,   417
Progress in Theoretical Chemistry and Physics 19, DOI 10.1007/978-90-481-2596-8_20,
© Springer Science+Business Media B.V. 2009

some other extensions have been done in order to introduce the concepts of "similarity" or "divergence" between two distributions, as comparative measures. Quantum similarity theory was originally developed in order to establish quantitative comparisons between molecular systems by means of their fundamental structure magnitudes: electron density functions. Applications of this important theory have been one of the cornerstones of recent chemical research in molecules [3–5].

Some pioneering efforts relating Information Theory to electronic structure and properties of molecules can be already found in the seminal papers by Daudel in the framework of loge theory [6, 7], subsequently followed by Mezey [8] and reexamined later by Nalewajski [9]. The studies of Mezey [10] and Avnir [11] on symmetry and chirality-related problems in molecules, and in other very diverse fields (e.g., image and texture analysis), are also examples of applications of informational measures on specific aspects of shape, disorder, and complexity.

This kind of measures and techniques, which in fact characterizes most of the information theory aims and tools, have been widely employed in recent years within the atomic physics framework. This work constitutes a survey of some of those applications for obtaining relevant information on different properties of atomic systems, including structural and experimental ones. The role played by the two conjugated variables, namely, position and momentum, appears fundamental for a complete description of the atomic information features. It is shown that, in spite of their simplicity among the many-body systems, the atomic ones possess a high-enough level of organization and hierarchy to be considered as an appropriate benchmark for the suggested complexity study.

The relevancy of the above concepts motivates the search for an appropriate quantification, giving rise to a variety of density functionals, each one with its own characterisitics and properties which make them more or less useful attending to the specific problem we are dealing with.

Diverse information measures of probability distributions have been widely applied with the aim of describing a great variety of systems or processes in many different scientific fields. One of the pioneering and most well-known of such measures is the variance, but later on many others have been also considered for these kind of applications. Among them, it should be emphasized the role played by the Shannon entropy S [12]

$$S(\rho) \equiv - \int \rho(\mathbf{r}) \ln \rho(\mathbf{r}) d\mathbf{r} \tag{1}$$

and the Fisher information I [13, 14]

$$I(\rho) \equiv \int \rho(\mathbf{r}) |\nabla \ln \rho(\mathbf{r})|^2 d\mathbf{r} \tag{2}$$

of a distribution $\rho(\mathbf{r})$. In fact, S is a basic quantity in statistical thermodynamics [15], and it is the essential tool on the application of the "Maximum Entropy" technique based on Jaynes' principle. More recently, Fisher information appeared as a fundamental magnitude for deriving important laws of, e.g., density functional theory [16, 17] or quantum mechanics [18] by means of the extremization Frieden principle [14]. The numerous applications of tools based on both S and I suggest the relevancy of using them in a complementary way, attending to their main characteristics and properties as will be described later.

1.1 Complexity: Meaning and Definitions

Another relevant concept within information theory, in some cases strongly related to the aforementioned measures, is the so-called 'complexity' of a given system or process. There is not a unique and universal definition of complexity for arbitrary distributions, but it could be roughly understood as an indicator of pattern, structure, and correlation associated to the system the distribution describes. Nevertheless, many different mathematical quantifications exist under such an intuitive description. This is the case of the algorithmic [19, 20], Lempel–Ziv [21] and Grassberger [22] complexities, as well as the logical and thermodynamical depths by Bennett [23] and Lloyd and Pagels [24], respectively, all them as others with many scientific applications.

Complexity is used in very different fields (dynamical systems, time series, quantum wavefunctions in disordered systems, spatial patterns, language, analysis of multielectronic systems, cellular automata, neuronal networks, self-organization, molecular or DNA analyses, social sciences, etc.) [25–27]. Although there is no general agreement about the definition of what complexity is, its quantitative characterization is a very important subject of research in nature and has received considerable attention over the past years [28, 29].

The characterization of complexity cannot be univocal and must be adequate for the type of structure or process we study, the nature and the goal of the description we want, and the level or scale of the observation that we use. In the same way it is interesting to combine the properties of the new proposals to characterize complexity and test them on diverse and known physical systems or processes. Fundamental concepts such as information or entropy are frequently present in the proposals for characterizing complexity, but some other ingredients that do not only capture uncertainty or randomness can also be searched. One wishes also to capture some other properties such as clustering, order or organization of the systems, or process. Some of the definitions and relations between the above concepts are not clear; even less so is how disorder or randomness takes part in the aforementioned properties of the system and vice versa.

The initial form of complexity is designed such that it vanishes for the two extreme probability distributions (little complex ones), corresponding to perfect order (represented by a Dirac-delta) and maximum disorder (associated with a highly flat distribution). Most of those definitions take into account elements of Bayesian and information theories. Some of the more recent ones consist of the product of two factors, measuring, respectively, order and disorder on the given systems or, equivalently, localization and delocalization [30, 31]. They will be referred to as product complexities.

These product-complexity measures have been criticized and consequently modified leading to powerful estimators successfully checked in a wide variety of fields [32–37]. Fundamental concepts such as entropy or information are frequently present in the proposals for characterizing complexity, but it is known that other ingredients capturing not only randomness are also necessary. In fact one would wish also to detect, for instance clustering or pattern.

Even restricting ourselves to the aforementioned factorization, there is no unique definition for complexity. The reason is that there exist different candidates for being one of the coupled factors which give rise to complexity. The most popular ones are well-known to play a relevant role in an information-theoretic framework. Among them, let us mention the Shannon entropy S, the disequilibrium D, the Fisher information I, and the variance V.

Much work has been done using those quantities as basic measures, not only for quantifying the level of spreading of distributions but also for many other applications, such as, for instance, maximum-entropy estimation and reconstruction of an unknown distribution from very limited information on it.

Other authors have recently dealt with some particular factors of the complexity measures. In particular, Shannon entropy has been extensively used in the study of many important properties of multielectronic systems, such as, for instance, rigorous bounds [38], electronic correlation [39], effective potentials [40], similarity [41], and minimum cross-entropy approximations [42].

More recently, Fisher information has been studied as an intrinsic accuracy measure for concrete atomic models and densities [43, 44] and also for quantum mechanics central potentials [45]. Also, the concept of phase space Fisher information, where position and momentum variables are included, was analyzed for hydrogenlike atoms and the isotropic harmonic oscillator [46]. The net Fisher information measure is found to correlate well with the inverse of the ionization potential and dipole polarizability [44].

Quantum similarities and self-similarities D for neutral atoms were computed for nuclear charges $Z = 1$–54 only in the position space [47, 48], but afterwards a more complete analysis including $Z = 1$–103 neutral systems and singly charged ions has been done in position and momentum spaces [49].

Some studies on atomic similarity, using magnitudes closely related to D or to relative Shannon entropies, have also been reported [50, 51]. Very recently a comparative analysis of I and D shows that they both vary similarly with Z within the neutral atoms, exhibiting the same maxima and minima, but Fisher information presents a significantly enhanced sensitivity in the position and momentum spaces in all systems considered [52].

1.1.1 LMC Complexity

Among the more recent and succesful definitions of complexity, usually built up as a product of two factors quantifying, respectively, order/disequilibrium and disorder/uncertainty, specially remarkable is the one provided by López-Ruiz, Mancini and Calbet [30], to be denoted by C(LMC) due to its pioneering authors, which satisfies as others do the condition of reaching minimal values for both extremely ordered and disordered limits. Additional relevant properties are the invariance under scaling, translation, and replication.

The initial definition of the LMC complexity has been criticized [28] and modified [35] in order to verify the aforementioned properties, giving rise to the expression

$$C(LMC) \equiv D \cdot e^S = D \cdot L, \tag{3}$$

of a distribution $\rho(\mathbf{r})$. It is built up as the product of two relevant quantities within an information-theoretic framework: the "disequilibrium" D [53],

$$D(\rho) \equiv \int \rho^2(\mathbf{r}) d\mathbf{r} \tag{4}$$

which quantifies the departure of $\rho(\mathbf{r})$ from equiprobability, and the aforementioned Shannon entropy S as measure of randomness or uncertainty on the distribution. The usefulness of C(LMC) has been shown in different fields, allowing detection of periodic, quasiperiodic, linear stochastic and chaotic dynamics [30, 36, 37].

1.1.2 Fisher–Shannon Complexity

It appears also interesting to look for statistical complexities involving a local information measure, as can be done by replacing one of the LMC global factors by a "local" measure of intrinsic accuracy. In this sense, the main properties of Fisher information I make this quantity to be an appropriate candidate with the aim of defining a complexity in terms of complementary global and local factors. Very recently, the Fisher–Shannon complexity C(FS) has been defined [52, 54] in terms of both Fisher information and Shannon entropy and, consequently, providing a measure combining the global and local characters, and also preserving the desirable properties for any complexity as previously described. The Fisher information I itself plays a fundamental role in different physical problems, such as the derivation of nonrelativistic quantum-mechanical equations by means of minimum I principle, as also done for the time-independent Kohn–Sham equations and the time-dependent Euler equation [17, 55].

The FS is defined in terms of the power Shannon entropy $J \equiv \dfrac{1}{2\pi e} e^{2S/3}$ and the Fisher information I as

$$C(FS) \equiv I \cdot J \tag{5}$$

where definition of J is chosen in order to preserve general complexity properties. As compared to LMC complexity, and apart from the explicit dependence on Shannon entropy, C(FS) replaces the disequilibrium global factor by the Fisher local one. The C(FS) expression arises from the isoperimetric three-dimensional inequality $I \cdot J \geq 3$ [56–58] providing a universal lower bound to FS complexity. Among the main applications carried out, it should be remarked those concerning atomic distributions in position and momentum spaces where FS complexity is shown to provide relevant information on atomic shell structure and ionization processes [52, 54, 59, 60].

1.1.3 Cramer–Rao Complexity

In this work, we will also analyze, apart from C(LMC) and C(FS), the "Cramer–Rao" complexity C(CR), also as the product of a local and a global measure, keeping the first one as the Fisher information I, and replacing the Shannon entropy exponential by the variance V, giving rise to

$$C(CR) \equiv I \cdot V, \tag{6}$$

product which has been considered in different contexts [61, 60, 59]. Specially remarkable is the existence of a lower bound, in spite of the factors being of very different origin as well as their definition in terms of the distribution, emphasizing again the strong connection between both the local and the global level of uncertainty.

1.2 Numerical Analysis

The main aim of this work is to analyze the above-defined LMC, FS, and CR complexities associated to the one-particle densities in conjugated spaces, namely, position $\rho(\mathbf{r})$ and momentum $\gamma(\mathbf{p})$ densities, as well as the product or phase-space distribution $f(\mathbf{r}, \mathbf{p}) \equiv \rho(\mathbf{r})\gamma(\mathbf{p})$, for a great amount of atomic systems including neutral atoms (Section 2), singly charged ions (Section 3), and isoelectronic series (Section 4).

Analyzing the main information-theoretic properties of many-electron systems has been a field widely studied by means of different procedures and quantities, in particular, for atomic and molecular systems in both position and momentum spaces. It is worthy to remark the pioneering works of Gadre et al. [62, 63] where the Shannon entropy plays a fundamental role, as well as the more recent ones concerning electronic structural complexity [27, 64], the connection between information measures (e.g., disequilibrium, Fisher information) and experimentally accessible quantities such as the ionization potentials or the static dipole polarizabilities [44], interpretation of chemical phenomena from momentum Shannon entropy [65, 66], applications of the LMC complexity [36, 37] and the quantum similarity measure [47] to the study of neutral atoms, and their extension to the FS and CR complexities [52, 60] as well as to ionized systems [39, 54, 59, 67].

The applications in this work on a global scale of 370 systems will be carried out in order to gain insight not only on the information content of those systems, but also to interpret the complexity values in terms of physical properties and processes, such as shell-structure and ionization. Also the associated informational planes sub-stended by the factors composing each complexity will allow to obtain relevant interpretations on the main physical processes and characteristics of the distributions here studied. In doing so, Near-Hartree–Fock wavefunctions [68, 69] will be employed to compute the densities and the associated information measures and planes as well as complexities. For atomic systems in the absence of external fields (as is the case of this work) it is sufficient to deal with the spherically averaged densities $\rho(r)$ and $\gamma(p)$. Main conclusions on the results will be given in Section 5.

2 Complexity and Atomic Shell Structure

Complexity studies for atoms have also been carried out, but most of them are only for $Z = 1$–54 [27, 64]. Recent complexity computations, using relativistic wavefunctions in the position space, were also done [70]. Some other complexity works simply take the position density, not the momentum one, as basic variable [71]. In this sense, it is worthy to point out the different behaviors displayed by some of these quantities in position and momentum spaces for atomic systems, as we have recently shown [50, 52].

In particular, it has been shown that it is not sufficient to study the above measures only in the usual position space, but also in the complementary momentum space, in order to have a complete description of the information theoretic internal structure and the behavior of physical processes suffered by these systems. Some other new proposals of product-type complexity measures (e.g., CR complexity) have been also constructed and computed for multielectronic systems [60].

This section is devoted to the analysis and interpretation, from a physical point of view, of the LMC, FS, and CR complexity values and information planes corresponding to all neutral atoms throughout the Periodic Table, within the range of nuclear charges $Z = 1$–103. Such a study is carried out in position, momentum, and phase/product spaces, which corresponding distributions and their complexities are obtained by means of the accurate wavefunctions provided in Ref. [68].

2.1 Comparison Between Atomic LMC and FS Complexities

First, let us compare the LMC and FS complexities for those systems, as done in Fig. 1 for position and momentum spaces (1(a) and 1(b), respectively). It is remarkable, attending to the curves displayed in these figures, the similar structure of LMC and FS complexities in both spaces, in spite of their strongly different definition, mainly due to the information measure accompanying the Shannon factor, namely, the "global" disequilibrium for LMC and the "local" Fisher information for FS. It is worthy to point out not only the almost identical magnitude orders of both complexities, but also the strong correlation between their structure, characterized by the number and location of extrema, and the shell-filling process as well as

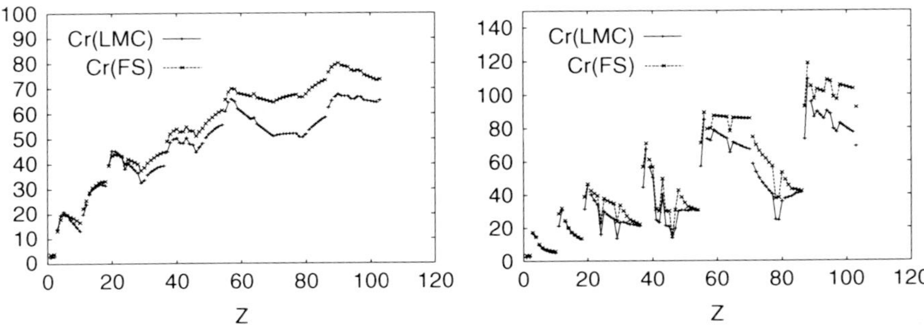

Fig. 1 LMC and FS complexities for neutral atoms with nuclear charge Z in position (*left*) and momentum (*right*) spaces. Atomic units (a.u.) are used

the groups the atoms belong to. Last comment is supported by the fact that both complexities in the two conjugated spaces display local minima for noble gases as well as for some atoms involved in the so-called 'anomalous shell-filling' (being specially relevant for the systems $Z = 24, 29, 46$). Similar comments can be done concerning maximal values.

2.1.1 LMC and FS Information Planes

Attending to the factors which compose complexities, it is also interesting to analyze the individual contribution of each one to the total complexity. For illustration, the "disequilibrium-Shannon plane" is shown in Fig. 2, drawn in terms of (D,L), as components of the LMC complexity, in position and momentum spaces (Figs. 2(a) and (b), respectively). Both figures again reveal the shell-filling patterns, much clearly in momentum than in position space. In fact, the different pieces of curves in momentum space belong to disjoint exponential entropy (L_p) values. Adding a new subshell makes L_p to increase, the disequilibrium D_p decreasing within each subshell. Opposite behaviors are displayed in position space concerning not only monotonicity, but also location of regions within the planes where heavy atoms

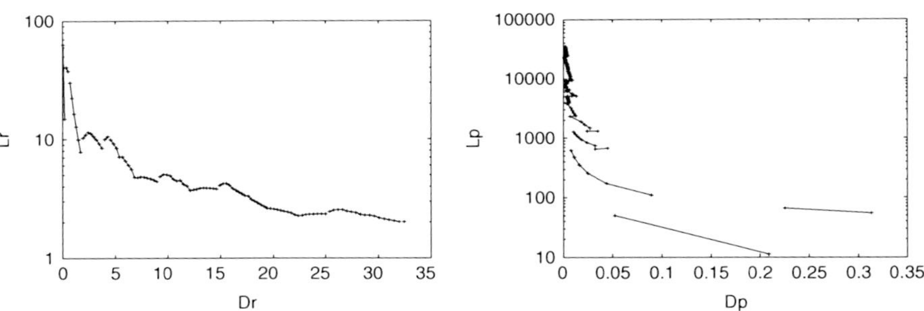

Fig. 2 Disequilibrium-Shannon plane (D,L) for neutral atoms with nuclear charge Z in position (*left*) and momentum (*right*) spaces. Atomic units (a.u.) are used

concentrate around: high disequilibrium in position space and high disorder (entropy) in the momentum one.

2.2 Atomic CR Complexity

Concerning CR complexity C(CR), main numerical results for atomic systems are displayed in Fig. 3 for position, momentum, and product spaces. In analyzing their structure as functions of the nuclear charge Z it is interesting to observe that most minima of $C_r(CR)$ and all of $C_p(CR)$ are the same of the LMC and FS complexities, previously specified. In fact, shell structure patterns are very similar for the three complexities, in spite of being determined by four quantities (S, D, I, and V) of very different character. The same also occurs for some of those isolated factors in all spaces, such as the exponential entropy L and the variance V, which figures are not shown for the sake of shortness.

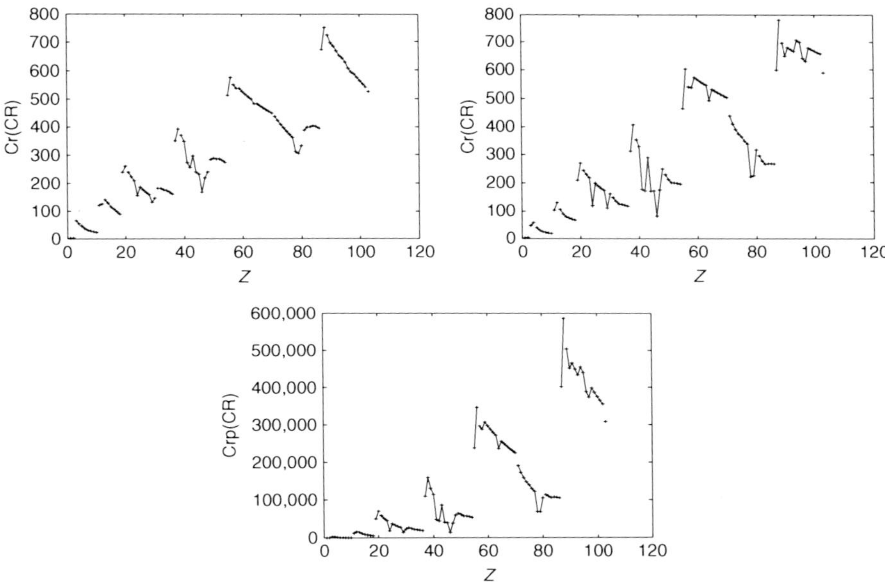

Fig. 3 CR complexity for neutral atoms with nuclear charge Z in position (*upper* left), momentum (*upper* right), and product (*lower*) spaces. Atomic units (a.u.) are used

2.2.1 CR Information Plane

The CR (I,V) information plane is shown in Fig. 4 for the two conjugated spaces, in order to check to which extent each composing factor is responsible for the

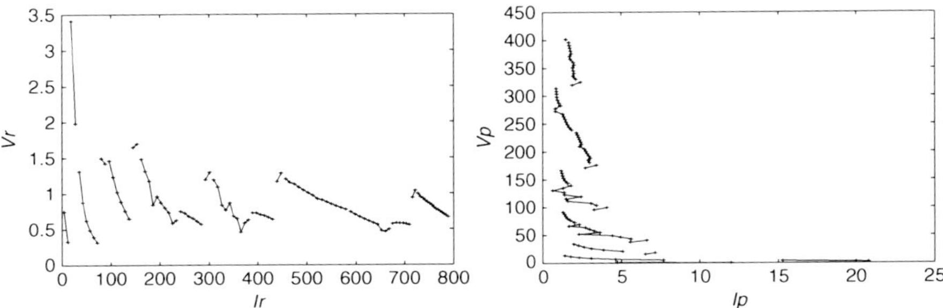

Fig. 4 Cramer–Rao plane (I,V) for neutral atoms with nuclear charge Z in position (*left*) and momentum (*right*) spaces. Atomic units (a.u.) are used

shell-filling pattern displayed. In position space (Fig. 4(a)), adding a new subshell makes Fisher information I_r to appreciably increase, its values belonging to disjoint intervals determined by the valence subshell. However, the variance V_r ranges over a unique interval for all systems without distinguishing their shell structure, but displaying a monotonically decreasing behavior (with few exceptions) within each specific subshell. Just the opposite behaviors for the corresponding momentum quantities I_p and V_p are observed in Fig. 4(b), in what concerns ranges of values and monotonicity.

It is worthy to notice how the three complexity measures here considered are able to provide information not only on randomness or disorder, but also on the structure and organization of the atomic systems. The same is not always true for the individual factors, appearing relevant to deal simultaneously with the localization and randomness factors, as well as the complementary conjugated spaces, in order to have a more complete description of the information content of atomic systems.

Summarizing the results of this section, (i) a complete description of the information-theoretic characteristics of atomic systems requires the complementary use of position and momentum spaces, (ii) LMC and FS complexities provide similar results (qualitatively and quantitatively) for all neutral atoms in both spaces, displaying periodicity and shell-filling patterns as also CR complexity does, and (iii) such patterns of the localization–delocalization planes in one space are inverse to those of the conjugated space.

3 Effects of the Ionization on Complexity

In this section the LMC, FS, and CR complexities are analyzed for singly charged ions with a number of electrons up to $N = 54$, that is, with a global charge $Q = Z - N = \pm 1$, Z being the nuclear charge. These quantities, together with the previously discussed values for neutral atoms within such N range, provide us with information on how complexity progresses in mono-ionization processes [54, 59]. In doing so, we are considering a global scale of 150 systems (53 cations, 43 anions, and 54 neutral atoms), the computations on ions being performed by employing the accurate wavefunctions of Ref. [69].

3.1 LMC and FS Complexities of Singly Charged Ions

A similar comparison between LMC and FS complexities as done previously for neutral atoms in both conjugated spaces has been also carried out for anions and cations in the two spaces. Conclusions raised by the analysis of these quantities for ions are almost identical to those provided when discussing Fig. 1 for neutral atoms, in what concerns similarity between C(LMC) and C(FS) values as well as their connection with the shell-filling process by means of the location of their extrema, most minima of complexity corresponding to noble gases, or the anomalous shell-filling set of atoms.

3.2 CR Complexity of Singly Charged Ions and Neutral Atoms

Concerning the C(CR) complexity, its evolution throughout the ionization is clearly displayed in Fig. 5, where its value is provided for the three considered species (anions, cations, and neutrals) in order to determine to which extent the ionization processes (by adding or removing electrons keeping fixed the nuclear charge Z) modify the atomic complexity. For illustration, this comparison is carried out for the C(CR) complexity $C_{rp}(CR)$ in the product space as shown in Fig. 5. Again, it is clearly observed the correlation of complexity with the atomic shell structure for all species. Additionally, it is appreciated that (i) complexity increases as the system loses an electron, and (ii) maxima are clearly associated to "s" valence subshells (those involved in ionization) while minima correspond to noble gases or some anomalous "d" subshells filling.

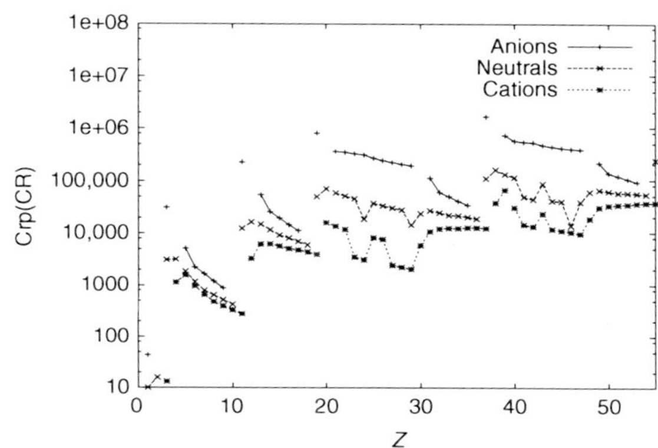

Fig. 5 CR complexity in product space neutral atoms and singly charged ions with nuclear charge Z. Atomic units (a.u.) are used

3.2.1 CR Information Plane for Monoionization Processes

The CR informational plane subtended by the constituing factors (I,V) also provides interesting results interpreted according to the atomic shell structure. Fig. 6 displays this plane in both conjugated spaces (6(a) for position and 6(b) for momentum) for the systems here considered. Apart from the faithful reproduction of shell structure, it is worthy to remark that, as shown in Fig. 6(a), systems of large Z are highly localized and organized in position space while the light ones appear much more delocalized. Location at the position (I_r, V_r) plane after an ionization process slightly changes for heavy atoms as to the light ones. Additionally, for fixed nuclear charge Z complexity $C_r(CR)$ decreases following the sequence anion–neutral–cation, that is, as losing electrons, the changes associated to "s" electrons being considerably higher to those of "p" or "d" subshells.

Exactly the opposite trends to those discussed in position space are observed in the momentum one, as shown in Fig. 6(b): large Z systems are now less localized and with a greater variance than the light ones, and losing electrons makes the variance to increase and Fisher information to decrease, just the reciprocal that happens in position space.

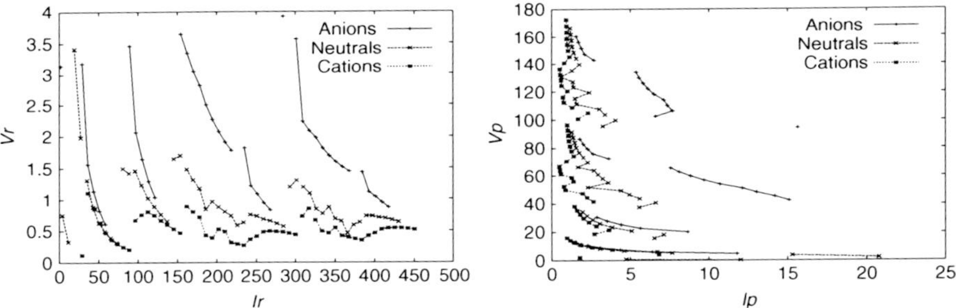

Fig. 6 Cramer–Rao plane (I,V) in position (*left*) and momentum (*right*) spaces, for neutral atoms and singly charged ions with nuclear charge Z. Atomic units (a.u.) are used

4 Isoelectronic Series: Dependence of Complexity on the Nuclear Charge

After carrying out the analysis of complexity dependence on the outermost subshells, as done in the previous section by considering ionization processes, let us now focus in the atomic core as source of the attractive forces and their effects on complexity values.

4.1 Composition and Number of Isoelectronic Series

We start by considering a neutral atom, that is, a system with identical values of the nuclear charge Z and the number of electrons N, from which we give rise to a set of cations by progressively increasing one-by-one the nuclear charge Z keeping fixed the number of electrons (or, equivalently, starting from a global charge $Q \equiv Z - N = 0$ until reaching a maximum positive value, being $Q_{max} = 20$ in the numerical application considered here). Such a set of cations together with the neutral atom is known as an "isoelectronic series," characterized by the fixed number of electrons N as well as the maximum value Q_{max}. Studying the previously considered complexity measures for a given isoelectronic series provides information on their dependence on the nuclear charge Z for fixed N electrons. In this section, such a study will be carried out for nine isoelectronic series, namely, those corresponding to $N = 2$–10, within a Hartree–Fock framework [69]. Each series consists of 21 members (a neutral atom and 20 cations), giving rise consequently to analyze complexities of a global of 189 atomic systems.

4.2 Complexities and Information Planes of Isoelectronic Series

As in Section 3, we consider here (i) LMC, FS, and CR complexities, (ii) the associated information planes (D,L), (I,J) and (I,V), and (iii) distributions in position, momentum, and product spaces.

4.2.1 LMC Complexity and Information Plane

In Fig. 7 the disequilibrium-Shannon plane (D,L) is shown in position, momentum, and product spaces (Figs. 7(a), (b) and (c), respectively) for the isoelectronic series $N = 2$–10. For the individual spaces (position and momentum), each series roughly follows a linear trajectory in a double logarithmic scale. In fact, the Helium series ($N = 2$) displays an almost constant $C(LMC) = D \cdot L$ line in both spaces, which means that increasing the nuclear charge produces, as should be expected, a higher localization D and a lower uncertainty, both effects compensating each other proportionally and providing an almost constant product which defines LMC complexity. Concerning product space, the corresponding Disequilibrium-Shannon plane (D, L) is shown in Fig. 7(c). It is worthy to notice the strong changes in the slopes of all series as compared with those of the isolated spaces. While product entropy does not suffer drastic changes, localization appears very different within each series. Additionally, shell-filling patterns are clearly displayed, with systems having 2s as valence subshell having a higher complexity than those filling the $2p$. It is also remarkable that the $N = 2$ series displays a very different behavior as compared with the other series. This can be interpreted by taking into account that those systems are the unique ones considered here consisting only on a core shell. From all these comments it should be concluded that the product space plane is relevant in order obtain an interpretation of the Disequilibrium-Shannon plane values in terms of shell structure.

In position space, systems with large nuclear charge Z for any isoelectronic series display a highly localized structure (large D) as shown in Fig. 7(a). In such a large D area, trajectories are almost linear which correspond to an almost constant product

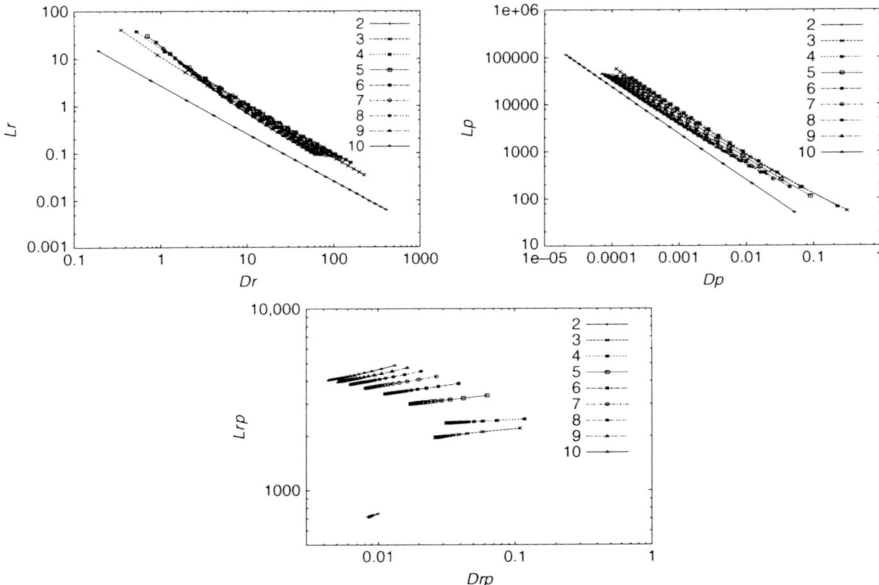

Fig. 7 Disequilibrium-Shannon plane (D, L) in position (*upper* left), momentum (*upper* right) and product (*lower*) spaces, for isoelectronic series with $N = 2 - 10$ electrons. Atomic units are used

measure. Deviations from this linear shapes are better observed for low nuclear charge systems, possessing a greater complexity. Biggest position space complexities correspond to neutral systems, with a relatively lower localization and greater uncertainty as compared with its cations. All those comments are just the opposite ones in momentum spaces, as can be readily realized by observing Fig. 7(b). Heavy systems are characterized by a low localization and high entropy in mometum space, and neutrals deviate from isoproduct lines as possessing a higher level of structure. It is worthy to remark also that spacing between consecutive systems within each isoelectronic series decreases as increases Z, because of a highers similarity between systems with large nuclear charge as compared with to those with low Z, which progressively separate among themselves.

4.2.2 FS Complexity and Information Plane

A similar analysis has also been carried out for the FS plane (I,J) in position, momentum, and product spaces (Fig. 8). It is worthy to remember the rigorous lower bound to the associated FS complexity $C(FS) = I \cdot J \geq$ constant (the constant being 3 for the conjugated spaces and $18\pi e$ for the product space) in order to verify such a bound for the systems considered here. The straight line $I \cdot J =$ constant drawn in the plane by using a double logarithmic scale divides it into an "allowed"

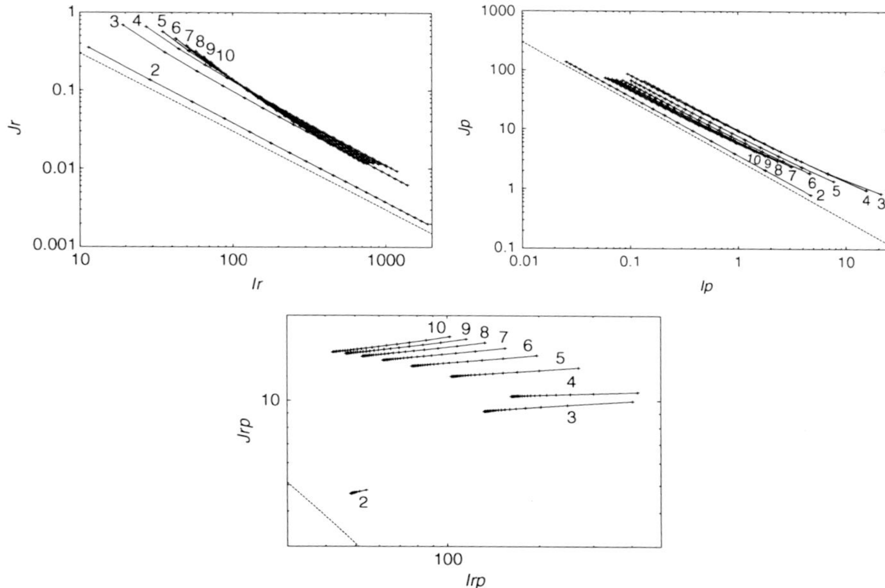

Fig. 8 Fisher–Shannon plane (I,J) in position (*u*pper left), momentum (*u*pper right) and product (*l*ower) spaces, for isoelectronic series with $N = 2 - 10$ electrons. Atomic units are used

(upper) and a "forbidden" (lower) parts. Parallel lines to that one represent isocomplexity points, and higher deviations from this frontier are associated to greater FS complexities. Such a parallel shape is roughly displayed by all isoelectronic series in both position and momentum spaces, as shown in Figs. 8(a) and (b), respectively. Similar comments to those provided on discussing Fig. 7 in what concerns location areas of systems at the plane, distances within a series between consecutive systems, deviation from minimal complexity values, and opposite behaviors in conjugated spaces are also valid for the position and momentum (I,J) planes as concluded by analyzing Figs. 8(a) and (b).

For the sake of briefty, results on the CR plane (I,V) are not displayed, but conclusions obtained from their values are the same as those just discussed for disequilibrium-Shannon and FS planes.

5 Concluding Remarks

Different information-theoretic quantities as well as complexities defined as the product of a couple of localization–delocalization factors have been shown to provide relevant information not only on the shell structure and organization of a great variety of atomic systems, but also on ionization processes and their dependence on both the nuclear charge and the number of electrons. In doing so, it appears necessary to deal simultaneously with the conjugated position and momentum space electron densities, being also important to consider the product space in order to get a more detailed and complete description of such systems. The method here employed for carrying out this study is also applicable to the analysis of additional multifermionic systems, as is the case of molecules and many others, as well as physical or chemical processes, such as reactions or polarization among others. Some of these subjects are now being studied and will be presented hopefully elsewhere.

It has also been shown the interest of studying the associated information planes substended by two information functionals, which for the atomic case clearly display the characteristic shell-filling patterns throughout the whole periodic table. It still remains open the question of the existence of additional functionals, planes and complexities providing further information on the atomic structure and the ionization processes, among others.

Acknowledgements This work was supported in part by the Spanish MICINN projects FIS-2008-02380 and FIS-2005-06237 (J.A.), and the grants FQM-1735 (J.C.A.), P05-FQM-00481 and P06-FQM-01735 of Junta de Andalucía. We belong to the Andalusian research group FQM-0207.

References

1. S. B. Sears, S. R. Gadre, J. Chem. Phys. **75**, 4626 (1981)
2. R. F. Nalewajski, R. G. Parr, J. Phys. Chem. A **105**, 7391 (2001)
3. R. Carbó-Dorca, X. Girones, P. G. Mezey, eds., *Fundamentals of Molecular Similarity* (Kluwer Academic, Dordrecht/Plenum, New York, 2001)
4. J. Cioslowski, A. Nanayakkara, J. Am. Chem. Soc. **115**, 11213 (1993)
5. R. Carbó-Dorca, L. Amat, E. Besalu, X. Girones, D. Robert, J. Mol. Struct. (Theochem) **504**, 181 (2000)
6. R. Daudel, C. R. Acad. Sci. (Paris) **237**, 601 (1953)
7. C. Aslangul, R. Constanciel, R. Daudel, P. Kottis, Adv. Quantum Chem. **6**, 94 (1972)
8. P. G. Mezey, R. Daudel, I. G. Csizmadia, Int. J. Quantum Chem. **16**, 1009 (1979)
9. R. F. Nalewajski, Chem. Phys. Lett. **375**, 196 (2003)
10. L. Wang, L. Wang, S.Arimoto, P. G. Mezey, J. Math. Chem. **40**, 145 (2006)
11. D. Avnir, A. Y. Meyer, J. Mol. Struct. (Theochem), **226**, 211 (1991)
12. C. E. Shannon, W. Weaver, *The Mathematical Theory of Communication* (University of Illinois Press, Urbana, 1949)
13. R. A. Fisher, Proc. Cambridge Phil. Sec. **22**,700 (1925)
14. B. R. Frieden, *Science from Fisher Information* (Cambridge University Press, Cambridge, 2004)
15. E. T. Jaynes, Phys. Rev. A **106**, 620 (1957)
16. A. Nagy, Chem. Phys. Lett. **425**, 157 (2006)
17. R. Nalewajski, Chem. Phys. Lett. **372**, 28 (2003)
18. M. Reginatto, Phys. Rev. A **58**, 1775 (1998)
19. A. N. Kolmogorov, Prob. Inf. Transm. **1**, 1 (1965)
20. G. J. Chaitin, J. ACM **13**, 547 (1966)

21. A. Lempel, J. Ziv, IEEE Trans. Inform: Theory **22**, 75 (1976)
22. P. Grassberger, Int. J. Theory Phys. **25**, 907 (1986)
23. C. H. Bennett, *Logical Depth and Physical Complexity*, pp. 227–257 in *The Universal Turing Machine: A Half Century Survey* (Oxford University Press, Oxford, 1988)
24. S. S. Lloyd, H. Pagels, Ann. Phys. NY **188**, 186 (1988)
25. C. R. Shalizi, K. L. Shalizi, R. Haslinger, Phys. Rev. Lett. **93**, 118701 (2004)
26. O. A. Rosso, M. T. Martin, A. Plastino, Physica A **320**, 497 (2003)
27. K. Ch. Chatzisavvas, Ch. C. Moustakidis, C. P. Panos, J. Chem. Phys. **123**, 174111 (2005)
28. D. P. Feldman, J. P. Crutchfield, Phys. Lett. A **238**, 244 (1998)
29. P. W. Lambert, M. P. Martin, A. Plastino, O. A. Rosso, Physica A **334**, 119 (2004)
30. R. López-Ruiz, H. L. Mancini, X. Calbet, Phys. Lett. A **209**, 321 (1995)
31. J. S. Shiner, M. Davison, P. T. Landsberg, Phys. Rev. E **59**, 1459 (1999)
32. C. Anteonodo, A. Plastino, Phys. Lett. A **223**, 348 (1996)
33. R. G. Catalán, J. Garay, R. López-Ruiz, Phys. Rev. E **66**, 011102 (2002)
34. M. T. Martin, A. Plastino, O. A. Rosso, Phys. Lett. A **311**, 126 (2003)
35. R. López-Ruiz, Biophysical Chemistry **115**, 215 (2005)
36. T. Yamano, J. Math. Phys. **45**, 1974 (2004)
37. T. Yamano, Physica A **340**, 131 (2004)
38. J. C. Angulo, Phys. Rev. A **50**, 311 (1994)
39. N. L. Guevara, R. P. Sagar, R. O. Esquivel, Phys. Rev. A **67**, 012507 (2003)
40. E. Romera, J. J. Torres, J. C. Angulo, Phys. Rev. A **65**, 024502 (2002)
41. M. Ho, V. H. Smith Jr., D. F. Weaver, C. Gatti, R.P. Sagar, R. O. Esquivel, J. Chem. Phys. **108**, 5469 (1998)
42. J. Antolín, J. C. Cuchí, J. C. Angulo, J. Phys. B **32**, 577 (1999)
43. A. Nagy, K. D. Sen, Phys. Lett. A **360**, 291 (2006)
44. K. D. Sen, C. P. Panos, K. Ch. Chtazisavvas, Ch. C. Moustakidis, Phys. Lett. A **364**, 286 (2007)
45. E. Romera, P. Sánchez-Moreno, J. S. Dehesa, J. Math. Phys. **47**, 103504 (2006)
46. I. Hornyak, A. Nagy, Chem. Phys. Lett. **437**, 132 (2007)
47. A. Borgoo, M. Godefroid, K. D. Sen, F. de Proft, P. Geerlings, Chem. Phys. Lett. **399**, 363 (2004)
48. F. de Proft, P.W. Ayers, K. D. Sen, P. Geerlings, J. Chem. Phys. **120**, 9969 (2004)
49. J. Antolín, J. C. Angulo, Eur. Phys. J. D **46**, 21 (2008)
50. J. C. Angulo, J. Antolín, J. Chem. Phys. **126**, 044106 (2007)
51. A. Borgoo, M. Godefroid, P. Indelicato, F. De Proft, P. Geerlings, J. Chem. Phys. **126**, 044102 (2007)
52. J. C. Angulo, J. Antolín, K. D. Sen, Phys. Lett. A **372**, 670 (2008)
53. O. Onicescu, C. R. Acad. Sci. Paris A **263**, 25 (1966)
54. K. D. Sen, J. Antolín, J. C. Angulo, Phys. Rev. A **76**, 032502 (2007)
55. A. Nagy, J. Chem. Phys. **119**, 9401 (2003)
56. T. M. Cover, J. A. Thomas, *Elements of Information Theory* (Wiley-Interscience, New York, 1991)
57. A. Dembo, T. A. Cover, J. A. Thomas, IEEE Trans. Inf. Theory **37**, 1501 (1991)
58. J. M. Pearson, Proc. Amer. Math. Soc. **125**, 3335 (1997)
59. J. Antolín, J. C. Angulo, Int. J. Quantum Chem. **109**, 586 (2009)
60. J. C. Angulo, J. Antolín, J. Chem. Phys. **128**, 164109 (2008)
61. J. S. Dehesa, P. Sánchez Moreno, R. J. Yáñez, J. Comp. Appl. Math. **186**, 523 (2006)
62. S. R. Gadre, Phys. Rev. A **30**, 620 (1984)
63. S. R. Gadre, R. D. Bendale, Curr. Sci. (India) **54**, 970 (1985) **17**, 138 (1985)
64. C. P. Panos, K. Ch. Chatzisavvas, Ch. C. Moustakidis, E. G. Kyhou, Phys. Lett. A **363**, 78 (2007)
65. S. R. Gadre, R. D. Bendale, S. P. Gejji, Chem. Phys. Lett. **17**, 138 (1985)
66. R. P. Sagar, N. L. Guevara, J. Chem. Phys. **124**, 134101 (2006)

67. E. Romera, J. S. Dehesa, J. Chem. Phys. **120**, 8906 (2004)
68. T. Koga, K. Kanayama, S. Watanabe, A. J. Thakkar, Intl. J. Quantum Chem. **71**, 491 (1999)
69. T. Koga, K. Kanayama, S. Watanabe, S. Imai, A. J. Thakkar, Theor. Chem. Acc. **104**, 411 (2000)
70. A. Borgoo, F. de Proft, P. Geerlings, K. D. Sen. Chem. Phys. Lett. **444**, 186 (2007)
71. J. B. Szabo, K. D. Sen, A. Nagy, Phys. Lett. A **372**, 2428 (2008)

Index